Time in Rutland

Thistleton
Market Overton
Clipsham
Teigh
Barrow
Stretton
Whissendine
Greetham
Ashwell
Cottesmore
Pickworth
Horn
Exton
Essendine
Langham
Burley
Ryhall
Barleythorpe
Little Casterton
Belmesthorpe
Whitwell
Tickencote
Oakham
Hambleton
Empingham
Great Casterton
Egleton
Stamford
Braunston in Rutland
Ketton
Brooke
Gunthorpe
Normanton
Martinsthorpe
Manton
Edith Weston
Tinwell
Ridlington
Lyndon
North Luffenham
Wing
Belton in Rutland
Preston
Pilton
Ayston
South Luffenham
Wardley
Glaston
Morcott
Tixover
Uppingham
Barrowden
Bisbrooke
Seaton
Stoke Dry
Lyddington
Harringworth
Thorpe by Water
Caldecott

N

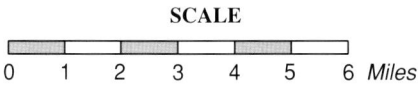

SCALE

0   1   2   3   4   5   6 Miles

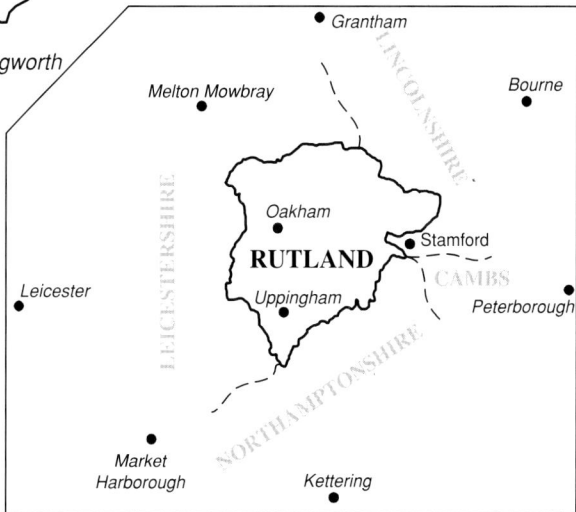

Grantham
Melton Mowbray
Bourne
LINCOLNSHIRE
Oakham
Stamford
RUTLAND
CAMBS
LEICESTERSHIRE
Leicester
Uppingham
Peterborough
NORTHAMPTONSHIRE
Market Harborough
Kettering

RUTLAND

A TIME FOR EVERYTHING

To every thing there is a season, and
a time to every purpose under the heaven:
A time to be born, and a time to die;
a time to plant, and a time to pluck up that which is planted;
A time to kill, and a time to heal;
a time to break down, and a time to build up;
A time to weep, and a time to laugh;
a time to mourn, and a time to dance;
A time to cast away stones, and a time to gather stones together;
a time to embrace, and a time to refrain from embracing;
A time to get, and a time to lose;
a time to keep, and a time to cast away;
A time to rend, and a time to sew;
a time to keep silence, and a time to speak;
A time to love, and a time to hate;
a time of war, and a time of peace.

Ecclesiastes 3, vv 1-8

(Benson 1875, iii)

Time in Rutland

A History and Gazetteer of the Bells, Scratch Dials, Sundials, and Clocks of Rutland

Robert Ovens & Sheila Sleath

[signature: Robert Ovens]   *[signature: Sheila Sleath]*

Rutland Local History and Record Society

Rutland Record Series No 4

2002

Registered Charity No 700273

RLHRS Rutland Record Series No 4

First published in 2002 by the Rutland Local History
& Record Society,
Rutland County Museum, Catmose Street, Oakham,
Rutland, LE15 6HW

The Society is grateful to the Heritage Lottery Fund
for a generous grant towards the cost of producing this
publication under the Millennium Festival Awards for
All scheme

ISBN 0 907464 30 0

British Library Cataloguing in Publication Data
A catalogue record for this book is available from the
British Library

Edited for the Society by T H McK Clough

Cover illustrations by Kitty Wigmore

Designed by John Robey, Mayfield Books,
Matherfield House, Church Lane, Mayfield,
Ashbourne DE6 2JR

Printed and bound in Spain by
Graficas Santamaria SA, Vitoria

Contents

	Foreword	6
	Preface	7
	Acknowledgements	8
1	Introduction to Time in Rutland	9
	1.1 Historical Introduction	9
	1.2 Church Bells	12
	1.3 Saxon Sundials	15
	1.4 Scratch Dials	16
	1.5 Scientific Sundials	19
	1.6 Clocks & Watches	21
	1.7 The Electric Telegraph & Standard Time	25
2	Bellfounders — The Founders of Rutland Bells	27
3	Clockmakers — Local Clock and Watchmakers & the Makers of Rutland Clocks	59
4	Gazetteer of the Bells, Scratch Dials, Sundials & Clocks of Rutland	93
	Select Glossary	347
	Bibliography	355
	Appendices:	
	1 Bellfounders' Marks, Devices, Decoration & Lettering on Rutland Bells	359
	2 Rutland Bells Scheduled for Preservation	374
	3 Rutland Ringing Customs	375
	4 Rutland Sundials	381
	5 William Potts & Sons' Specification & Tender of 1919 for Seaton Church Clock	386
	6 Thomas North FSA	388
	Index	394

Foreword

Time in Rutland represents the culmination of a project which began modestly enough, as one might have thought befitted England's smallest historic county. From that modest beginning it has become a major work of reference, having grown as a sturdy oak from an acorn — a figure of speech which seems appropriate because, after all, the acorn is included in Rutland's coat of arms.

In *Time in Rutland* we have a detailed account of the devices for the measurement of time, from simple sundials to complex clocks, and of the signalling of time by the ringing of bells, for the whole of the county. The descriptions of the clocks and dials and bells in the Gazetteer form a comprehensive statement of what is known today. They supersede the only other such account, that of Rutland's church bells and bellringing customs by Thomas North published in 1880. One might have expected there could be little to add to that work, beyond the occasional correction and the addition of more recent material, and indeed the authors rightly acknowledge their debt to that earlier publication. In fact, they have gone to some pains to discover more about North, a man whose work might be better recognised than it is. However, this volume covers a wider brief than he did in including everything to do with the measuring of time. Not only does it set out information about the clockmakers and bellfounders whose work survives in Rutland, it also includes many references to and quotations from primary archive sources, such as churchwardens' accounts, which help to document and authenticate the history of the bells and clocks described. Inevitably there remain some unanswered questions, particularly relating to the history of bellfounding dynasties and the attribution of bells and clocks to their makers. However, far from diminishing the value of the work, this will encourage future researchers to pursue such matters further.

Time in Rutland has become a definitive corpus and, I believe, a work of sound academic value which will itself stand the passage of time. In achieving this, the authors would probably admit that they have taken themselves slightly by surprise, but it is true. As the Society's editor, I have seen the fruits of their research incorporated into the final text and I know how much sound original work it contains. They deserve full credit for this, just as they have themselves acknowledged the contributions by many others which have made the completion of the book possible. Those interested in the history of Rutland, and in horology and in campanology in general, have good cause to be grateful for its publication.

T H McK Clough
Curator, Rutland County Museum

Preface

The inception of *Time in Rutland* grew from our article entitled 'Keeping Time in Rutland' in *A Celebration of Rutland* (Ovens & Sleath 1994, 57-63). It was apparent at this time that we had only scratched the surface of what was a fascinating aspect of the county's heritage and we decided to extend our research. Over the following years we continued to look for and list additional scratch dials and sundials found on Rutland buildings. The locations of church and secular clocks were also recorded, but little attempt was made to gain access to see clock movements. By 1998 we had built up an interesting database of information, much of which, as far as we knew, was not available elsewhere. At about this time the Rutland Local History & Record Society was looking for a suitable Millennium project and we offered our research for publication to fulfil this need. Time was obviously a very relevant subject for such a project, our research was strictly limited to Rutland, and it met the Society's objective of recording Rutland.

The adoption of the project meant that every scratch dial, sundial and clock had to be visited and recorded accurately and in detail, especially following our successful application for grant aid from the Heritage Lottery Fund's Millennium Awards for All programme. Very early on it was realised that there were other elements of 'time in Rutland' which had to be researched if our work was to be complete, especially church and secular bells because of their role in announcing time. In addition to the physical details we also wanted to include historical background, and relevant details of bellfounders and clockmakers.

Thus began the systematic visiting of every parish in the county, looking for and recording in detail anything of relevance. Almost every bell, belfry, bellcote, scratch dial, sundial, clock dial and clock movement was photographed and in most cases measured. Plaster casts were made of bellfounders' marks, decorations and letter styles, and a rubbing made of every scratch dial. Sufficient details enabling us to draw a plan of each church and bellframe were also noted, and photographs of any relevant memorials and notices taken. Another important component of our visit was to talk to churchwardens and local historians who often provided or directed us towards very interesting and useful information.

Our work on the historical background for this project involved many hours spent on searching parish documents at the Record Office for Leicestershire, Leicester and Rutland. Here, the Churchwardens' Accounts proved invaluable, as did correspondence, faculty documents and vouchers. Supplementary information was also found in other local record offices, museums and libraries. We also consulted experts on bells, scratch dials, sundials and clocks and we are particularly grateful for their help. They guided us with enthusiasm and contributed freely to our investigations.

This book has fulfilled its primary objective of thoroughly recording another aspect of Rutland's heritage. It is the only detailed record of scratch dials and pre-1900 sundials in the county, it updates and extends the work started by Patrick Hewitt in recording Rutland turret clocks, and is a long overdue revision of Thomas North's *Church Bells of Rutland*, published in 1880. It will therefore, we hope, be of interest to campanologists, horologists, scratch dial and sundial enthusiasts, local and family historians, and many others, particularly those resident in the parishes of this small but historic county.

Robert Ovens & Sheila Sleath
February 2002

Acknowledgements

Time in Rutland is the result of five years of detailed research and survey work. It could not have been published in this time and in such detail without the aid of personal computers and, more importantly, the help of many organisations and individuals. Without exception, those we turned to for help provided it willingly and with enthusiasm. The list includes churchwardens, tower captains, village historians, specialists in bells, clocks, scratch dials, sundials and coins, societies, record offices, libraries and museums. The interest in our project was overwhelming and a particular 'thank you' must go to the following:

Our mentor Tim Clough, Curator of Rutland County Museum and Honorary Editor of Rutland Local History and Record Society, for advice, answering our many questions, editing our script and providing access to the museum's sources and resources; George Dawson, bell historian and adviser, for information on bells and bellfounders, and checking the relevant script; Sue Howlett, for Stretton local history and reading through the final version; John Robey, Mayfield Books, for guiding us through the publication of this, our first book; the staff at Rutland County Museum, at the Record Office for Leicestershire, Leicester and Rutland, at Northamptonshire Record Office, at the University of Leicester Library, at Newarke Houses Museum (Leicester City Museums Service), at Stamford Museum, and to the following other individuals and organisations who helped in so many different ways:

John Ablott (clockmaker), Jane Allsop (Belton in Rutland), Janice Atkinson (Tower Captain, Uppingham), Dr Ray Ayres (clocks by Thomas Eayre), Ray Bailey (Tower Captain, Langham), Mrs A M Bardwell (Pickworth), John and Connie Beadman (Braunston in Rutland), Mr & Mrs Birch (Wing), David Bland (Greetham), Robert Boyle (Bisbrooke), Trustees of Browne's Hospital, Stamford (archives), Andrew Butterworth (Rutland clockmakers), Ian & Rosemary Canadine (Lyddington), David Carlin (drawing of Greetham House), Michael Clayton (Cottesmore Hunt), Yolanda Courtney (Leicester City Museums Service), the late Ralph Cox (Greetham clock), Ron Creese (Preston), Hilary Crowden (Seaton and Uppingham), Tony Cutting (Bisbrooke), Jonathan Dent (Dent family history), John C Eisel (Central Council of Church Bell Ringers), Joe Ecob (Melton Mowbray clockmakers), Betty & George Finch (Ayston, Thistleton and Wardley), Geoff Fox (Ketton), Mr Frearson (Bisbrooke), Charles Gilman (Pickworth), Steve Gluning (Market Overton), Brian Gooch (Tinwell), Michael Gray (Ridlington), David Griffiths (Belton in Rutland), Sir David Davenport-Handley (Clipsham), Miranda Hall (Hambleton), Stephen Hargraves (North Luffenham), Peter Hayward (Hayward Mills Associates — bell hangers), David Heasum (Uppingham School Estate), Christine Hill (Oakham Library), Raymond Hill (Burley), Mr P E Hodgson (Normanton Church bell), Mrs Hoy (Oakham bells), Mr & Mrs P Johnstone (Ketton Grange Stables), Ioan Jones (Lyddington), Harold Killingback (Brooke), Les Kirk (John Smith & Sons, Derby — turret clocks), the late Miss Verona Kitson (The Pastures, North Luffenham), Roy Knight (Head of Cultural Services, Rutland County Council), Peter Lane (Uppingham), Michael Lee (John Watts clocks and Henry Penn bells), Sue Lee (Glaston), Richard Lees (Cottesmore), the Editor, *Leicester Mercury* (photographs), Ian Lyon (Empingham), Heather McGuire (Aris family history), Chris McKay (Antiquarian Horological Society Turret Clock Group), Nicholas Meadwell (church restoration), Warwick Metcalfe (Uppingham School history), Malcolm and Esme Mottram (Egleton), Michael Moubray (Ridlington), the late David Nettle (turret clocks), Brian Nichols (photograph), Ron Pace (Thomas Eayre II and Rutland watches), David Parkin (Egleton and St John & St Anne's Hospital, Oakham), Denis Pearson (bell historian), John Pearson (The Vaults, Uppingham), Pam Phillips (Thomas North), Paul Phillips (Whissendine), Christopher Pickford (bell historian), Andrew Pilbeam (Stretton), Ken Popple (Ryhall), Michael Potts (Potts' turret clocks), Canon John R Prophet (church drawings), David Rippon (Ketton Handbell Ringers), Professor Alan Rogers (Uppingham), Royal Cornwall Museum (*The Ringers of Launcells*), Gerry Rudman (Uppingham School Archives), Bob Salmon (Stoke Dry), Bill Sewell (North Luffenham), Paul Sibbering (Lyndon), Margaret Sillett (Wardley), John Smith (lately Curator, Stamford Museum), Alan Snodin (Uppingham), Philip Snowden (Stamford clocks), Brian Sparkes (John Watts' clocks), Brian Stokes (Lyddington), John Taylor Bell Foundry (archives), Auriol Thomson (Glaston), David Thomson (Latin translation), David Thorne (Seaton), Dick Tidd (former clock winder, Ashwell Church), the late Dr E C Till (Norris and Watts research, Stamford), Marilyn Tomalin (Tower Captain, Oakham), Peter Tomalin (Lyddington), Margaret Towl (Burley), the Rev Peter Townsend (Greetham), Tony Traylen (Rutland photographs), Sarah Tuck (Wing), Jean Turner (Belton in Rutland), Alan Walker (Belton in Rutland), Bernadette Wallace (Exton), Michael Ward (Great Casterton), Ken Weatherhogg (Oakham), Walter Wells (sundials and scratch dials), Whitechapel Bell Foundry (archives), Kitty Wigmore (cover illustrations), John Williams (Morcott), Anthony Woodburn Ltd (John Watts musical clock), Mr & Mrs Wright (Caldecott), Linda & Graham Worrall (Barrowden). There were many others who assisted with this project and to them we also offer our thanks.

Finally, we must thank those responsible for awarding a substantial grant towards the cost of producing this publication from the Heritage Lottery Fund under the Millennium Festival Awards for All scheme.

Robert Ovens & Sheila Sleath
February 2002

AWARDS
FOR ALL

Chapter One

Introduction to Time in Rutland

1.1 Historical Introduction

Modern technology makes it possible to know the time instantly and accurately anywhere, and it is difficult to understand how lives were conducted without this knowledge. However, up to the end of the nineteenth century and beyond, the main occupation in Rutland was farming and there was little need for accurate timekeeping. People were well aware of dawn, dusk, lunar cycles, the seasons and other natural phenomena. The pattern of their everyday lives was so attuned to these aspects of Mother Nature that they possessed a sense of time which was quite sufficient for most of their needs.

From the beginning of the seventeenth century the sundial or church clock more than adequately satisfied their minimal 'accurate' timekeeping requirements. Even the introduction of low-priced mass-produced clocks and watches from America and Europe in the second half of the nineteenth century did little to alter the daily activities of the majority of Rutlanders. Agricultural wages were very low and a high percentage of the workforce could not consider buying a clock even if they were able to find a use for one. Nevertheless, there was a demand for better timekeeping, especially from those who could afford to buy their own clocks and watches, and this was satisfied by the clock and watchmakers of Oakham, Uppingham, Stamford and other local towns. The building of railways in the mid nineteenth century led to the introduction of 'railway time' throughout the whole country. This standard time became known as Greenwich Mean Time, and by 1880 Wright's *Directory of Leicestershire and Rutland* records that it was being transmitted by electric telegraph to six post offices in Rutland. This was the beginning of a time revolution which now dominates our lives.

The earliest time-measuring device recorded in Rutland, and now in Rutland County Museum (OS 89) was found *circa* 1908 at Market Overton during the excavation of a Saxon cemetery. The following report gives more details:

> We now come to the most unusual and interesting relic in the collection. This is a small saucer-shaped vessel 4 inches [102mm] in diameter made of bronze, the bottom of

which is pierced with a small hole. This is no doubt a water clock, of a kind employed by the early Britons. The bowl is placed on the surface of the water and allowed to fill through the perforation; on sinking in a definite time it is emptied and replaced. Mr. Wing made a series of experiments with it and found the average time for filling and sinking to be 62.9 minutes; the longest time recorded being 72 and the shortest 56 minutes. An interesting feature connected with this object was a clay vessel in which it was found, and into the bottom of which it exactly fitted. Whether this was merely a chance association or a case to preserve the somewhat fragile bronze bowl, is a matter which will be left for the experienced archaeologist to decide (Phillips 1911-12, 166).

William Wing, FSA, a man of antiquarian interests, was a resident of Market Overton. He and his family donated the church clock in 1912.

The Saxons placed some remarkable sundials on church walls and a number have survived (see Chapter 1.3 — Saxon Sundials). Although no such dial has been recorded on a Rutland church, *Victoria County History* (**II**, 1935, 134-5) notes that during alterations in 1929 to the Ram

The bronze dish found at Market Overton circa 1908 (Rutland County Museum OS 89)

The early sundial at the Ram Jam Inn, on the Great North Road near Stretton

Jam Inn, in the parish of Greetham, 'A Saxon sundial, found in excavating the new foundations, is inserted in the walling ...'. This imperfect dial, which is difficult to date (see Chapter 1.3 — Saxon Sundials), can still be seen over a doorway on the east side of the building.

Crude medieval sundials can be found on the south-facing walls of many Rutland churches. They are known as mass clocks, mass dials or scratch dials, and their main function was to indicate the time when the priest should ring a bell to call people to worship. The number of lines scratched onto the dials varied according to the individual priest's requirements. Service times, and perhaps hours designated for prayer, would have been regulated in this way and the numerous examples found in Rutland are indicative of their importance. Scratch dials are not found on every church, but this is not surprising considering the extensive alterations and restorations carried out since their introduction in the eleventh or twelfth century. Scratch dials were obviously ineffective on sunless days. Before clocks became available the only alternatives were hour glasses, 'sinking bowl' water clocks and candle clocks. Any one of these could have been used to measure the time from dawn to determine when bells should be rung. More complicated water clocks were available but their high cost meant that they were only found in the wealthier monasteries. It is unlikely that any were used in Rutland as the only full monastery in the county was a small Augustinian priory at Brooke.

From the eleventh century, church bells played an increasingly important role in communicating time and announcing events within Rutland's towns and villages. Their prime religious function, as in Saxon times, continued to be as a call to worship. They were also rung at

death, at funerals, and to celebrate the great religious festivals. There were many secular reasons for ringing bells. They included, for example, the pronouncement and commemoration of national events, the celebration of local festivals, the marking of times for gleaning and curfew, and as a warning in times of danger. Each bell had a specific function, and the way it was rung communicated a particular message which was understood by everyone in the parish. There must have been periods when the communities of Rutland were dominated by the sound of bells.

All recorded bellringing customs in Rutland are described in Appendix 3 — Rutland Ringing Customs. It is important to note that through the centuries, many of these customs were revised or renamed as a result of political, ecclesiastical and social changes. Nearly all had ceased by the early twentieth century, partly due to the general availability of accurate time. Nevertheless, a recent survey of Rutland parishes carried out by the authors reveals that a number of bellringing customs are still followed and the messages they convey are just as important as those of the past.

Determined fund-raising efforts and the availability of substantial grants have ensured that Rutland bells and belfries have never been in a better condition. However, because individual incumbents are responsible for an increasing number of parishes, services are now conducted on a rota basis. This inevitably removes the regular weekly 'call to worship' in many villages. Although congregations no longer rely on church bells to summon them to services their sound still continues to communicate a timely message. **CUM VOCO AD ECCLESIAM VENITE** (When I call come to church) is a seventeenth-century inscription retained on the recast treble bell at Manton.

Scientifically constructed sundials, capable of indicating the time with reasonable accuracy, were introduced into this country towards the end of the fifteenth century, but it was probably another hundred years or more before they started to appear in Rutland. Their use continued alongside that of clocks until well into the nineteenth century. Vertical dials were usually placed in the gable of the south porch of a church, although some were located on the south wall of the tower, nave or chancel. As well as serving the purpose of the former scratch dial, they also indicated local time to the community and provided a means of regulating the church clock. Early records of church sundials include seventeenth-century vertical dials at Ashwell and Morcott. Evidence for these is provided in the Churchwardens' Accounts. Another early example is the cuboid sundial over the porch at Wardley. It is the only dial of its kind on a Rutland church. It is dated 1694 and displays the motto memento mori (remember death). The only known example of a horizontal or table sundial in a Rutland churchyard was that at Ridlington dated 1614.

During the seventeenth and eighteenth centuries it became fashionable to include scientific sundials as architectural features on the façades of houses, stables and public buildings. Engravings in the *History and Antiquities of the County of Rutland* (Wright 1684, 49 & 90, Additions 1788, 7) provide evidence of early examples on three former mansions: Exton House (Exton Old Hall), Martinsthorpe House and Luffenham House. Examples of dials on smaller seventeenth-century domestic buildings and early eighteenth-century inns can be seen in Uppingham (*see* Chapter 4 — Gazetteer, Uppingham). Table dials became fashionable for the wealthy during the same period and in Rutland the earliest known dial of this type was installed in the grounds of Normanton Hall during the late eighteenth century. Such dials would have been used by their owners and servants for regulating household clocks and watches as well as serving as an ornamental centrepiece for the garden. At least twenty-five table sundials are shown in the gardens of rectories, vicarages and other larger houses on the early 1900s Ordnance Survey Second Edition 25 inch maps of Rutland. It appears that only one, that in the east garden of Clipsham Hall, has survived in its original location.

Over ninety pre-1900 horizontal and vertical sundials have been recorded in Rutland. Unfortunately more than one third no longer exist. Of those that remain, although there are some notable exceptions, many are in a sad and neglected state, having long since lost their ability to indicate the time.

The first church clocks were made in the thirteenth century but they probably did not reach Rutland for another three hundred years or more. They had a crude form of escapement known as a 'verge and foliot'. This resulted in poor timekeeping and even the better clocks would gain or lose thirty minutes or more in a day, hence the need for regular checking against a sundial. However, this inaccuracy was of little consequence at the time as almost everyone in the parish would have been using the same clock. They were probably grateful for a device which told the time whether or not the sun was shining. The earliest known clocks installed in Rutland churches were at Wing, Uppingham, Exton and Ridlington, all of which were working by 1618. Almost one third of the county's churches possessed a clock by the end of the seventeenth century.

None of these early church clocks survive in Rutland,

but details of some have been gleaned from surviving documents, particularly Churchwardens' Accounts. They did not have dials and their movements would normally have been located at the foot of the towers. They communicated the time by striking a bell, and this was sometimes a small dedicated clock bell rather than one of the larger bells hanging in the belfry. There were clock bells at Burley, Glaston, North Luffenham and Morcott. Although some of these early clocks were later converted, the majority were eventually replaced by pendulum clocks with single-handed, and later, two-handed dials. These clocks were placed higher in the tower so that their dials could be seen from afar. The first turret clocks with a pendulum and anchor escapement were introduced by Joseph Knibb at Oxford *circa* 1669 and this development resulted in a dramatic improvement in timekeeping accuracy. A working example of Knibb's work, dated 1678, remains in the tower of the church at Burley. Four late seventeenth-century pendulum controlled turret clocks installed in Rutland churches by

This 'Patent Brass' American wall clock by Ansonia was supplied by James Sparkes of Uppingham. It was owned in the late 1800s by James Marlow, an agricultural labourer, and his wife Elizabeth of Belton in Rutland. This clock is still in good working order today

John Watts have also survived, although all have been removed from the towers.

By the end of the nineteenth century many of the larger houses in Rutland had become country retreats for the aristocracy who came to the county for the hunting. Stables were built or converted for this purpose and a clock placed over the entrance to the stable yard was a decorative and functional feature. All were installed between 1850 and 1912 and would have been of particular importance to the estate workers. Many of these houses also had a house bell which was often installed on a wall near the kitchen.

Inevitably many of the wealthy within the county would also have acquired one or more domestic clocks. The earliest were like small turret clocks, and these, from their shape, were known as lantern clocks or sheep's head clocks. They had an escapement which was very similar to the verge and foliot, the foliot bar being replaced by a balance wheel. By 1680, weight and spring driven pendulum clocks were being made in England. They almost immediately took the well-known forms of longcase (grandfather) and bracket clocks. Many such clocks, as well as watches, were made by local clockmakers, but this domestic industry had virtually ceased by the middle of the nineteenth century owing to factory production methods and low cost imports. Even so most of the population of Rutland would have still relied on the church clock or the ringing of a bell for their time needs.

By the close of the nineteenth century owning a clock became a little more likely for the average Rutlander. Mantle clocks would have been more of a feature within the homes of the early twentieth century, and the man of the house may have possessed a pocket watch, but wrist watches were by no means a common possession until the 1950s. The progressive introduction of the electric telegraph, the telephone, the wireless and the electricity grid (synchronous electric clocks) were other innovative mediums which brought more accurate and more available time to Rutland people.

In the latter decades of the last century and into the new Millennium, digital clocks and watches, the Speaking Clock, radio and television ensure that everyone has instant and affordable access to very accurate time information. However, the bells, scratch dials, sundials and clocks of Rutland are an important part of the county's heritage and it should never be forgotten that throughout the ages they played a significant part in ordering the lives of Rutlanders.

1.2 Church Bells

The church bells we are familiar with today were evident in England as early as the seventh century, but most were installed in the larger religious institutions. Directives were issued in 977 by St Dunstan with regard to ringing canonical hours within the monasteries (Walters 1977, 8-13). However such rulings would have had little influence on the activities of parish priests within the locality of Rutland.

By the tenth century the building of churches and the founding of bells in England had been encouraged by a decree. This stated that a thane's rank could be obtained by a Saxon churl or franklin if he possessed about five hundred acres of land, and had a church with a bell tower on his estate (North 1880, 6). The effect of this appears to have been minimal in Rutland because the Domesday Survey of 1086 only records eight churches in the county. Although by this time the art of bellfounding in England was well established, it is quite likely that handbells were used to summon Rutland congregations to worship.

The law of curfew was partially implemented by King Alfred and later fully implemented as a result of a decree by William I. It could not have been efficiently enforced in communities which did not have a church with a large bell, and this may have been the general position in Rutland. However the Norman period saw an increase in church building, and the taxation returns of Pope Nicholas IV in 1291 indicate that there were, by then, forty-five churches and at least thirteen parochial chapels in the county, a very different picture compared to two centuries earlier (*Victoria County History* I, 144-5).

The majority of Rutland's church towers and bellcotes were built in the thirteenth and fourteenth centuries and none is earlier than the twelfth century. This does not, however, eliminate the possibility that some of the more important churches in Rutland may have possessed bells, other than handbells, before this date. In the absence of a tower, bells could have been housed in a separate structure detached from the church, hung in trees or fixed to purpose-built frames. An early reference to bell ringing can be found on one of the late Norman pillars supporting the chancel arch in Stoke Dry Church (*see* Chapter 4 — Gazetteer, Stoke Dry). This church, built in the twelfth century, originally consisted of a small nave and chancel and a century later may have incorporated a bellcote over the west wall of the nave. This was probably the style of many of the churches in Rutland at this time. Six double bellcotes of thirteenth-century origin remain in Rutland. They are at Essendine, Little Casterton, Manton, Pilton, Stretton and Whitwell. Similar bellcotes are known to have existed at Bisbrooke, Great Casterton, Tickencote, Wing and possibly at Ridlington and Stoke Dry.

Early bells in England were regarded with such importance that they were baptised before being hung in the belfry. They were often dedicated to a saint, and the inscription ᴵᴺ ⱧOᴺOᴿE SᴬᴺᴄTI EIᴄDII (In honour of St Ægidius [St Giles]) on an early fifteenth-century bell at Whitwell and Sancta Fides Ora Pro Dobis (Saint or Holy Faith pray for us) on the bell of similar date at Tixover are examples of this practice.

In medieval Rutland, and right through to the twentieth century, agriculture was the main occupation, with self-sufficiency being the main concern. Life was hard for the inhabitants and their main source of comfort would have been a shared Christian belief. Daily life would have been set within a religious framework. A regular pattern of services and prayers in the individual parish churches would have been set, the bells playing an important part in the liturgy. In addition to their religious functions, the bells were important to the laity, for they relied upon them for many secular reasons. The bells would have been used to warn of danger from attack, fire and impending storms. In general, the belief held by the parishioners that the ringing of bells had the power to quench fires, dispel storms, stop disease and drive away plagues verged on the superstitious. They were convinced that the main disrupters of their well-being were evil spirits or demons and that they could be warded off by the sound of the bells. This is illustrated in the grovelling figure below the bellringer carved on the Norman pillar at Stoke Dry, which is thought to be a representation of Satan trying to shut out the sound of the Sanctus Bell.

The Constitution of Archbishop Winchelsea of 1300 mentions bells amongst the essential church ornaments to be provided by parishioners. However a church was only required to provide one bell which would serve the duty of ringing for services and for providing a death bell (Walters 1977, 115). The likelihood is that by about 1550 most churches in Rutland, with their added towers or bellcotes, would have had at least two bells and possibly a 'little bell' or Sanctus Bell. In Rutland today, only six of the county's forty-nine parish churches have only one bell.

Research by the authors has revealed that just under a quarter of the churches in Rutland had an early Sanctus Bell, but of these only one, that at Preston, cast by John Barber of Salisbury *circa* 1400, still exists. The Sanctus Bell was often hung in a bellcote on the gable at the east end of the nave. Today, only those bellcotes at Manton and Market Overton remain. The Sanctus bellcote at Stretton is shown in a drawing of *circa* 1793 (Rutland County Museum F10/1984/51), but it had been removed by 1839. That at Caldecott was taken down in 1976. The Sanctus Bell was rung at the beginning of and during Mass so that all the villagers, even if they were at work or at home, could join in the Holy Song of Adoration.

By the Reformation, local bellringing customs, both religious and secular, were firmly established and the ringing of bells would have been an invaluable source of communication for the Rutland population. Everyone would have known the messages given by the way particular bells were rung. More importantly they marked the time, thus answering the purposes of a clock. A comment by Bishop Latimer in 1552 confirms that by then most parishes had a church with at least one bell, thus emphasising their importance:

> ... if all the bells in England should be rung together at a certain hour, there would be almost no place but one bell might be heard there ... (Corrie 1844, 498).

What immediate impact the Reformation made on the use of church bells in Rutland is not entirely known. The Prayer Book of 1552 directed that the curate should say morning and evening prayer in the church every day, preceded on each occasion by the tolling of a bell to summon the parishioners. The new book dispensed with most of the ceremonial of the medieval church, and with the passing of the Act of Uniformity in 1558 the liturgical use of bells finally ceased. The Canons of 1603 directed that:

1. The Litany was to be said on Wednesday and Friday, and that the parishioners should be informed of this by the tolling of a bell.
2. A Passing Bell was to be tolled when anyone was dying. After the party's death there was to be rung no more than one short peal, with another before and one after the burial.
3. The superstitious use of bells on unlawful festivals and other occasions was forbidden.

Although such directives removed the excessive ringing of bells that had previously regulated the daily round of prayer, in general even the religious and political upheavals of the sixteenth and seventeenth centuries did little to diminish the overall use of them. Even if the religious significance of a particular ring was no longer relevant, a secular function gradually took its place, for 'old habits die hard'. In fact there was an ever-increasing demand for the bells to be rung.

It was, and had been since medieval times, the duty of a church officer, usually the parish clerk, to ring a bell for services and at the death of a parishioner. His religious duties would have included ringing the Angelus, Shrive and Sermon Bells. After the Reformation many of his former religious bellringing duties remained but others took on a secular guise and were given a different name. Two such examples are the Pancake Bell and the Early Morning Bell. Churchwardens' Accounts show that extra payments were made for ringing on such occasions. Over the next few centuries the Curfew, Gleaning and Dinner Bells were the more important 'time markers' rung in Rutland (*see* Appendix 3 — Rutland Ringing Customs). Disputes inevitably arose as a result of requests for extra payment, and there is a good example in the Uppingham records of 1628. The parish clerk firmly stated that he was not bound by law to ring the Curfew Bell. He argued that its purpose was secular and was no part of his ministerial duty!

After the Reformation it became traditional for all the bells in the belfries to be rung on an increasing number of special occasions. This required a group of men, one for each bell, to perform the task. As a consequence the hobby of 'pleasure ringing' was born. Bells were rung to announce the great religious feasts and festivals, to celebrate weddings, and to commemorate local and national events. Loyal peals abounded on the occasion and anniversary of coronations, royal marriages, and the birthdays of monarchs and their offspring. The sound of these peals would have done much to lighten and brighten the days of toil in this agricultural county.

The introduction of 'change-ringing' in the early seventeenth century encouraged the continuation of this enjoyable pastime, which was initially taken up by young gentlemen, who tended to regard it as a sport. More of the laity became interested in the art and many of the locals were eager to participate, for the increased demand for such ringing presented the opportunity of earning extra money. Gradually the laity took over the running of the belfries from the clerics, and societies of lay ringers were formed. As a consequence, during this century and the next, much attention was directed towards the recasting and augmentation of the bells, made possible due to support from the gentry. By 1880 two-thirds of the church bells in Rutland were either newly cast or had been recast during the seventeenth and eighteenth centuries.

Many of the available church accounts include payments to 'ringers', and these were sometimes in kind, such as ale, cheese and tobacco. At Ashwell in 1708 payments were made to village ringers as well as to bands from Cottesmore, Langham, Oakham and Stapleford (Leicestershire). During the latter part of the seventeenth and into the eighteenth century, Ashwell ringers sounded the bells regularly for 'Gunpowder Day', for the birthdays of Royalty, for Coronation Day, for peace, for victorious battles, and on the Archdeacon's Visitation. These examples are typical of ringing in many of the villages of Rutland, reminders of events both past and present.

The nineteenth century saw the fabric of the majority of the churches in Rutland in a very poor state. Major programmes of restoration took place in many of them and it was inevitable that the towers and bells received attention. This revival of interest in the belfry led to the laity losing control of bellringing. Many diocesan and county societies for bellringing were founded, the emphasis being placed on ringing for Sunday services. The list of rules for bellringers on a board at Ashwell Church

includes: *to be present at Morning, or Evening service, every Lords day, & on great Festivals, or Fasts, such as Xmas day. Good Friday ... only to Ring by leave of the Rector ... not to be wilfully nor habitually guilty of swearing or drunkenness: any Member guilty of these to be excluded from the society ....* There are also 'ringer's rules', both with similar wording, on a board at Ryhall and painted on the ringing chamber wall at Tinwell. Lyddington and Oakham also have ringers' rules.

Many of the traditional and secular reasons for ringing bells had ceased by 1880. By this date over half of the parish churches in Rutland had a clock and this had a bearing on the demise of some of these customs. Fortunately, a detailed record of Rutland bells and their 'inscriptions, traditions and peculiar uses' was made and published in 1880 by Thomas North, FSA, in *The Church Bells of Rutland*. This was at a time when a number of the 'peculiar uses' of church bells were still alive. In order to collect this information as well as the physical and historical details of the bells, their inscriptions and founder's marks, Thomas North enlisted the help of local people, many of them being the incumbents (*see* Appendix 6 — Thomas North). *The Church Bells of Rutland* has been a major source of information and inspiration in producing *Time in Rutland*, and full credit is due to Thomas North and his assistants for publishing such a remarkably accurate survey. Full details of religious and secular bellringing customs, both in 1880 and at the present time, are included in Chapter 4 — Gazetteer, and Appendix 3 — Rutland Ringing Customs.

Despite the ecclesiastical, political and social changes over the centuries the use of church bells has endured right through to the present day, both for their original intent and for other traditional reasons. Their usefulness has never been in doubt and the fact that during the nineteenth century bells were placed on new schools, railway stations and large houses emphasises their importance and worth as instruments of time.

In 1940, Parliament ordered that all bells remain silent during the Second World War. They could only be rung as an alarm signal in case of invasion. At the end of the war the bells throughout the country rang out in victory, a parallel with those customs that spread good news in this way centuries earlier.

The results of a survey, completed by the incumbents, churchwardens and tower captains of Rutland, reveal that today the bells continue to be rung for the announcement of all Sunday services. They are chimed for periods that vary from between two to thirty minutes but at Great Casterton, Greetham, Empingham and Uppingham all or the majority of the bells are rung full circle. At Greetham they are also rung after services. At Belton, after two of the smaller bells have chimed for fifteen minutes, one of the bells, referred to as the Priest's Bell, is then chimed for three minutes prior to the start of the service. At Uppingham the second bell is chimed at the Elevation of the Host for the entire town to hear. At Oakham the bells are rung on All Saints Day and Ascension Day. If requested the bells are tolled for funerals at almost a third of Rutland's churches. They are usually tolled if the deceased was a ringer, a person of local note or of national importance. At Stoke Dry the bell is still tolled when a parishioner dies. At Stretton on 1 April 1997 the bells were rung when Rutland regained its county status. These examples reflect traditional customs of long ago.

Other occasions when the bells are rung may not be as numerous as those in centuries past, but their sounds still act as a reminder of local and national events. If requested, and if there is a band of ringers available, the bells will ring out to celebrate the occasion of a wedding in almost half of the churches. They are also rung in some communities for New Year, harvest festivals, village fêtes, visiting preachers, birthdays and anniversaries. At Oakham they are rung for civic and memorial services and at Uppingham on the occasion of royal weddings, birthdays and funerals. Twenty Rutland churches have their own band of ringers and celebrations to mark the new Millennium helped to create additional interest in bellringing.

Although many bellringing customs have become extinct other than for the call to worship and to celebrate the main Christian Festivals, a twentieth-century innovation prevails. The ringing of muffled or half-muffled bells on Armistice Day or Remembrance Sunday is a custom which continues in many Rutland parishes.

1.3 Saxon Sundials

The origin of the sundial has not been recorded but early man must have noticed that the direction and lengths of shadows varied according to the time of day and the time of year. A pole driven vertically into the ground may have been the first rudimentary sundial. The earliest sundials were Egyptian, about 1300 BC. The earliest surviving sundials in Britain are of Saxon origin and the oldest of these is at Bewcastle Cross in Cumbria, dated *circa* AD 685. Well-preserved examples can also be seen at Kirkdale, North Yorkshire, Daglingworth, Gloucestershire and at Barnack, Lincolnshire

The sundial, either circular or semi-circular, would have a style or gnomon hole at its centre. From this would be a series of downward radiating lines extending to the circumference. The gnomon was placed perpendicular to the dial face and its shadow indicated the time

The well-preserved Saxon sundial over the south door of St Gregory Minster, Kirkdale, Yorkshire. It is considered to be circa 1065. The dedication is on panels on either side of the dial. Part of the inscription translates as:
+ THIS IS THE DAY'S SUN MARKER + AT EVERY TIDE + AND HAWARD MADE ME AND BRAND PRIESTS

The Saxon sundial on the tower at Barnack Church, Lincolnshire, drawn on 7 March 1885 by J T Irving (Gatty 1890, 482). *The decoration is similar to that found on other Saxon sundials*

from sunrise to sunset. It is understood that the Saxons divided the twenty-four hours into eight tides, each of three hours. The Christian church had a strict system to mark the regular times of prayer and these church offices, held at three-hourly intervals, were known as 'canonical hours'. They therefore fitted in with the 'tide system' of time measurement. The lines that mark the canonical hours on the dial, indicating the present day equivalent of 9am, noon and 3pm, were of special significance, and were usually marked with a cross at their extremity. The nearest Saxon sundial to Rutland survives on the south face of the tower at Barnack Church, near Stamford, Lincolnshire.

Although archaeologists have found considerable evidence of Saxon occupation in Rutland (Clough et al 1975), no similar dial has been recorded in the county. However an unusual dial was found in 1929 during site excavations at the Ram Jam Inn on the Great North Road near Stretton. It is said (*Victoria County History* II, 135) to be a Saxon sundial but there is considerable doubt about this (*see* Chapter 1.1 — Historical Introduction).

1.4 Scratch Dials

The careful inspection of the south-facing walls of many churches will often reveal marks consisting of radial lines around a small hole. We refer to these as scratch dials, and many are so weathered and covered in lichen that they can only be seen when the wall is illuminated by the early morning or late afternoon sun. They are called scratch dials because many of them have been scratched, rather than incised, into the surface of the stone. They are clearly related to Saxon sundials although significantly inferior in construction and considerably more variable in design. They are also known as primitive sundials, medieval sundials, mass dials and mass clocks. It is generally agreed that they are of medieval origin and were a type of shadow dial made by priests to indicate specific times when bells should be rung. They were, however, incapable of telling the time as we know it today as only specific event lines were drawn.

The basic scratch dial consists of a small hole, usually between one and two metres above ground level, into which a straight stick was placed, perpendicular to the wall. This stick is referred to as a gnomon. On days when the sun was shining, this stick would create a shadow. At noon this shadow would be vertically below the gnomon. Hence, a line scratched vertically down from the hole would be the noon marker. This is the only occasion when a scratch dial is at all accurate as a time indicator, and is only correct for a direct south facing wall, but slight deviations were of little consequence in medieval times. Additional radial lines were the markers

for mass, other services and times of prayer. Scratch dials are generally found where the footpath meets or used to meet the church on the south side, and so the majority are located near the south door or near the Priest's door.

There are no known references to scratch dials in contemporary documents, and until recently, there has been no serious attempt to record them in any detail on a national basis. One of the first writers on this subject was Sir Henry Dryden who made detailed drawings of many Northamptonshire scratch dials. In December 1896 he presented a paper to the Northampton and Oakham Architectural Society under the title *Squints and Dials*. His comments on the dials he surveyed apply equally well to the scratch dials found on Rutland churches today. It is interesting to note that at this stage he had no idea what their function was:

> The sort of dials to be described must not be confused with ordinary Sundials, either upright or horizontal. They are, with rare exceptions, upright upon the walls and formed by incisions in common building stones, often on quoins and buttresses. Their use and the reason for their formation have not been determined. They are usually circles or parts of circles from 3 inches [75mm] to 10 inches [250mm] in diameter, formed by grooves about ⅛ inch [3mm] wide and the same deep, placed from 4 feet to 7 feet [1.2m to 2.1m] from the ground. They have a central hole of ½ an inch [13mm] or ¾ inch [19mm] in diameter and from ¾ inch to 1 inch [19mm to 25mm] in depth, and lines or rows of small holes radiating from the centre, and in some instances small dots or cavities in the periphery of the circle. Some have rays or dots only in the lower half of the circle. There are many varieties. The number of rays differs much, and they are usually more or less irregular. Rays in the upper part of the circle were useless for dial purposes. None contain figures.

His paper was subsequently printed in the *Transactions* of the Worcestershire Architectural Society (Dryden 1897, 354).

At this time there were a number of other researchers who presented papers on their theories as to the possible function of these dials, but Dom Ethelbert Horne (1917

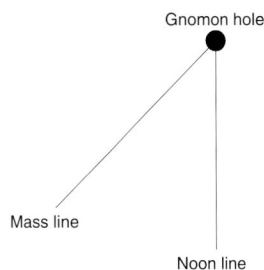

Gnomon hole

Mass line

The basic scratch dial   Noon line

& 1929) was the first author to publish books devoted specifically to the study of scratch dials and much of what he said then is in line with current thinking. The present interest is due mainly to the national survey being conducted by the Mass Dial Group of the British Sundial Society.

The only published source which refers to scratch dials in Rutland is *Victoria County History* (**II**, 1935) where they are recorded at Braunston, Ketton, South Luffenham, Lyndon, Stretton, Stoke Dry and Whitwell. However, a detailed survey conducted by the authors between 1998 and 2000 reveals that there is in fact a total of at least sixty-two surviving scratch dials on twenty-eight of Rutland's parish churches, and one on a secular building. All but two are illustrated in Chapter 4 — Gazetteer. This total probably represents less than 30% of the original as many have been lost as a result of weathering and extensive restoration work, particularly during the late Victorian period. It is thought that by the end of the fifteenth century every Rutland church would have had at least one scratch dial, and in some cases there would have been many more. At Egleton twelve have survived, and there are five at Caldecott and four at both South Luffenham and Whitwell. These may possibly be explained by the fact that each succeeding priest made his own scratch dial. Changes in service times, fabric alterations and the ever-growing churchyard yew tree, which put the dial into shadow, are other possible reasons for a multiplicity of scratch dials on any one church.

As there are no contemporary records which refer to scratch dials, it is necessary to consider the dating evidence provided by the walls on which they have survived. Most church walls have been dated and this provides the earliest possible date for a dial scratched onto a particular wall. Fourteen Rutland scratch dials are inside porches, to the side of, or above, the main door. They are in permanent shadow and are therefore incapable of performing their intended function. They were obviously made before the porches were built, thus providing the latest possible date for when they were made. The dating evidence for these scratch dials is given in the following table. All dates are from *Victoria County History* (**II**, 1935):

Church	Scratch Dial Ref (as in the Gazetteer)	Date of Doorway	Date of Porch
Braunston	1	early 13th C	15th C
Egleton	4	12th C	14th C (original porch)
Lyndon	2	13th C	modern
Lyndon	3	13th C	modern
North Luffenham	1	13th C	modern
Preston	1	early 14th C	14th C
Seaton	1	12th C	14th C
Stretton	1	12th C	13th C
Stretton	2	12th C	13th C
Tixover	1	early 13th C	15th C
Whissendine	1	13th C	14th C
Whissendine	2	13th C	14th C
Whitwell	3	possibly 12th C	13th C
Whitwell	4	possibly 12th C	13th C

All the above scratch dials are of similar form. From the detailed survey, although there are many varieties, it is possible to divide Rutland scratch dials into two groups. However, the loss of lines due to erosion makes this classification imprecise. Group A, which includes all the above, consists of scratch dials with lines radiating from a central hole, the majority being below this hole. These scratch dials possibly date from the twelfth to the fourteenth century. 68% of all recorded scratch dials in Rutland fall into this group.

Group B consists of what appear to be more developed scratch dials. On these there are lines both above and below the central hole and there is often a circum-

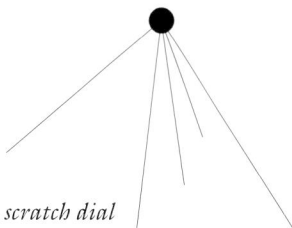

A drawing of the Group A scratch dial to the west of the main door inside the porch at Braunston Church

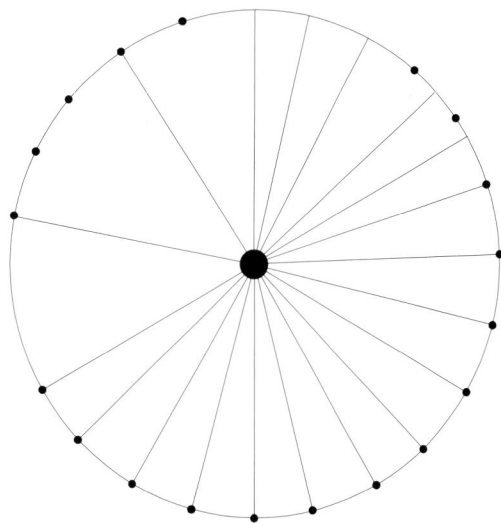

A drawing of the Group B scratch dial on the west front of the south porch at Manton Church

scribed circle. In some cases there is a ring of dots, known as 'pocks', instead of or as well as the circle. The majority of these are thought to be of fourteenth and fifteenth century origin. Caldecott 2, Manton 1 & 2, Preston 2, South Luffenham 1 & 2, Stoke Dry 1 & 2, Whitwell 1 and Wing 1 (*see* Chapter 4 — Gazetteer for scratch dial reference numbers and details) are in this group and appear to be misguided attempts to divide the day into twenty-four hours. Preston 2 is the most accurately laid out dial in this respect but it would be a very poor time keeper with a horizontal gnomon. The lines in the top half of the dial would of course serve no useful function.

Although there is a great variety of scratch dials in Rutland, there is a number of common features which apply to both groups. All have a style or gnomon hole, although some have now been filled with mortar. Some are very shallow and in this case the stick gnomon was held in place by hand. None has the remains of a gnomon. 84% have a 'noon line' within 10 degrees of vertical and 76% have a 'mass line' which would indicate a time of between approximately 9am and late morning. Most dials have a multiplicity of lines, possibly indicating different or changed mass and service times. Thirty-nine scratch dials (62%) have lines which would indicate afternoon service times.

Ten Rutland scratch dials have been relocated as a result of restoration work and are now inverted, on walls which do not face south, or in shadow from the church structure. They are Ayston 1, Burley 1, Clipsham 2, Egleton 6 & 12, Greetham 1, 2 & 3, Lyddington 1 and Teigh 1. Another is on the south elevation of the cottage at 20 St Mary's Road, Manton, and is the only scratch dial on a secular building in Rutland. The cottage is dated 1733 and the dial is probably on a stone reclaimed following a restoration of Manton Church, or salvaged from a demolished medieval church, the nearest being Martinsthorpe.

A scratch dial was obviously only effective on days when the sun was shining. Although water clocks, candles and hour glasses were used in monasteries to record the passage of time, it is very doubtful if such sophistication was available at parish church level. No doubt the priest relied on his body clock to know when to ring the bell for services on days when there was no sun. The scratch dial was a very simple but highly effective instrument. Its importance is proven by the great number that survive today.

There is little doubt that the ability of a scratch dial to indicate noon fairly accurately was used as a means of regulating early church clocks, and this practice may have carried on well into the seventeenth century. They were, however, gradually replaced by scientific sundials, with gnomons parallel to the earth's axis. These could indicate local time quite accurately any time that the sun was shining on the dial. Between the two there are some sundials which are probably better described as transitional scratch dials and there is a good example in the gable of the thirteenth-century south porch at Stretton, the only one in Rutland (*see* Chapter 4 — Gazetteer, Stretton). It is crudely marked on a limestone block set into the wall and originally had an angled gnomon. Roman numerals can be seen at the outer ends of the lines and it has been suggested that this was an original scratch dial, removed from its former location, with the gnomon and numerals added at a later date (information from Walter Wells). It is not shown on early drawings and there is no record of a clock at this church.

One other unusual scratch dial is that on the south aisle wall at Clipsham. This is very carefully constructed with straight lines incised into the stone, all meeting at the centre of the gnomon hole (*see* Chapter 4 — Gazetteer, Clipsham). It is thought to be a late scratch dial and was no doubt used to regulate the clock installed by John Watts in 1688.

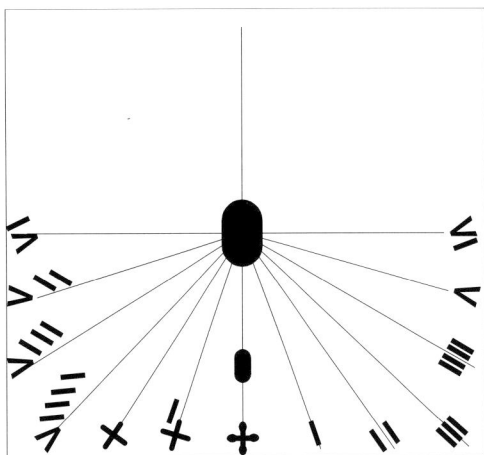

The transitional sundial in the gable of the south porch at Stretton Church

The scratch dial on the south aisle wall at Clipsham Church

1.5 Scientific Sundials

The Saxon sundial and the medieval scratch dial were unable to show the time as we know it today as both had gnomons which were perpendicular to the vertical dial face. They would indicate noon quite accurately but the time taken for the shadow to traverse the dial between sunrise and sunset varied considerably between winter and summer. Equal division of the dial between these extremes does not result in equal divisions of time throughout the year unless the dial is on the equator. In Rutland the length of time between sunrise and sunset varies between about 7.75 hours in winter and about 16.5 hours in summer so summer hours would be more than twice the length of winter hours. The change between the two would of course be gradual and of relatively little significance.

The scientific sundial, however, with its gnomon parallel to the earth's axis, is able to indicate equal hours throughout the year. It is thought to have been developed by the Arabs and possibly first came to Europe as early as the twelfth century. But its adoption was very gradual and it was not until the fifteenth century, by which time the division of the day into twenty-four equal hours had come into common use, that the art of dialling became established in Britain as an important mathematical subject. By this time mechanical clocks were becoming established and they also recorded equal hours. The scientific sundial was therefore a very convenient means of checking their regulation. The alternative was to use the noon line on a scratch dial.

For four hundred years scientific sundials and clocks flourished side by side. The sundial performed the function of clock regulator and public timekeeper on sunny days, and the clock a public timekeeper throughout the day and night and in all weathers. Initially, however, sundials were much cheaper than clocks and consequently were by far the most widely employed of all time indicating devices. Many were set up in public places, and portable sundials, which could be folded up and put in the pocket, became very popular. The usefulness of the sundial began to wane in the early nineteenth century, with the greatly increased accuracy of clocks.

The use of a sundial for checking a clock is not quite as straightforward as it might seem. If it is compared to an accurate clock over a period of a year it will be found that the sundial is slow by about fourteen minutes in mid February, fast by about sixteen minutes at the beginning of November, and accurate on 16 April, 15 June, 1 September and 25 December. The sundial indicates apparent solar time whereas the clock indicates mean time, the difference between the two being known as the equation of time. The equation of time has two causes. The first is that the plane of the Earth's Equator is inclined to the plane of the Earth's orbit around the Sun. The second is that the orbit of the Earth around the Sun is an ellipse and not a circle. The total of these two effects gives the equation of time. Early foliot clocks were not accurate enough to distinguish between apparent and mean time and the equation of time was only of academic interest until accurate pendulum clocks were introduced towards the end of the seventeenth century. The equation of time was systematically worked out by the astronomer John Flamsteed (1646-1719) between 1665 and 1670. It was published in almanacks and included on some sundials so that owners and keepers of

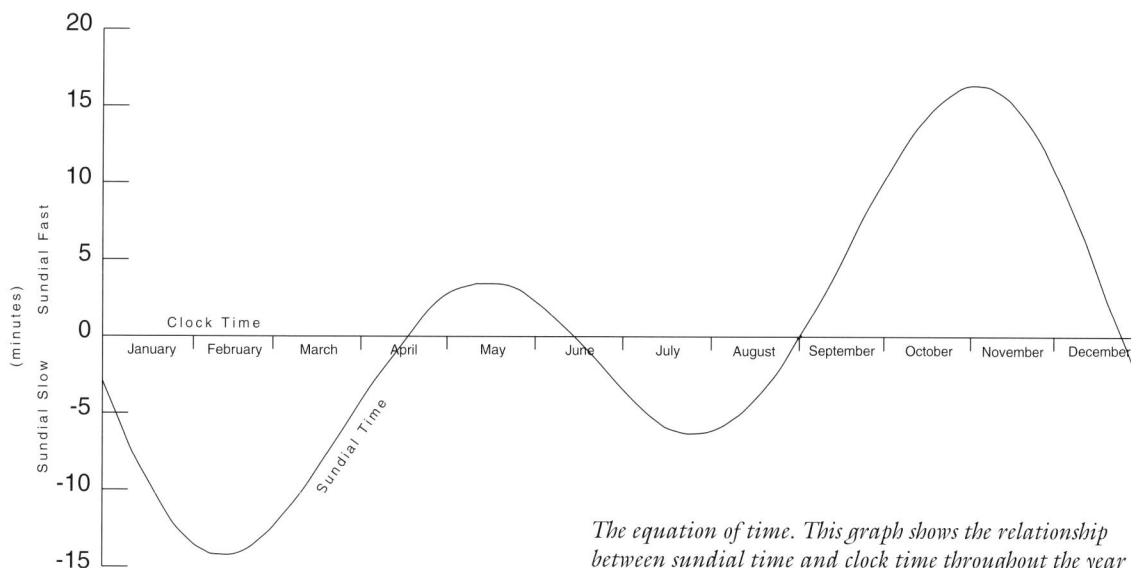

The equation of time. This graph shows the relationship between sundial time and clock time throughout the year

clocks and watches could obtain an accurate time check. Today, a correction also has to be made for British Summer Time. It must also be remembered that any sundial erected before the adoption of Greenwich Mean Time will have been designed to indicate local time, although in Rutland the correction for this is of little consequence (*see* Chapter 1.7 — The Electric Telegraph and Standard Time).

There are many types of sundial and these are adequately described in a number of sources (*see*, for example, Daniel 1986 and Waugh 1973). Primary dials include those drawn on a horizontal plane, known as horizontal or table sundials, and vertical dials which are those drawn perpendicular to the horizon and directly facing north, south, east or west. Secondary dials are those which are drawn on other planes. A vertical dial facing due south is known as a direct south dial. Very few walls exactly face a cardinal compass point and consequently most vertical dials decline and are therefore classed as secondary dials. A dial which declines from the south towards the west is known as a south-west declining dial.

Almost a hundred existing and former sundials have been recorded in Rutland (*see* Chapter 4 — Gazetteer and Appendix 4 — Rutland Sundials). During the eighteenth and nineteenth centuries there were undoubtedly many more, all of which have been lost without trace. This is supported by a note in *Rutland Magazine* (Phillips 1911-12, 72) which states that in Caldecott 'Sundials were formerly rather numerous ...'. The remains of an early sundial on a building attached to Mill Garage is the only one known to have survived.

Most of the county's lost sundials have been identified from the *circa* 1793 (Rutland County Museum) and *circa* 1839 (Uppingham School Archives) series of drawings of Rutland churches, and the early 1900s Ordnance Survey Second Edition 25 inch maps of Rutland which show table sundials in many of the larger houses of the county. Others have been noted on old photographs and the engravings in James Wright's *History and Antiquities of the County of Rutland* (1684).

The majority of Rutland's collection of sundials are vertical south decliners. Many are in poor condition, and at the time of writing, few are capable of indicating the time. Some, however, have been restored to good working order. Examples are the 'Tempus Fugit' sundial at 13 High Street, Oakham, the sundial at Sundial House, Wing, and the sundials over the church porches at Belton in Rutland and Preston. The best examples of the unrestored sundials still in working condition include those on the manor houses at Preston and Langham, Sundial House in High Street East, Uppingham, and City Yard House, Wing. Perhaps the most interesting of the vertical dials are those over the porch at Wardley Church and over the Buttercross in Market Place, Oak-

ham. At the time of writing both were in need of major restoration. Sundials with mottoes include the table sundial at The Cottage in Nether Street, Belton, and vertical dials at Caldecott Church and Hambleton Hall. There are eight vertical and two horizontal sundials with mottoes in the county, and full details of these are given in Appendix 4 — Rutland Sundials.

The loss of so many of Rutland's sundials is unfortunate. In the case of the Victorian table dials it is hoped that more have survived than the authors' survey suggests. The county town's most recognised sundial in High Street aptly displays the motto 'Tempus Fugit' (Time Flies), reinforcing the message that if sundials are not identified, recorded and preserved, yet more of our heritage will soon be lost forever.

Although sundials are no longer prime 'time markers', table sundials in particular remain popular as functional garden ornaments. Two such sundials were installed in Rutland in the Millennium year. They are at Clock House Court, Barleythorpe, and on Masons' Lawn at Uppingham School (*see* Chapter 4 — Gazetteer, Barleythorpe and Uppingham).

A new horizontal or table sundial was presented to Uppingham School by the praepostors [prefects] in the Millennium year and erected on Masons' Lawn, near the School Library

1.6 Clocks & Watches

Continental words for bell include *clocke* (Dutch), *glocke* (German), and *cloche* (French), thus highlighting the connection between clocks and bells. The very first mechanical clocks were used in religious houses to indicate when bells should be rung to mark the start of one of the many daily observances. The next step was to modify the design so that the clock could lift a hammer to strike the bell automatically. Thus the turret clock came into being.

Geared mechanisms were understood and used by the Greeks more than two thousand years ago but they had no mechanical method of turning these devices continuously, and at a uniform speed, in order to measure the passing of time. Before the advent of mechanical clocks, water clocks, known as *clepsydrae*, were used. These took many forms, including graduated vessels from which the water was allowed to escape slowly, and the 'sinking bowl' type, an example of which was found at Market Overton (*see* Chapter 1.1 — Historical Introduction).

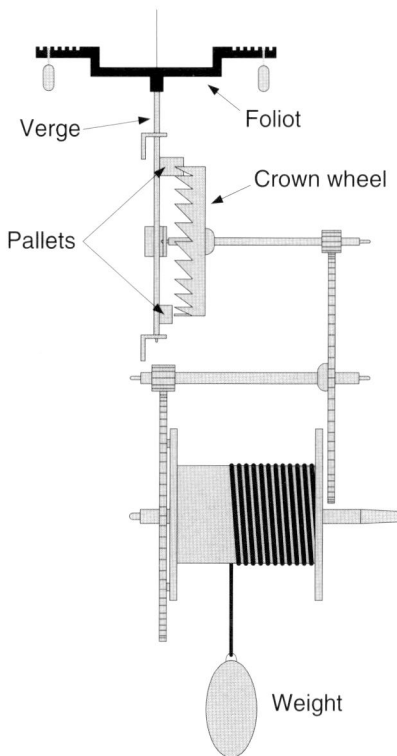

The verge and foliot escapement. The upper beam, or foliot, which carries adjustable weights, is attached to the verge which has two pallets. These engage alternately with the teeth of the crown wheel and as a result the foliot is pushed backwards and forwards in a horizontal plane about its centre. The positions of the weights on the arms of the foliot determine the period of oscillation

For mechanical clocks to be developed, an escapement and a constant motive force were required. The latter was provided by a lead, iron or stone weight. The earliest escapement was the verge and foliot and it was first applied to a mechanical clock in the last quarter of the thirteenth century. This type of escapement can be seen on the former Edith Weston church clock, now in Rutland County Museum (L1975.11) (*see* Chapter 4 — Gazetteer, Edith Weston). It was the only form of escapement used on clocks until the invention of the anchor escapement nearly four centuries later.

Although adequate for most of the period in which it was used, the verge and foliot escapement was a poor timekeeper by modern standards. This was mainly due to variations in friction in the gear train, and the best that could be expected was a gaining or losing rate of twenty to thirty minutes per day. Some adjustment was possible by moving the weights on the foliot.

According to Beeson (1977, 6-9) the earliest recorded escapement-controlled clock, in this country or on the Continent, was working in the Priory of the Augustinian Canons at Dunstable, Bedfordshire, in 1283. Other records of the same period include references to a clock at Exeter Cathedral in 1284, Old St Paul's Cathedral in 1286, Westminster Abbey in 1288, Norwich Cathedral Priory in 1290, and the Benedictine Abbey at Ely in 1291.

All the early clocks were turret clocks with striking trains. A few also had chiming trains, but dials were very rare and these were limited to internal astronomical dials. Skilled blacksmiths copied the basic design once it had been established. These men gradually became known as clocksmiths, and eventually watch and clockmakers. The installation of turret clocks gradually spread from the monasteries to cathedrals and to larger churches in towns, but it was probably not until the middle of the sixteenth century that they were first installed in country parishes. These were essentially similar to the early clocks in operation and construction, but they became smaller, and sometimes wooden frames were used instead of iron. The relative positions of the gear trains also varied, from end to end (fieldgate) to side by side (birdcage) or one above the other (doorframe). The form and function of these and later turret clocks were studied and analysed in great detail by C F C Beeson who published his findings in *English Church Clocks 1280-1850* (Beeson 1977).

There are no surviving records which might indicate when the first church clocks were installed in Rutland. Neither do we have any details of their movements or who made them. Irons' Notes (*see* Chapter 4 — Gazetteer Introduction) provide the earliest confirmation of working clocks in Rutland churches:

Wing

1605 There was a clock in the church on the death of Mr [Robtus] Cooke [*circa* 1602] the late parson & now taken away by Mrs Cooke the relict of the said Mr Cooke
(MS 80/1/3)

Uppingham

1618 April 11 Wilbron (Everard, Guard) [Everard Wilbron, churchwarden] to certify that the dyall goeth well
(MS 80/1/3)

Exton

1618 April 11 The clock there goeth not right but maketh a greate rumbling when it goeth   (MS 80/5/24)

Ridlington

1618 The clock & chimes are out of repair and do not go   (MS 80/1/3)

The earliest surviving Churchwardens' Accounts at Lyddington (DE 1881/41) (*see* Chapter 4 — Gazetteer Introduction for source references) show that there was a church clock there by 1626, and in 1629 'Bouth [Roger Booth] ye clock keper' was paid 6s 8d for 'kepinge ye Clock' (DE 1881/41). A little later, Preston Churchwardens' Accounts provide the earliest record of a clock being installed:

1656-57 ffor a Clocke and setting it up   £3 5s 0d
(DE 2461/39)

An agreement to maintain the clock was made with Henry Nicholls of Glaston and it is quite likely that he was also the supplier and installer (*see* Chapter 4 — Gazetteer, Glaston). Another early clockmaker, or clocksmith, was William King who was paid 18s for mending Uppingham church clock in 1633 (DE 1784/17), and a Glebe Terrier of the same date (MF 495) confirms that there was also a clock at Belton.

Churchwardens' Accounts and other documents show that by the end of the seventeenth century nearly one third of all Rutland churches had a clock. Ownership may well have been higher than this but accounts have not survived in many parishes.

Weight-driven domestic clocks were available from about 1400 onwards but until about 1550 they were great rarities and only the very wealthy could afford to own one. They had iron frames and, except in size, they were very similar to turret clocks. The movement usually consisted of going and striking trains, and often an alarm. The clock stood on a wall bracket with the weights hanging below and the escapement was a modification of the verge and foliot, with the foliot being replaced by a balance wheel. After 1550 brass was used extensively in the construction and they became known, from their shape, as lantern clocks, or sheep's head clocks. They were very popular and were being made, with a verge escapement and short pendulum, until well into the 1700s. A lantern clock by John Watts of Stamford is illustrated in Chapter 3 — Clockmakers, John Watts. The manufacture of portable clocks and watches was

made possible when a spring was used instead of a suspended weight as the motive force. The first watches were made in this country after about 1580. The first spring-driven clocks were introduced about 1630. They were in the shape of a drum with the dial uppermost, but by the late 1600s they had taken the well-known form of the bracket clock. Early continental spring-driven clocks and watches were even more inaccurate than weight-driven clocks because of the reducing torque exerted by the spring as it wound down. This was solved by the introduction of the fusee, a device consisting of a cone with a spiral groove cut into it. This evened out the torque, and was fitted to English watches, bracket clocks and wall clocks from the start.

The greatest revolution in timekeeping came as a result of a discovery by Galileo in 1582. He found that the time of swing of a pendulum was almost independent of its amplitude. The next step came when a clock mechanism was used both to impel the pendulum (to keep it swinging) and to count the swings. Galileo and his son, Vicenzio, worked on this and their drawing of such a clock mechanism has survived, but there is no record of them actually constructing a working version. Christiaan Huygens, a Dutch mathematician, was the first to construct a pendulum clock with a verge escape-

The anchor escapement of the John Watts clock at Clipsham

ment in 1656. He published his design in *Horologium* in 1658 and Ahasuerus Fromanteel introduced it into this country in the same year. The verge escapement and short pendulum was first used in domestic clocks, and despite the invention of the more accurate anchor escapement about 1668, it was used in lantern, bracket and wall clocks well into the next century.

The anchor escapement allowed the use of a longer pendulum with smaller oscillations, and the Royal Pendulum, beating seconds, became the norm for turret and longcase clocks. This development resulted in a dramatic improvement in timekeeping accuracy, and longcase clocks began to be produced in large numbers.

Joseph Knibb was very interested in the work of Ahasuerus Fromanteel and he built the first turret clock with an anchor escapement about 1669-70 for Wadham College, Oxford. However, it is not clear whether it was he, Robert Hooke or William Clement who was responsible for inventing or perfecting it. Rutland is fortunate in that it has a working example of a Knibb clock, dated 1678, in the tower of the church at Burley (*see* Chapter 4 — Gazetteer).

One local clockmaker who adopted this invention for his turret clocks was John Watts of Stamford. Only sixteen years after Joseph Knibb built the first pendulum controlled turret clock in Oxford, Watts was making his own version in a timber frame and they were subsequently installed in a number of local churches (*see* Chapter 3 — Clockmakers and Chapter 4 — Gazetteer, Burley). He also made lantern and longcase clocks.

With the invention of the anchor escapement and the introduction of the longcase clock, domestic and turret clocks developed along different lines.

Early watches were considered more as jewellery than as serious timekeepers. They were usually worn on a cord round the neck, to be seen and admired as an item of rarity and high cost. This was to change by the end of the seventeenth century. Both Christiaan Huygens and Robert Hooke had made or commissioned watches with verge escapements and spiral balance springs by 1675. Thomas Tompion was also involved in this development work and he invented a device to alter the effective length of the balance spring and so provided a means of easily adjusting the timekeeping of watches. The improvements made by these and other inventions were so great that from then on watches were provided with two hands and their dials were subdivided into minutes.

There were many other developments, including the introduction of jewels to reduce friction, and temperature compensation. However, the verge pocket watch was to become the standard which was to last until nearly 1850 when it was replaced by the detached lever watch. By this time watches were being mass produced in Swiss and American factories, and traditional English clock-making was at an end.

The only known verge watch by a Rutland watchmaker in a public collection is at the Newarke Houses Museum, Leicester (L.H292.1960). It is signed John Simpson, Oakham, and the London hallmark on the case has the date letter for 1795. A number of others by Rutland makers are known to be in private collections. It is very unlikely that any of these were made in Rutland. Most watches of this period were manufactured by a group of watchmakers, each craftsman specialising on one particular part. One important watchmaking centre was the area around Liverpool and Prescott, Lancashire, where retailing watchmakers could acquire finished and unfinished watches. For further information on the history and development of pocket watches see, for example, *The Pocket Watch Handbook* (Cutmore 1985).

According to Owen and Bowen's *A Map of Rutland-shire* of 1720 (*see* page 58) there were then 48 parishes and 3,263 houses in the county. In 1801 the population was 16,300, rising to 19,720 by 1901 (*Victoria County History* I, 231). The market for clocks and watches was therefore never more than that of an average-sized town, and probably less due to the predominance of low-paid agricultural labourers and the distributed nature of the population. Local clockmaking had virtually ceased by 1780 owing to the availability of ready-made components and complete movements, and later, competition from low-cost imports. But during the hundred years prior to this clocks were made by a number of locally based craftsmen, examples of whose work survives in the area, in museums and in private ownership. The predominant makers during this period include the following (earliest and latest dates when known to be working as a clock or watchmaker are shown as, for example, [1745-75]):

Oakham:
BLACKBURN [*circa* 1720 to *circa* 1750], Stephen BLACKBURN [before 1731-71], John WILKINS [1741-51]

Uppingham:
William ARIS I [1753-98], John FOX [1744-1802], Robert FOX [1707-50], William FOX [1666-1703], William WATTS [*circa* 1750]

Harringworth, Northamptonshire:
Richard HACKETT [1741-81]

Stamford:
Boniface BYWATER [1689-1752], Thomas RAYMENT [1760-92], John WATTS [1661-1719], Robert WATTS [*circa* 1719-59], Ralph WILSON [1767-1801]

Biographical details of these and other local watch and clockmakers are given in Chapter 3 — Clockmakers. For further details on the history and development of domestic clocks see, for example, *Britten's Old Clocks & Watches & Their Makers* (Clutton 1982) and *Country Clocks and their London Origins* (Loomes 1976a).

Rutland has a good collection of turret clocks by a wide variety of makers, and together they demonstrate well the development of this type of clock from the late 1600s. After the anchor escapement, the next important step was the invention of the deadbeat escapement by George Graham in 1715. It eliminated the recoil action of the anchor escapement, which turns the train backward slightly after each beat, and was soon adopted for domestic regulator clocks and better quality turret clocks. This was too late for John Watts' turret clocks, all of which have anchor escapements. But Thomas Eayre II of Kettering had adopted this escapement by 1750 and many of his clocks were installed in the East Midlands area, in churches, country houses and public buildings. None of his Rutland clocks have survived, but a clock made by him and installed by Stephen Blackburn of Oakham in 1754 has been preserved at Barkby Church, Leicestershire, and this has a deadbeat escapement.

In 1676 a new form of striking, known as rack striking, was invented by the Rev Edward Barlow. Prior to this, countwheel striking had been used since the appearance of the first striking clocks. Rack striking has the advantage that the hands always stay in phase with the number of hours struck and from this date on it became the standard system for eight-day longcase, bracket and other better quality domestic clocks. Countwheel striking was generally retained for cheaper thirty-hour longcase clocks and turret clocks, although there are some notable exceptions. The clock at Stocken Hall Coach House signed by John Walker is one of only two turret clocks in Rutland with rack striking, the other being the clock installed by Joseph Wilson at Lyndon Hall stables.

Over the next hundred years there was a steady development as a result of the increased understanding of physics and improved engineering skills. Throughout this period turret clocks were made with side-by-side trains, mainly in wrought-iron birdcage frames. The exceptions to this are the extended barrel birdcage frames made by Bosworth of Nottingham and Whitehurst of Derby. There is a redundant clock of this type preserved in the tower at

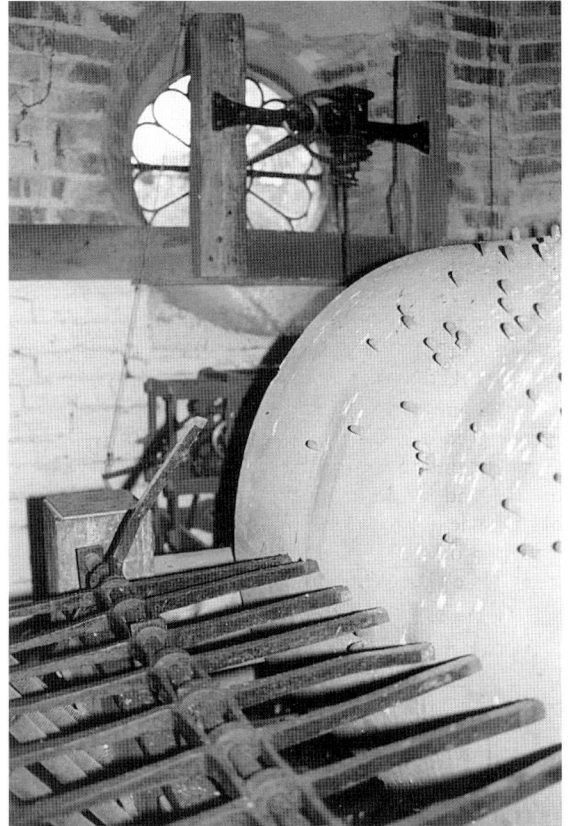

(above) The six-tune chime barrel by Edward Arnold at Stapleford Church, Leicestershire. It has twelve chime levers which are connected to bell hammers in the belfry by wires and bellcranks and each bell has two bell hammers. The names of the tunes are recorded on a brass plaque on the frame and the barrel is still used today for weddings and concerts

(left) Another view of the chime barrel at Stapleford Church. The barrel is 1625mm (64in) long and 800mm (31in) diameter and its substantial oak frame has 125mm (5in) square corner posts

Ashwell, although it is thought to be a later conversion (*see* Chapter 4 — Gazetteer, Ashwell). Improvements in gear-cutting and bearing design, the availability of better quality materials, the use of nuts and bolts instead of wedges in slots, and the adoption of temperature compensation are examples of developments which together resulted in more reliable and more accurate clocks. Clocks were made, or converted, to drive dials with two hands, and their greater efficiency resulted in smaller movements and a move towards eight-day running.

A further development during the second half of the eighteenth century was the installation of separate chime barrels, controlled by the church clock. These normally played a tune on the church bells every three hours, the tune being changed every twenty-four hours. Records show that there were chime barrels of this type at Cottesmore and Oakham churches, the former removed in 1867 and the latter in 1858. From the few details that survive it seems that they were similar to the chime barrel which is still in working order at Stapleford Park Church, Leicestershire, some eight miles north of Oakham. It was made and installed in 1785 by Edward Arnold of Leicester.

By the end of the first half of the nineteenth century the demand for turret clocks was such that they were being made by specialist manufacturers and exported all over the world. This buoyant market was to last until the start of the First World War, with a short revival afterwards to satisfy the demand for memorial clocks. The cast-iron flatbed frame, which was also introduced at this time, is very rigid and access for assembly, maintenance and repair is much easier than with a birdcage frame. The earliest clock of this type in Rutland is that by Frederick Dent which was installed in Oakham Church in 1858. This clock has a single four-legged gravity escapement designed by E Beckett Denison (*see* Chapter 3 — Clockmakers and Chapter 4 — Gazetteer, Oakham). He also developed the double three-legged gravity escapement which became the standard for better quality turret clocks. The clocks at Lyddington Church, South Luffenham Church and Stocken Hall Coach House have this form of escapement.

Over the next sixty years two- and three-train flatbed turret clocks were installed in Rutland churches, stables and a school by many of the well-known makers including: G & F Cope & Co of Nottingham; W F Evans & Son of Handsworth, Birmingham; James Gent & Son of London; Gillett, Bland & Co of Croydon; William Potts & Sons of Pudsey and Leeds; John Smith & Sons of Derby; John Smith & Sons of St John's Square, Clerkenwell, London; Thwaites & Reed of Clerkenwell, London; Elisha Tucker of London; John Walker of London. Fifteen turret clocks were installed in Rutland by John Smith & Sons, Derby, the last being the synchronous electric clock and striking unit at Greetham Church in 1967.

1.7 The Electric Telegraph & Standard Time

Local noon at Ryhall, the moment when the sun is directly overhead, is about one minute before local noon at Belton in Rutland, but the difference between Great Yarmouth and Land's End is half an hour. This time difference was of little consequence when communication was either person to person, or by the occasional letter, and travel was little faster than walking pace. The requirement for standard time throughout the country, and an integrated time service to match, was recognised when fast co-ordinated stage coach services were introduced in the early nineteenth century, but it was not achieved until the late 1840s when it was forced on the nation by the rapidly expanding railway network.

The Royal Observatory at Greenwich was founded by Charles II in 1675 and it came to be established as Britain's point of zero longitude. By the middle of the eighteenth century the techniques for establishing the exact local time by astronomical observation were well known, and there was a network of amateur observers throughout the country providing this service. Also by this time, the country had been surveyed well enough for every town and city to know its distance and hence time difference from the Greenwich meridian. By the 1840s this multiplicity of local time zones was hampering the development of the railway system. Another problem was that with trains travelling at up to sixty miles per hour, the railways needed some form of instant communication system for safety reasons. Both problems were solved by the introduction of the electric telegraph. The first telegraph link available for public use was commissioned in 1843. It was built alongside the Great Western Railway between Paddington and Slough, a distance of some 31 miles.

Aided by the convenient routes provided by the railway network, the electric telegraph expanded rapidly, providing an instant communication system for the railways, for the public and for businesses, and for this reason it has been referred to as the 'Victorian Internet' (Standage 1999). In 1852 it was connected to the electric time system at Greenwich Observatory and by 1870 the electric telegraph was providing time signals to all major cities and most towns throughout the country. By now most of the country was operating on 'railway time', although for a period after 1852 there was a great deal of confusion as 'local time' and 'railway time' were used concurrently. In fact some public clocks were fitted with an extra hand to show 'railway time'. As Rutland is very close to the line of zero longitude it did not have this problem. In 1877 the Standard Time Company was established and signals were available via the electric telegraph to strike a bell one blow every hour. In August 1880 Greenwich Mean Time became the legal standard as a result of the Statutes (Definition of Time) Act, and within a few years most countries had adopted this standard.

In Rutland, the first access to the electric telegraph was probably provided at Oakham Station. It was transferred to the new Post Office in Market Street when it was opened in November 1869, being connected to the telegraph system by wires from Brooke Road. In 1871 the telegraph wires were extended to Empingham Post Office where telegraphic business commenced on 13 December. In 1872 telegraph offices were opened at North and South Luffenham on 1 September and at Ketton on 1 December, followed by Cottesmore on 1 June 1884. Postal and Telegraph offices opened at Barrowden in 1893 and at Upper Hambleton in 1898. The date of arrival of the electric telegraph at Uppingham is not recorded but the office was originally at what is now 18 High Street West. On 12 July 1901 the Postmaster General applied to Rutland County Council to erect telegraph poles at the side of the main road from Uppingham to Belton (Traylen 1982b, 150 & 158). In 1925 Kelly's *Directory* recorded that over 50% of Rutland parishes had telegraph offices.

W Potts and Sons' 'Instructions for Winding and Regulating Clock' which they supplied with the Lyddington (1890) and South Luffenham (1907) church clocks, and the clock at The Pastures, North Luffenham (1902), include:

> WINDING ... It should also be done as near the same time each week as possible - say every Monday morning - preferably directly after the Attendant has got the current Telegraph time from some reliable source. (Correct time is sent every day from Greenwich, to all Post Offices and Railway Stations, by Electric Telegraph at 10am.) Postal Telegraph is best.

By the end of Victoria's reign in 1901 the electric telegraph's greatest days were behind it and the telephone was in its ascendancy. Even by 1886, ten years after its invention, there were more than a quarter of a

Hambleton Post and Telegraph Office is now closed but the clock and sign remain

million telephones in use world-wide, and in Rutland, by 1906, there were 43 subscribers on the Oakham and Langham Telephone Exchanges. One of these was Robert Corney, Jeweller and Watchmaker, whose number was Oakham 9 (Traylen 1982b, 181). The first Speaking Clock was introduced in the USA in 1927 and in Britain in July 1936, but only for subscribers to the London Telephone Service. For the first time it was possible to obtain the time, to an accuracy of 0.1 of a second, at any time of the day or night and nearly thirteen million calls were registered in the first year. It became a national service in 1942 and today's Speaking Clock is accurate to within 0.05 of a second.

Another medium which was to bring a time service to many Rutland people was the wireless. The first licence for the reception of 'Wireless Telegraphy' in Rutland was issued to Captain Thompson of Station Road, Morcott in 1905 (Traylen 1982b, 177). The first 'six pips' time signal was heard on crystal sets and primitive wireless sets in 1924, and in the same year, on 17 February, the first broadcast of 'Big Ben' chiming and striking the hour was heard.

Chapter Two

Bellfounders —
The Founders of Rutland Bells

There is no record of a bell foundry in Rutland, the nearest being the Norris foundry at Stamford. All but five of Rutland's church bells have been ascribed to a particular founder, and details of these are given below, grouped together according to the location of their foundries. At the time of Elizabeth I, bellfounding was still recovering from the effects of the Reformation, and Walters (1977, 179) was of the opinion that it was 'at this period that the itinerant element [was] most marked in the history of bell-founding'. As several of the medieval foundries had ceased production, it was the opportunity for the itinerant journeyman founder to procure business.

Richard Holdfield was an itinerant founder before he settled in Cambridge *circa* 1599. So too was Matthew Norris, who, having initially worked in Leicester, may have been instrumental in establishing the Stamford foundry *circa* 1603. Richard Holdfield is believed to have had connections with the Oldfields of Nottingham, and Matthew Norris probably worked with the Newcombes of Leicester (see under Cambridge and Stamford). Both are believed to have cast bells for Seaton Church in 1597 (George Dawson). A bell of the same date at Edith Weston has the same inscription as one of the bells at Seaton. This and a bell cast in 1597 for Lyndon, since recast, have been ascribed to Matthew Norris (George Dawson).

These four bells may have been cast in or near the churchyards concerned but no evidence of this has been found. The only evidence of this activity in Rutland is at Empingham, where a mass of bell-metal 'clearly in a state of fusion' (North 1880, 11) was found *circa* 1876 in the churchyard.

BUCKINGHAM

RICHARD BENTLEY

From the middle of the sixteenth century Buckingham had a foundry of some importance. It was originally run by John and George Appowell, and then taken over by Bartholomew Atton in 1585. Anonymous bells of about this time with lettering [87] are found in the locality of Buckingham, and it is believed that they came from this foundry. Bartholomew Atton had learned his trade with the Newcombes at Leicester where he was admitted to the Merchants' Guild in 1581/82. He was described as 'tanner and bellfounder, p. [apprentice] of Thomas New-com, tanner and bellfounder decd' (Hartopp 1927, 85). The Buckingham business continued under the owner-ship of Robert Atton until the foundry closed in 1633.

Little is known of Richard Bentley, other than his family may have lived in the parish of All Saints, Leicester, where the Newcombes had their bell foundry. The parish registers record a marriage between a Richard Bentley and Jone Browne in 1571, the baptisms of three of their children in the 1570s and that of another in 1585.

The only bell in Rutland by Richard Bentley is the third at Seaton. The reverse inscription, using lettering [87], together with the initial cross [25] and blocks of ornate decoration [88], is ꝶꬲꝺꝺꝴꝯꝹꝼ ꝲꝲꬲꞆ ꬲꝴꝲꞆꬲꝺꬲꝯ ꬲꝺꝶꝯꝺꝺꬲꝯꝴꝶ (Ryecharde Benetlye Bell Fovndder). The use of these distinctive letters is a positive indication that there is a connection between him, Bartholomew Atton and the Newcombe family. Although the location of Bentley's foundry is not known it is possible that he worked alongside Bartholomew Atton at Buckingham.

[87]

[88]

[25]

A bell dated 1585 at Passenham, Northamptonshire, which is only a few miles from Buckingham, has the same cross and lettering, and this provides an indication of the date of the Seaton bell.

It is interesting to note that decoration [88] incorporates the band decoration [39] as used by Toby Norris I of Stamford on three Rutland bells all dated 1610 (see Stamford, Toby Norris I). Matthew Norris, the presumed father of Toby I is believed to have been associated with the Newcombes of Leicester (see under Stamford). This decoration was also used by Hugh Watts II of Leicester.

CAMBRIDGE

RICHARD HOLDFIELD

Richard Holdfield was an itinerant founder who later settled in Cambridge, and cast bells there from *circa* 1599 until *circa* 1612.

He may have been related to the Oldfield bellfounders of St Peter's, Nottingham, and was possibly the son of Reginald Oldfield, potter. He generally used similar lettering to the Nottingham foundry (information from George Dawson) although his gothic capitals are not so ornamented and the decoration is different.

On this basis the fourth bell at Seaton dated 1597 has been ascribed to Richard Holdfield (George Dawson). The inscription CELORVM CHRISTI PLATIAT TIBE REX SONVS ISTI is in lettering like [125]. The initial cross [8] on this bell is very similar to the [72] used by the Oldfields of Nottingham and to the [10] used by the Norrises of Stamford. This bell may have been cast in association with Matthew Norris (see under Stamford).

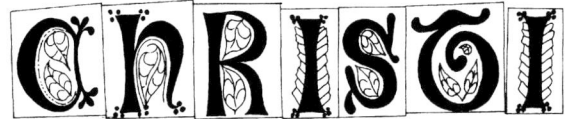

[125]

[8]   [72]   [10]

CHACOMBE

HENRY BAGLEY

The Chacombe bell foundry near Banbury in Oxfordshire was casting bells with few interruptions for a hundred and fifty years. It was owned by members of the Bagley family and was one of the most important foundries in the second half of the seventeenth century. The first owner was Henry Bagley I, who had learned the bellfounding trade when working with the Attons of Buckingham and from whom he derived some of their stamps. His career spanned almost half a century from 1631, and during the later stage of his career his son Henry Bagley II went into partnership with him. After the retirement of Henry I *circa* 1679, Henry II was joined by his brothers William and Matthew.

Five years after the death of their father in 1682, Henry II and Matthew set up their own foundries elsewhere, but bells continued to be cast at Chacombe by William and other members of the family until 1712. The foundry then seems to have been quiet for just over a decade, but after 1726 Bagley bells continued to be supplied right through to the early 1780s.

The gravestone of Henry Bagley I, which has been restored on three occasions in order 'To perpetuate ye memory of an ingenious bellfounder ...', can be found to the south-east of the church in Chacombe churchyard.

The only known Bagley bell in Rutland is the present treble at Seaton, which was cast by Henry II in 1684. Although he set up a new foundry at Ecton in Northamptonshire, at this time he would have been supplying bells from Chacombe. The inscription on the Seaton bell is in plain roman lettering and makes a bold statement giving the founder's name and the date. The Bagleys invariably used rich decoration and [100] is placed between the words of the inscription and in a complete band immediately below.

[100]

CROYDON

GILLETT & JOHNSTON

William Gillett, a clockmaker, established a company in his own name at Croydon in 1844. A decade later Charles Bland joined him and the firm then traded under the name of Gillett & Bland, with the manufacture and installation of public clocks and carillon machines forming an ever-greater part of the business (*see* Chapter 3 - Clockmakers). From about 1879 the company traded as Gillett, Bland & Co until 1884, Gillett & Co until 1887, then as Gillett & Johnston of Croydon until 1958. For a short period, from 1925 until 1930, it was also known as the Croydon Bell Foundry.

In the early clockmaking days they purchased their bells from other founders, mainly John Taylor, but from 1877 when Arthur Johnston became a partner in the

business they started to cast their own. By the end of the nineteenth century Johnston had taken full charge of the foundry and under his guidance and that of his son Cyril, who joined him in 1902, the bellfounding side of the business rapidly expanded. When the company went into receivership in 1958 all bellfounding ceased. The clock-making side of the business was acquired by Synchronome and moved to Wembley. However, the Gillett and Johnston name was later re-established by Cyril Coombes and his son and moved back to Croydon where it continues today.

There are twenty-four bells in Rutland supplied by this foundry and with the exception of the School Bell at Uppingham School, all form complete rings. The earliest are the eight at Oakham cast in 1910, although two of these were recast in 1924, and the ring of six at Belton cast in 1911. A ring of five was supplied to Greetham Church in 1923 and a new treble added in 1949. The three bells hanging in the tower of Uppingham School Chapel were installed in 1929.

Nearly all the Rutland bells display the name of the founder and the foundry's location, and those at Belton, Oakham and Uppingham have band decoration [58] below the inscription. On the treble at Greetham, the foundry placed shield [74] and incorporated the device [76] on the five earlier bells, which also have continuous decoration [75] below the inscription band. [76] is also placed on the seventh bell at Oakham. This device includes the initials of Cyril F Johnston.

The recast Oakham tenor being tuned at Gillett & Johnston's Croydon foundry in 1910 (Phillips 1911-12, 41)

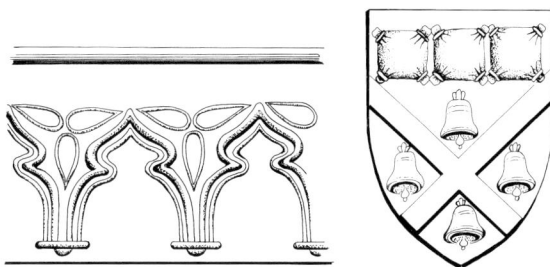

[58]   [74]

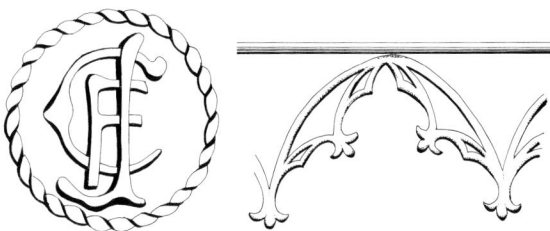

[76]   [75]

KETTERING

THOMAS EAYRE I

By the beginning of the eighteenth century, Thomas Eayre I of Kettering was a renowned clockmaker (*see* Chapter 3 — Clockmakers), and his son and grandson of the same name were to continue in this trade. Although Thomas Eayre I, unlike his son and grandson, is not generally believed to have been involved in bell founding, for the purposes of this book, which deals with the history of the clockmakers as well as the bellfounders, it is convenient to group them together and to distinquish them numerically. Tilley and Walters (1910, 80) state 'the initials **T. E.** appear on the 2nd [bell] at East Farndon, Northants, with the date 1710, which seems to suggest that this Thomas Eayre [Thomas Eayre I who died in 1716] had tried his hand at bell-founding'. However this is an error as the bell is actually inscribed with the initials **T. C.** for Thomas Clay of Leicester (information from George Dawson). Thomas Eayre I was succeeded by his son, Thomas II.

THOMAS EAYRE II

Thomas Eayre II, had decided by 1717 to diversify by opening a bell foundry in Kettering. However by 1719 this foundry appears to have been in the sole manage-

ment of Thomas II who ultimately cast many good bells within the surrounding area. Seven of his bells, cast between 1726 and 1748, are still hanging in the church towers of Rutland. Chronologically, they are the tenor at Morcott, the second and fourth at Market Overton, the fifth at North Luffenham, the treble and second at Teigh and the treble at Ketton. An eighth bell, that hanging in the turret of St John and St Anne's Chapel in Oakham and dated 1744, was also cast by him. With the exceptions of the Ketton bell, which is obviously a recast of a 1640 bell (where Thomas II reproduced the old inscription), and the bell at Oakham Chapel, all his Rutland bells have one or more of his favoured Latin inscriptions:

OMNIA FIANT AD GLORIAM DEI
GLORIA DEO SOLI
GLORIA PATRI FILIO ET SPIRITUI SANCTO
IHS NAZARENE REX IUDÆORUM FILI DEI
~ MISERERE MEI

The lettering on all Eayre bells generally consists of neat roman capitals. On the bells at Market Overton, Teigh, North Luffenham and Oakham Thomas II used various combinations of dots for plain stops. Decorative stops used on Rutland bells are [51] on the bell at Oakham Chapel, the fifth at North Luffenham, and the

---

The Eayre family of bellfounders and clockmakers

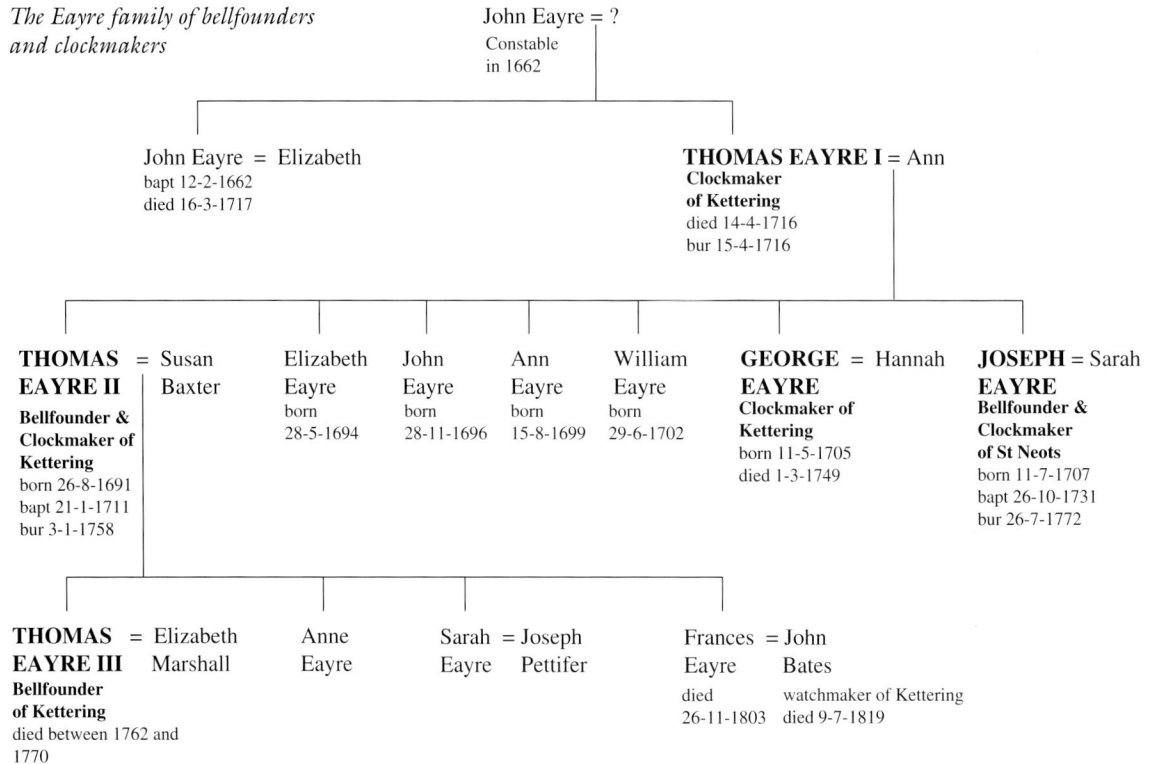

John Eayre = ?
Constable in 1662

John Eayre = Elizabeth
bapt 12-2-1662
died 16-3-1717

THOMAS EAYRE I = Ann
Clockmaker of Kettering
died 14-4-1716
bur 15-4-1716

THOMAS EAYRE II = Susan Baxter
Bellfounder & Clockmaker of Kettering
born 26-8-1691
bapt 21-1-1711
bur 3-1-1758

Elizabeth Eayre
born 28-5-1694

John Eayre
born 28-11-1696

Ann Eayre
born 15-8-1699

William Eayre
born 29-6-1702

GEORGE EAYRE = Hannah
Clockmaker of Kettering
born 11-5-1705
died 1-3-1749

JOSEPH EAYRE = Sarah
Bellfounder & Clockmaker of St Neots
born 11-7-1707
bapt 26-10-1731
bur 26-7-1772

THOMAS EAYRE III = Elizabeth Marshall
Bellfounder of Kettering
died between 1762 and 1770

Anne Eayre

Sarah Eayre = Joseph Pettifer

Frances Eayre = John Bates
died 26-11-1803
watchmaker of Kettering
died 9-7-1819

[51]

[117]

[99]

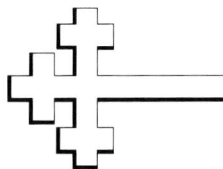
[49]

[104]

[128]

THOMAS EAYRE III

Thomas II was buried at the beginning of January 1758, and although his son Thomas III continued working from the foundry, within three years he was bankrupt and the business was closed. The only bell in Rutland supplied by him is that at Stoke Dry which is dated 1761. It bears his name, **THO^S : EAYRE**, place of work, date and the inscription **OMNIA FIANT AD GLORIAM DEI: LAUDATE ILLUM CYMBALIS SONORIS**. He used the decorative stop [101] and arrangements of dots as plain stops.

second and fourth at Market Overton and [117] on the treble at Teigh. [99] is used on both Eayre bells at Teigh and [49] is placed on the tenor at Morcott.

The Langham treble, originally a Thomas Eayre II bell of 1754, was recast by John Taylor in 1900. The inscription was retained and stops [49], [99], [104] and [128] were retained together with an arrangement of dots as a plain stop.

Thomas II used a variety of forms when placing his name on his Rutland bells, including **T E**, **T EAYRE** and **THO EAYRE**.

[101]

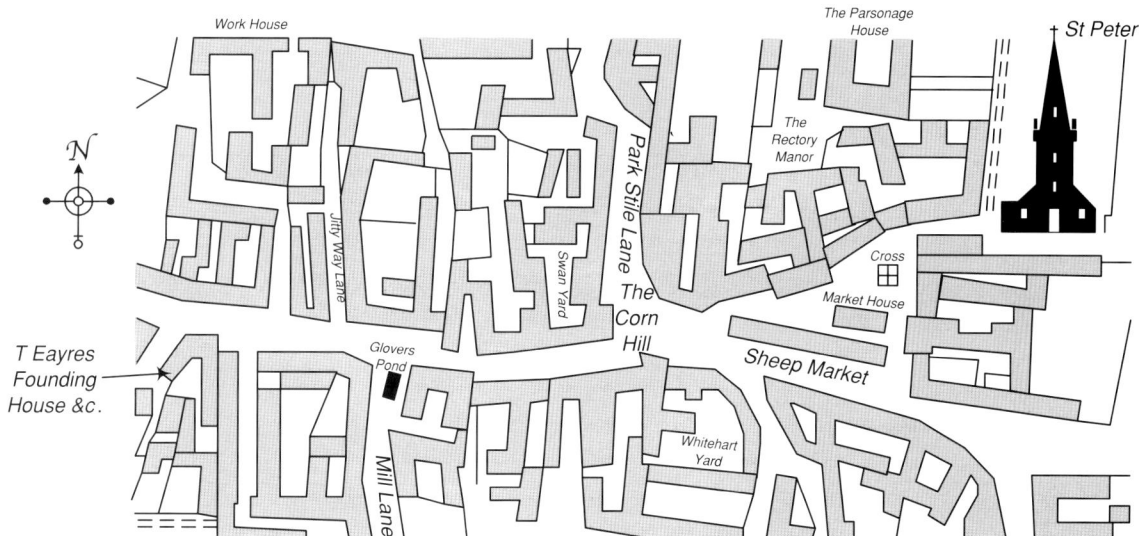

The location of the Eayre bell foundry based on 'A Plan of the Town of Kettering in Northamptonshire' drawn by Thomas Eayre II circa 1745

LEICESTER

Bellfounders are known to have been casting bells in Leicester from the early fourteenth century. Three who were admitted Freemen of the town at this time were 'Rog. le Belleyetere' 1308/9, 'Steph. le Belleyetere' 1328/9 and 'Tho. de Melton, belmaker' 1368/9 (Hartopp 1927, 27, 33 & 46). The fifth at Preston is the earliest example in Rutland of a bell cast by a Leicester founder. It has the initial cross [33] and GⱯBRIET in large decorated capitals as illustrated in [89]. Tilley and Walters (1910, 17) recorded that a bell cast for Mancetter in Warwickshire had the same cross and inscription, noting that the lettering was of the same type. They assigned the bell to a Leicester founder of *circa* 1400. In the light of this information the bell at Preston has been similarly ascribed (Robert Ovens & Sheila Sleath).

[33]

[89]

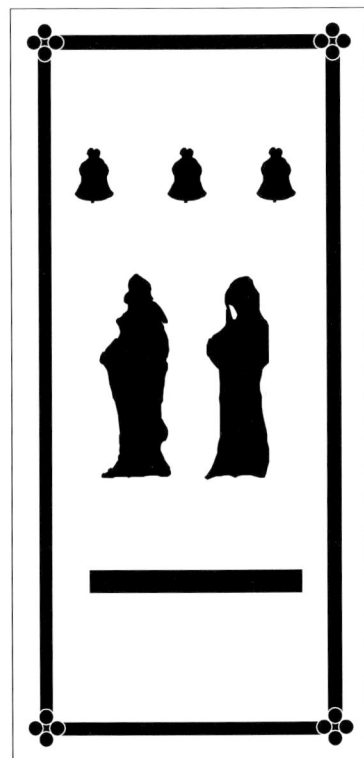

A drawing of the memorial to Thomas Newcombe I which was originally in the floor of the nave at All Saints' Church, Leicester

If any other Rutland bells were cast in Leicester during the fifteenth century, none remain. From *circa* 1529 until 1618 Thomas Bett and members of the Newcombe and Watts family are known to have provided Rutland churches with twenty-one bells, and of these, sixteen still exist. Whether the Watts founders had their own foundry in Leicester has not been proven and it is possible that they shared premises, sited next to All Saints' Church, with the Newcombes. However Thomas North noted, 'Their house, and most probably their foundry, being in the Gallowtree Gate, they would be in St Martin's parish' (North 1880, 55).

There are several bells in Rutland which bear the characteristic marks of both Newcombe and Watts, and it is assumed that they worked in partnership for some time. As it is impossible to determine, with complete certainty, which individual member of either family was involved, the Rutland bells considered to have been cast under such an alliance are here assigned to both Newcombe and Watts.

THOMAS NEWCOMBE I

Little is known of Thomas Newcombe I but he was described as 'Thomas Newcom of Leic., yeoman' when he was admitted as a Freeman of Leicester in 1507/8

(Hartopp 1927, 61). Thomas described himself in his will as 'fusor campanarius' [bellfounder] (North 1880, 49). He died in 1520 and was buried at All Saints' Church, Leicester. His inlaid slate memorial is preserved in this redundant church, and although the brasses have been removed, the impressions of three bells, signifying his trade, are apparent. His descendants continued in his chosen occupation for over a century.

THOMAS BETT

On the death of Thomas Newcombe I the foundry passed to his wife Margery. She subsequently married Thomas Bett who became associated with the business. He was admitted a Freeman of Leicester in 1521/2 (Hartopp 1927, 65) and became Mayor in 1529 when he was described as 'Bell Founder in All Saints' (Hartopp 1936, 57). The only bell in Rutland cast by him is the

[77]

[2]

[108]

[24]

[78]

[48]

second at Wardley (Robert Ovens & Sheila Sleath). Upon this he placed his foundry shield [77] which incorporates his initials, and cross [2]. The inscription reads TḫOMA and the lettering is like [108].

ROBERT NEWCOMBE I

On the death of Thomas Bett in 1539 the founding business passed to his stepson Robert Newcombe I. Robert, the son of Thomas Newcombe I, married his stepsister Katherine, the daughter of Thomas Bett by a previous marriage. He was admitted a Freeman of Leicester in 1536/7 (Hartopp 1927, 68) and his foundry is known to have been next to All Saints' Church. He was a prosperous man and was elected Mayor in 1550. His three sons, Thomas II, Robert II and Edward, all practised the bellfounder's craft.

The undated clock bell above the stable block at The Pastures in North Luffenham has on the inscription band ROBAḊE ḊEꝂCOME with stamps [11] and [2], and this is one of the bells originally supplied to the church at Little Bowden in Leicestershire (*see* Chapter 4 — Gazetteer, North Luffenham). The lettering is like [108] and this bell has been attributed to Robert Newcombe I *circa* 1550 (Robert Ovens & Sheila Sleath).

Stamps [11] and [24] are found alongside two others on the treble at Ayston. The well-defined cross on this bell closely resembles [78] and [48] almost certainly depicts a dog. The pictorial interpretation of [48] fits in well with information supplied by Tilley and Walters (1910, 33), who provide other examples of bells with this stamp. One of these, cast for Haddon, Huntingdonshire, was dated 1568, indicating that Robert Newcombe's son Thomas II was the supplier of that particular bell.

The presence on this bell of cross [78] and stamps [11] and [24], together with lettering [108] for the inscription AMBROSE, proves unquestionably that the treble at Ayston was cast at a Leicester foundry. However, although it is difficult to allocate this bell

[11]

positively to a particular founder, it is not beyond the bounds of reason to assign it to Robert Newcombe I (Robert Ovens & Sheila Sleath), as the personal shield [27] of Thomas Newcombe II (see below) is not included on the inscription band.

The second bell at Ayston is attributed to John Rufford of Toddington, *circa* 1365 (Pearson 1989). However, seventeenth-century records transcribed *circa* 1940 by Canon Aldred (Lane 1999, 18) indicate a date of 1541 for this bell. It is thought that a mistake may have been made in the seventeenth-century records or in the transcription and that this date is probably that of the treble. If this is the case it would be within the period when Robert Newcombe I was casting bells.

The former tenor at Manton has been ascribed to Robert Newcombe I *circa* 1550 (Robert Ovens & Sheila Sleath). Although this bell was recast by John Taylor in 1920 the initial cross [2], stamp [11] and the inscription were reproduced in facsimile.

THOMAS NEWCOMBE II

Thomas Newcombe II was the eldest son of Robert Newcombe I and he was admitted a member of the Merchants' Guild of Leicester in 1567/8. He carried on with the bell founding business and used many of the stamps of the earlier Newcombes and former Leicester founders. He frequently utilised the initial cross [2] and a personalised stamp [27], which was either an adaptation of Thomas Bett's shield [77] or perhaps originally the property of his grandfather Thomas Newcombe I.

Both of these marks and lettering [108] are on the third bell at Braunston and the tenor at Wing. Thomas Newcombe II usually inscribed his bells with the name of a saint and the bell at Braunston conforms with this, the dedication being to **S THOMA** (St Thomas). The tenor at Wing is dedicated to **S TADDEE** (St Thaddæus). According to Walters (1977, 270 & 271) this bell is dedicated to one 'of the rarer saints', and the bell at Wing is unique in this respect.

Thomas II used his shield [27] with cross [22] on bells in Leicestershire (North 1876, 48). The fifth bell at Barrowden incorporates [22] with the initial cross [2] and the inscription, which is part of the alphabet, uses lettering like [108]. All three bells have been ascribed to Thomas Newcombe II (Robert Ovens & Sheila Sleath).

[27]   [22]

[106]   [103]   [30]

NEWCOMBES OF THE 16TH CENTURY

The third bells at Morcott and North Luffenham are difficult to assign to a particular founder and date. They have, however, been generally assigned to the Newcombes of the sixteenth century, based on the following observations.

The Morcott bell has shield [9] as used by Robert Mellour of Nottingham (see under Nottingham) and stop [13], which was originally used by the Newcombes and later appropriated by the Watts of Leicester. The letters in the inscription **S MARIA** are like [108] as used consistently by the Newcombes. It is believed that [9] was a poor recast of the Mellour stamp which was in the hands of the Leicester foundry (information from George Dawson).

It would be interesting to see if a firm family connection could be established between the Nottingham Mellours and William Mellars. William was admitted to the Leicester Merchants' Guild as a bellfounder in 1499/1500. He died in 1506 and his wife, Margery, married Thomas Newcombe I, possibly explaining the use of a Nottingham stamp by the Newcombe foundry. On being widowed a second time she married Thomas Bett.

The third bell at North Luffenham displays the very ornate cross [106] and shield [103], both of which have lost much of their detail. [106] has been found on bells in Leicestershire with and without Thomas Newcombe II's shield [27] (North 1876, 50-1).

The Newcombes unquestionably had a hand in the casting of the North Luffenham bell. [103] may have been an earlier Newcombe stamp (possibly used by Thomas Newcombe I, who had three bells signifying his trade placed on his memorial slab). Alternatively it may have been a very poor casting of [3] (see Newcombe and Watts). The words in the inscription Melodie Geret Domen Campana are in black-letter with gothic capitals. Unfortunately they are too indistinct to illustrate.

The sixteenth-century fourth bell at Wing was recast

by John Taylor in 1903 but the inscription was retained using a 'mixed gothic' style of lettering. The initial cross [30], which is thought to have originated with the Newcombe foundry, was also retained.

ROBERT NEWCOMBE II

On the death of Thomas Newcombe II in 1580, his brothers Robert II and Edward continued casting bells. Prior to his death in 1580, Thomas Newcombe II was a tanner as well as a bellfounder (see under Buckingham), and the Newcombe family continued to trade as bellfounders and tanners into the next century.

Robert Newcombe II went into occasional partnership with Francis Watts. His brother Edward Newcombe and his sons appear to have worked in partnership with both Francis and other members of the Watts family, both at Leicester and at a foundry in Bedford. The fact that Robert Newcombe II married Helen, the daughter of Francis Watts, must have played a part in this business arrangement. Tilley and Walters (1910, 33) expressed the opinion that 'we may perhaps assign the period of that partnership to the closing years of the sixteenth century'.

The following extract from Loughborough Churchwardens' Accounts (North 1876, 230) implies that the two families were working together at this time:

> 1585 Item pd to ffrauncis Watts and Mr. Newcome the
> Bellfounders of Leic. for one half of the payment
> for Castinge the great Bell   iiij li [£4]

NEWCOMBE AND WATTS

As it is impossible to identify particular family members or to be precise about the dating of their various partnerships, three bells in Rutland have been ascribed to Newcombe and Watts (Robert Ovens & Sheila Sleath).

These bells, the undated tenor at Preston, the fourth at Ketton and the tenor at Glaston (both dated 1598), have similar characteristics. All three display the ornate cross [23] and the Watts foundry shield [3]. [23] is recorded as being used independently by Robert Newcombe II on a bell at All Saints' Church, Leicester in 1586 (North 1876, 200). A bell he cast in 1598 for Gloucester Cathedral confirms that he was working up to the year he died. He could therefore have been involved in casting the Ketton and Glaston bells.

[9]   [13]

[23]

[3]

[16]

[17]

[110]

[111]

[31]

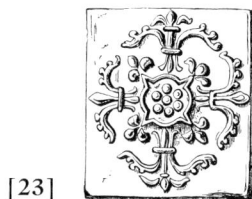

Francis Watts and other members of his family generally used the beautifully decorated capitals like [16] and [17] for their inscriptions, and lettering of this style is found on all three bells. They obviously did not have a complete alphabet as the **w** and **y** [110] used on the Ketton and Glaston bells and the **z** [111] on the Preston bell are of a completely different style. There is no w, y or z in the Latin alphabet and this implies that the Watts set of decorated capitals was originally made for Latin inscriptions.

The inscription on the tenor at Preston is GOD SAVE OUR QUEENE ELIZABETH which obviously dates it to 1603 or before. COELORUM CHRISTE PLACEAT TIBI REX SONUS ISTE is on the tenor at Glaston and is one of the Watts' favourite inscriptions. The inscription on the fourth bell at Ketton is ME ME I MERELY WILL SING and it has blocks of decoration [40] after the words and date. Below this is a complete band of decoration [119].

EDWARD NEWCOMBE

Edward Newcombe, the brother of Robert II, is known to have been working from 1570 and, as noted previously, worked at some time in partnership with the Watts family. He received his freedom to trade in Leicester as a 'tanner and bellfounder' in 1567/8 (Hartopp 1927, 78). At least three of Edward Newcombe's sons were connected with the bellfounding trade and all were known to be casting by 1611.

At the commencement of the seventeenth century,

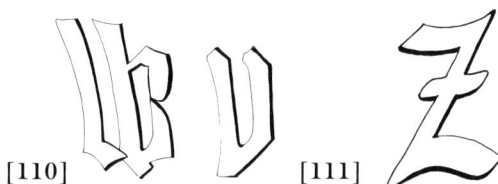

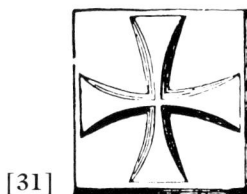

[40]

[119]

the Newcombes virtually ceased to use their former initial crosses, stamps, decorations, sets of letters and forms of inscription. Instead they employed plain roman capitals using the same inscription, or a shortened version, alongside the initial cross [31]. An example of this can be seen on the tenor at Ketton dated 1606 which was undoubtedly cast by Edward in association with his sons:

BE YT KNOWNE TO ALL THAT DOTH ME ~ SEE THAT NEWCOMBE OF LEICESTER ~ MADE MEE

After the death of Edward in 1616 the foundry appears to have been taken over by Hugh Watts II, the son and successor of Francis Watts (see below).

WATTS

The genealogy of the Watts family of bellfounders in the sixteenth century has not been fully researched. Until a detailed analysis has been completed, in connection with their founding activities in both Leicester and Bedford, it is difficult to determine the exact classification and date of some of their bells. Their involvement with the Newcombes has already been noted.

The Watts family were renowned for supplying good quality bells. They generally used the fine ornamented letters and stamps that had been in the hands of the Brasyers of Norwich up to the beginning of the sixteenth century. Stamps [3], [13] (as noted previously) and [37] were amongst those owned by the Norwich foundry, and having been used by the Newcombes, they were later appropriated by the Watts for their own trademarks.

[37]   [38]

It is fortunate that the Newcombes favoured lettering [108] for in many cases this helps to differentiate their bells.The gothic lettering [38] (North 1880, 58 & 59) was also used by the Watts founders but no bell in Rutland has an inscription using this style.

FRANCIS WATTS

As noted previously (see Robert Newcombe II) Francis Watts was in partnership with the Newcombes and, although it is difficult to determine the full extent of the Watts' involvement, it is believed that they were closely associated, particularly during the last decade of the sixteenth century. 'Francis Watts, tanner' was admitted a Freeman of Leicester in 1566/8 (Hartopp 1927, 77). He died in 1600.

The following bells have been credited to Watts in general but it is possible that they could have been cast by Francis.

The second bell at Morcott has part of the alphabet for the inscription but it has no date or founder's marks. Although the Newcombe foundry had supplied two alphabet bells to Rutland the lettering on this bell is like [16] and [17], implying that it had been cast by a member of the Watts family (Robert Ovens & Sheila Sleath). John Taylor recast the tenor at Ridlington in 1903 and the inscription, using lettering [16] and [17], was copied from the former bell. The familiar Watts shield [3] was also reproduced.

As already noted, [13] came to be recognised as a Watts stamp, and this is placed on the fourth bell at Barrowden with foundry shield [3]. Before being recast by John Taylor in 1915, there was another bell in the same tower bearing these same stamps. Both bells when cast were dated 1595 and the lettering was of the black-letter type. The original Watts bell at Barrowden has as its inscription god saue the queene. The inscription on the recast third is a facsimile copy and reads cum cum and preau, an inscription favoured by the Watts founders. Both of these bells cast for Barrowden have been ascribed to Watts (Pearson 1989).

HUGH WATTS I

Hugh Watts I and William Watts are presumed to be the brothers of Francis Watts. William is believed to have worked entirely at Bedford (Walters 1977, 242).

The fifth bell at Ketton, dated 1601, has been attributed to Hugh Watts I (Pearson 1989), likewise the undated fifth at Braunston (Robert Ovens & Sheila Sleath). Both bells display [3] and the large decorated gothic lettering [16] and [17] characteristic of this family. The inscriptions on these bells, SARUE THE LORDE and PRAISE THE JORDE, were frequently used by the Watts family.

A Hugh Watts placed the date, his foundry shield [3] and stamp [13], along with the reverse inscription ƎM Ɔ꩜ꞪM ꙅꞱꞱꙅ꙰ Ꞇ꙰Ə (Hew Watts made me), in plain gothic capitals on the treble at South Luffenham. Unfortunately some of the letters are inverted and reversed and the whole inscription reads backwards. The 9 and 3 of the date 1593 are so placed that they can easily be misinterpreted as 1563. There seems to be some confusion as to the true identity of this Hugh Watts. The fact that the lettering used was not from the ornamental set as consistently used by the Watts at this time implies that this 'Hew Watts' may not be the brother of Francis. The inscription was placed in a poor manner, suggesting perhaps that this could have been a younger member of the family learning the trade.

HUGH WATTS II

Hugh Watts II, the son of Francis, was believed to be apprenticed to his uncle, William Watts, at Bedford. He was a bellfounder of repute and is known to have been casting bells both at Bedford and Leicester. He was admitted a Freeman of Leicester in 1611/12: 'Hugh Wattes, 2nd s. of Francis Wattes, bellfounder, decd.' (Hartopp 1927, 106). When he took sole charge of the Leicester foundry he introduced new letters, ornaments and styles of inscriptions. He generally used no other mark than his foundry shield [3].

The only bell known to have been supplied to Rutland by Hugh Watts II was the former sixth at Oakham cast in 1618. The inscription **IH'S : NAZARENVS REX : IVDEORVM FILI : DEI MISERERE : MEI** was so frequently used by him that bells bearing this inscription became known as 'Watts Nazarenes'. This inscription using roman lettering always had the first S placed in reverse. [142] is an example of the first word that would have been placed on the bell at Oakham.

Hugh Watts II died in 1643. Although his son

[142]

inherited all the tools of the trade he does not appear to have entered the founding business. The foundry continued for a while under a George Curtis, who died in 1650, but it seems to have closed by 1655 and some of the foundry equipment found its way into the hands of the Nottingham founders.

The earliest date of the appearance of the 'Nazarene' inscription on Nottingham bells is 1656 on the disused treble at St Nicholas, Leicester, where George Oldfield I wished, perhaps, to emphasise his take-over of the Leicester foundry (information from George Dawson).

EDWARD ARNOLD

Edward Arnold, bellfounder and clockmaker, became joint owner of the foundry at St Neots following the death of Joseph Eayre in 1772 (see under St Neots and Chapter 3 — Clockmakers). He first came to Leicester in 1784 when he set up a bell foundry near the South Gates. According to North (1880, 62) it was in Hangman's Lane, now Newarke Street. He announced his arrival in Leicester by an advertisement in the *Leicester and Nottingham Journal* of Saturday 17 July 1784 (No. 2644, 1):

BELL-FOUNDING
EDWARD ARNOLD, BELL-FOUNDER and BELL-HANGER, from ST. NEOTS, in the County of Huntingdon, is removed to, and has erected, a New BELL FOUNDRY, Near the South-Gates, in Leicester, where he proposes carrying on the BELL-FOUNDING BUSINESS in all its Branches, and flatters himself he shall by his Diligence and Care, continue to deserve that favourable Attention the Publick have heretofore bestowed upon him.

He continues to make upon the best capital principles, repairs, and cleans Church Clocks and Chimes, Turret and other Clocks, Watches &c .... The Bell Founding Business only will still be continued at St. Neots by the said Edward Arnold.

Apart from a few bells cast by a Thomas Clay of Leicester between 1711 and 1715, Edward Arnold's bells were virtually the first to be cast there in any number after the closure of the Watts foundry *circa* 1650. It has been said that Arnold himself knew little of the art of bell founding and consequently had to depend upon the skill of his foreman Islip Edmonds.

In addition to the third at Brooke which was cast at the St Neots foundry, there are four more bells in Rutland cast by Edward Arnold: the treble at Ryhall, the third at Whissendine, the fifth at Ashwell, and the Priest's Bell at Barrowden. They date from 1785 to 1790. On each of his Rutland bells, he placed his name as **EDWD ARNOLD**, the foundry location, **FECIT** and the date. The Priest's Bell at Barrowden, presumably for lack of space, omits the location.

Edward Arnold used the same type of roman lettering as the Eayres and occasionally used their scroll-patterns

[49]

[46]

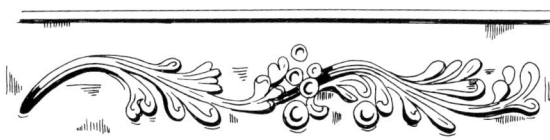
[45]

for ornamental purposes. On the Priest's Bell at Barrowden he used stop **[49]** as applied by the Eayres, and on the others he used either decoration **[45]** or **[46]**.

He also included the impression of the obverse and reverse of a coin on the inscription bands of the Brooke and Ashwell bells.

Arnold set up an iron-foundry in 1792 and this is recorded in his advertisement of 9 May in the *Leicester and Nottingham Journal*, part of which is transcribed here:

EDWARD ARNOLD
BELL-FOUNDER
Begs leave to inform the Public, that, in addition to his present business, he has erected an IRON-FOUNDRY near the West-Bridge where he intends perfecting the branch of CASTING IRON to its utmost extent.

A business of this Nature, he doubts not, will prove highly advantageous and convenient to the town and county, from the very great expense of carriage that attends the articles, (which the Purchaser will save) and from the advantage of its being situated on the spot, in executing orders with exactness and despatch.

This business is said to have been bankrupt by 1795, but he denied this in an another advertisement (*Leicester and Nottingham Journal*, No 6320, 21 March 1795). It seems, however, that both foundries had closed by 1800. The bellhanging side of the business continued for a few more years and he is known to have hung bells cast by Thomas Mears in the belfry at St Mary's Church, Stamford, in 1802 (Ketteringham 2000, 291). On the death of Arnold, his foreman Islip Edmonds went to work for John Briant, a bellfounder at Hertford.

The headstone of Mary Ann Arnold can be found in the churchyard of St Mary de Castro, Leicester. She was the second daughter 'of Edward Arnold Bellfounder by Mary his wife' who died on 7 June 1794 at the age of 20 years.

LONDON

Of seven bells ascribed to medieval founders in Rutland, three, that at Tixover, the fourth at Langham and possibly the former tenor at Tickencote, now on display in the church, were cast in a London foundry. All three are scheduled for preservation by the Council for the Care of Churches.

No other Rutland bells are known to have been cast in London until the late eighteenth century, and from then until the present day they have been cast by two different foundries. The following bells, all from the Whitechapel Bell Foundry, have since been recast. The bell number is as recorded by Thomas North in 1880:

> Uppingham fifth, 1772, Pack & Chapman
> Oakham fourth and fifth, 1858, George Mears
> Empingham treble, second, third, fourth and fifth, 1859, George Mears

The fate of the former Barrow Chapel bell, cast by Thomas Mears II of the Whitechapel Bell Foundry *circa* 1830, is not known. It was removed in 1970 when the chapel was demolished.

RICHARD HILLE

Very little is known of this London founder who was casting from *circa* 1423 to 1440, but many of his bells bear the initial cross and letters used by the Burfords, also of London. His foundry shield [29] is a very distinctive trademark showing a bend between a cross and annulet. The crosses [12] and [32] seem to have been used regularly by him and all three of these stamps are found on the bell at Tixover, the only Rutland bell assigned to him (Pearson 1989). The period when Hille was founding was a transitional stage, during which inscriptions were no longer placed exclusively in gothic capitals. The bell at Tixover is dedicated to St Faith and the lettering is of a type termed 'mixed gothic'. This bell is scheduled for preservation. According to Walters (1977,

270-1) this bell is dedicated to one 'of the rarer saints [who appears] on ... two bells at most'. The other was cast for Higham Church, Suffolk.

JOHN DANIEL'S SUCCESSOR

John Daniel of London, who was casting between *circa* 1456 and 1470, and his unknown successor both used the Royal Arms [35], but John Daniel did not use the trademark [36] on his bells, which included only the arms and his initials **I D**. The implication is that his successor, who probably worked alongside him, actually introduced mark [36]. Because [36] was first found on a bell at Brede in Sussex this mark is often referred to as a 'Brede mark'. There are probably ten bells cast with this mark in Sussex and the inscriptions on all of them begin with [1] and terminate with [36] and [35] (Elphick 1970, 58). The fourth bell at Langham displays these marks in the same order about the inscription Sit Nomen Domini Benedictum and has therefore been attributed to a London foundry *circa* 1480 (Council for the Care of Churches).

[35]   [36]   [1]

JOHN WARNER

In 1762, brothers Tomson and John Warner succeeded to their father's bell and brass founding business which was sited in Cheapside, London. The following year, the business moved to Cripplegate. In the early 1780s the brothers dissolved the partnership and John moved to Fleet Street where he set up an independent bellfounding business, trading under his own name. In 1850 this business moved from Fleet Street back to premises in Cripplegate.

The few bells cast by John Warner prior to 1850 did not exceed 450mm (18in) in diameter. However from that date larger bells were supplied by the company, and it built up a worthy reputation. The foundry trademark became a bell set within a crescent and it came to be known as the Crescent Foundry.

The foundry was transferred to Spelman Street, Spitalfields *circa* 1910 but the offices remained at Jewin Crescent, Cripplegate. It continued to be administered by members of the Warner family until it closed in 1921.

The Royal Arms can be found on some of Warner's

[29]   [12]   [32]

Sancta Fides Ora pro Nobis

The inscription on the bell at Tixover. The letters do not appear to be decorated

[135]   [140]

[134]

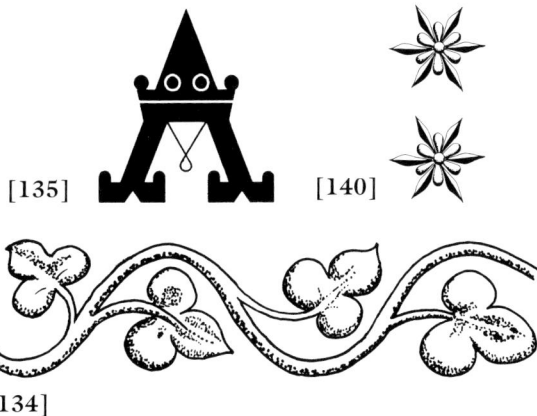

bells, for they had been appointed by Royal Warrant as bellfounders to Queen Victoria, Edward VII and George V. Perhaps their greatest achievement was when they secured the contract to cast the bells for the Clock Tower at The Palace of Westminster.

The second bell at Whissendine is the only church bell in Rutland to be supplied by Warners. It has no decoration and the inscription reads **CAST BY JOHN WARNER & SONS L**^TD^ **LONDON 1897**.

There are two other Rutland bells from this foundry. The bell retained in the bellcote of the former school at Stretton has the date 1879 on the waist. Below this is [135], possibly used to identify the second Lord Aveland of Normanton Hall, who paid to have the school built. The founder's name is on the soundbow.

There is large clock bell in the turret at the converted Stocken Hall Coach House near Stretton. It was cast by the Warner foundry in 1914 for Major Charles Hesketh Fleetwood-Hesketh. It has a double inscription band. The upper band displays the initials of the Major and his wife and incorporates stop [140]. The lower band has **WARNER. LONDON**. Both bands are filled with decoration [134]. The date and STOCKEN HALL are placed on the waist.

WHITECHAPEL FOUNDRY

Robert Mot, who died in 1608, was the first of a long line of master founders at the Whitechapel Bell Foundry, which is still in business today. Only those master founders connected with Rutland bells are noted here.

The earliest Whitechapel bells in Rutland were cast by Pack and Chapman in the eighteenth century. Thomas Lester, one of the foundry's less successful masters, ran the business between the years 1738 and 1752 but things steadily improved when in the latter year he took Thomas Pack as his partner. After Lester's death in 1769, William Chapman joined the firm and the Whitechapel Foundry began to enjoy almost a monopoly in the trade, mainly due to the gradual disappearance or absorption of

[97]

London and provincial foundries. Another factor which helped the firm was the popularity of the art of change ringing in the early eighteenth century. This brought increasing orders for new bells, and by 1800 the Whitechapel Foundry had achieved the international fame which it still enjoys.

Between 1770 and 1780 Pack and Chapman cast important rings of bells for London and provincial towns. The first bells known to have come to Rutland from this foundry were the eight sent to St Peter & St Paul's Church at Uppingham in 1772 and 1773. Six of these bells still hang in the tower today.

The Whitechapel Foundry used decoration sparingly in the latter half of the eighteenth century but [97], usually known as the 'Whitechapel Pattern', was introduced by Thomas Lester and used for nearly a century. This design was employed on at least five of the Uppingham bells. It was also a practice of the foundry to add the churchwardens' names by incising or engraving them onto the bell after it was cast. This can be seen on the tenor at Uppingham with the date included in the same manner. One of their favourite inscriptions is also found on the tenor:

YE RINGERS ALL THAT WHO PRIZE YOUR ~ HEALTH AND HAPPINESS BE SOBER MERRY ~ WISE AND YOU'LL THE SAME POSSESS

The Mears family enjoyed a long association with the Whitechapel Bell Foundry. It began when William Mears joined William Chapman following the death of Thomas Pack in 1781.

Three bells almost certainly cast by the Whitechapel Foundry under the management and ownership of Thomas Mears I and his son, Thomas II, are the two at Essendine dated 1805 and 1823 (George Dawson), and the single bell at Pickworth dated 1821 (Pearson 1989). The tenor at Essendine is ascribed to Thomas I and only displays the date, as does the Pickworth bell which is ascribed to Thomas II. The treble at Essendine, also ascribed to Thomas II, has the date and the name of a churchwarden. All three bells are without decoration. The Priest's Bell at Oakham dated 1840 may also have been cast by the Whitechapel Foundry.

In the nineteenth century the foundry, under the management of Thomas Mears II, cast some very large bells, the most notable being 'Big Ben' in 1858. It also ventured into the field of bellhanging for the first time.

On his death in 1844, Thomas Mears II was succeeded by his sons Charles and George Mears. The inscription on the treble supplied by them to Ashwell

The Whitechapel Foundry, London, from Cassell's Magazine of Art, **II**, *1854* (Whitechapel Bell Foundry Ltd)

The tuning room at the Whitechapel Foundry, London, in 1854 from Cassell's Magazine of Art, **II**, *1854* (Whitechapel Bell Foundry Ltd)

Church in 1850 was in Latin and in gothic capitals. So also was C ET G MEARS LONDINI FECERUNT. They utilised the initial cross [**44**]. Although Charles died about 1855 the foundry continued to use his name for a further two years. **G. MEARS FOUNDER LONDON** was placed on the former fourth and fifth at Oakham dated 1858 and the former five bells at Empingham dated 1859.

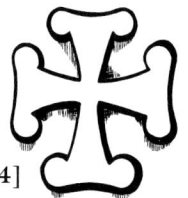

[**44**]

C et G Mears Londini Fecerunt

Mears' lettering on the treble at Ashwell Church

In 1865 the foundry passed into the hands of Robert Stainbank and it continued to trade as Mears and Stainbank for over a century.

The treble and second at Braunston cast in 1967 have **MEARS**, [59] and the date on the waist. [59], and [93] on the treble at Great Casterton, are both adaptations of the early stamp used by Robert Mot, who is traditionally believed to have begun the business at Whitechapel *circa* 1570. The initials stand for Albert A Hughes, William A Hughes and Douglas Hughes who were all master founders.

The foundry now trades under the name of Whitechapel Bell Foundry Ltd. The principal director is Alan Hughes. Along with [93] and the date 1990 the bell at Great Casterton has **WHITECHAPEL** on the waist. Decoration [102] is found on the inscription band along with the dedication to St Peter.

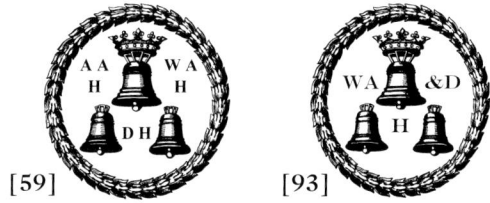

[59]   [93]

[102]

LOUGHBOROUGH

Robert Taylor laid the foundations for his descendants to set up a successful bellfounding business at Loughborough, Leicestershire (see under St Neots). The original Loughborough foundry of Robert's son, John Taylor, was located at Pack Horse Lane, and by the end of the 1850s the benefits of his move from Oxford became apparent. At his death in 1858 the management of the foundry passed to his son John William, and the tradition of this craft was further continued by members of this family for over a century. From 1859 casting was gradually run down at the Pack Horse Lane foundry and all activities transferred to a new foundry erected on the Cherry Orchard site. This foundry became well established and today is one of only two bellfounding busi-

nesses remaining in the United Kingdom. It has traded under various names and the company is now known as John Taylor Bellfounders Ltd. It continues to supply good quality bells, both nationally and internationally.

Nineteenth-century correspondence between Taylors and their prospective customers reveals their businesslike approach in their efforts to obtain orders. The following extracts are taken from letters sent to Mr J T Hollis, churchwarden of Cottesmore in 1885, when the tenor needed recasting and the fourth bell rehanging. The closing remarks of a letter dated 8 December read: '... again soliciting the honour of your commands which should have our prompt & careful attention, We remain Dear Sir Your obedt Servant John Taylor & Co' (DE

This envelope sent to the vicar of Lyddington in 1860 depicts two of Taylor's successful projects

1920/79/69, *see* Chapter 4 — Gazetteer, Introduction, for source references).

During the negotiations Mr Hollis made a query regarding payment for the work and Taylors replied on 15 December: 'We will agree to the terms you propose though it will be an expense to us to wait so long for the balance, as we are not capitalists, and in such an enviable position as to be able to allow credit' (DE 1920/79/71).

Of the 213 bells hanging in the towers and turrets of Rutland Parish Churches sixty-one (almost twenty-nine per cent) have been supplied by Taylors of Loughborough. Their Rutland bells cast between 1861 and 1886 display the founder's name on the inscription band in one of the following ways:

J TAYLOR & CO on the bell at Bisbrooke, the fourth at Ayston, the third at Market Overton, the Cottesmore tenor, and the second and third at South Luffenham.
JOHN TAYLOR & C⁰ on the Hambleton fourth, Whissendine fifth, Tinwell second and fourth, and the fifth at Cottesmore.
TAYLOR & C⁰ on the fourth at Lyddington.

All these bells are dated, and with the exception of that at Lyddington, they include **LOUGHBOROUGH**, the location of the foundry. Of the twelve, six include **FOUNDER** and two **BELLFOUNDER** as part of the inscription.

From 1887 until the present date, all but seven of the Taylor bells cast for the county display a foundry trademark. Some of these incorporate the 'lamb and flag' design. Nine different devices are used and they fall into the following date periods:

[120]	1887-1888	Hambleton treble and tenor. Market Overton treble and tenor
[121]	1889	Lyndon treble and second
[81]	1895-1897	Empingham treble, second, third, and fourth. Exton treble and tenor. Ketton third
[107]	1900-1903	Langham treble and second. Wing fourth
[105]	1903	Ridlington second, third and fourth.
[94]	1908	Preston second and third
[122]	1911	Ridlington treble
[56]	1915-1920	Barrowden second, third and tenor. Manton treble and tenor.
[47]	1931-1935	Glaston second. Tickencote second. Cottesmore treble
[47]	1977	Lyddington treble
[56]	1985-1993	Caldecott treble and fourth. North Luffenham treble and tenor. Brooke treble and tenor. Barrowden treble. Uppingham Angelus

The trebles at Glaston and Preston, and the treble, second and third at Edith Weston, all cast between 1937 and 1964, do not have a Taylor device. Instead, **JOHN TAYLOR & CO** and foundry location were placed on the inscription band, as on those cast for Rutland prior

[120]   [121]   [81]

[107]   [105]   [94]

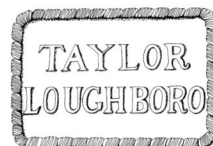

[122]   [56]   [47]

to 1887. Of the ring of six cast in 1895 for Empingham Church the fifth and sixth have no device or founder's name.

All the twenty-seven Taylor bells dated between 1861

Decoration, inscriptions and foundry marks are impressed on the inside of the cope [outer mould] whilst the special clay is still soft. New stamps were created for bells to be cast during the Millennium year and some examples are shown here, together with Taylor's 'lamb and flag' device [56] and decoration [54]

[91]   [92]   [85]

[54]

[65]   [95]

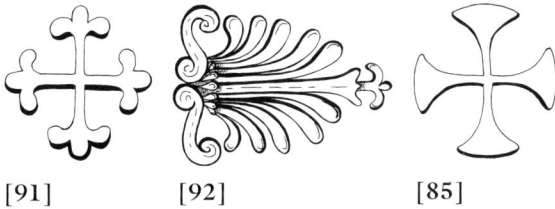

and 1895, with the exception of four, bear no decoration or decorative stop. Of these four, the fourth at Hambleton displays cross [91] and decoration [92], the tenor at Market Overton uses cross [85], as does the second at Lyndon and the treble at Exton. The latter bell also includes a Toby Norris III decoration [52] (see under Stamford). With the exception of the Market Overton bell these bells are recasts and the Lyndon and Exton bells retain the former inscriptions.

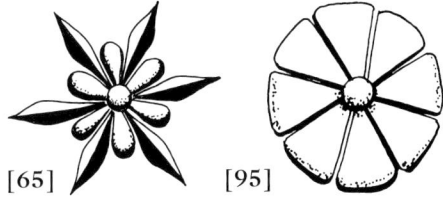

Some of the later recast bells also retain earlier inscriptions. Founders' marks and decorations from former bells are retained on the following:

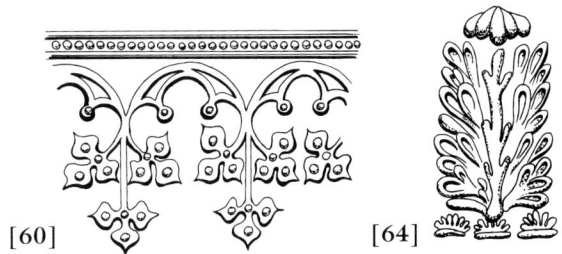

Ketton third (1897)	[50] (see under Peterborough)
Langham treble (1900)	[99], [49], [128] and [104] (see under Kettering)
Wing fourth (1903)	[30] (see under Leicester)
Ridlington second (1903)	[6], [21] and [9] (see under Nottingham)
Ridlington tenor (1903)	[3] (see under Leicester)
Ridlington third (1903)	[10] and [52] (see under Stamford)
Barrowden second and tenor (1915)	[14] and [57] (see under Stamford)
Barrowden third (1915)	[3] and [13] (see under Leicester)
Manton second (1920)	[2] and [11] (see under Leicester)
North Luffenham tenor (1989)	[7], [34] and [42] (see under Nottingham)

The majority of the forty-three Rutland bells cast after 1895 use one or other of the following: [54], [65] and [95]. [54], a band decoration, is the most frequently used and is found on seventeen of the bells. It is invariably used with stop [65] which is applied throughout the twentieth century. Stop [95] is used on Glaston treble.

Decorations [60], [64], [66], [71], [86], [98], [126] and [127], are used sparingly and only eleven of the bells cast between 1895 and 1993 display any of these marks. The ornate cross [55] is placed on the treble at Barrowden.

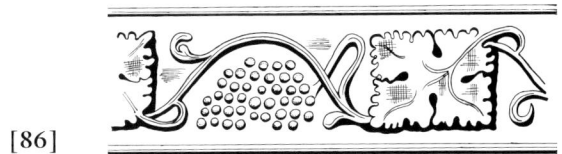

Roman lettering is used for the majority of the inscriptions on Taylor bells. However some of the letters in the alphabet used on all six bells at Empingham and the treble and tenor at North Luffenham have slight variations, as illustrated in [137]. Five recast bells, the third at Barrowden (1915), the second at Manton (1920), the second and tenor at Ridlington (both 1903), and the tenor at North Luffenham (1989) are in gothic lettering

[60]   [64]

[66]

[71]

[86]

[98]

[126]

[127]

[55]

reproduced in facsimile. That at Barrowden retains black-letter type, and the tenor at Ridlington the gothic capitals as used by the Watts family of Leicester. The others, at Manton, Ridlington and North Luffenham, use the capitals as employed by Robert Newcombe I of Leicester, and Robert Mellour and Henry Oldfield II, both of Nottingham.

Both the cross [55] and gothic lettering [130] and [131] placed by John Taylor on the treble at Barrowden were copies made from alphabets originally used in the fifteenth century by John Smith of Louth (Ketteringham 2000, 205). The capitals used by the Taylor foundry in a retained Watts inscription on the fourth bell at Wing are from the same set of decorated capitals [131], but the lower case letters used are as illustrated in [113].

The undecorated capitals [130] can be found on the second, third and tenor at Ridlington and upon the Preston second and third. At Preston they are used with the lower case letters [82]. Another combination of gothic type lettering is used on these two bells at Preston, with capitals like [61] and lower case letters like [113]. Lettering like [109] is used on the trebles at Ridlington and Whissendine, and the fourth at Wing.

The Uppingham School clock bell cast by John Taylor has not been included in the above analysis. Physical inaccessibility prevented a full survey of this bell.

[137] # AND EVERY TONGUE

The foundry area of John Taylor at Loughborough, Leicestershire, where a bell is being cast. The mould is buried in the sand floor to ensure the metal cools slowly. The molten bell-metal, an alloy of 77% copper and 23% tin, is being poured into a header box which acts as a reservoir as the bell cools. The man on the left holds an ash stick which he pumps up and down to ensure that there are no gas pockets in the casting, and he will continue with this long after the metal has been poured. The outer mould of the bell, known as the cope, is created inside a cast-iron case, a number of which are seen in the foreground

[130]

[131]

[113]  [61]

a f g m r s v

[82]

AECMST

[109]

The early method of tuning bells was to chip metal away with a chipping hammer until the right pitch was obtained. Today bells are tuned by turning on a vertical lathe and the harmonic frequency is checked electronically. This bell has been tuned by both methods

NOTTINGHAM

By the late fifteenth century, and probably well before, there was an active group of bellfounders in Nottingham.

RICHARD AND ROBERT MELLOUR

The Mellours were one of the last families casting in Nottingham before the Reformation. Richard, who died about 1508, was succeeded by his son Robert who continued to cast bells until 1525. The Mellours used the shield [9] as their foundry mark, and this is found on two bells hanging in the belfries of Rutland, although one of them is a recast with the former inscription and stamps reproduced.

Shield [9] appears on the tenor at South Luffenham with cross [6] and stamp [21]. The gothic capital S [124] found on this bell is characteristic of that used by the Mellour family (information from George Dawson). All three stamps and the letter S are also found in facsimile on the second bell at Ridlington which was recast by Taylors in 1903. Both bells have been assigned to Robert Mellour *circa* 1510. The repetition of the letter S probably indicates that the bell was used as a Sanctus Bell. This letter on the former Mellour bell at Ridlington, as noted by North (1880, 149), is placed on its side.

HENRY OLDFIELD I

An important Nottingham foundry of the sixteenth century was run by the Oldfields, and the first founding member of this family was Henry Oldfield I who was casting from *circa* 1540 until 1590. As his trademark he used a shield with a saltire cross together with his initials [118]. The tenor at Teigh has been ascribed to Henry Oldfield I and is an early example of his work (information from George Dawson). It has this shield alongside stamps [18] and [19] which are known as the 'Royal Heads', the images of which represent Edward III and Queen Philippa. They are thought to have originally belonged to the Ruffords of Toddington during the fourteenth century (see under Toddington). The inscription on the bell at Teigh is IN NOIE IHS MARIA, with abbreviation marks above the E and S, and part of this is shown in [136]. It also has four coin impressions, two of the obverse and two of the reverse of a silver groat on the soundbow. This is dated 1526-40 and confirms the dating of the bell as *circa* 1540.

HENRY OLDFIELD II

Henry Oldfield II, the son of Henry I, known to have

[9]

[6]

[21]

[124]

[118]   [18]   [19]   [136]

been casting from *circa* 1590 to *circa* 1619, had a good reputation as a founder, and four of his bells remain in the church towers of Rutland. Bells by this founder can be identified by his foundry shield [7] which displays a calvary cross between his initials with a crescent and star above. This is found on the third bell at Cottesmore dated 1598, the fourth and tenor at Whissendine and the second at Ketton, all dated 1609. This stamp was often accompanied by either cross [34], found at Ketton and on the fourth at Whissendine, or cross [72] as on the bell at Cottesmore.

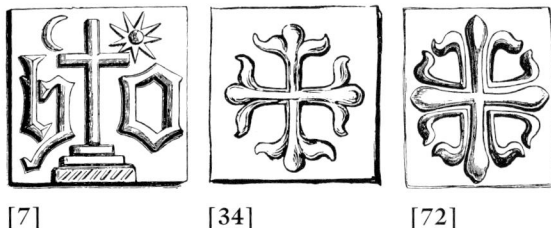

[7]   [34]   [72]

The Nottingham founders used several band ornaments to fill the spaces between the words of the inscription, and Henry Oldfield II occasionally used the grotesque pattern [73] evident on the Cottesmore bell. The significance of the letters included in this pattern is not known. This bell also incorporates the Royal Arms [68].

[73]

[68]

The lettering used by Henry Oldfield II for his inscriptions is varied. That at Ketton, I sweetly toling men do call to taste on meate that feeds the soole is in black-letter type [114], but the first letter of the sentence [84] is a very ornate capital. Lettering [114] is also used for the inscription my roaring sound doth warning geve that men cannot heare always luve on the tenor at Whissendine. He favoured both of these inscriptions and the words used were placed in blocks of lettering. Another of his favoured inscriptions, GOD SAVE HIS CHVRCH, is found on both the fourth at Whissendine and third at Cottesmore. It is in large gothic capitals [116] with minimal decoration.

The tenor at North Luffenham was recast in 1989 by John Taylor. The original bell was by Henry Oldfield II and dated 1619. The inscription, decoration [42] and stamps have all been retained.

The Oldfield foundry continued to supply bells for at least one hundred and twenty years after the death of Henry Oldfield II in 1620. The dynasty came to an end in 1741 with the death of George Oldfield III, the great-grandson of Henry II. Although there are no Rutland bells supplied by any of Henry's descendants there are some that nevertheless came from this Nottingham foundry under the name of William Noone.

[114]

[84]

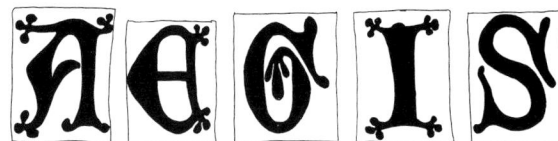

[116]

[42]

WILLIAM NOONE

When George Oldfield I retired *circa* 1678 William Noone took charge of the business. William cast four of the bells now hanging in Rutland belfries. Evidence that William Noone was casting bells in Nottingham at the beginning of the eighteenth century is given by entries made in the Churchwardens' Accounts of St Martin's Church, Leicester (North 1876, 209):

1700 The fifth bell was recast by William Noone of Nottingham ...

1704 Mr. Noone of Nottingham was engaged to recast the tenor bell ...

The full extent of his relationship with George Oldfield I has not been determined (it is said that he was a cousin), but the fact that he used many of the Oldfields' ornaments is an indication of his involvement with the Nottingham foundry. Noone supplied three bells to Ashwell Church in 1708 and virtually all of the ornamentation on these bells is known to have been used by the Oldfields. A typical example is the band ornament [40] found on the inscription band of the third, fourth and tenor. This was originally derived from Newcombe and Watts of Leicester, and so also was the distinctive band decoration [42] below the inscription on the fourth and tenor. Both of these decorations were also placed on the fourth bell at Braunston which was cast by Noone in 1710. The interesting and unusual stop [41] is also placed on the third and fourth at Ashwell.

The inscriptions used by Noone on his Rutland bells are in plain roman capitals of two sizes except for the Ashwell tenor. This particular inscription, Dec Campana Sacra Fiat Trinitate Beata, had been used by earlier Nottingham founders including Richard Mellour *circa* 1470-1508, possibly implying that this was a recast bell. None of the Noone bells in Rutland display his name or initials. He died in 1731.

[40]

[41]

THOMAS HEDDERLY I

With the death of George Oldfield III in 1741 the Nottingham foundry in Bellfounder's Yard, Long Row, passed to the Hedderly family. Thomas Hedderly I supplied two bells to Rutland. He supplied the second bell to Ashwell in 1760 and included decoration [42] as used by Noone on the fourth and tenor. There are two features on this bell that are found on no other bell in Rutland: he applied a decoration (albeit indistinct) around the lip of the bell, and an adornment [43] between the canons. The Latin inscription is in roman capitals and alongside the churchwarden's name he placed his own name: **THO HEDDERLY FOUNDER**.

Eleven years later he cast the fifth bell for Langham but this time the inscription, again in roman capitals, is in English. The location of the foundry, **NOTTING**[M], was included and his Christian name in full. This is one of the most ornamented of the Rutland bells. There is a band of decoration [42] around the collar and directly above the soundbow, and a band of [129] below the inscription. The king's head [133] from a second set of 'Royal Heads' as used by John Rufford (see under Toddington) in the fourteenth century is placed in a gap in the decoration immediately above the soundbow. It appears that these 'Royal Heads' had not been used by the Nottingham foundry for at least two centuries.

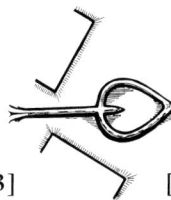

[43]   [133]

[129]

GEORGE HEDDERLY

After the death of Thomas Hedderly I in 1778 his sons Thomas II and George succeeded to the foundry. However with the death of his brother in 1785, George only continued with the business until 1793 when, following its failure, he emigrated to America. He supplied the bell at Thistleton in the same year. It has the plain inscription **W: FREAR CH. WARDEN G. H. NOTT**[N] **1793** (**W: FREAR** is incised). The remaining space on the inscription band indicates that an inscription and decoration has been removed, implying that originally this bell had been cast for another church.

PETERBOROUGH

HENRY PENN

Henry Penn was casting bells between the years 1703 and 1729. Before 1703 he worked as an apprentice with his uncle, Henry Bagley II, whose foundry was at Ecton in Northamptonshire. Thomas Penn, the father of Henry Penn, had married Henry Bagley's sister Sarah.

Henry Bagley II worked alongside Thomas Franklin who was a blacksmith. Thomas was uncle to Benjamin Franklin, the American statesman. There is also some speculation that Henry Penn the bellfounder was related in some way to William Penn, who founded Pennsylvania, one of the original thirteen states in the United States of America.

On the deaths of Thomas Franklin in 1702 and Henry Bagley II in the following year, Henry Penn took over the Ecton foundry. He continued to use Bagley's tools,

moulds, ornaments and other equipment essential to the trade. By 1708 Penn is known to have worked from a foundry in Bridge Street in Peterborough, having been commissioned to cast the cathedral bells. It may be that for an intervening period he worked from both Peterborough and Ecton. However by 1714 he was living in Peterborough and had set up a permanent business there.

Henry Penn is known to have cast fourteen bells for Rutland churches, of which ten remain today. They date from 1710 to 1723 and it is assumed that they were all supplied by his Peterborough foundry. The earliest is the treble hanging in the bellcote at Stretton. The inscription reads **HENRY PENN FVSORE** and also includes the date 1710 together with his well-known trademark, a rose [96].

[96]

Penn was very sparing with his inscriptions and decoration, and the third bell at Lyndon only has the date **1716** and the name of the donor. The fourth at Preston just has **1717**. The tenor in the bellcote at Pilton of *circa* 1720 is devoid of any inscription, decoration or date but it has been ascribed to Penn, mainly because of the wide inscription band, a characteristic typical of his bells (information from George Dawson). Five bells at Great Casterton were supplied by Penn in 1718 and they represent one of the few complete rings by this founder to have survived. He placed the date on all these bells but, as was often the practice in a full ring, his name appears on just one of them, the then treble: **HENRY PENN FOVNDER 1718**. The tenor has the names of the serving churchwardens and the impression of an unidentifiable coin. Otherwise the bells are blank.

Around 1722, Penn began to use more decorative work and the lettering was bolder. This is evident on the tenor at Edith Weston, cast in 1723. He used decoration

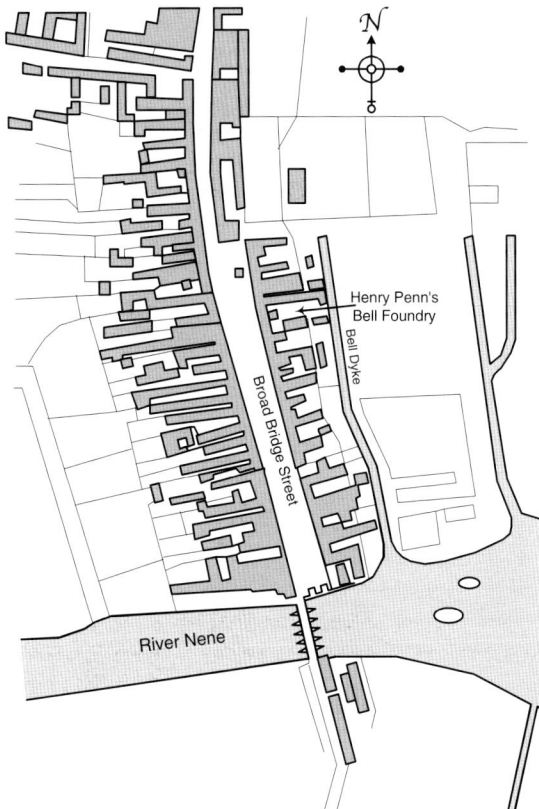

Possible location of Henry Penn's house, workshop and foundry in Peterborough, based on a map of 1723 (Lee 1988, 4). Bell Dyke and remains of pits from bellfounding both existed in 1850 and confirm that the foundry was in this vicinity. Bell Dyke was filled in circa 1875 (North 1880, 73-4)

[80]

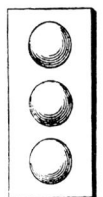

[50]

[80] as a filler between the end and beginning of the inscription and as a complete band on the shoulder. Stop [50], as used by the very early founders, is placed between the words of the inscription.

The end of Henry Penn's life was tragic. The parishioners of St Ives, Cambridgeshire, were dissatisfied with the bells he had cast for them and they brought a lawsuit against him. The case was tried in 1729 at the assizes in that town. Although the verdict was in Penn's favour, the strain of the proceedings had affected his heart and he died immediately after the trial. The saddest thing about this episode is that the burial entry in the Huntingdon St Mary & St Benedict church register describes him as a 'stranger'. It is therefore fitting and fortunate that his bells, many of which are still in use today, are able to serve as a more apt and lasting memorial.

ST NEOTS

JOSEPH EAYRE

Joseph Eayre was the son of Thomas Eayre I and brother of Thomas Eayre II of Kettering. About 1735 he built a brick foundry in the shape of a bell in the Priory at St Neots and he ran a flourishing business there until 1772. Three of his Rutland bells survive. His bell at Normanton Church, inscribed **CVM VOCO VENITE JOSEPH EAYRE ST. NEOTS FECIT 1766** was removed when the church was deconsecrated in 1970 and it now hangs, without its canons, above the roof at St Jude's Church, Peterborough. It has decoration [45] after the Latin inscription and after the date. The treble in the bellcote at Whitwell has the simple inscription **J EAYRE ST NEOTS 1749** together with two [49] stops. That in the bellcote at Exton Hall Chapel is inscribed **JOSEPH EAYRE ST NEOTS FECIT 1771**. From a ground survey it appears that it has no stops or decoration. This must have been one of the last bells that he cast. The tenor at Exton Church was the only other Rutland bell cast by Joseph. It was recast by John Taylor in 1895 but the full inscription was retained, a credit to those who commissioned the replacement.

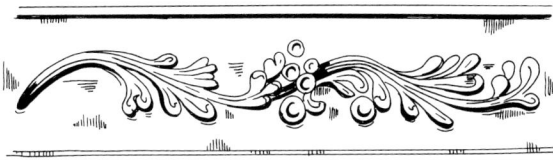

[45]

[49]

EDWARD ARNOLD

On the death of Joseph Eayre in 1772, the foundry at St Neots was held jointly for a time by his nephew Edward Arnold and his late foreman Thomas Osborn. This partnership was soon dissolved but Edward continued with the business until 1784, when he migrated to Leicester, where he erected a bell foundry in Hangman's Lane, now Newarke Street, near to the South Gates. He ran both foundries together for a short time, referring to himself as of St Neots and Leicester. From his St Neots foundry he supplied the present third at Brooke in 1780 (see under Leicester).

ROBERT TAYLOR

Robert Taylor had been an apprentice with Edward Arnold. When the latter moved to Leicester, the foundry at St Neots was left in the hands of Robert and by 1786 he was in full ownership of the St Neots business. He became the first in a long line of this family who were to follow in the founder's craft.

There are five bells in Rutland by Robert Taylor: the treble, second and third at Wing, all cast in 1789, the fourth at Uppingham cast in 1804, and the fifth at Brooke dated 1811. The lettering is plain roman capitals and he used either the initial **R** or the shortened form, **ROB**^T for his christian name. With the exception of the treble at Wing he included **ST. NEOTS** in the inscription. All have blocks of the attractive decoration [67] in the inscription band.

A fire at his foundry in 1821 compelled Robert Taylor to find a new location, and new premises were set up in Oxford. On his death in 1830 at Oxford, he was succeeded by his sons William and John. By this time John had set up a foundry at Buckland Brewer in Devon, but he returned to Oxford in 1835. In 1839 John and his son John William travelled to Loughborough to recast the bells for All Saints' Church, a stipulation of the order being that the bells had to be cast in the town. They found the area very suitable for business and as a consequence John and his family took up residency and set up a permanent foundry there (see under Loughborough). The foundry at Oxford continued to supply bells under the auspices of John's brother William until the latter died in 1854.

[67]

SALISBURY

JOHN BARBER

John Barber was a bellfounder working in Salisbury. He died in 1403 but little else is known of him. The Sanctus Bell at Preston has been ascribed to him and dated *circa* 1400 (Pearson 1989). The inscription **S MARI** is in small, crowned gothic capitals [115] and includes the hexagonal stamp [15], the initial cross [132] and the impression of an unidentified lead token (information from Yolanda Courtney, Leicester City Museums Service). It is the only medieval Sanctus Bell to survive in Rutland. With the bell being so far from its foundry, it is possible that it was used elsewhere before being hung at Preston.

[115]

[15]   [132]

STAMFORD

There was a bell foundry at Stamford, Lincolnshire, throughout most of the seventeenth century, and it supplied bells to many Rutland churches. The foundry was owned by at least three generations of the Norris family and their successor Alexander Rigby.

Toby Norris I first appears in Stamford documents when he obtained his Freedom in 1607. There are, however, bells in the Stamford area in the Norris style from as early as 1597. Since Toby I was probably only eleven or twelve years old at this date (see below), there must have been an earlier member of the family casting bells.

Little is known of the origins of Toby I, but recent research by George Dawson has established that he was possibly the son of Matthew Norris, bellfounder, known to have been of Leicester in 1575/6. The following notes on Matthew Norris are based on the results of this research.

North (1876, 217-18) quotes two sixteenth-century references which associate a Mr Norris with one Thomas Newcombe in connection with bells and possibly bell-founding. The first is in 1563 when the Leicester Hall Book records that 'Mr. Norys, ... & Thomas Newombe wth others' witnessed the weighing of the bells from the former church of St Peter, Leicester. The second is from the Chamberlains' Accounts which records the sale of one of St Peter's bells:

> **1563-64** It. Receyved of Thomas Newcombe & Mr. Norys for j [1] bell of sent Peters church weying xj[c]. xvj [li] [11cwt 16lb]
> xiiijli . xvjs . viijd. [£14 16s 8d]

'Mr Norys' in both instances may have been Thomas Norris, father of Matthew. Alternatively, the 'Mr Norys' mentioned in 1563 could have been William, who was buried in the churchyard of All Saints' Church, Leicester. Many facts were recorded about him on 'a large wooden framed tablet, hung in the pew next the May-

or's, on the South' (Nichols 1971, 553), not least that 'Twice was he grac'd with serving twice The office of the maioraltye'. The fact that he was a man of distinction may account for his presence at the bell-weighing ceremony.

The Thomas Newcombe is likely to have been Thomas Newcombe II although at this time he had not been admitted a Freeman of the town (see under Leicester).

MATTHEW NORRIS

The marriage of a Matthew Norris is noted in 1571/72 in the *Records of the Borough of Leicester* (Bateson **3**, 137). When he obtained his freedom (to trade) at Leicester in 1575/6 he was described as 'bellfounder, 2nd s. of Thomas, late of Leic., decd.'. Matthew's first son was John Norris who became a tanner, and he obtained his Feedom at Leicester in 1604/5 as the 'p. [apprentice] of Mr. William Norice, tanner' (Hartopp 1927, 81 & 101). Further research may reveal that this William was a brother of the Thomas noted above. The *Register of the Freemen of Leicester* (Hartopp 1927) lists many Norrises, confirming that the family was an important one in Leicester throughout the sixteenth century. Two, including the William Norris already mentioned, were Mayors during this period.

Matthew's son John is the important link between the Norrises of Stamford and Leicester, for when Toby I died in 1626 he left a will (Lincolnshire Archives, LCC Wills 1626/603) which states that if his son Thomas 'shall refuse or not satisfie and paie unto my said Children the said severall sumes of monie and porcons according and at the time or times limited and sett downe' he gave 'full power & lawful authoritie unto my beloved brother John Norris of Leicester, gentleman to sell, convey and assure my said house ...'. This leads to the

The Norris family of bellfounders

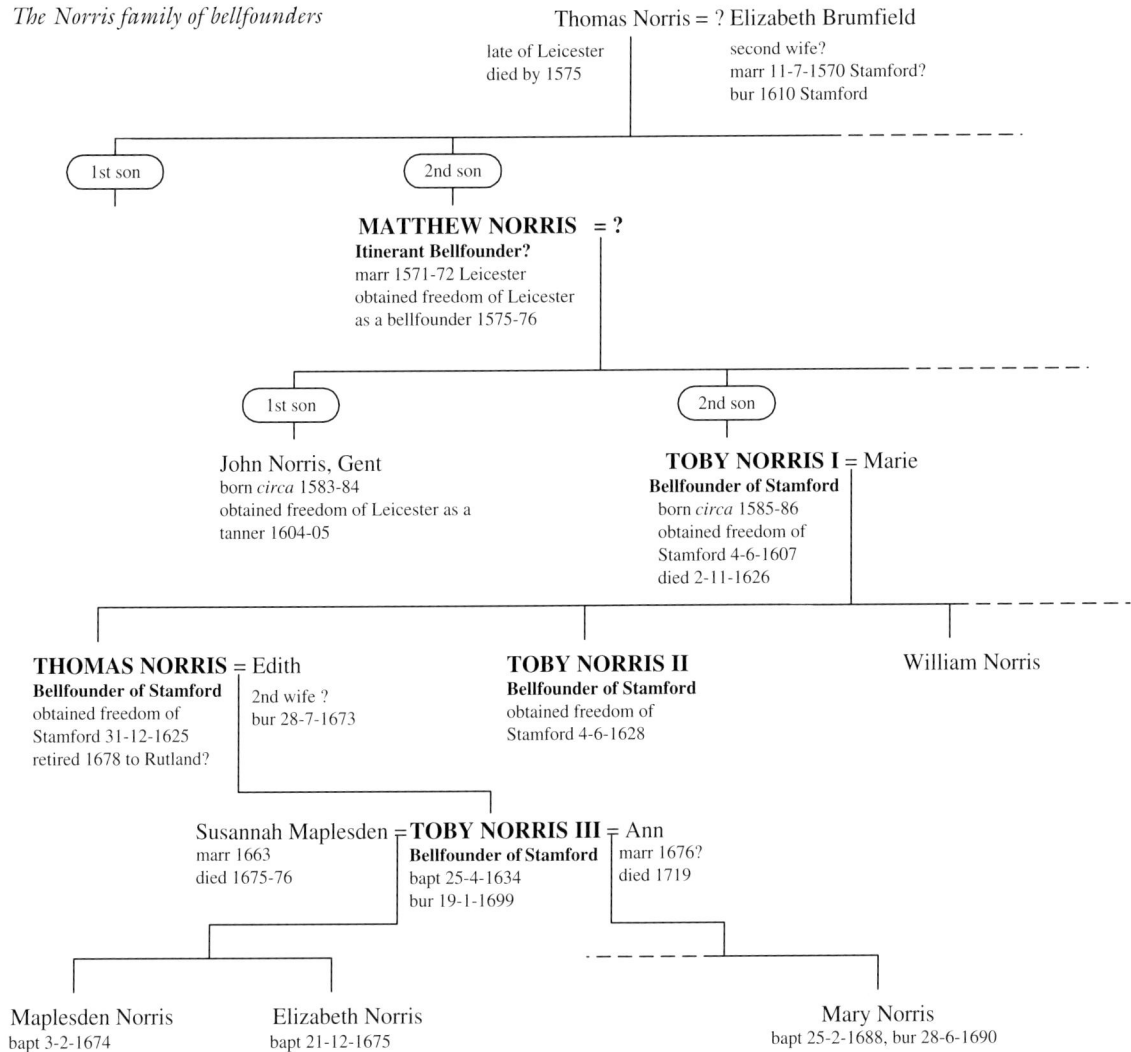

Thomas Norris = ? Elizabeth Brumfield

late of Leicester
died by 1575

second wife?
marr 11-7-1570 Stamford?
bur 1610 Stamford

1st son

2nd son

MATTHEW NORRIS = ?

Itinerant Bellfounder?
marr 1571-72 Leicester
obtained freedom of Leicester
as a bellfounder 1575-76

1st son

2nd son

John Norris, Gent
born *circa* 1583-84
obtained freedom of Leicester as a
tanner 1604-05

TOBY NORRIS I = Marie

Bellfounder of Stamford
born *circa* 1585-86
obtained freedom of
Stamford 4-6-1607
died 2-11-1626

THOMAS NORRIS = Edith

Bellfounder of Stamford
obtained freedom of
Stamford 31-12-1625
retired 1678 to Rutland?

2nd wife ?
bur 28-7-1673

TOBY NORRIS II

Bellfounder of Stamford
obtained freedom of
Stamford 4-6-1628

William Norris

Susannah Maplesden = **TOBY NORRIS III** = Ann

marr 1663
died 1675-76

Bellfounder of Stamford
bapt 25-4-1634
bur 19-1-1699

marr 1676?
died 1719

Maplesden Norris
bapt 3-2-1674

Elizabeth Norris
bapt 21-12-1675

Mary Norris
bapt 25-2-1688, bur 28-6-1690

conclusion that Toby I originated in Leicester and that he was probably the second son of Matthew Norris.

Toby I was admitted a Freeman of Stamford on 4 June 1607. If both John and Toby I obtained their Freedom at the normal age of twenty-one, John would have been born in 1583/4 and Toby in 1585/6.

It is apparent that Matthew Norris must have worked initially with the Newcombes at Leicester and had been one of their collaborators. The fact that both the Norris and Newcombe families were involved in tanning as well as bellfounding supports this connection. By the 1590s the Newcombes were becoming associated with the Watts family, both possibly using the same bell foundry near All Saints' Church. It may well be that Matthew saw the writing on the wall as far as his future employment prospects were concerned, and decided that a new loca-tion was needed, although it seems that he was probably an itinerant founder before settling in Stamford.

What is certain, however, is that Matthew formed an association with another founder whose initials were RO. In 1595 RO and MN cast a bell for the church at Terrington St Clement, Norfolk. This fine bell has the founder's mark [138] which includes their initials, and the initial cross [8]. Other bells using the same lettering [125] and initial cross, but not the RO/MN mark are: the fourth of five bells at Newton, Lincolnshire, dated 1596, the third of three bells at Swaton, Lincolnshire, dated 1596, and the fourth of five bells at Seaton, Rutland, dated 1597.

The initial cross [8] is almost identical to cross [72] as used by the Oldfields of Nottingham except for the small dots between the arms.

  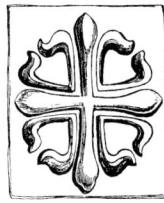

[138] [8] [72]

[125]

From this it can be concluded that RO was probably related to the Oldfield family of bellfounders. On the evidence of the Churchwardens' Accounts at St Margaret's Church, Kings Lynn, where there are payments in 1595 to 'rychard howlfeld, a belfounder', it is surmised that RO must be this Richard (H)Oldfield. It would appear that he moved to Cambridge *circa* 1599, casting bells there until 1612. He is known to have used lettering [125].

Until the appearance of the RO/MN mark no bells had been assigned to Matthew Norris, but this founder's mark makes it clear that the Terrington St Clement bell was the joint work of Matthew Norris and Richard Holdfield. It is likely that they worked with each other on the 1596 and 1597 bells mentioned above, but there is a possibility that the partnership had already broken up and they are the sole work of one of them. The fact that at Seaton, Rutland, there are two bells dated 1597, the fourth using decorated gothic capitals [125], and the second small plain roman capitals, suggests that Richard Holdfield and Matthew Norris cast these bells alongside

each other before parting company. At nearby Edith Weston the fourth bell has identical lettering and the same inscription as the second at Seaton: ƧVM ROƧA PVLSATA MVИDIA MARIA VOCATA 1597. Thomas North (1880,128 & 151) suggested that both of these bells were recasts, and that the inscription was 'probably an attempted copy of that on the ancient bell'. The treble at Lyndon, Rutland, of the same date was inscribed NUNC MARTIИE EGO CAИA VOBIS ORE IVCVNDO REMMEDG HVИTE 1597 before being recast by John Taylor in 1889. Note that in all three cases some of the Ss and Ns are reversed. On this evidence these three bells have been ascribed to Matthew Norris possibly in association with Richard Holdfield (George Dawson). As there is no record of a local foundry at this date it is very likely that both Matthew Norris and Richard Holdfield were itinerant founders at this time and it is possible that all four Rutland bells were cast in Seaton churchyard.

At the end of this period Matthew Norris disappears. No will or date of death has been found. However on the assumption that Toby I was born *circa* 1585, some of the early bells which had previously been assigned to him in the period 1603 to 1607 would have been cast when

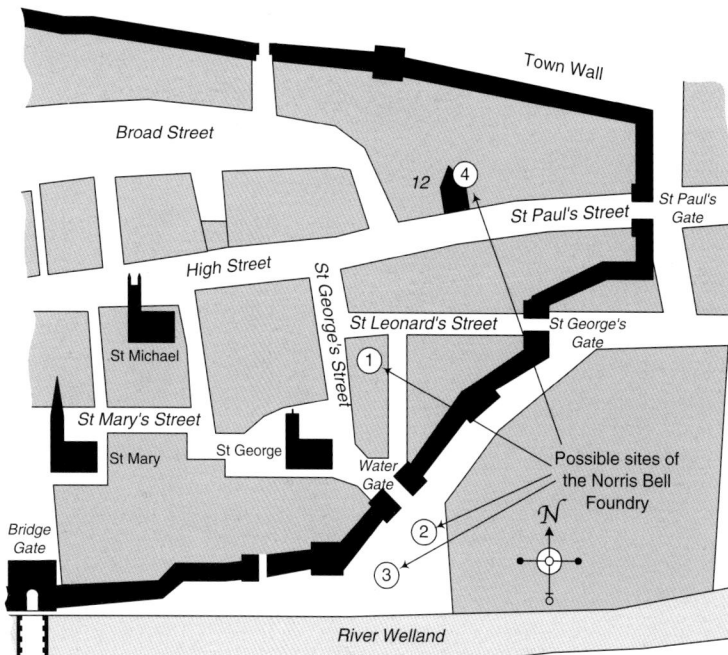

An interpretation of Speed's 1611 map of Stamford showing the possible locations of the Norris foundry. 1 is to the west of Gas Lane on a site occupied by the Pick Motor Company in 1903, but shown as gardens in 1833 (Smith, M, 1992, 44). 2 and 3 are both suggested by North (1880, 68): 2 is the site of the former gas works which was built in 1824-5, and 3 is the site of Grant's Brass and Iron Foundry, built in 1845, later to become J M Blashfield's terracotta works. 4 is 12 St Paul's Street which was the home of the Norris founders. Members of the family were tenants here until the building was purchased by Toby Norris I in 1617. A lozenge-shaped plaque in the gable above the east bay window is inscribed T N S 1663. The initials refer to Toby III and Susannah Norris who were married in 1663

he was in his teens. This seems unlikely and it is assumed therefore that Matthew was still alive and had started the Stamford foundry business by 1603.

At the church of St John the Baptist in Stamford there used to be a Priest's Bell inscribed **CVM VOCO VENITE 1605**, and the Church Books record the following entry as noted in *The Church Bells of the County and City of Lincoln* (North 1882, 671-2):

> 1605-06 Itm paid to Tobye Norrysh for our bell castine xvijs [8s]

The above is a favoured inscription of Toby I and being a small bell, this could have been the first he cast in his own name at the end of his apprenticeship. The earliest bell which has the name or initials of Toby I is the third of four at Orton Waterville, near Peterborough. It is dated 1606. Perhaps his father, Matthew, had recently died and the young Toby wished the world to know that he was working on his own.

TOBY NORRIS I

Toby Norris I was one of the 'Capital Constables' of Stamford in the years 1607 and 1621-2, and a warden of St George's Church, Stamford, in 1613-14 (North 1880, 63). He is known to have cast fourteen, possibly fifteen bells in Rutland between 1608 and 1626.

The site of the Norris foundry at Stamford has not been recorded in any detail. From the few clues that have survived it appears to have been just outside the town wall, either near to the sites of the later Blashfield terracotta works or gas works, or inside the wall near St Leonards Street and Gas Lane. A fourth possibility, suggested by Ketteringham (2000, 268), was that the foundry was on the same site as the Norris family home at 12 St Paul's Street.

Toby I used the initial cross [10], but not on any Rutland bells, the earliest being the eleventh of twelve at Surfleet, Lincolnshire, dated 1607. It is almost identical with [72] which was used by the Oldfields. [10] was frequently used on bells in Rutland and elsewhere by Toby's successors. A particular characteristic of his bells is that the **N**, as on the bells ascribed to his father Matthew, is almost always cast in reverse, presumably as the result of the matrix used to make the wax letters for the inscription being incorrectly made. Toby I's incriptions sometimes included a reversed **S**.

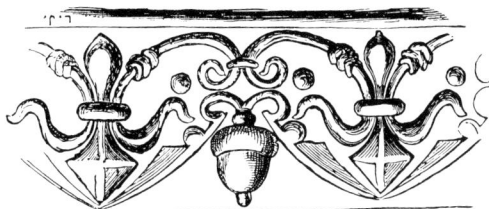

Part of the inscription on the third bell at Ayston. Note the reversed **N** and **S**

His nine surviving Rutland bells date from 1608 to 1626 but only the Ayston third carries his name. The rest are identified by the initial cross [14] or [26].

The lettering used by Toby I and his successors was generally plain roman capitals, the only exceptions being the tenor at Little Casterton, which is in gothic capitals, and the third at Ayston which has a mixture of both. All of his bells in Rutland, apart from that at Ayston, include one of the following favoured inscriptions:

CVM VOCO VENITE on the second at Little Casterton and Brooke.
OMNIA FIANT AD GLORIAM DEI on the second at Hambleton, and the fourth at Glaston and North Luffenham.
NON CLAMOR SED AMOR CANTAT IN AVRE DEI on the third at Hambleton and the fifth at Glaston and Edith Weston.

The decoration used on these bells was either [39], [53], [79] or [90] together with stops [5] or [62]. The fourth at Glaston is the only Norris bell in Rutland to bear the Royal Arms [83].

Toby Norris I died in 1626 and appropriately all the bells in Stamford rang out as a mark of respect for him.

[10]

[14]

[26]

[39]

[53]

[79]

[90]

[5]

[62]   [83]

The bell metal memorial to Toby Norris I in St George's Church, Stamford

He was buried in the north transept of St George's Church, and his sons Toby II and Thomas cast a special plaque in bell metal to his memory. This was originally on the flagstone floor immediately above his tomb, but it is now on the nearby east wall.

The business then passed into the hands of his eldest son Thomas. Toby, a younger son, took up his freedom in 1628 and was described as 'Toby Norris of Staunford bellfounder' in 1638. However, it is believed that he occupied 'a subordinate position in the foundry' (North 1880, 65) and there are no bells specifically assigned to him.

The following is a list of all the definite and possible [*] bells cast by Toby Norris I which have since been recast. The average lifespan of these former bells was 272 years. Five of the six were recast between 1859 and 1889.

Empingham fifth	1611-1859*	(248 years)
Hambleton tenor	1611-1887	(276 years)
Lyndon second	1624-1889	(265 years)
Manton treble	1610-1920	(310 years)
S Luffenham third	1618-1886	(268 years)
Tinwell tenor	1620-1883	(263 years)

THOMAS NORRIS

Thomas succeeded to the Stamford bellfounding business on the death of his father, having taken up his freedom as a bellfounder on 31 December 1625. The will of Toby Norris I dated 11 October 1626 (Lincolnshire Archives, LCC Wills 1626/603) includes, 'I give unto my said sonne Thomas Norris all and singular such Tooles and Instruments belonging to my trade as are

now remayninge & being in my Workhouse ...'. Following his father's example, Thomas became a prominent figure in the parish of St George and a well-respected bellfounder. He held several important positions within Stamford including that of Chamberlain and Alderman. He was also at some time churchwarden and constable for the parish of St George, Stamford (North 1880, 65). Seventeen of the known twenty-nine bells he sent to Rutland between 1626 and 1671 have survived.

Like his father he used roman capitals for his lettering. There is only one small exception which can be seen on the tenor at Langham where the letters **IH** in gothic were used.

He placed **THOMAS NORRIS MADE ME** on many of his Rutland bells, the surviving examples being the fifth at Ryhall (1633), the third at Langham (1636), the treble at Morcott (1637), and the third (1639) and treble (1654) at Tinwell. However the third at Ryhall (1626) and the treble at Tickencote (1630) have **CAST** and not **MADE**. The tenor at Clipsham (1657) has **MEE** as do the second and fourth bells at Cottesmore, the tenors at Braunston and Langham (all 1660) and the tenor (1663) at Stretton. When the second at Langham was recast by John Taylor in 1900, **THOMAS NORRIS MADE MEE 1660** was retained. Similarly, the fifth at Greetham, recast by Gillett & Johnston in 1923, retains this same inscription from Thomas' bell cast in 1650. The reversed letters, characteristic of all early Norris bells, were not used after 1630.

Thomas used the initial cross [10] on five of his bells in Rutland together with crosses [14] and [26] which had been used by his father.

Only two of his early bells in Rutland display one of his father's favoured inscriptions, these being the second (1627) and third (1626) at Ryhall. Of the decorations used by Thomas Norris on his Rutland bells, those prior to 1637 are [53] and [90], as used by his father, and [69]. After this date the decoration [52] is used exclusively. The [63] used on the fourth at Brooke is the centre of decoration [52]. Both the third at Langham and the tenor at Clipsham have impressions of coins on the soundbow.

Thomas Norris is known to have been casting bells

[69]

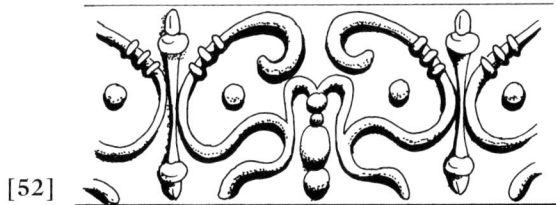

[52]

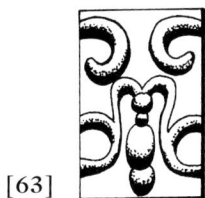

[63]

until 1678 and it is at about this time that he left Stamford to live with relatives at either Barrowden or Tickencote.

The following is a list of all the definite and possible [*] bells cast by Thomas Norris which have since been recast. The average lifespan of these former bells was 184 years.

Belton third	1664-1730*	(66 years)
Belton tenor	1660-1911	(251 years)
Cottesmore fifth	1660-1886	(226 years)
Empingham third	1648-1859*	(211 years)
Empingham fourth	1661-1859	(198 years)
Greetham fourth	1658-1923	(265 years)
Greetham fifth	1650-1923	(273 years)
Ketton treble	1640-1748*	(108 years)
Langham second	1660-1874	(214 years)
Market Overton third	1658-1885	(227 years)
Ridlington third	1671-1903	(232 years)
Ryhall treble	1633-1790	(157 years)
Uppingham unknown	1662-1772*	(110 years)
Uppingham unknown	1662-1772*	(110 years)
Uppingham unknown	1666-1772*	(106 years)

The following extract from the Churchwardens' Accounts of All Saints' Church, Cambridge, provides evidence that Thomas utilised local waterways as a convenient means of transporting bells (Till 1990, 92-3):

1632

For our charges in going for a bellfounder	6s 0d
For taking down the bell by Mr. Mane	3s 0d
For carryedge to the bridge and waying	2s 6d
For the carryedge of it by water	6s 0d
It. for bringing it bake from Stamford to the bridge	6s 0d
It. for waying it and bringing from the water home	3s 0d
It. Paide to Thomas Noris, the bellfounder	£6 14s 4d

It would be easy to assume that the route from Stamford to Cambridge by water would involve the use of the nearby River Welland. However, although the river had been navigable by the 1570s, it had silted up by the early 1600s and was not opened to water traffic again for at least another seventy years. It is more likely therefore that Thomas Norris used the River Nene, the nearest access point being at Wansford. The route to Cambridge would then be via Peterborough, the fen waterway system, the River Ouse and the River Cam.

TOBY NORRIS III

The successor to Thomas Norris was his son Toby Norris III who was baptised at St George's Church, Stamford, on 25 April 1634. Like his father and grandfather before him, he was well connected within the town. He was casting bells until his death in 1699 and sixteen of at least twenty-five bells he sent to Rutland remain in the county, their dates ranging from 1669 to 1696.

The tenor at Seaton, dated 1669, is his earliest Rutland bell. Five bells, the second, third, fourth and fifth at Exton together with the second at Clipsham, were all cast in 1675. Upon all of these, with the exception of the second at Exton, he included his name: **TOBIEAƧ ИORRIƧ CAƧT ME**. Note the re-introduction of the reversed **S** and **N**. He placed his name in two other ways on his Rutland bells. At Lyddington the tenor of 1694 has **TOBYAS NORIS** but the second, dated one year later, has **TOBY NORRIS**, and this latter form is used on the fifth at Caldecott in 1696. The fourth at Caldecott, originally a Toby Norris III bell, was recast by John Taylor in 1985. The retained inscription reads **TOBY NORRIS 1696**.

The Seaton tenor (1669), the treble at Wardley (1677) and all his Exton bells display the loyal inscription **GOD SAVE THE KING**.

In general Toby III's lettering is uniform but that on the Lyndon tenor of 1687 is poorly formed and irregularly spaced. It is the only Norris bell in Rutland to have the date incised into the bell metal and this is probably the result of an error or omission.

The crosses [10] and [26] and decorations [69] and [70] were employed on the second at Clipsham (1675), the tenor at Lyndon (1687), the four bells at Lyddington (1694/5) and the four at Caldecott (1696).

[70]

[70] is incorporated in the band decoration [79] as found on the fifth bell at Edith Weston cast by Toby Norris I in 1621. Prior to 1687 he used the crosses [26] and [14] and decorations [69] and [52]. With the exception of decoration [70], the other crosses and decorations used by Toby III were included on Rutland bells by his father and grandfather.

The following is a list of all the known Rutland bells cast by Toby Norris III which have been recast. The average lifespan of these former bells was 198 years.

Belton fourth	1681-1911	(230 years)
Belton fifth	1695-1911	(216 years)
Caldecott fourth	1696-1985	(289 years)
Cottesmore tenor	1699-1885	(186 years)
Empingham treble	1695-1859	(164 years)
Exton treble	1675-1895	(220 years)
Lyddington fourth	1694-1861	(167 years)
Oakham third	1677-1910	(233 years)
Oakham tenor	1677-1875	(198 years)

ALEXANDER RIGBY

Alexander Rigby had connections with the Stamford foundry, possibly as foreman, for some years before the decease of Toby Norris III in 1699. In 1880 Thomas North recorded four Rigby bells in Rutland but only that at Burley remains today. On this bell, which is believed to be the largest of his surviving bells, he used the bold statement **ALEXANDER RIGBY MADE ME** alongside the date **1705**, and the initial cross [14] which was

[57]

consistently used by the Norris family. He also included **BVRLEY IN RVTLAND** in the inscription. This is one of only three examples in the county of the village being named on a bell. A scroll ornamentation [57] was inserted between the words of the inscription, initial crosses and date.

Three of Alexander Rigby's Rutland bells have been recast:

Barrowden second	1706-1915	(209 years)
Barrowden tenor	1706-1915	(209 years)
Greetham third	1703-1923	(220 years)

The inscription has been retained on all three and the Barrowden bells include Rigby's cross [14] and decoration [57].

Following the death of Alexander Rigby in 1708 the Stamford foundry closed, bringing to an end just over a century of bellfounding in the town. He was buried at St Martin's Church, Stamford, and the entry in the register reads '1708 Alexander Rigby, bellfounder, bur. Octr. 29' (North 1880, 69).

TODDINGTON

JOHN RUFFORD

John Rufford was a founder at Toddington, Bedfordshire, in the middle of the fourteenth century. He was appointed Royal Bellfounder in 1367, and it was possibly this honour which allowed him to introduce stops depicting 'Royal Heads'. The images are believed to represent Edward III and Queen Philippa and there are two varieties of each head.

John Rufford had died by 1390 and a William Rufford, possibly his son, continued in the trade for a short while. John Rufford's 'Royal Heads' are known to have found their way to Leicester and Nottingham foundries by the end of the sixteenth century and examples from both sets can be found on the tenor at Teigh and the fifth at

Langham (see under Nottingham, Henry Oldfield I, and Thomas Hedderly I).

The second bell at Ayston has been ascribed to John Rufford (Pearson 1989) and dated *circa* 1365. It has as its inscription AVE REX GENTIS ANGLORVM and the lettering is of the type [112]. The bell also displays the initial cross [4] and stop [50].

[112]

UNKNOWN FOUNDERS

There are only five church bells in Rutland which have not been ascribed to a particular founder. Four of these, the trebles at Pilton and Little Casterton, and the Priest's Bell and treble at Egleton, are plain bells. All four are thought to be of eighteenth-century origin and are totally devoid of any foundry mark, decoration, inscription or other identifying mark.

The fifth is the former clock bell at Morcott which is retained in the nave of the church. Until 1940 it was hung externally on the south-west corner of the tower.

It has as its inscription ᴀᴅ ᴍᴏᴅᴇᴏ ᴄᴠᴍ ᴍᴏᴠᴇᴏ. The lettering is plain gothic and there are small blocks of decoration [123] after each word except ᴀᴅ.

Thomas North (1880, 144) reported that the chapel bell of St Peter's College, Cambridge, had the same inscription and that it was dated 1622. On this basis the Morcott bell has been dated *circa* 1620.

[123]

The inscription on the Morcott clock bell

UNKNOWN LOCATION

JOHANNES DE COLSALE

Johannes de Colsale was an itinerant founder (George Dawson). Little is known of him but it is believed that he may have hailed from near Nottingham.

He is known to have cast bells for Milwich, Staffordshire, and Beckingham, Nottinghamshire, and on both he placed cross [28], his name and the date 1409. This cross is also known to have been used on two Sanctus Bells in Northamptonshire, one at Walgrave and the other at Harringworth. As cross [28] is found on the tenor at Whitwell this bell has been ascribed to Johannes de Colsale (George Dawson). The inscription in gothic

[28]

capitals is ɪɴ ʜᴏɴᴏʀᴇ ꜱᴀɴᴄᴛɪ ᴇɪᴅɪɪ. The intended dedication is Saint Ægidius which is the Latin for St Giles. Although there are many churches with this dedication, it is very seldom found on bells.

A performer playing a 'carillon of five bells', from a manuscript said to be of the ninth century (North 1880, 36)

This map of Rutland was first published by John Owen and Emmanuel Bowen in 1720. At this time the county was very much an agricultural community consisting of 3263 houses in 48 parishes and two small market towns - Oakham and Uppingham. The market for clocks and watches was therefore never more than that of an average-sized town, and probably less due to the predominance of low-paid agricultural labourers and the distributed nature of the population. Local clockmaking, which began around the beginning of the eighteenth century, had virtually ceased by 1780 owing to the availability of ready-made components and complete movements. The predominant makers during this period included the Blackburns of Oakham, the Fox family of Uppingham, Richard Hackett of Harringworth, and John Watts, Boniface Bywater, Robert Watts, Thomas Rayment and Ralph Wilson of Stamford

Chapter 3
Clockmakers
Local Clock & Watchmakers and the Makers of Rutland Clocks

Nearly all the clock and watchmakers recorded in Rutland were based at either Oakham or Uppingham, the only two towns in the county. The few clockmakers that lived in the villages tended to be early craftsmen who were probably blacksmiths as well. They are referred to here as clocksmiths. Clock and watchmakers in the nearby larger towns of the surrounding counties also supplied clocks and watches to Rutland, and maintained and repaired some of the county's turret clocks. Stamford was the most important in this respect, but the clock and watchmakers of Melton Mowbray and Market Harborough, and possibly Leicester, Kettering, Grantham and Bourne, also benefited from trade with Rutland people. One important clockmaker from a village just outside the county who worked in Rutland was Richard Hackett of Harringworth, Northamptonshire.

There are no records of any turret clockmakers in Rutland. The nearest was John Watts of Stamford [working 1661-1719] who probably only made church clocks for his local area. Thomas Eayre II of Kettering [working 1717-57] supplied clocks to local clockmakers. Most of the county's present turret clocks were manufactured and installed by established and well-known makers between 1858 and 1932.

The criteria for inclusion in this section are therefore:
• Clock and watchmakers based in or just outside Rutland.
• Clockmakers located outside Rutland who are recorded as having maintained or repaired turret clocks in the county.
• Clockmakers who sent turret clocks to a Rutland church or secular building.
Also included are any others associated with timekeeping in Rutland.

Rutland clock and watchmakers have been recorded previously by Britten (1932), Daniell (1975), Baillie (1976), Loomes (1976b), and Hewitt (1992 & 1994). Some Stamford clock and watchmakers have been researched by Tebbutt (1975) and Wilbourn and Ellis (2001). Current clock and watchmakers were also recorded in nineteenth and early twentieth-century directories. All these sources, and more, have been used to generate the list below. Research into parish records,

Churchwardens' Accounts in particular, as well as other archives, has produced a great deal more information on clock and watchmakers from Rutland and its bordering counties. A number of previously unrecorded clockmakers have also been identified.

The terms 'clockmaker' and 'watchmaker' are somewhat misleading when applied to many such local craftsmen after about 1780, as by then most were assemblers, repairers and retailers rather than makers. Ready-made components, movements, cases and complete clocks and watches for retailing were obtained from workshops in Birmingham, Lancashire, London and elsewhere, and after about 1850 many were imported from America and the continent. It was common practice for local clockmakers to add their names to the dials in the same way as if they had actually made the clock. Such craftsmen are, however, included in the following list, with a cut-off date of *circa* 1920.

Where a clockmaker is noted as having supplied, maintained or repaired a Rutland turret clock, further details may be found in Chapter 4 — Gazetteer.

Dates shown in the first line of each entry represent date of birth to date of death, for example (1825-85), and when known to be working as a clock or watchmaker, for example [1845-75]. Names in bold in the text of an entry, for example, **William Aris II**, indicate that there are biographical notes on this maker or company elsewhere, either in this chapter, or in Chapter 2 — Bellfounders. Abbreviations are given in the introduction to Chapter 4 — Gazetteer.

ADAMS, Clement, of Stamford [1762-77]
The *Lincoln, Rutland and Stamford Mercury* of 11 November 1762 reported that Clement Adams was apprenticed to Samuel Haselwood, watchmaker, and Baillie (1976, 2) records that he was working as a clockmaker in Stamford before 1777.

ARIS, Thomas, of Uppingham (1803-76) [1831-76]
Thomas Aris was the son of **William Aris II** and he was born in Uppingham in 1803. He married Catherine Tyler at St Peter & St Paul's, Uppingham, on 9 October 1827 and he succeeded his father as parish clerk in 1830,

a post he was to hold until 1842. He is first noted as a watch and clockmaker in Pigot's *Directory* of 1831, working in Market Place, Uppingham. The court rolls of the Manor of Preston with Uppingham record that in fact he was living at the Tap to the Swan Inn in Market Place. Sometime after this he acquired premises in Horn Lane, now known as 4 Queen Street (information from Peter Lane), but the Census Returns for 1841 and later show that he was living on the north side of High Street, at or near what is now 41 High Street East. At the 1841 Census he was living with his wife, Catherine, four children, and his father, William Aris II. In all, he had eleven children, eight of these being sons, but none became clockmakers, most of them moving to London to become plumbers and carpenters.

Thomas was responsible for Uppingham church clock. He maintained and repaired it from 1830 until 1876 and many of his detailed vouchers for work carried out on the clock are in the parish archive. He also worked on a number of other local church clocks, including Edith Weston from 1842 to 1847, Preston from 1842 to 1848, Belton in Rutland from 1859 to 1876, North Luffenham in 1861, Caldecott in 1864 and 1865 and Lyddington from 1865 to 1867. He died in 1876 and is buried in Uppingham churchyard.

ARIS, William I, of Harringworth and Uppingham (1736-98) [1753-98]
William Aris I was born in 1736 (information from Heather McGuire) and the first record of him as a clockmaker may be in the Churchwardens' Accounts of South Luffenham:

| 1753 | pd Mr Ayres Bill | £1 3s 6d |
| 1755 | paid Mr Ayres for cleaning the Clock | 5s 0d |

If this is William Aris I then he would only have been 17 years old and still serving his apprenticeship, probably with **Richard Hackett** at Harringworth, as explained below. The next record of him is in the militia ballot lists of 1762 (NRO, Corby Hundred — Militia Papers): 'Wm Ares Clockmaker'. It is almost certain that he was then working with **Richard Hackett** who was a witness when he married Ann Bradshaw at Harringworth on 27 September 1765. 'Wm Aries' and 'Wm Tiplee' are included in the 1771 militia ballot list as clockmakers (NRO, Corby Hundred — Militia Papers). In the 'Town Rents' recorded in the Harringworth Churchwardens' Accounts (NRO 156P/84) it appears that **Richard Hackett** was paying the rent on the house occupied by William Aris I:

1773	Mr Hackett Rent £3 10s 0d for Wm Aires House agreed to this 28 June 1773	
1774	Mr Hackett for Shop & Wd Simsons house	6s 6d
	Do for Aires House	£3 10s 0d

A settlement certificate dated 18 May 1785 (NRO,

Thomas Aris clockmaker of Uppingham circa 1875 (John Pearson)

Biographical Notes — Aris Family) states: 'That William Aris Watch Maker, and Ann his wife, and 5 Children to be our Inhabitants legally settled in the parish of Harringworth in the County of Northampton aforesaid ...'. This allowed William Aris and his family to move to Uppingham in that year. The Harringworth Churchwardens' Accounts (NRO 156P/84) show that **Richard Hackett** had been responsible for the maintenance and repair of the church clock until 1781. William Aris I took over these duties following the death of Hackett in 1782, and continued until 1796, travelling from Uppingham after 1785. He took over the same duties at South Luffenham Church, continuing until 1798, and at Glaston Church. An eight-day arch-dial longcase clock with moon and tidal work and signed 'Wm Aris Harringworth' was reported to Leicestershire Museums some years ago (Hewitt 1992, 310) (see **Richard Hackett**). When William Aris I and his family moved to Uppingham they lived at what is now 10 High Street East, where he continued as a clock and watchmaker until his death in 1798. He was succeeded by his son, **William Aris II**.

ARIS, William II, of Horn Lane, Uppingham (1774-1842) [1799-1842]
William Aris II was born in 1774 at Harringworth where his father, **William Aris I**, was a clockmaker working

The dial of a thirty-hour longcase clock by William Aris I, circa 1790 (Andrew Butterworth)

The dial of an eight-day longcase clock by William Aris II. The dial surround is in gold leaf on a black background, circa 1810 (Andrew Butterworth)

with **Richard Hackett** and others. He continued to service Harringworth church clock after his father's death and the Churchwardens' Accounts (NRO 156P/84) contain relevant entries from 1799 to 1816. He also worked on many Rutland church clocks including Lyddington in 1801, Hambleton from 1810 (when the clock was transported to his workshop for repairs) to 1822, North Luffenham from 1810 to 1842, Edith Weston from 1818 to 1842, Belton in Rutland in 1826 and Preston from 1831 to 1842. He also travelled to Gaulby in Leicestershire where his signature can be seen in the clockroom window reveal. By 1820 he was the Uppingham clockmaker responsible for the church clock and he was also the parish clerk. In 1830 his son, **Thomas Aris**, took over both roles.

William Aris II initially worked from what is now 10 High Street East, his parents' home. On 11 September 1796 he married Frances Lupton of Wing at St Peter & St Paul's, Uppingham. In 1821 they were living in a house owned by John Wadd near the Swan Yard and by 1824 they were living in Fisher's (now Printer's) Yard, behind 10 High Street East (information from Peter Lane). Pigot's *Directory* of 1835 records him as working in Horn Lane (now Queen Street), where he had been from 1827 (information from Alan Rogers), and the 1841 Census Returns show that he was living with his son, **Thomas Aris** and his family in High Street, Frances having died in 1839.

There are a number of surviving longcase clocks by William Aris II in local private ownership. Another, in Rutland County Museum signed 'W Aris Uppingham'

(1971.53), is a small oak-cased eight-day clock of *circa* 1820 with a painted square dial showing seconds.

William Aris II died at the age of 68 years and was buried in Uppingham churchyard on 31 October 1842.

ARNOLD, Edward, of St Neots and Leicester [1772-1802] Following the death of **Joseph Eayre** in 1772 '... the business at S. Neots was held jointly for a short time by his late foreman Thomas Osborn, and his cousin Edward Arnold. After they dissolved the partnership Edward Arnold held the foundry at S. Neots ...' (North 1882, 134). In 1784 Edward Arnold established new workshops in Hangman's Lane, now Newarke Street, Leicester, leaving the St Neots foundry in the hands of his former apprentice Robert Taylor. At Leicester he set up a new foundry to cast bells and 'make upon the best capital principles ...', repair, and clean '... Church Clocks and Chimes, Turret and other Clocks, Watches &c ...'. He also made 'upon the most approved Plans ... Machines for weighing Carriages, and Engines for extinguishing Fire ...' (*Leicester and Nottingham Journal* 17 July 1784, No 2644, 1).

Although there are no records of any of his clocks being installed in Rutland, he did repair the chime barrel at All Saints' Church, Oakham, in 1793, and he, or **Joseph Eayre**, may have been the original supplier. The six-tune chime barrel at Stapleford Church, some eight miles due north of Oakham, is signed 'Edwd. ARNOLD. Leicester fecit. 1795'. This chime barrel is almost exactly the same as the nine-tune chime barrel at Kings Norton in Leicestershire built in 1765 by **Joseph Eayre**. In 1801 '... it was ordered [by Stamford Corporation] that Edward Arnold of Leicester be employed to repair the chimes ...' at St Mary's Church, Stamford (North 1882, 677). This eight-tune chime barrel had originally been installed by **Joseph Eayre** in 1770 at a cost of £40.

Hewitt (1994, 19) records that Edward Arnold supplied a new clock to St Martin's Church (now Leicester Cathedral) in 1787 and that in 1795 he was a founder member of the Society of Leicester Clock and Watchmakers. His last known bell is dated 1798 (Pickford 1998) and his last recorded work in the area was to hang two new trebles at St Mary's Church, Stamford, early in 1802. (*See also* Chapter 2 — Bellfounders, Leicester and St Neots).

BATES, John, of Uppingham [1846-48]
John Bates is described in White's *Directory* of 1846 as a watch and clockmaker with premises in Market Place, Uppingham. The following extract from the *Lincoln, Rutland and Stamford Mercury* of October 1851 may refer to him: 'The tradesman in the clock and watch business who left Uppingham for America some three years ago without previously informing his creditors, and who had posted on his shop shutters "wound up and the main spring broken" last week sent over the needful to supply his creditors in full demand' (Traylen 1982c, 19).

BILLINGTON, Everard, of Market Harborough, Leicestershire [1738-70]
Everard Billington supplied a new clock to Lyddington Church in 1738 at a cost of £10 10s. Bearing in mind the close proximity of Kettering to Market Harborough and that **Thomas Eayre II** was foremost turret clock maker in the area at this time, it is quite likely that the Lyddington clock came from this source (*see* Chapter 4 — Gazetteer, Lyddington). One of Everard Billington's oak-cased thirty-hour longcase clocks is in the Newarke Houses Museum, Leicester (H105.1948). It has a brass dial with a single hand. The centre of the dial has elaborate scroll engraving and it is signed on the chapter ring 'Everard Billington HARBOROUGH'. The clock originated from Tur Langton, Leicestershire (Daniell 1975, 16).

BIRD, William (senior and junior), of Seagrave, Leicestershire [*circa* 1741 to *circa* 1780]
Several members of the Bird family of Seagrave, Leicestershire, were noted clockmakers and watchmakers. Many examples of their work have survived, particularly thirty-hour longcase clocks, and they cover the period from *circa* 1741 to *circa* 1780. The collection at the Newarke Houses Museum, Leicester, includes a brass dial thirty-hour longcase clock (H1079.1951) and a silver-cased verge pocket watch of 1785, numbered 303 (H75.1949). An eight-day longcase clock of *circa* 1760, with an arch

dial and moon-work, is one example known to be in private ownership.

The clock at Ryhall Church, signed 'William Bird Seagrave 1771', is the only known turret clock by this family. It is therefore possible that it was acquired by them from a specialist maker. A slate headstone in Seagrave churchyard records that William Bird senior died at the age of 52 years, four years prior to the date on the Ryhall clock. He was succeeded by his sons, William and Richard. It appears that William junior carried on the business until about 1780, concentrating mainly on watchmaking and retailing.

BLACKBOND, Peter, of unknown location [1710]
Peter 'Blackbond' was paid 6d in 1710 for 'mending ye clock' at Teigh. He may be the Blackburn of Oakham referred to below, or a relative.

BLACKBURN, of Oakham [*circa* 1720 to *circa* 1750]
Blackburn is the earliest recorded domestic clockmaker in Rutland. He was working in Oakham between 1720 and 1750. His Christian name is not known. A number of Blackburns are included in the parish registers at this time but none is described as being a clockmaker. Several Blackburn clocks have survived and one is exhibited in Rutland County Museum (H258.1952). It is a *circa* 1720 thirty-hour longcase clock in a narrow elm case with a 200mm (8in) single-handed square brass dial inscribed 'Blackburn Oakham No. 1058'. All other known clocks by this maker also have a serial number and they include another thirty-hour clock of *circa* 1720 (No 1066) and an eight-day longcase of c*irca* 1740 (No 1208). Although it is quite usual for watches to be numbered in this manner, it is rare on longcase clocks. His numbering system probably started at 1000.

It may have been this Blackburn who repaired Ashwell church clock in 1729: 'Charges Mr blackburn when he came about the Clock 1s 6d'.

BLACKBURN, Stephen, of Oakham [before 1731-71]
Stephen Blackburn was probably the son of the **Blackburn** noted above. He was working in Oakham from before 1731 and several of his clocks are known. A tall eight-day arch dial longcase clock of *circa* 1770 in a mahogany case with columns to the trunk and signed 'Stephen Blackburn, Oakham' is exhibited in Rutland County Museum (H149.1987). It was bequeathed to Leicestershire Museums by the late L W H Paybury. A similar clock in private ownership was reported in 1975 at Stanmore, Middlesex.

William Pochin MP, of Barkby Hall, in his notes on Barkby Church, Leicestershire, records 'Oct. 3, 1754, I gave a clock to Barkby church: it was made by Steven Blackburn of Oakham; and I paid him for it £25' (Nichols 1800, **III**, Pt 1, 62). In the same reference, Nichols adds

(left) The elm hood and 200mm (8 inch) brass dial of Blackburn's clock, circa 1720, in Rutland County Museum (H258.1952)

The movement of Blackburn's thirty-hour longcase clock in Rutland County Museum (H258.1952)

a note about the church: 'It holds a peal of five bells, and a good parish-clock with an excellent dial of Swithland Slate'. Although it was replaced in 1898 by a flatbed movement by Smiths of Derby, the original clock is retained in the church.

It was in fact manufactured by **Thomas Eayre II** of Kettering in 1754 (Hewitt 1994, 25). Stephen Blackburn presumably acquired it from this source and carried out the installation work. He also maintained and repaired another clock by **Thomas Eayre II** at Whissendine Church between 1746 and 1771. Here, the accounts of 1770 show that he had at least one employee: 'Pd Mr Blackburns man for raiseing ye dyal Board and makeing ye hand go 3s 6d'. He also worked on the Hambleton church clock from 1730 to 1767.

BOYFIELD, Richard and Thomas, of Great Dalby and Melton Mowbray, Leicestershire [1750 to after 1801] Richard Boyfield senior, clockmaker of Great Dalby, was working from 1750 until *circa* 1780 and one of his thirty-hour longcase clocks is in the collection at the Newarke Houses Museum, Leicester (H1088.1951). His sons, Richard junior and Thomas, both became watch and clockmakers (*Universal Directory* 1791), working in

Great Dalby and later Melton Mowbray. Watches and longcase clocks by both have been recorded. A Mr Boyfield, no doubt one of these brothers, was paid for work on the clock at Whissendine Church for three consecutive years until 1801.

BRADLEY, Simpson, of Oakham [1841] Simpson Bradley is recorded as a clockmaker in the 1841 Census Return for Oakham. No other details are known.

BRITTON, John, of Stamford, Lincolnshire [1872-1900] John Britton's repair signature has been noted on a clock by **Robert Broughton Haynes** of Stamford (Tebbutt 1975, 46) and he is recorded as having repaired Ryhall church clock in 1887 for £1 5s, and in 1889 at a cost of 12s.

BROOKS, Charles, of Stamford, Lincolnshire [1810-53] Charles Brooks, clockmaker and silversmith, had premises in St Mary's Street, Stamford. He was engaged by Stamford Corporation to maintain the clock and chimes at St Mary's Church and a receipt dated December 1844 in the Churchwardens' Accounts indicates that he was paid £6 per annum for this duty (Tebbutt 1975, 46). For this salary he also had to toll a bell for fifteen minutes before Court Sessions at the Town Hall and on 'Bull-Running day' to warn the timid to stay off the streets (Tebbutt 1975, 28).

BRUMHEAD, John (senior), of Stamford, Lincolnshire (1762-1809) [1780-1809]
The Churchwardens' Accounts of St Mary, Clipsham, show that John Brumhead maintained the clock there in 1807 and 1808.

He was working in the Butter Market, Stamford, until he died on 15 December 1809 at the age of 47. About two months before his death, possibly due to his illness, he advertised for a 'Journeyman Clock-maker' in the *Lincoln, Rutland and Stamford Mercury*: 'A steady man and a good workman may have constant employ by applying to J. Brumhead' (Tebbutt 1975, 47). The business was taken over by his wife, Mary, who advertised herself as 'watch and clock-maker, silversmith etc'. She employed **William Hickman** as a journeyman clockmaker until he left in 1811. In 1825, as noted below, she advertised that she had relinquished the business to her son, **John Brumhead (junior)**.

BRUMHEAD, John (junior), of Stamford, Lincolnshire [1825-57]
John Brumhead (junior) is recorded by Loomes (1976, 33) as working from 1828 to 1835. However, in 1825 he placed the following advertisement in the *Lincoln, Rutland and Stamford Mercury*:

> JOHN BRUMHEAD, Clock and Watch maker, from London, respectfully informs his friends and the public that he has commenced the above business on the premises in the occupation of his Mother, MARY BRUMHEAD, Red Lion Square, Stamford, where he hopes by his attention to business to merit their future favours (Tebbutt 1975, 48).

He is the first recorded clockmaker to work on the clock at Ryhall Church in 1838 and he, and possibly his son William, continued to repair and maintain this clock until the 1850s. White's *Directory* of 1826 records that by that date John (junior) was working in St John's Street and the Poll Book of 1857 gives his address as St Leonard's Street.

BYRON, Thomas, of Stamford, Lincolnshire [1847]
The Poll Book of 1847 records that Thomas Byron, watchmaker, had premises in St Peter's Street, Stamford (Tebbutt 1975, 48).

BYWATER, Boniface, of Stamford, Lincolnshire (*circa* 1675-1752) [1689-1752]
Boniface Bywater was a well-known clockmaker, whitesmith and gunsmith. Stamford Town Hall Books record

that he was admitted to the freedom of Stamford in 1696 following his seven-year apprenticeship with **John Watts**, and later he was a Capital Burgess and an Alderman of Stamford. The Old Swan, at what is now 10-12 St Mary's Street, Stamford, was altered in 1702 by Boniface Bywater to make a shop, although little of this building has survived (Smith M 1992, 94). He later moved to the south side of St Mary's Street, near St John's Passage. The parish registers of St Mary's Church record the baptisms of two of his sons, Major in 1700 and Boniface in 1707, as well as his own burial on 13 May 1752. Although he is not recorded as working on any Rutland church clock, it seems quite possible that he helped with the construction and installation of some of them when he was apprenticed to **John Watts** between 1689 and 1696.

A brass plate attached to the former clock at St Mary's Church, Stamford, recorded that it and the chimes were the gift of the Honourable Charles Cecil and the Honourable Charles Bertie, and that they were 'performed' [made and installed?] by Boniface Bywater in 1709.

A rubbing of the brass plate which was attached to the former clock and chimes at St Mary's Church, Stamford (Michael Lee)

After this date he constantly figures in the Corporation Hall Book for carrying out repairs to both (Lee 2000, 5). They replaced a clock which was working in 1623 and chimes which were referred to in the Town Hall Books in 1683 (North 1882, 677). The clock remained in service until 1890 when it was removed to make way for a new clock by **William Potts** of Leeds (Tebbutt 1975, 26-28). The chimes were replaced by **Joseph Eayre** in 1770 at a cost of £40 (North 1882, 677).

It appears that Boniface Bywater also made domestic clocks although few have been recorded. A lantern clock by him was reported to Stamford Museum in 1975 (Lee 2000, 5).

COOKE, John, of Oakham (*circa* 1823-?) [1841-81]
John Cooke was the son of **Thomas Cooke** and he was born in Rutland *circa* 1823. He is listed in the 1841-81 Censuses as a watch and clockmaker in Oakham. White's *Directory* of 1877 records that his business was in High Street. Churchwarden' Accounts show that he repaired the clock and chimes at Cottesmore Church in 1861.

COOKE, Thomas, of Uppingham and Oakham (*circa* 1793-1883) [1821-71]
Thomas Cooke was born *circa* 1793 in Leicester and was

working as a clockmaker in Uppingham from 1821-23. By 1828 he had moved his business to High Street, Oakham, and White's *Directory* of 1846 records that he was a clockmaker and silversmith. The Churchwardens' Accounts show that he repaired the clock at Langham Church in 1836. He worked with his son **John Cooke**, who succeeded him, from about 1841. Thomas Cooke retired in 1871 and died in 1883.

COPE, G & F, & Co, of Nottingham [1845 to after 1928]

G & F Cope was formed in 1845 when George Cope and his brother Francis took over the turret clock business of Reuben Bosworth in Nottingham. The business prospered and they became well known throughout the world for their tower clocks and carillons.

William Cope became an indentured apprentice in 1874 when, at the age of 14 years, he joined his uncles' business. When George and Francis died just before 1900, he took control. He died in 1922 at the age of 52 years, and was succeeded by his son William W Cope (Mather 1979, 31). The company workshops were in Holden Street, Radford, Nottingham, and they supplied the two-train flatbed turret clock to Clipsham Hall stables in 1904, and then, in 1928, the Edith Weston Church clock. The small flatbed turret clock at Tinwell church was installed by William Cope a year later. The clock at Edith Weston, and possibly the Tinwell clock, were not actually manufactured in the Cope workshops. Apparently, William W Cope occasionally purchased clocks from continental makers in order to study the mechanisms. One such clock was that at Edith Weston which was made by Phil Horz of Ulam-am-Donau, Germany (*see* Chapter 4 — Gazetteer, Edith Weston).

G & F Cope's letterhead of 1928

CORNEY, Mrs Mary Ann, of Oakham [1925-32]

Mary Ann was the wife of **Robert Corney** and she took over his clock and watchmaking business when he died in 1925, continuing until 1932.

CORNEY, Robert, of Oakham [1895-1925]

Robert Corney frequently advertised his business in Matkin's *Oakham Almanack* from 1895, describing himself as 'Manufacturing Jeweller, Silversmith, Watch and Clock Maker'. His shop, which he called 'The Clock House', was in Bank Buildings, High Street, Oakham, adjacent to what was then the Post Office. A photograph of his shop is included in *Oakham in Rutland* (Traylen 1982a, 40). Braunston Churchwardens' Accounts show that he carried out repairs on the church clock there for many years from 1901. Other customers included Lord Lonsdale, with whom he had an annual clock-winding contract, Earl Cowley of Cold Overton Hall, and the Duke of Rutland. Outside his business one of his interests was playing bowls: 'Mr Bob Corney, watchmaker, was a bowls fanatic; when playing the game he was apt to become excited and would twist himself into all shapes and postures as the "woods" went up the green in a vain attempt to influence their course' (from the memories of Mr L Wakefield *circa* 1897) (Traylen 1982a, 24). One of his apprentices was **Charles Hetterley**. When Robert Corney died in 1925 the business was taken over by his wife, **Mary Ann Corney** until 1932. Entries in Braunston Churchwardens' Accounts show that the business continued until at least 1940.

CRANE, of unknown location [1925]

Langham Churchwardens' Accounts record that a Mr Crane was paid £5 17s 6d for repairing the church clock in 1925. No other details are known.

CURE, William, of Oakham [1863]

White's *Directory* records that William Cure was a clockmaker in Oakham in 1863. No other details are known.

DENISON, Edmund Beckett (Lord Grimthorpe), of St Albans (1816-1905)

Edmund Beckett Denison qualified as a barrister and was called to the bar in 1841. On the death of his father in 1874 he succeeded to the Beckett Baronetcy and dropped the Denison name, his grandmother's surname, to become Sir Edmund Beckett QC. In 1886 he was raised to the Peerage and became the first Baron Grimthorpe. He was deeply interested in bells, clocks, watches and architecture, and made a thorough study of the theory of clockmaking and watchmaking. He published *A Rudimentary Treatise on Clocks, Watches, & Bells for Public Purposes* in 1850 and a later edition was used as the basis for the horological articles in the eighth and ninth editions of *Encyclopædia Britannica*. He became president of the British Horological Institute in 1868, where his blunt views on many traditional practices raised many hackles with the professionals. He held this position until his death in 1905.

His horological interests led him to take a fundamental role in the design and construction of the new clock

A caricature of Lord Grimthorpe as depicted in Vanity Fair. Date unknown (Chris McKay)

and bells for the Palace of Westminster, often referred to as 'Big Ben'. This resulted in a close association with clockmakers **Edward and Frederick Dent**, who made the clock.

Denison recognised the benefits of the gravity escapement for turret clocks and considered using a design by Thomas Bloxham for the Westminster clock. However, he eventually rejected this because it was unreliable under certain conditions. Instead he developed Bloxham's escapement and eventually perfected it into the double three-legged gravity escapement. Thereafter it became the standard escapement for all better quality turret clocks (Rawlings 1993, 125-126). He also developed the single four-legged gravity escapement and this was fitted to Oakham church clock which was installed in 1858 by Frederick Dent.

In his preface to the eighth edition of *A Rudimentary Treatise on Clocks, Watches, & Bells for Public Purposes* published in 1903, Denison writes:

> The book led to my designing, either directly or indirectly, not only the Westminster and St. Paul's clocks, and the great peal of bells there, but those of many other Cathedrals and Churches, as well as Town-hall, Railway-station, and others in several of our Colonies, by special request. As I did all that work gratuitously, I have no means of tracing them, or probably remembering the names of them all. I know that I once counted above forty.

His architectural work included the major restoration of St Alban's Abbey at his own expense (information from Chris McKay).

C A Stevens, a former curate at Oakham, describes his own role in the restoration of the church and comments:

> I should much like to know whether the clock, which was of special construction recommended by Mr E Beckett Denison (Lord Grimthorpe), has done justice to its inventor and maker during its 45 years use (Stevens 1905-06, 25).

DENT, Edward, of London (1790-1853) [1811-53]
DENT, Frederick William Rippon, of London (1808-60) [1831-1860]
DENT, E & Co, of London [1864-*circa* 1976]
Edward Dent (born 1790) was a celebrated watch and clockmaker working in London on his own account, and then with Edward Arnold from 1830-40. After making a clock for the Royal Exchange he eventually received the order for the Palace of Westminster clock. When he died in 1853 the great clock was almost complete. The business was inherited by his nephew Frederick Rippon, a condition of the inheritance being that he used Dent as his surname.

Frederick was a clockmaker in his own right, having gained the Freedom of the Clockmakers Company in 1831 at the age of 23 years. He worked with Edward Dent and was particularly interested in the inventive side of the business, his expertise being in chronometers and watches. The Westminster clock, although completed by 1858, was still in the Dent workshops when the Oakham church clock was being assembled. **Sir Edmund Beckett Denison**, Lord Grimthorpe, was technical adviser to the Dents and the large three-train quarter-chiming clock at Oakham Church was made to his design. Frederick died in 1860, but the business continued in the hands of other members of the Dent family under the name of E Dent & Co.

This company supplied a small two-train turret clock to the Earl of Lonsdale for his new stables at Barleythorpe Hall in 1871. The serial number 239 is stamped on the pendulum cock bar. The Dent catalogue refers to it as a 'No 1 clock'. Oakham church clock is referred to as a 'No 4 clock' in the same catalogue.

EAYRE, Joseph, of St Neots, Cambridgeshire (1707-72) [1731-72]
Joseph Eayre was the brother of **Thomas Eayre II**. He was also a bellfounder and clockmaker and he set up his foundry and workshops at St Neots. Although he supplied no clocks to Rutland, he (or **Edward Arnold**) may have been the maker of the chime barrel formerly at Oakham Church, and possibly that at Cottesmore Church (*see* Chapter 4 — Gazetteer). In 1770 he supplied a new chime barrel to St Mary's Church, Stamford, at a cost of £40 (North 1882, 677), so his work was known in the area. The St Mary's chime barrel played tunes at 3, 6, 9 & 12 o'clock day and night. A different tune was played each day and the tunes available were engraved on a brass plate on the frame:

108 Psalm	Lodging on the ground
General Toast	God save the King
Tight Little Island	Highland Laddie
Gramnocree Molly	145 Psalm

He also supplied the chime barrel at Kings Norton Church, Leicestershire, in 1765.

On the death of Joseph Eayre in 1772, the St Neots foundry was taken over by **Edward Arnold** who, as well as founding bells, continued to supply and repair clocks and chime barrels, initially from St Neots, and later from Leicester (*see* Chapter 2 — Bellfounders, St Neots).

EAYRE, Thomas II, of Kettering, Northamptonshire (1691-1758) [1717-57]

Thomas Eayre I of Kettering was, by the beginning of the eighteenth century, a well-known clockmaker with a good reputation. Both his son and his grandson, both named Thomas, continued in this trade. Thomas Eayre II opened a bell foundry in Kettering in 1717 and he cast a great number of bells, some of which were sent to Rutland churches. He was born on 26 August 1691 and married Susan Baxter, by whom he had four children. **Thomas Eayre III** was his only son. Thomas II was the most outstanding member of the talented Eayre family (King 1952, 11) and he is probably remembered more for his bellfounding activities than any other. However, apart from also being a clockmaker, he was a surveyor and cartographer and produced the first large-scale map of Northamptonshire *circa* 1745.

He made a large number of turret clocks, including, for example, a new clock for Wellingborough Church in 1750 and also a new 'Town Clock' for Kettering. He was also regularly employed by the third Earl of Cardigan to look after the clocks at Deene Park, Northamptonshire.

In Rutland, the new clock installed at Glaston Church in 1739 was by Thomas Eayre II. The Churchwardens' Accounts of 1739 include 'for going to Kettering to fetch ye Clock 10s 0d', and before it was replaced in 1905 it was recorded that 'The inner dial bears the name of Thomas Eayre of Kettering ...'. A new clock for Barkby Church, Leicestershire, was acquired from Thomas Eayre II by **Steven Blackburn** of Oakham in 1754. Although it was replaced in 1898 the original clock is retained in the church. The Glaston clock may have been very similar to this. At Whissendine the following items in the Churchwardens' Accounts of 1746 point to the installation of a new clock by Thomas Eayre II: 'pd Mr Eayrs his bill £20 2s 11d' and 'pd George Snodin for going to 3 times to Kettering for Mr Eayr 12s 0d'.

The turret clock movement, and a bell by Thomas Eayre II dated 1753, from Loddington Hall stables, Leicestershire, were acquired by Mr A T Ringrose, the proprietor of Belton Garage (*see* Chapter 4 — Gazetteer, Belton in Rutland). Further details of this clock and bell are given in the Addenda to *Thomas Eayre of Kettering and other Members of his Family* (King 1953, 10):

Mr Ringrose writes of the bell, 'it is shaped like a bowl weighing about ¾-cwt. and not the usual church bell type ... the clock is of earlier date than 1753 and was originally fitted with a "foliot and verge escapement". It was probably bought by Eayre ... and converted by him to anchor escapement and 1½ second pendulum added. Another interesting feature is no centre gear train for driving hands and ... this points to it having been made before dials were used'.

In 1969 the movement, but not the bell, was donated to the Rutland County Museum where it is now on display (1969.428).

A number of other hemispherical clock bells by Thomas Eayre II and his successors have been noted. The former clock bell at Glaston Church (*see* Chapter 4 — Gazetteer, Glaston) was of this type, and two such bells can be seen in the museum at Wollaston, Northamptonshire.

Although Thomas II is only recorded as having installed two clocks in Rutland, those at Glaston and Whissendine churches, there are several other clocks which he probably installed, either directly, or through other clockmakers. These include the clocks at Caldecott Church and Normanton Park Hotel, and the former clocks at North Luffenham and Lyddington Churches. Two other turret clocks, one definite and one possible, were supplied to Rutland clockmakers for installation out of the county. One was installed at Barkby Church, Leicestershire in 1754, by **Stephen Blackburn** of Oakham, and the other may have been the clock installed in 1745 at Gretton Church, Northamptonshire, by **Robert Fox** of Uppingham.

His specification for a new clock at Stoke Doyle Church, near Oundle, Northamptonshire, is given in a letter to the churchwardens dated 27 November 1726. It provides an interesting insight into the design and construction of his clocks:

As to a new Church Clock I would advise you to have it made strong and plain and thereto would make it go 30 hrs. from winding up to winding up again to be made in a strong iron frame with brass wheels and steel pinions etc., with a pendulum of about 8 or 9 feet long and the pendulum bob about 50 or 60 pounds weight (the longer and heavier the pendulum the more certain is the going of the clock). The Clock is to turn one hand which shall on the dial show the hours and quarters. As to the painting of the dial, the index and figures and circles of gold their ground black and the middle a smalt blue. Such a clock with the painting of the dial will be well worth £30 at which price I will undertake it at. I am, Sir, your most humble servant, Tho. Eayre.

The final accounts show that he charged £40 for the complete installation. The accounts also include: 'For a horizontal dial on the post at the Church 3s 0d', showing that he also supplied a sundial, presumably for regulating the clock (information from Dr Ray Ayres).

Thomas Eayre II was buried on 3 January 1758, and was succeeded by his only son, the bellfounder **Thomas Eayre III** (*see* Chapter 2 — Bellfounders, Kettering).

ECOB, William, of Whissendine (1836-1913) [1892-1900]
'Clockie' Ecob had premises in The Nook, Whissendine. Matkin's *Oakham Almanack* from 1892 to 1900 records that he was a watch and clock cleaner and the *Melton Mowbray Year Book* of 1900 records him as a watch and clock repairer. He was baptised on 17 January 1836 and died on 21 October 1913 at the age of 77 years (information from Joe Ecob).

ESAM, of Stamford, Lincolnshire [mid eighteenth century]
Only one clock by Esam of Stamford has been noted. It is a *circa* 1760 thirty-hour longcase clock with a brass dial in an oak case. He may have been a journeyman clockmaker, or the clock may have been made for him. No biographical details of Esam have been found. He is not noted in Tebbutt (1975) or in any of the other lists of clock and watchmakers. However, amongst the vouchers for work carried out at Browne's Hospital, Stamford, are a number of receipted bills signed by John Essom, a plumber, confirming that a family of similar name was living in or near Stamford in the mid 1700s.

EVANS, W F (senior), of Handsworth, Birmingham (1819-99) [1832?-99]
EVANS, W F & Son, of Handsworth, Birmingham [1843 to after 1904]
William Frederick Evans (senior) succeeded to the clockmaking business of John Houghton on the latter's retirement in 1843. The Soho Clock Factory was established in Soho Road, Handsworth, Birmingham, and he went into business with his son William F Evans (junior) trading as W F Evans & Son. The company manufactured turret clocks for installation in their own name and for the trade, and they also exported clocks in large quantities to the United States, Canada, Australia and South America (McKenna 2002). William (senior) died in 1899 aged 81years and was succeeded by his son. Following the death of William (junior) in 1904, the business was taken over by his two sons. The small two-train flatbed turret clock with a deadbeat escapement and countwheel hour striking at Cottesmore Hunt Kennels, Ashwell, was supplied by this company in 1890. It was installed by **Thomas Large** of Melton Mowbray and his name is on the setting dial.

FLINT, Mark, of Uppingham (1829-88) [1851-75]
Mark Flint was born in 1829 at Teigh. He was working as a clockmaker in Melton Mowbray, Leicestershire, in 1851. By 1855 he had moved to Uppingham, taking

The dial of a thirty-hour longcase clock signed by Esam of Stamford, circa 1760 (Philip Snowden)

The rear view of Esam's thirty-hour longcase clock (Philip Snowden)

over the premises in High Street West, now known as No 7, which were vacant as a result of the death of **John Houghton**. From the beginning of his time in Uppingham he worked with **David Robinson**, and they were often referred to as Robinson and Flint. Mark Flint is recorded in the Census Returns of 1861 as a master clockmaker aged 32 years. Living with him were his wife, Eliza, two daughters, a house servant, and **Samuel North**, then a seventeen-year-old apprentice watchmaker.

W F Rawnsley ['An Old Boy'] in *Early Days at Uppingham under Edward Thring* (Rawnsley 1904, 122-3) described some of the 'Old Folk' he had noticed in the town during his days at the School in the 1850s:

> To begin with the tradesmen. There were two watchmakers, whose names, oddly enough, were Flint and Sparkes. Of these, Flint wore a white tie, was an elder of the Nonconformist church, and spoke with a most approved snuffle, and when you asked him what you had to pay for the new glass or hands to your watch, he always answered, 'I sh'll charge you a shull'n'; whilst Sparkes, who looked of a sporting turn, wore a black beard and big gold watch-chain, fastened always to the bottom button of his waist-coat, and always had his hat on in the shop, stammered so badly that his answer to the same question as to price invariably was, 'A sh-, a sh-, a sh-, *fifteen-pence*'.

Mark Flint is recorded in Churchwardens' Accounts as working together with David Robinson on Belton in Rutland church clock in 1855, and on North Luffenham church clock between 1860 and 1864. Mark Flint remained in Uppingham until about 1875 when he moved to Tunstall, Staffordshire. He died in 1888. His shop in Uppingham was subsequently taken over by **Francis Pinney** of Stamford.

William R Flint is included in the 1871 Census Return for Uppingham as a clockmaker aged 14 years, so presumably he was an apprentice. He was possibly a relative of Mark Flint but he is not included in the 1861 Census Return.

FOX, George, of Uppingham [1682]

An entry in the Preston Churchwardens' Accounts shows that George Fox repaired the church clock in 1682 at a cost of 5s 0d. He was probably a relative of **William Fox**. No other details are known.

FOX, John, of Uppingham (?-1802) [1744-1802]

John Fox, whitesmith and clockmaker of Uppingham, was the son of **Robert Fox**. A longcase clock with brass dial by him of *circa* 1775 has been recorded (Daniell 1975, 23). He lived at Cranwell's Cottage in Horn Lane, now Queen Street; the site of Cranwell's Cottage is thought to be at 6 and 8 Queen Street. As well as this property he owned the Pump Inn in London Road from 4 October 1770, 2 Leamington Terrace from the same date, 28 and 30 High Street East, land in Oakham Road and a blacksmith's shop tenanted by James Sneath (information from Peter Lane). In 1784 William Goodman of Seaton was apprenticed to John Fox in the trade of whitesmith (DE 2417/22). John Fox maintained and repaired Lyddington church clock from 1744 until 1802, and Uppingham church clock from *circa* 1750 to *circa*

1775. He died on 11 April 1802 and his estate was inherited by his son, Robert Breton Fox, who did not continue the family tradition of clockmaking (information from Peter Lane).

FOX, John, of Great Easton, Leicestershire [*circa* 1825-65]

Caldecott Churchwardens' Accounts show that John Fox maintained and repaired the church clock from 1834 until 1865. A longcase clock of *circa* 1825 has also been noted (Daniell 1975, 23). He was probably the son of **John Fox** of Uppingham. The house where he lived and worked still exists and can be recognised by the clock on the front elevation. He was succeeded by his son, Thomas Henry Fox, watch and clockmaker, who is recorded by Loomes (1976b, 81) as working at Great Easton in 1876, and in Kelly's *Directory* as still working there in 1900.

FOX, Robert, of Uppingham [1707-50]

Robert Fox was probably the son of **William Fox** and the father of **John Fox**, all Uppingham clockmakers. An oak-cased, thirty-hour longcase clock by Robert Fox has been noted (Andrew Butterworth).

Churchwardens' Accounts show that Robert Fox maintained and repaired a number of church clocks in Rutland. At Glaston he probably installed the new clock supplied by **Thomas Eayre II** of Kettering in 1739, and he continued to look after it until at least 1754. He took over responsibility for the clock at Uppingham Church in 1707 and continued with this duty until at least 1745, the year in which he supplied a new clock to Gretton Church and also the year in which he became a church-warden at Uppingham. For some of this time he was also the clock winder. He also worked on the church clocks at Preston (1738), Lyddington (1742-43) and South Luffenham (1750).

A two-train turret clock signed by Robert Fox is

Robert Fox's signature on the dial of a single-handed thirty-hour longcase clock (Andrew Butterworth)

Gretton church clock, signed by Robert Fox of Uppingham (Andrew Butterworth)

(below) Robert Fox's plaque on the church clock at Gretton, Northamptonshire (Andrew Butterworth)

displayed in the church at Gretton, Northamptonshire. It was the church clock until 1896 and a plaque on the clock frame is engraved:

Mr. Tho⁵ Boon
and Mr Robᵗ Laxton
Church Wardens 1745
Robᵗ Fox
Uppingham
Fecit

The movement, which appears to be the work of **Thomas Eayre II** of Kettering, has a wrought-iron birdcage frame with brass and iron wheels, anchor escapement and countwheel striking. Marks on the rear of the name plaque indicate that it was repaired by **David Robinson** of Uppingham in 1855.

FOX, William, of Uppingham [1666-1703]
William Fox was probably the father of **Robert Fox**. Little is known of him other than the details contained in local Churchwardens' Accounts. At Uppingham Church he was responsible for repairing and maintaining the church clock from 1666 until 1698 and he probably installed a new clock in 1686: 'paid to Wm ffox in part of the Clock £5 10s 0d'. At Preston entries in the accounts show that he repaired and maintained the church clock from 1676 to 1684. He repaired the clock at Morcott in 1689 and he was paid 6s 8d for mending the church clock at Lyddington in 1703.

FROMANT, of Stamford, Lincolnshire [1911]
Fromant was the last recorded clockmaker to repair the **John Watts** clock at Edith Weston Church in 1911.

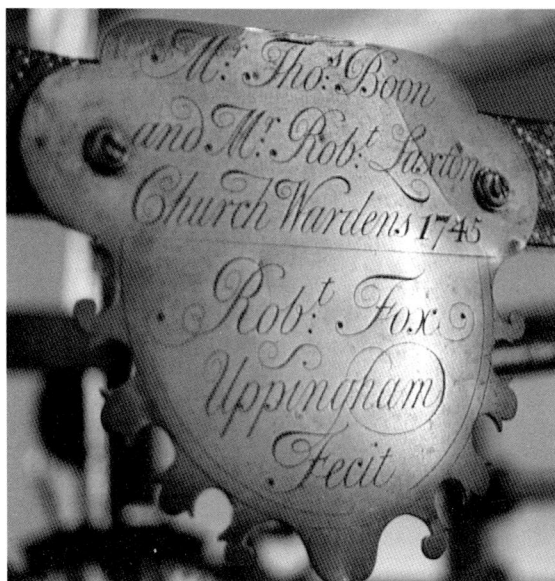

FURNISS, Joseph, of Uppingham (?-1804) [1776-95]
Joseph Furniss is recorded by Baillie (1976, 116) as a watchmaker in 1795, and a longcase clock with a brass dial of *circa* 1785 is noted by John Daniell (1975, 24). Other thirty-hour painted dial and eight-day brass dial longcase clocks have been noted in local auctions.

From 1781 he lived at what is now 25 South View, Uppingham, and also owned the Malsters Arms (information from Peter Lane).

When John Whitehurst I of Derby supplied a new

clock for Uppingham Church in 1776, Joseph Furniss was involved with the installation. This included 'figuring the dial with gold letters' for which he was paid £6 4s 6d. In October of that year, a Vestry Meeting

> Resolved that Mr Joseph Furniss, clockmaker be appointed to the care and management of the Parish clock, in building up and accepting always repairs of lines, springs, hammers, hammerwork and unforeseen accidents, keeping the same in good repair, and that this said Joseph Furniss be allowed the sum of £2 12s 6d yearly for the said allowance to commence at Michelmass 1776.

This was increased to three guineas at a Vestry Meeting on 15 April 1777 where he agreed 'to Wind up and keep the Town Clock in all repairs'. His involvement with the church did not stop with the clock as two years later he became the church organist.

Elsewhere, he repaired Hambleton church clock in 1789, which involved the clock being removed to his Uppingham workshop, and another repair was carried out in 1791. At South Luffenham he cleaned and repaired the clock on several occasions between 1782 and 1790 and this clock was taken to his workshop in 1792 for repairs which cost five guineas. John Furniss was buried in Uppingham churchyard on 4 September 1804.

GENT & Co, of Leicester [1872 to late 1970s]
The company was founded by John Thomas Gent in 1872 and became known throughout the world for its electric clock systems. Their first commercial electric master clock, known as the Thornbridge Transmitter, was introduced in 1905. It was further developed to become the *Pul-syn-etic* system, and many thousands had been made by the time it ceased production in the late 1970s (Hewitt 1994, 19-20). One of these electric master clocks is installed in the physics laboratories at Oakham School. It has two slave dials, one of which is over the western entrance to the former wharf buildings, now known as Old Stables.

GENT, James & Son, of London [1912]
In 1912 James Gent & Son supplied and installed the two-train flatbed turret clock at Market Overton Church. It is the only clock by this maker in Rutland.

GILLETT, BLAND & Co, of Croydon [1844-87]
(*see* Chapter 2 — Bellfounders, Croydon)
William Gillett was established as a clockmaker in Croydon by 1844. Some ten years later he was joined by Charles Bland and they traded under the name of Gillett & Bland, and from 1879 as Gillett, Bland & Co (Pickford 1995, 81). The manufacture and installation of turret clocks and carillon machines became a major part of the thriving business. After Arthur Johnston joined the company in 1877 a bell foundry was set up at the Croydon works and as **Gillett & Johnston** they sent a number of

bells to Rutland. The two-train flatbed clock at Braunston Church was supplied by Gillett & Bland in 1879. As Gillett & Co the company quoted unsuccessfully for the new clock at Thistleton in 1887.

GILLETT, William, of Uppingham [1900]
Kelly's *Directory* of 1900 records that William Gillett was working as a watchmaker in High Street, Uppingham. No other details are known.

GOODMAN, of unknown location [1680-81]
Preston Churchwardens' Accounts record that a Mr Goodman repaired the church clock in 1680 and 1681. He was probably a local clocksmith.

GRIMADELL, Peter, and Samuel (senior and junior), of Stamford [1757-1857]
Peter Grimadell was engaged by the Marquess of Exeter to repair clocks at Burghley House between 1757 and 1759; a longcase clock by Peter Grimadell is also recorded. Samuel Grimadell is noted by Tebbutt (1975, 49) at Burghley Lane, St Martins, Stamford, between 1760 and 1800. His son, also Samuel Grimadell, was at the same address in 1836 and 1857. A *circa* 1760 eight-day longcase clock with a square brass dial in an oak case by Samuel Grimadell of Stamford is illustrated by Tebbutt (1975, 135).

GRIMTHORPE, Lord, see **DENISON, E Beckett**

HACKETT, Richard, of Harringworth, Northamptonshire (1714-82) [1741-81]
Richard, the son of John and Jane Hackett, was baptised in Harringworth on 11 August 1714. He married Eleanor Hackett at Harringworth on 4 July 1754. He is first recorded as a clockmaker in the Churchwardens' Accounts for Harringworth when in 1741 he repaired the church clock, and he continued to be responsible for this work for the next forty years.

It has been suggested (Tebbutt 1975, 78) that Richard Hackett had a clock factory at Harringworth, the site of which is now known as Limes Farm. No documents have been found which confirm this but it seems certain that he had a number of clockmakers working with or for him over a period of time. The first to be noted is Joseph Beaumont of Harringworth, whose occupation was given as 'clockmaker' when he married Elizabeth Elliot at Harringworth 14 May 1755. Richard Hackett was a witness at the marriage of **William Aris I**, clockmaker, to Ann Bradshaw at Harringworth in 1765 and he paid the rent of their house in 1774. In 1771 the militia ballot list for Harringworth (NRO, Corby Hundred — Militia Papers) includes 'Wm Aries' and 'Wm Tiplee', both described as clockmakers.

A number of Hackett clocks have been noted. One is described and illustrated in *Grandfather Clocks and their*

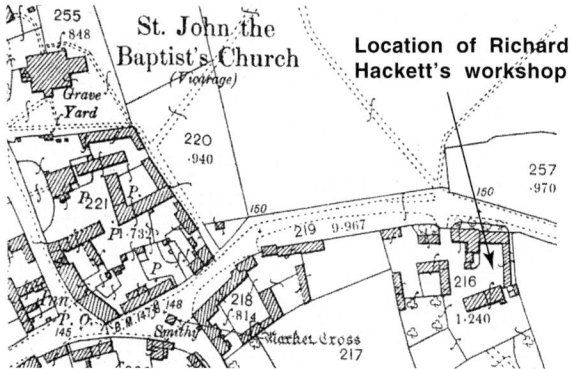

The site of Richard Hackett's workshop in Harringworth, Northamptonshire (early 1900s OS Second Edition 25 inch map)

The dial of a Richard Hackett eight-day longcase clock with centre seconds and date hands (Philip Snowden)

The dial of Richard Hackett's eight-day longcase clock which has a moon with tidal readings for six different ports. Note the centre seconds and date hands (Loomes 1985, 180)

Cases (Loomes 1985, 177-9). It is an eight-day clock of *circa* 1755 with moon dial and incorporating tidal readings for six different ports. The dial is 325mm (13in) wide and has centre seconds and centre calendar work. The oak case, which is 2.13 metres (7ft) high, retains its original caddy top and has a starburst inlay in the door. This is a really high quality clock and demonstrates the type of work he was capable of. This was possibly a special order from someone connected with the Navy, as most of the ports mentioned were naval bases. Brian Loomes (1985, 179) comments on Hackett: 'A further puzzle is the maker himself, who seems misplaced in a tiny out-of-the-way village like Harringworth if he was making such sophisticated clocks as this one. This is not at all the type of clock we expect from a rural maker'. A similar clock, but without tidal readings to the moon, is in a private collection.

An eight-day arch dial longcase clock with moon and tidal work and signed 'Wm Aris Harringworth' was reported to Leicestershire Museums some years ago (Hewitt 1992, 310). Perhaps the clockmakers of Harringworth concentrated on this type of high quality work, but it does pose the question as to where they received their training. A number of movements have been seen with 'H' on the front plate and other castings, many of these being on thirty-hour clocks, and it has been suggested that these are Hackett clock movements made for the trade (Tebbutt 1975, 78). Richard Hackett was also involved in turret clocks and his work on the Harringworth clock has already been mentioned. Although there is no record of a complete clock by him there are two examples in Rutland where he converted turret clocks. Edith Weston Churchwardens' Accounts of 1775 include 'Mr Hacket's Bill for repairing the clock £13 12s 6d'. This was for providing the clock with a single hour hand, and the work included fitting two brass wheels in the going train, a new escape wheel and probably a longer pendulum. Although now removed they are displayed alongside the clock which is in Rutland County Museum. One brass wheel is engraved **R**^D **HACKETT HARRINGWORTH 1775**. In 1772 he was engaged to repair or modify the clock at South Luffenham which involved its removal to Harringworth. The cost of this work was £3 17s 6d. He also worked regularly on Glaston Church clock between 1770 and 1781.

Richard Hackett is mentioned in an advertisement in the *Lincoln, Rutland & Stamford Mercury* of 15 August 1771 where he appeals for the return of a lost watch (No 69) (Tebbutt 1975, 78), but there is no other indication that he was involved in watchmaking.

Richard Hackett was buried on 13 January 1782 and his table tomb can be seen in Harringworth churchyard near the south-west corner of the tower:

HERE
lieth the Body of
Richard Hackett
Who died the [11th?] Day of January
1782
AGED 69 YEARS

HERE
lieth the Body of
ELEANOR the Wife of
RICHARD HACKETT
Who died the [30th?] Day of
September 17[84?]
AGED [70?] YEARS

HARDING, William Pilkington, of Uppingham [1850]
William Harding, clockmaker, watchmaker and silversmith is recorded as working in Uppingham in Slater's *Directory* of 1850. No other details are known.

HARTSHORN, John Stanley, of Badby, Northamptonshire [1969-80]
John Hartshorn installed synchronous electric movements, bell strike units and automatic winding units, all of his own design, in many churches throughout the Midlands, including an electric movement at St Peter & St Paul's, Market Overton, in 1969.

HASLEWOOD, Samuel, of Stamford, Lincolnshire (?-1744) [1744]
Samuel Haslewood was a clock and watchmaker, probably in St Mary's Street, Stamford, until he died in 1744. He was a freeman of the town but the date of his admission is not known (Tebbutt 1975, 49). Tebbutt's notes suggest that there were two clock and watchmakers of this name, probably father and son, the son dying about 1772. Two longcase clocks by Samuel Haslewood were noted in the Stamford area in 1975.

HAYNES, Robert Broughton, of Stamford, Lincolnshire [1828-59]
Robert Haynes was the son of **Thomas Haynes**. They worked together as Haynes & Son from about 1828. Robert Haynes took over the business *circa* 1831 and he continued working from their premises in Stamford. Although now no longer occupied by a clockmaker, the shop still exists, and is located at 9 Red Lion Square, as does the clock dial installed by Thomas Haynes which overlooks the square. The clock movement, however, has been replaced by a synchronous electric unit. Robert Haynes advertised as a goldsmith, silversmith, optician, and watch and clockmaker. It is thought that he only assembled clocks from bought-in movements and cases, many of the cases having the trade labels of Wilcox of Dyke (near Bourne) and Oliver of Spalding. Twenty-three longcase clocks and two wall clocks by Haynes, **T Haynes**, Haynes & Son and R B Haynes were known in the Stamford area in 1975 (Tebbutt 1975, 51, 79-81). Robert Haynes maintained and repaired the clock by **John Watts** at Clipsham Church from the early 1830s until 1847, and he was also responsible for Tinwell church clock for about 25 years until 1859, both duties which he took over from his father.

HAYNES, Thomas, of Stamford, Lincolnshire [1799-1832]
Thomas Haynes was a nephew of **James Wilson** and they were in partnership as watch and clockmakers from 1799. Thomas Haynes took over the business when James Wilson retired in 1803. Their shop and workshop were located in the parish of All Saints, Stamford. Although his premises were at what is now known as 9 Red Lion Square, he sometimes advertised his address as the Butter Market, also in Red Lion Square, probably because it was a well known nearby landmark. On 9 September 1810 he advertised in the *Lincoln, Rutland & Stamford Mercury* for a journeyman watchmaker: 'a good hand may have constant employment and good wages',

and a journeyman clockmaker, to join the business (Tebbutt 1975, 50). Thomas Haynes maintained and repaired the clock at Clipsham Church from 1809 until about 1832, and he was also responsible for Tinwell church clock for a similar period, both of which duties he took over from James Wilson. He married a Miss Hodges of Stamford on Tuesday 6 September 1803, and their son, **Robert Broughton Haynes**, took over the business in the early 1830s.

HEDGES, Frederick, of Oakham [1895-98]
Frederick Hedges is recorded as working in Oakham from 1895 to 1898. His shop in Mill Street was taken over by **John Knight** in 1898. No other details are known.

HEDLEY, Amos, of Stamford [1746-51]
Amos Hedley moved from London to the Bull Gate in St Mary's Road, Stamford, in 1746. He was admitted to the freedom of Stamford as a watchmaker on 20 May 1747 and moved to Ironmonger Street in the same year (Tebbutt 1975, 51).

HETTERLEY, Charles, of Oakham [1906-61]
Charles Hetterley was apprenticed to **Robert Corney** of Bank Buildings, High Street, Oakham:

> Although Mr Hetterley senior has spent over 55 years on horological work of one kind or another, his interest in timekeepers remains keen, and even increases with every new development he meets. His horological work began when, as a boy of 14, he was apprenticed to Mr Corney, in the quiet county town of Oakham, Rutland. At first, he was paid one shilling a week, but towards the end of his apprenticeship, he was considered good enough to receive seven shillings a week. Whilst with Mr Corney, he assisted with a lot of silversmith's work for Lord Lonsdale, and another of his jobs was to wind Lord Lonsdale's clocks. Other customers of the firm included Earl Cowley and the Duke of Rutland, and work on their clocks added to the experience of the young apprentice. It also taught him to appreciate the work of the old masters of the craft whose clocks were included in the collections he helped to care for, and his admiration for such things has remained with him ever since (*Horological Journal* **103**, No 1239, 1961, 788).

HICKMAN, William, of Stamford, Lincolnshire [1809-47]
William Hickman worked for Mary Brumhead, widow of **John Brumhead (senior)**, as a journeyman clockmaker until 1811 and he was admitted to the freedom of Stamford as a watchmaker on 26 August 1813. His premises were in St John's Street and later, according to the Poll Book of 1836, in St Mary's Street (Tebbutt 1975, 53).

HICKS, John, of Stamford, Lincolnshire [1862-1900]
John Hicks' shop was at 4 Ironmonger Street, Stamford. Churchwardens' Accounts record that he worked on a number of Rutland church clocks including Tinwell from 1862 until at least 1884, North Luffenham in 1868,

Ryhall in 1880 and South Luffenham from 1881 to 1888.

HINDS, Joseph, of Stamford, Lincolnshire [1825-1833]
Joseph Hinds was a nephew of **William Hickman**. He was admitted to the freedom of Stamford as a watchmaker on 10 October 1833. In the *Lincoln, Rutland and Stamford Mercury* of 6 May 1825 he announced that he had opened a shop in High Street, Stamford, '... at Mr Goodwins (next door to Mr. White, Draper) ...' and he also listed some of his prices: 'Eight day clock cleaning 1s 6d, Thirty hour ditto 1s 0d, New mainsprings to watches 3s 0d, Watches cleaning 1s 0d' (Tebbutt 1975, 53).

HOLMAN, John, of unknown location [1869 to *circa* 1879]
Churchwardens' Accounts record that John Holman was the clockmaker responsible for the clock at Caldecott Church from 1869 to *circa* 1879. No other details are known.

HORZ, Phil, of Ulam-am-Donau, Germany [1928]
Phil Horz, of Ulam-am-Donau, Germany, manufactured the clock which was installed by **G & F Cope** at Edith Weston Church in 1928.

HOUGHTON, John, of Uppingham (1789-1854) [1810-51]
John Houghton was born in Little Bowden, Leicestershire, in 1789 and was working as a watch and clockmaker in Uppingham from *circa* 1815 to 1851. His shop was in High Street (court rolls of the Manor of Preston with Uppingham) at what is now 7 High Street West where, according to the 1841 Census Return, he lived with his wife, Susannah, and two children. In the 1851 Census Return he is described as a master watchmaker. A number of his longcase clocks have been recorded, mainly eight-day and thirty-hour painted dial clocks. One thirty-hour clock noted at auction in 1997, signed 'John Houghton, Uppingham', had a round dial with roman numerals, stamped brass hands and false winding squares. The slender oak case had swan neck pediments to the hood, and a shell inlay at the centre of the trunk door. Churchwardens' Accounts record that he cleaned, maintained and repaired the church clock at Belton in Rutland from 1826 until 1848. He is also recorded as having worked on North Luffenham church clock in 1810 and 1822, and the clock at Lyddington in 1848. In May 1830 'After a spate of robberies including one on John Houghton, the watchmaker ... the town decided promptly to set on a watchman by voluntary subscription' (Traylen 1982c, 16). He died on 26 July 1854 at the age of 65 years and his premises were taken over by **Mark Flint** in the following year. His headstone, in the shape of a pinnacle, can be seen on the western edge of Uppingham churchyard.

Matthias Houser's advertisement of 1932 (Matkin's *Oakham Almanack* 1932, 61)

HOUSER, Matthias (Matthew), of Leicester and Oakham (1838-?) [1861- after 1871]
White's *Directory* of 1877 records that Matthew Houser was a watchmaker in Northgate Street, Oakham. He was born in Germany in 1838 and worked in Leicester from 1861 to 1867 before moving to Oakham in 1871. He and his successor advertised in Matkin's *Oakham Almanack* from 1923 to 1932.

HUBBARD, Edward, of Oakham and Melton Mowbray (1744-1810) [1773-1810]
Edward Hubbard was working as a clockmaker in Oakham from 1773 until after 1777 and in Melton Mowbray from before 1804 until he died in 1810. Surviving clocks include an eight-day longcase with an arched brass dial and moon work of *circa* 1780, and a thirty-hour longcase with a brass dial, of *circa* 1775 (Daniell 1975, 28).

JACKSON, John, of Oakham [1846]
White's *Directory* of 1846 lists John Jackson as a watch and clockmaker in Melton Road, Oakham. No other details are known.

JOYCE, J B & Co, of Whitchurch, Shropshire [1883 to present]
The Joyce family were involved in clockmaking from 1690, initially at Wrexham and then in Whitchurch from about 1780. Thomas Joyce (1793-1861) was recognised as a fine craftsman and his association with **Edmund Beckett Denison** (Lord Grimthorpe) ensured the success of the company. John Barnett Joyce took over the company in 1883 and it became known as J B Joyce & Co. Under his leadership the company continued to prosper, sending clocks to all parts of the world (Pickford 1995, 119). The company was acquired by **John Smith & Sons, Derby,** in 1966, but the name was retained for the Whitchurch business. J B Joyce & Co quoted unsuccessfully for the new clock at Thistleton in 1887.

KING, William, possibly of Uppingham [1633-36]
Uppingham Churchwardens' Accounts record that 'Wily' King repaired the church clock in 1633 and 1634-35.

He was also paid 1s 1d in 1636 for 'worke about the fore-bell'. He is not recorded as working on any other clock in Rutland and he was probably a clocksmith.

KNIBB, JOSEPH, of Oxford and London (1639-1711) [*circa* 1655-1711]
Joseph Knibb, one of a very famous family of clockmakers, was born at Claydon, Oxfordshire, in 1639. Samuel Knibb, Joseph's cousin, set up his clockmaking business at Newport Pagnell, and it was here that Joseph, and John his younger brother, served their apprenticeships. In about 1662, at the end of his seven-year apprenticeship, Joseph established a workshop in Holywell Street, Oxford. His cousins John and Peter joined him as apprentices in 1664 and 1668. In 1670 he became free of the Clockmaker's Company and in the same year he moved his business to Fleet Street in London, at the sign of 'The Dyal' near Serjeant's Inn. In 1693 he moved to 'The Clock Dyal' in Suffolk Street, near Charing Cross. He was a contemporary of Thomas Tompion: they were born in the same year and both were in business in Fleet Street. These two eminent makers had a profound influence on the development of clockmaking during the latter part of the seventeenth century.

Joseph Knibb was very interested in the work of Ahasuerus Fromanteel who had introduced the verge escapement and vertically oscillating pendulum into this country in 1658. It had been developed by Christiaan Huygens, a Dutch mathematician, who described it in his *Horologium* of that year. Some ten years later the anchor escapement appeared. This allowed the use of a longer pendulum with smaller oscillations, and the Royal Pendulum, beating seconds, became the norm for turret and longcase clocks. Over two hundred of Joseph Knibb's clocks survive and it has been suggested that this represents about fifty per cent of his total output (Jagger 1983, 28). Only five of these are known to be turret clocks. He is credited with making the first turret clock with a pendulum and anchor escapement. He made this in 1670 for Wadham College, Oxford, the benefactor being Sir Christopher Wren (Beeson 1977, 79). He also made a similar turret clock for another Oxford building in the same year. A later turret clock was installed above the State Entrance to the quadrangle at Windsor Castle. It had the signature 'Joseph Knibb Londini 1677'. It was taken down in 1829. It has been suggested that the clock installed in the church at Burley by the Earl of Nottingham was its twin (Jagger 1983, 28-9). Although it is signed 'Joseph Knibb London 1678' it was not installed until *circa* 1703 (*see* Chapter 4 — Gazetteer, Burley). A fifth turret clock by Joseph Knibb was installed in St John's College, Oxford, in 1690. In 1697 he moved to Hanslop in Buckinghamshire where he died in 1711.

KNIGHT, John James, of Oakham [1898-1931]
John Knight is first noted as working as a clockmaker in Sileby, Leicestershire, in 1898 and he moved to Oakham in that year, taking over the shop of **Frederick Hedges** at 1a Mill Street. A photograph of his shop is included in *Oakham in Rutland* (Traylen 1982a, 41). Advertisements for J J Knight 'Watchmaker, Jeweller &c' appear in Matkin's *Oakham Almanack* almost continuously from 1898 until 1931. A late nineteenth-century American Wall Clock by the Ansonia Clock Co of Brooklyn, New York, in Rutland County Museum (H563.1982) has 'Knight Oakham' on the dial.

LARGE, Thomas, of Melton Mowbray, Leicestershire (1823-?) [1855-90]
Thomas Large, watch and clockmaker, is first noted as working in Market Place, Melton Mowbray, in 1855 (*Post Office Directory* 1855). In the Census Returns of 1881 he is described as a watchmaker, aged 58 years, living with his wife, aged 33 years, two children, and William Mee, an apprentice watchmaker. They were living in King Street, Melton Mowbray. He supplied and installed the small two-train flatbed turret clock at the Cottesmore Hunt Kennels, Ashwell, in 1890. The movement was made by **W F Evans & Son**, Handsworth, Birmingham.

LARRATT, George, of Uppingham (1801-?) [1867-71]
George Larratt was born in Uppingham in 1801 and was working as a clockmaker there from 1867 to 1871 (information from Newarke Houses Museum, Leicester).

LINE, John, of unknown location [1770-75]
Hambleton Churchwardens' Accounts record that John Line repaired the church clock three times between 1770 and 1775. No other details have been found. He may have been a local blacksmith or clocksmith.

MONCK, Thomas, of Stamford, Lincolnshire [1790-1857]
Edith Weston Churchwardens' Accounts record that Thomas Monck repaired the old timber-framed church clock at his workshop in 1809 at a cost of £15 16s. He continued to service and repair the clock until 1817. Thomas Monck was a gunsmith, clockmaker and watchmaker, and at this time had a shop in High Street, Stamford. He was in business with his brother Edmund who, until 1790, was a journeyman clockmaker to **James Wilson**. John Barker was a journeyman clockmaker to Thomas Monck, but apparently he absconded in 1796, leaving his tools and working clothes (Tebbutt 1975, 54). Monck's advertisement in the *Lincoln, Rutland & Stamford Mercury* of 13 October 1809 included: 'Guns bored to short thick and strong. Superfine powder and patent shot. House and Church clocks cleaned and repaired'. In 1817 Monck moved to Red Lion Square,

An American wall clock signed by John Knight in Rutland County Museum (H563.1982)

near the Crown Inn, and later had premises in St John's Street and St Mary's Street (Tebbutt 1975, 54).

NEWTON, Sir Isaac, of Woolsthorpe Manor, near Colsterworth, Lincolnshire (1642-1727)
Isaac Newton was born 25 December 1642 at Woolsthorpe Manor, near Colsterworth, Lincolnshire. He was the son of Isaac and Hannah Newton. Isaac's father died in 1641 and his mother married Barnabas Smith, the rector of North Witham, in 1646. Isaac was left in the care of Margery Ayscough, his maternal grandmother, who moved into Woolsthorpe Manor leaving her husband James to look after their home in Market Overton. Isaac's early education was at schools in Skillington and Stoke Rochford, Lincolnshire. By the time he was ten his stepfather had died and his mother had returned to Woolsthorpe Manor together with Isaac's half-sisters Marie and Hannah and his half-brother Benjamin. Isaac's grandmother returned to her home in Market Overton at the same time. Two years later, in 1655, at the age of twelve, Isaac went to King's School in Grantham, where he boarded with a family friend.

In *Memoirs of Sir Isaac Newton's Life* (White 1936) William Stukeley relates that Isaac was adept with his hands and that he made a model windmill and a water clock for his bedroom, as well as other mechanical contrivances. He also writes: 'Besides the clocks which Sir Isaac made, he showd another method of indulging his curiosity to find out the sun's motions, by making dyals of divers forms and constructions every where about the house, in his own chamber, in the entrys and rooms wherever the sun came in'.

Sir Isaac Newton, when a boy, painted a dial on a ceiling in his grandmother's house at Market Overton. It is thought that the sunlight was reflected there by a piece of glass. When this house was demolished the piece of plaster with the dial was saved, and kept in the new house built on the site. Sir Isaac Newton also carved two sundials on the south wall of the Manor House at Woolsthorpe. Under one of these he carved his name. The stone was removed in 1844 and presented to the museum of the Royal Society. When the Newton Chapel at Colsterworth Church was rebuilt in 1876-77, Sir William Erle offered to present a copy of the dial in the Royal Society Museum to the church. The Rev John Mirehouse at first accepted this offer, but decided to make a search at Woolsthorpe Manor to see if Newton's second dial could be located. It was found covered by a small coalhouse which had been built against the wall. It was given to the church by the owner of Woolsthorpe and erected on the north wall of the Newton Chapel. The dial, which unfortunately was mounted upside down, is accompanied by the inscription: 'Newton: aged 9 years cut with his penknife this dial: The stone was given by C. Turner, Esq., and placed here at the cost of the Rt. Hon. Sir William Erle, a collateral descendant of Newton. 1877' (Gatty 1890, 484-5 & Mills 1992a, 126-39).

The two seventeenth-century sundials abutting each other on the south-west corner of the tower at Market Overton Church are said to be the gift of Sir Isaac Newton. There are, however, no primary sources which support this statement.

NICHOLLS, Henry, of Glaston [1656-69]

Preston Churchwardens' Accounts of 1656-57 record: 'ffor a Clocke and setting it up £3 5s 0d'. They also record that on 30 March 1657 an agreement was made between the inhabitants of Preston and Henry Nicholls of Glaston to 'keepe the clocke in good and sufficient repaire and find all materials belonging to it ...' at a wage of five shillings a year. He may also have supplied the clock.

Later accounts show that he attended the clock each year until 1669. 'Henry Nicolls' is included in the Glaston Hearth Tax returns of 1665 (Bourn & Goode, 1991, 13) as having one hearth 'not chargeable', and the baptisms of his children are recorded in the parish registers (information from Auriol Thomson).

NORTH, Samuel, of Uppingham (1843-?) [1861]

Samuel North, born at Frisby, Leicestershire, is recorded in the Uppingham Census Returns of 1861 as a seven-teen-year-old apprentice watchmaker living with **Mark Flint**. No other details are known.

NORTON, Robert, of Stamford, Lincolnshire [1841 to *circa* 1870]

Robert Norton's premises were in High Street, Stamford. He is recorded by Loomes (1976b, 173) as a watch-maker and silversmith. His advertisement in Dolby's *Old Moore's Illustrated Almanack* of 1869 (Tebbutt 1975, 57) includes:

R.N., in again returning his grateful acknowledgements to the Inhabitants of STAMFORD, and particularly to his numerous Friends in the adjacent Villages, begs leave to inform them he has lately made a very extensive and splendid addition to his Stock of Gold and Silver Watches, Eight-day Clocks, Timepieces, Skeleton ditto, and every other article connected with the trade ....

He also lists the prices of watches, including: 'Gentle-men's Gold Watches' £6 6s, 'Gentlemen's Silver Watches' £3 5s, 'Gold Lever Watches' £12 and 'Silver Lever Watches £5. American clocks were offered at 'TWENTY SHILLINGS EACH'

Robert Norton was succeeded by his son Henry.

NUTT, Thomas Cornelius, of Uppingham [*circa* 1760 to *circa* 1784]

Thomas Cornelius Nutt is not recorded in any of the published lists of clock and watchmakers. Uppingham parish records confirm that a Thomas Cornelius Nutt was buried in Uppingham churchyard on 19 March 1784. The manorial records show that he was a tenant of Thomas Knight and that his occupation was that of cutler (information from Peter Lane).

A silvered and skeletonised dial plate, without a movement or chapter ring, signed 'Thos Cornel^s Nutt UP-PINGHAM' above and below the centre, has been re-corded. It has no winding holes and is therefore assumed to be from a thirty-hour longcase clock of *circa* 1760 (information from Ron Pace). The engraved centre of the dial plate is similar in style to that seen on clocks signed by **Richard Hackett** of Harringworth and they may have worked together.

OGDEN, W, of unknown location [1877]

Empingham Churchwardens' Accounts record that W Ogden renovated the old church clock in 1877 and he carved his name and the date upon the new frame. This clock, by **John Watts**, is now in Stamford Museum. No other details are known.

PARMITER, John F, of Oakham (*circa*1859-?) [1881]

Census Returns show that John Parmiter was born at Bourne, Lincolnshire, *circa* 1859 and that he was work-ing as a clockmaker in Oakham in 1881. No other details are known.

PAYNE, Charles, of Oakham [1867-1928]
Charles Payne was working as a clock and watchmaker in Catmose Street, Oakham, in 1870 according to Harrod's *Directory*. By 1891 he had moved to Market Place, and he was still there in 1928 (Kelly, *Directories*). Churchwardens' Accounts show that he repaired the 'clock and chimes' at Cottesmore on several occasions between 1867 and 1874, attended the church clock at Langham between 1900 and 1905, and wound Oakham church clock from 1910 until at least 1913.

PAYNE, William, of Wymondham, Leicestershire [1847-76]
William Payne was the father of **Charles Payne**. Surviving accounts show that he repaired the clock and chimes at Cottesmore between 1847 and 1849. No other details are known.

PHILLIPS, Richard, of unknown location [1698]
Ashwell Churchwardens' Accounts show that Richard Phillips went 'three times to mend ye clock' in 1698. No other details are known.

PHILLPOT, of unknown location [1710]
Teigh Churchwardens' Accounts of 1710 include: 'p'd to Phillpot for mending ye Clock & spout 3s 6d'. He may have been a local blacksmith or clocksmith. No other details are known.

PINNEY, Charles, of Uppingham (*circa* 1837-?) [*circa* 1855-1908]
Details of Charles Pinney and his family are given in the Census Returns for Uppingham of 1891. He was born in Stamford *circa* 1837 and was resident in High Street, Uppingham, with his wife and four daughters. It is known that his actual address was what is now 7 High Street West (information from Peter Lane) and that his father was **Francis Pinney**, watch & clockmaker of Stamford. In fact they were in business together as Francis Pinney & Sons from *circa* 1855 in Broad Street, and later at 21 High Street, Stamford, opposite St Michael's Church. They also had branch shops at 21 Narrow Bridge Street, Peterborough, High Street, Oakham and Market Place, Uppingham before Charles moved to Uppingham. A new clock by **John Smith & Sons** of Derby was installed at Belton in Rutland Church in 1887. The setting dial confirms that it was supplied and installed by Pinney & Son, Uppingham. The Churchwardens' Accounts show that Pinney serviced and repaired the clock until at least 1908.

PINNEY, Francis, of Stamford, Lincolnshire [1844-77]
Francis Pinney was probably the son of **Richard Matthew Pinney** of Stamford. The following is from Francis Pinney's advertisement in *Johnson's Household Almanack*

A circa 1860 two train mahogany bracket clock by Francis Pinney of Stamford (Phillip Snowden)

& Year Book of Useful Knowledge and Local Compendium of 1871 (Tebbutt 1975, 61):

ESTABLISHED 1844
FRANCIS PINNEY & SON
Watch & Clock Makers,
SILVERSMITHS, JEWELLERS, OPTICIANS
AND ENGRAVERS,
HIGH STREET, STAMFORD,
Opposite St. Michael's Church.

Branch Shops:- Market Place, Uppingham,
open every Wednesday:
and High Street, Oakham, open every Saturday

The court rolls of the Manor of Preston with Uppingham record that on 20 November 1877 Francis Pinney, watchmaker, was admitted to 7 High Street West, Uppingham, on the surrender of **Mark Flint**, and that he was succeeded by his son **Charles Pinney** to the same property (information from Peter Lane).

Ryhall Churchwardens' Accounts record that Francis Pinney and later Francis Pinney & Sons were responsible for the repair and maintenance of the church clock from the 1850s. A voucher for some of this work is illustrated in Chapter 4 — Gazetteer, Ryhall.

PINNEY, Richard Matthew, of Stamford, Lincolnshire [1830-57]
Richard Pinney is recorded by Loomes (1976b, 186) as a watchmaker working in Stamford from 1830 to 1857. His premises were in St John's Parish in 1830 and in Empingham Road in 1857 (Tebbutt 1975, 60). He was probably the father of **Francis Pinney**.

POTTS, William & Sons Ltd, of Pudsey and Leeds [1833-1935]

This branch of the Potts family originated in the north-east of England and became involved in clockmaking when Robert Potts was apprenticed to a clockmaker in Darlington in 1790. His son, William Potts (1809-87), served his apprenticeship with William Smith of Keighley and then started his own clockmaking business at Pudsey, West Yorkshire, in 1833. After initially making domestic clocks in his shop at Chapeltown, the business expanded into the manufacture of turret clocks and he moved to Guildford Street, Leeds, in 1862. A new clock factory was opened later in Cookridge Street, Leeds, and mass-produced wall clocks for schools, offices and railways were made in the workshops behind the Guildford Street shop in Butts Court.

Three of William's sons eventually joined their father in the business and in due course it became known as William Potts & Sons Limited. A further generation joined the business after World War I but their time with the company was short-lived. Tom Potts left in 1928 to operate on his own and Charles Potts left in 1930 to set up his own turret clock business as Charles H Potts & Co at Marshall Mills in Leeds where it remained until the 1950s, latterly under the control of Anthony Potts. The former business of William Potts & Sons Limited was sold to **John Smith & Son** of Derby in 1935 and has remained a subsidiary company in Leeds to the present day (information from Michael Potts).

The following turret clocks and dials by William Potts & Sons have been installed in Rutland:

Greetham, The Old School
Small single-train movement in brass plates with anchor escapement and one dial; installed on 3 August 1911 to commemorate the Coronation of that year; Potts' Order Number 2074

Lyddington, St Andrew
Two-train flatbed turret clock with double three-legged gravity escapement and one dial; installed in August 1890; Potts' Order Number 506

North Luffenham, The Pastures
Two-train flatbed turret clock with deadbeat escapement; installed in 1903; made for William Potts & Sons by Haycock of Ashbourne; dial by the architect C F A Voysey; Potts' Order Number 1539

South Luffenham, St Mary the Virgin
Two-train flatbed turret clock with double three-legged gravity escapement and one dial; installed in 1903; Potts' Order Number 1491

Additional details of these clocks are given in the Gazetteer. William Potts & Sons of Leeds also tendered for the supply and installation of a clock for Seaton Church in 1919. A transcript of the unsuccessful hand-written tender and specification is included as Appendix 5.

RAYMENT, Thomas (senior), of Stamford, Lincoln-shire [*circa* 1760-92]

Thomas Rayment's shop and workshop was at what is now 10 High Street, Stamford (Smith, M 1992, 47). He frequently advertised his business in the *Lincoln, Rutland and Stamford Mercury*, referring to himself as watch-maker and silversmith. In 1762 he was asked to investi-gate an unspecified problem with the clock at Uppingham Church, but the outcome is not recorded. In 1791, his son, Thomas Rayment (junior), set up in business in High Street as a clock and watchmaker. He was also a dentist. By 1792 Thomas Rayment (senior) was bank-rupt and his stock was sold by auction at the Town Hall

William Potts circa 1867 (Michael Potts)

(Tebbutt 1975, 63). His shop became the town Post Office, but by the end of the nineteenth century it was a draper's shop (Smith M 1992, 47). It is now Walker's bookshop, but 'T. RAYMENT . 1788' can still be seen on one of the beams in an attic room (information from John Smith).

RICHARDSON, Ebenezer, of Uppingham [1891]
The 1891 Census Returns for Uppingham includes Ebenezer Richardson, aged 65 years, 'Watch & Clock Maker', who was living in High Street with his wife, Elizabeth. No other details are known.

ROBINSON, David, of Uppingham [1855-56]
David Robinson worked with **Mark Flint** and they were known as Robinson and Flint. Their shop was at what is now known as 7 High Street West, Uppingham. David Robinson worked on Belton in Rutland church clock in 1855-56. He also repaired the clock supplied by **Robert Fox** at Gretton Church in 1855. This is recorded on the rear of the name plaque:

Robinson
Uppingham
Rep'd this clock all my [own work?]
1855

RODELY, Stephen Simpson, of Market Place, Oakham [1854-63]
White's *Directory* of 1863 records that Stephen Simpson Rodely was a watch and clockmaker with a shop in Market Place, Oakham. The Churchwardens' Accounts record that he repaired and cleaned the church clock and chimes at Cottesmore in 1854 and 1855. He also repaired the chiming train of the new Dent clock at Oakham Church in 1861 (*see* Chapter 4 — Gazetteer, Oakham).

SHARMAN, John, of Melton Mowbray, Leicestershire (1776-1849) [1800-49]
John Sharman was baptised at Greetham in 1776. He was the son of Thomas Sharman and Mary (née Warren) and married Eleanor Laxton at Melton Mowbray in 1806. His occupation at this time is given in the parish registers as ironmonger (Sharman 1984, 132-3), but he is recorded by Loomes (1976b, 210-11) as a clock-maker, working from 1800 to 1835, and in 1849 as Sharman & Son. He is noted in Pigot's *Directory* in 1828 and 1840, where his address is given as Sherrard Street. A thirty-hour painted dial longcase clock in a slender oak case and signed 'John Sharman, Melton Mowbray' was noted at a local auction in 1997.

SHARPE, Hugh, possibly of Lyddington [1706-38]
Hugh Sharpe was responsible for maintaining and repairing Lyddington church clock from 1706 until 1738.

He also wound the clock, repaired locks and carried out work on the bells, and may therefore have been a village blacksmith who had a good working knowledge of clocks. This is confirmed by entries in the Churchwardens' Accounts.

SIMPSON, John, of Oakham [*circa* 1770-95]
John Simpson was working in Oakham from before 1770 until at least 1795. **Stephen Simpson,** possibly his son, had premises in Market Place and he may have taken these over from John. The first mention of John Simpson is in the Churchwardens' Accounts for Ashwell where he is noted as being paid for mending the church clock in December 1783. The last reference to him in these accounts is in 1792. He also repaired the church clock at Hambleton in 1788 and that at Oakham in 1793. A thirty-hour longcase movement with a 250mm (10in) square painted dial, but without a case, was noted at a local auction in October 1997. A thirty-hour brass dial longcase clock by John Simpson of Whissendine was noted recently at an auction in Melton Mowbray. It had a rope weight line, a separate silvered chapter ring, a tall oak case

Pocket watch signed by John Simpson (Leicester City Museums Service L.H292.1960)

and was dated *circa* 1770. The collection at the Newarke Houses Museum, Leicester, includes a silver-cased pocket watch with a verge escapement by J Simpson of Oakham. It is numbered 5045 and the London hallmark on the case has the date letter for 1795 (L.H292.1960).

SIMPSON, Stephen, of Market Place, Oakham [1828-67] Stephen Simpson of Market Place, Oakham, is known to have been working from 1828 to 1867. He was probably the son of **John Simpson**.

An eight-day longcase clock, originally made for Oakham Union Workhouse, was presented to Rutland County Museum by Rutland County Council in 1969 (1969.413). The painted dial, which has a false-plate by Walker & Hughes, Birmingham, includes a fairly accurate colour painting of the workhouse in the arch and a representation of the common seal of the County of Rutland in each spandrel. Although the dial is unsigned, the maker is revealed in the following minute in the Oakham Union Minute Book for 1836-38 (DE 1381/401): 'A Clock or Timepiece to be provided by Mr Simpson of Oakham for the Workhouse under the direction of Mr Morton and Mr Stimson'. John Morton and Henry Stimson were members of the Board of Guardians representing Egleton and Oakham Lordshold respectively (Clough 1981, 82-3).

Stephen Simpson is recorded as working on a number of Rutland church clocks including Langham from 1848 to 1867, Oakham in 1853 and 1854, and the clock and chimes at Cottesmore from 1849 to 1857.

SIMMONS (SIMMONDS), Alexander Sadler, of Warwick (1809 to *circa* 1882) [*circa* 1835 to *circa* 1882] Alexander Sadler Simmons was born at Henley-in-Arden in September 1809 and was apprenticed to his father, John Simmons, in 1823. He worked as a watch and clockmaker in West Street, Warwick, from *circa* 1835 to *circa* 1882. He also had a workshop in Regent Street, Leamington, from 1866 to 1870. Alexander had two brothers who were also clockmakers. John worked in Henley-in-Arden and Charles worked in Redditch, and all three had their mother's maiden name of Sadler as a middle name.

Turret clocks supplied by Alexander Simmons were installed in the Warwickshire churches of Hampton Lucy (1855), Wellesbourne (1858), Sherbourne (1864), St Nicholas, Kenilworth (1865), and Henley-in-Arden (1868) (McKenna 2001). Churchwardens' Accounts show that he maintained and repaired clocks in many other local churches in the Warwick area (Seaby 1981, 51). His only clock in Rutland is at Ketton Grange stables. It is a two-train posted-frame movement with a pinwheel escapement. It is dated 1855. The wooden pendulum rod, which has a cast-iron bob, is two metres long and hangs in a separate lath and plaster case (*see* Chapter 4 — Gazetteer, Ketton).

SMITH, John & Sons, of St John's Square, Clerkenwell, London [*circa* 1844 to mid twentieth century] John Smith & Sons were important Clerkenwell clockmakers in the late nineteenth century. They made all types of clocks but they were particularly well known for their turret clocks, which were exported to all parts of

The dial of Stephen Simpson's Oakham Workhouse clock (RCM 1969.413)

The Rutland seal on Stephen Simpson's Oakham Workhouse clock

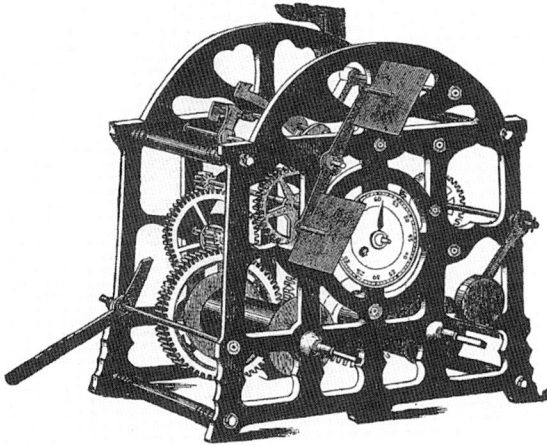

Lyndon Hall stables clock as shown in the catalogue of John Smith & Sons, Clerkenwell (McKay 2001, 37)

the world, and which they often supplied to other clock-makers for selling on. The company is no longer in-volved in clockmaking but it still exists as part of the Delta Metals Group. There is no connection between this company and **John Smith & Sons of Derby** who remain well known for their turret clocks. The two-train birdcage turret clock movement with deadbeat escape-ment and rack striking at Lyndon Hall stables is signed WILSON STAMFORD on the setting dial and was supplied and installed by **Joseph Wilson** of Stamford. The undated movement is *circa* 1860 and it was manu-factured by John Smith & Sons of Clerkenwell.

SMITH, John & Sons, of Derby [1850 to present]
The Smith family originated from Hognaston, Derby-shire, and the clockmaking business was founded by John Smith (1813-86). He served his apprenticeship as a clockmaker with **John Whitehurst II** from 1827 to 1834, continuing to work for him until he left in 1850 following an argument with the foreman (Craven 1996, 174-5). He set up on his own to make turret clocks and noctuaries in the Whitehurst tradition, and from 1870 onwards, and supported by Lord Grimthorpe, the com-pany became very successful, supplying large numbers of turret clocks to customers all over the world. It was known as John Smith & Son from 1868 and John Smith & Sons from 1873. After John's death in 1886, the company was run by brothers John Henry Smith and Frank Symon Smith. The old clockmaking firms of **William Potts & Sons** of Leeds and **J B Joyce & Co** of Whitchurch were acquired in 1933 and 1965 respec-tively. Today the company is still regarded as the leading specialist in public clocks and remains in the control of the Smith family (Pickford 1995, 143).

The following turret clocks by John Smith & Sons have been installed in Rutland. All have a pinwheel escapement:

Belton in Rutland, St Peter
Two-train flatbed; installed in 1887; one Smith's skeleton dial

Cottesmore, St Nicholas
Two-train flatbed; installed in 1909; one Smith's skeleton dial

Empingham, St Peter
Two train flatbed; installed in 1895; one Smith's skeleton dial

Glaston, St Andrew
Two-train flatbed; installed in 1905; Smith's automatic winding units installed in 1989; original wood dial re-tained and overlaid by a convex copper dial

Greetham, St Mary the Virgin
Synchronous electric movement and electric bell strike unit; installed in 1967; convex dial

Hambleton Hall Stables
Two-train flatbed; installed in 1897; movement replaced by synchronous electric motor; four illuminated copper dials

Langham Hall Stables
Single-train flatbed; installed in 1929; two small convex copper dials and two wind direction dials

Morcott, St Mary
Three-train flatbed; installed in 1921; one Smith's skel-eton dial

Seaton, All Hallows
Two-train flatbed; installed in 1919; one Smith's skeleton dial

Thistleton, St Nicholas
Two-train flatbed; installed in 1887; one Smith's skeleton dial

Uppingham, St Peter & St Paul
Three-train flatbed; installed in 1898; train wheels re-moved and replaced by electric motors and gearboxes; one Smith's skeleton dial

Uppingham School, School House
Two-train flatbed; early automatic winding; installed in January 1930; one dial

Uppingham School, 'Upper' Cricket Pavilion
Single-train plate and spacer frame; deadbeat escapement; installed in 1897; one dial

Whissendine, St Andrew
Two-train flatbed; installed in 1911; original dial retained

Wing, St Peter & St Paul
Two-trains; installed in 1920; Smith's automatic winding units installed in 1997; one Smith's skeleton dial

Further details of these clocks are given in Chapter 4 — Gazetteer.

In addition to those noted above, automatic winding units by John Smith & Sons have also been installed on the clocks at:

Burley, Holy Cross	installed in 1976
Edith Weston, St Mary	installed in 1987
Langham, St Peter & St Paul	installed in 1986
Ryhall, St John the Evangelist	installed in 1979

SPARKES, James, of Uppingham (?-1875) [1839-67]
James Sparkes is recorded in White's *Directory* (1863) as
a watch and clockmaker in Market Place, Uppingham.
The site of his shop in 1839 was next to the Town House
(demolished in 1960) and is now occupied by a Doctor's
Surgery. In 1862 he purchased four tenanted houses in
an auction at the Falcon Hotel, Uppingham. At this time
he is recorded as being an innkeeper at the George and
Dragon, which was in London Road, near the Market
Square, Uppingham, having married Rebecca Worsdale,
the widow of the former landlord. He made a will in
1870 in which he left the newly erected house '... in
which I live' to his wife and legacies to his mother and
sisters. The will also mentions eight properties in Uppingham and four in Bisbrooke which he owned (information from Alan Rogers). The newly erected house
mentioned in his will, which he was still living at in 1873
was the property at or adjacent to the east side of what is
now 4 Stockerston Road, Uppingham (information from
Peter Lane). This is referred to as West End in Harrod's
Directory of 1870. He made a codicil to his will on 4
November 1875 in which he mentions another new
house which he had erected since 1870.

He is recalled by an old boy of Uppingham School:

... Sparkes, who looked of a sporting turn, wore a black
beard and big gold watch-chain, fastened always to the
bottom button of his waistcoat, and always had his hat on
in the shop, stammered so badly ... (see **Mark Flint**)
(Rawnsley 1904, 122-3).

The clock by **Thomas Eayre II** of Kettering at Glaston
Church was fully restored in 1867. The work, the details
of which are recorded in the Churchwardens' Accounts,
included removing the clock and transporting it to James
Sparkes' workshop in Uppingham, fitting a new tooth to
the escape wheel, repairing the escapement pallets, altering the wheels to 'act on a different part', as well as
cleaning and painting the assembly. He also painted and
gilded the dial and hands, the total being £10 8s 0d.

James Sparkes died on 26 November 1875, and his
wife, Rebecca, took over the business.

SUTTON, The Reverend Canon Augustus, of Cottesmore Hall, Rutland, and West Tofts, Norfolk (1825-85)
[1844-85]
Augustus Sutton, born 13 January 1825, was the fifth
son of Sir Richard Sutton, 2nd Baronet, of Norwood
Park, Nottinghamshire, and owner of Cottesmore Hall.
It appears that Augustus resided, for some time at least,
at Cottesmore Hall. In 1849 he became Rector of West
Tofts, Norfolk. He married Charlotte Carter of
Northwold on 2 October 1851 and they subsequently
had five sons (*Burke's Peerage* 1967, 2428-29). He had a
particular interest in horology and the vestry minutes of
St Lawrence's Church, Norwich, record that he was 'a
very clever mechanic, his speciality being the reconstruc-

James Sparkes' label found inside an American wall clock (David Griffiths)

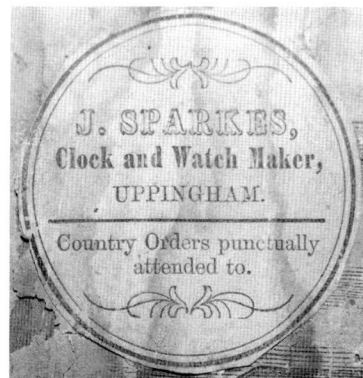

tion and repairing of old disused church clocks'. He
presented at least thirteen of these clocks to Norfolk
churches whilst at West Tofts (Bird & Bird 1996, 166).

This interest was evident during the time he was at
Cottesmore Hall, as confirmed by the following extract
from White's *Directory* (1846, 613) concerning Greetham
church clock: 'The clock, after being useless for half a
century, has recently been repaired at the expense of A.
Sutton, Esq.'. He also worked on Cottesmore church
clock. When Richard Coverley, the village carpenter,
submitted a bill in 1844 for making a new bellframe and
rehanging the bells at Cottesmore, he also included for
'assisting Mr Sutton with clock'.

An item in *Some Recollections of Exton* (Wilton Hall
1913, 11) refers to Exton church clock: 'An enthusiastic
Curate from Cottesmore who "understood these things",
having obtained Mr Knox's permission to overhaul the
clock, took it to pieces'. The 'enthusiastic Curate' may
well have been Augustus Sutton, although by this time
he had moved to West Tofts. Perhaps he returned to
Rutland for this purpose (*see* Chapter 4 — Gazetteer,
Cottesmore, Greetham and Exton).

He was succeeded at West Tofts in 1885 by his eldest
son, Arthur Frederick Sutton, who had a similar interest
in church clocks.

THWAITES & REED, originally of Clerkenwell, London, now at Hastings, East Sussex [1740 to present]
The business was started in 1740 by Aynsworth Thwaites.
One of the first clocks he made was a turret clock for
Horse Guards Parade in London. It continues working
today and is maintained by the company. Aynsworth was
elected Freeman of the Clockmakers' Company in 1751,
and was succeeded by his sons. They established a partnership with George Jeremiah Reed, a relative, and the
company traded as Thwaites & Reed from *circa* 1826.
They manufactured a wide range of domestic and turret
clocks, and in addition to making clocks for many churches
and public buildings they also supplied to the trade
(information from Thwaites & Reed). The company

records, covering the period from 1780 to 1955, are held in the Guildhall Library (MS 6788-6808) (Pickford 1995, 41). Although the two-train flatbed turret clock with deadbeat escapement at Ashwell, St Mary, installed in 1890, has no signature or other identifying marks, it has been positively identified as being made by Thwaites and Reed (information from Les Kirk). For additional details of this clock see Chapter 3 — Gazetteer, Ashwell. It is the only turret clock in Rutland by this company.

TILLEY, Samuel, of Uppingham [1891]
The 1891 Census Returns for Uppingham include Samuel Tilley, watchmaker, who was living in Hope's Yard with his wife, Sarah, and their son of one month. No other details are known.

TUCKER, Elisha, of Gray's Inn, London [1849-75]
The two-train flatbed turret clock with deadbeat escapement and one dial at the church of St Peter & St Paul, Langham, was supplied in 1875 by Elisha Tucker of 42 Theobalds Road, Gray's Inn, London. The cost was £100 and it was donated by the Rev John Mould, then vicar of the parish. Elisha Tucker advertised in London directories as a clock and watchmaker (Clutton 1982, 628).

VINES, John, of unknown location [1797-1804]
Hambleton Churchwardens' Accounts record that John Vines cleaned and repaired the church clock each year between 1797 and 1804. No other details are known.

WALKER, John, of London (1836-80) [1862-80]
John Walker was a chronometer maker, and inventor and manufacturer of the crystal case watch, for which he won prize medals in 1862 and 1867. He also invented and manufactured a railway guard's watch (Clutton 1982,

634). The company of this name continued after John Walker's death and is recorded as makers of turret clocks in 1890 (Loomes 1976b, 243). They were responsible for the supply and maintenance of wall clocks for a number of railway companies and were sole suppliers to the London & South West Railway (Parr 1997, 338). The large two-train flatbed turret clock with double three-legged gravity escapement and rack striking at Stocken Hall Coach House, near Stretton, was installed in 1914. It is signed 'John Walker, 1 South Molton Street, London', and was one of the first turret clocks to be supplied from these workshops. Some turret clocks signed by John Walker are known to be the work of other makers (information from Chris McKay).

WALKER, Lewis, of Stamford, Lincolnshire [1720]
Lewis Walker is recorded as a watchmaker having been admitted to the freedom of Stamford on 30 April 1720 by payment of £5. He was the son of Lewis Walker, a miller of Hudd's Mill, Stamford (Tebbutt 1975, 64).

WARREN, Henry, of Stamford, Lincolnshire [1857-71]
Henry Warren's shop was in St Mary's Street, Stamford. According to his advertisement in *Johnson's Household Almanack & Year Book of Useful Knowledge and Local Compendium* of 1871 (Tebbutt 1975, 66) he was a 'Watch Maker, Working Jeweller, and Practical Optician'. He was selling 'SILVER WATCHES from 20s each', 'SILVER ENGLISH LEVERS from £3 15s to £6 10s', 'GOLD ORIZONTAL WATCHES from 56s each' and 'GOLD ENGLISH LEVERS from £8 to £20'. A watch by Henry Warren is illustrated in Tebbutt (1975, 107). Ryhall Churchwardens' Accounts record that he attended the church clock in 1864 and 1865 at an annual cost of 10s 6d.

A pocket watch by Henry Warren of Stamford (Philip Snowden)

The movement of Henry Warren's pocket watch (Philip Snowden)

WATTS, Charles, of Stamford and Bourne, Lincolnshire (1722-98) [mid 1700s]

Charles Watts of Stamford, possibly the son of **Robert Watts**, was baptised 17 December 1724 at St John's, Stamford. An oak-cased thirty-hour longcase clock of *circa* 1760 in a private collection is the only known example of his work.

WATTS, John (senior), of Stamford, Lincolnshire (?-1719) [1661-1719]

John Watts, maker of turret and domestic clocks, and gunsmith, was working in Stamford during the latter part of the seventeenth and the early part of the eighteenth century. Little is known of his early life, or where he received his training. An early reference to a John Watts is included in Lyddington Churchwardens' Accounts where he is recorded as being paid for 'mending ye clock' five times between 1661 and 1670. The following entry suggests that he then lived in Caldecott:

1662 Spent with John Watts when he come to ye clok   6d
payd to Steeven Manton for carrying ye clock and fetching of it hom from Colldicoat [Caldecott]   3s 0d
payd to John Watts for mending ye clock   £2 3s 0d

The Stamford Town Hall Books record that in 1682 John Watts, whitesmith, paid £3 6s 8d to the Chamberlain to purchase his freedom of the town. It appears that he had at least three apprentices and another entry in 1696 records that one of them was **Boniface Bywater** who '... having served John Watts for seven years is to become free'.

His son, **John Watts (junior)**, was also an apprentice

and he received his freedom in 1704. The baptisms of seven children of a John and Elizabeth Watts between 1682 and 1693 are recorded in the parish registers of St John's, Stamford, but there is no reference to their father being a clockmaker (information from Michael Lee).

A number of late seventeenth-century timber-framed turret clocks signed on the frame 'I W' have survived in the Stamford area, the 'I' being the initial for 'Iohannes', the Latin form of John. The following clocks have been attributed to John Watts of Stamford:

Clipsham Church, Rutland. Restored and now displayed in the church; signed '16 + IW + 88'
Empingham Church, Rutland. Restored and now in Stamford Museum; signed 'IW 1686'
Apethorpe Church, Northamptonshire. Restored and still working in the church; signed 'IW 1704'
Nassington Church, Northamptonshire. Restored and now displayed in the church; signed 'IW 1695'

John Watts also installed a single-train turret clock at Peterborough Cathedral in 1687 and the former timber framed clock dated 1692 at Kings Cliffe Church is believed to have been his work. There is also some evidence to show that he installed turret clocks in other Rutland churches including Ashwell, Belton, Exton, Morcott, North and South Luffenham, Oakham, and Tinwell.

Two other Rutland clocks, Greetham church clock, removed in 1967 and exported to the United States, and Edith Weston church clock, now in Rutland County Museum (L.1975.11), have also been identified as John Watts clocks. Neither is signed or dated but this information may have been lost due to part or all of the frame being replaced. All these clocks originally had anchor

A conjectural family tree of the Watts clockmakers of Stamford (based on information provided by Michael Lee)

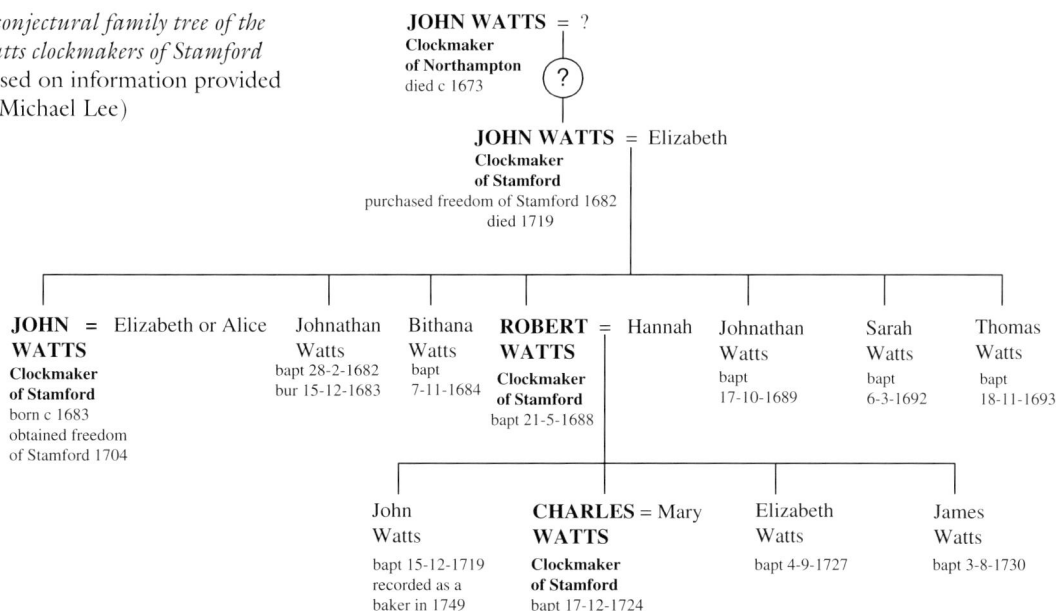

JOHN WATTS = ?
Clockmaker
of Northampton
died c 1673

?

JOHN WATTS = Elizabeth
Clockmaker
of Stamford
purchased freedom of Stamford 1682
died 1719

JOHN WATTS = Elizabeth or Alice	Johnathan Watts	Bithana Watts	**ROBERT WATTS** = Hannah	Johnathan Watts	Sarah Watts	Thomas Watts
Clockmaker of Stamford born c 1683 obtained freedom of Stamford 1704	bapt 28-2-1682 bur 15-12-1683	bapt 7-11-1684	Clockmaker of Stamford bapt 21-5-1688	bapt 17-10-1689	bapt 6-3-1692	bapt 18-11-1693

John Watts	**CHARLES WATTS** = Mary	Elizabeth Watts	James Watts
bapt 15-12-1719 recorded as a baker in 1749	Clockmaker of Stamford bapt 17-12-1724	bapt 4-9-1727	bapt 3-8-1730

∞∾J6 + I +W + 88 ∞∾

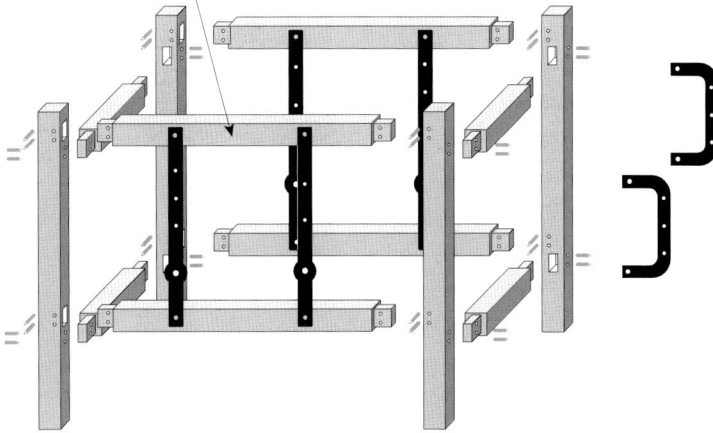

The frame components of a typical turret clock by John Watts. The frames were
made of oak and the mortise and tenon joints were fixed with two pegs. The
clock could therefore be assembled in the workshop and then easily dismantled
for transporting to site. Some clocks had extended legs. This was either to lift
the movement to align the clock with the dial, or to accommodate a longer
pendulum as a result of a later conversion

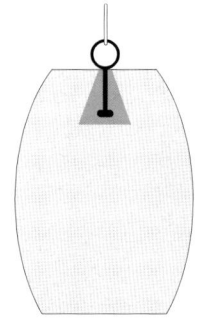

A cross-section of a John Watts clock
weight. After being fashioned from a
block of limestone a recess in the shape
of a truncated cone was cut in the top.
The weight eye is then cast in place
with molten lead. Although
dimensions and shapes varied
considerably, the going train weight
was typically 14kg (30lb) and 200mm
(8in) in diameter, and the striking
train weight was typically 32kg (70lb)
and 250mm (10in) in diameter

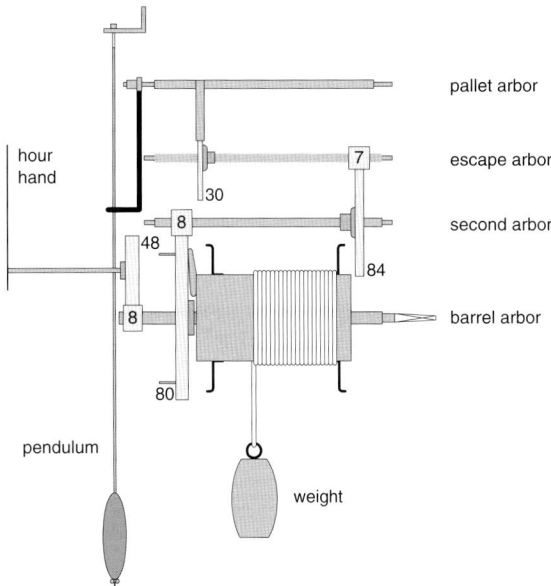

pallet arbor

escape arbor

hour
hand

7

30

second arbor

8

48

84

barrel arbor

8

80

pendulum

weight

The going train of a typical John Watts turret clock. With a 30-tooth
escapement wheel and a seconds pendulum the barrel arbour revolves once
every two hours. The great wheel therefore has two pins to trip the strike
train every hour. If the clock was to have a dial, a pinion of 8 was fixed on
the barrel arbor and this engaged with a wheel of 48 on the leading-off
shaft to the hand. This 6:1 reduction ratio ensured that the hand revolved
once in 12 hours. This also moved the leading-off shaft to the side thus
allowing the pendulum to swing about the centre line of the train. Clocks
with longer pendulums generally have an escape arbor pinion of 8

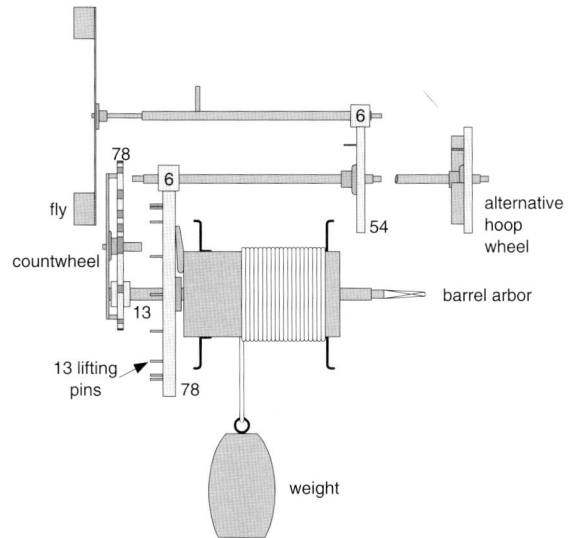

6

78

6

fly

54

alternative
hoop
wheel

countwheel

13

barrel arbor

13 lifting
pins

78

weight

The striking train of a typical John
Watts turret clock. Clocks with a hoop
wheel and a pinned countwheel are later
conversions. The countwheel on the
original clocks was a slotted disk
attached to a wheel with internally cut
teeth

A typical John Watts dial consisted of four planks with four lengths of edge moulding. These would be assembled and decorated in the workshop and, like the clock frames, could be dismantled for easy transport

*A close-up view of the strike train of the former Empingham church clock, now in Stamford Museum. It is signed **I W 1686** indicating that it was made in 1686 by John Watts of Stamford* (Lincolnshire County Council: Stamford Museum 31/2/26)

escapements with thirty-tooth escape wheels, seconds pendulums, countwheel striking and stone weights. Some were later converted to longer pendulums, requiring the frame legs to be extended. The earliest of these clocks is dated only sixteen years after **Joseph Knibb** built the first pendulum-controlled turret clocks in Oxford (see under **Joseph Knibb**) and it seems that John Watts saw the commercial potential for these more accurate time keepers and sold them to many local churches. Details of the Rutland clocks are included in Chapter 4 — Gazetteer.

Evidence of the cost of these turret clocks is provided by an entry in the 'Disbursments Extraordinary' for the year 1684 for a new clock at Browne's Hospital, Stamford: 'It. paid fr a clock cum ptinent [complete] £10 0s 0d'. It was replaced in the late 1800s and the support frame for the new clock was constructed using the timber frame from the old one. This still exists and has been identified as the frame of a clock made by John Watts (information from Michael Lee).

Further evidence is provided by an entry in the Church-wardens' Accounts of All Saints' Church, Stamford (North 1882, 668):

1705 Paid Mr Watts for the new Clock as p. acquittance
£10 0s 0d

Beeeson (1977, 120) refers to there being a timber-framed clock at Castle Bytham, Lincolnshire, which was similar to the former Edith Weston church clock. The following extract may also refer to this clock:

Here also ought to be remembered the Remarkable Charity of Mr. Endymion Canyng, an old Cavalier, and a Captain of Horse in the Service of King Charles I: of ever Blessed Memory. After the Civil Wars, he lived for many years at Brook in this Country [Brooke, Rutland], in the Family of the Right Honourable Julian, late Viscountess Campden, to which Lady he was Steward, and her Principal Servant. He Died a Bachelor at Brook in the year 1683. And by his Will, ... gave to Pious Uses as follows ... To the Town of Castle-Bytham in the County of Lincoln, to buy a Clock for the use of the Town 6l [£6] (Wright 1684, Additions 9-10).

Several domestic clocks by John Watts have also survived. There is a tall eight-day longcase clock with a black lacquered case of *circa* 1690 in Stamford Town Hall, signed 'John Watts Stamford'. A musical longcase clock of *circa* 1690 with a massive three-train posted frame movement playing a tune every three hours on eight bells was sold at auction in 2000 for £48,000. It is in a walnut and marquetry case and signed 'John Watts Stamford'. A marquetry bracket clock with a turn-table base, and a lantern clock, both signed by John Watts, have also been noted.

John Watts died in 1719 at Boston, Lincolnshire. 'A true and Perfect Inventory of the Goods Chattells and Creditts of John Watts late of Boston in the County of Lincolne Clock maker deceased ...' dated 10 March 1719 includes:

In the Shop	
Two Clocks and three cases	£6 10s 0d
Three Old Muskets	7s 6d
The Shop Tools at	15s 0d
Jn [Journal] Book Debts	£1 0s 0d

The massive posted frame musical longcase clock movement of circa 1690 by John Watts. The train arbors are pivoted between pairs of vertical plates which are held in position between the top and bottom plates by wedges as in a traditional lantern clock. The chiming train countwheel has three slots and this train plays every four hours via a pin barrel and fifteen hammers on eight bells. The movement has an anchor escapement with the anchor arbor pivoted above the top plate. Note that the suspension block can be adjusted laterally to bring the clock into beat (Anthony Woodburn Ltd)

The total value of his estate was £12 13s 6d (Lincoln-shire Archives Admon 1719/156).

At the time of writing, Michael Lee of Wansford was researching the life and work of this interesting clock-maker. The results to date of this continuing research have been published in *John Watts Stamford Clockmaker 1686-1704* (Lee 2000).

WATTS, John (junior), of Stamford, Lincolnshire (1683?-?) [1704-?]
John Watts (junior) was an apprentice to his father **John**

The walnut and marquetry case of John Watts' musical longcase clock. The 300mm (12in) square dial is signed 'John Watts Stamford' on the silvered chapter ring. The dial has a matted centre, seconds ring, decorated calendar aperture, ringed winding holes, cherub spandrels, and original pierced steel hands (Anthony Woodburn Ltd)

A circa 1700 lantern clock by John Watts (senior) of Stamford (Philip Snowden)

A rare transitional lantern clock of circa 1730 by Robert Watts of Stamford (Philip Snowden). *Like his thirty-hour longcase clocks, it has an anchor escapement, long pendulum and posted frame movement. However, it was made to look like a lantern clock by the use of a lantern type dial and a single front fret. It is designed to sit on a wooden wall bracket. The use of iron for the bell standard and the vertical frame bars, and the deletion of expensive brass finials, side frets, side doors and feet, made this a very much cheaper clock. This clock is described by Brian Loomes in* Clocks (2002, 16-19)

Watts (senior) and he received his freedom as a clockmaker in 1704. On the assumption that he received this freedom on his twenty-first birthday, this gives his birth date as *circa* 1683. Little is known of him otherwise, and no clocks have been attributed to him.

WATTS, Robert, of Stamford, Lincolnshire (1688 to *circa* 1760) [1719-59]
Robert Watts of Stamford was the son of **John Watts**

(senior), clockmaker and gunsmith. He was born in 1688 and, like his brother, **John Watts (junior)**, he may have served his apprenticeship with his father. He married Hannah, and the baptisms of four of their children are recorded in the parish registers of St John's, Stamford.

A number of his longcase clocks have been recorded. One of *circa* 1750, in private ownership at Braunston in Rutland, has a 250mm (10 in) brass dial with a silvered chapter ring signed 'Robt Watts Stamford', and a single hour hand. It has a posted frame thirty-hour movement and is housed in a small, plain oak case.

The dial of a circa 1750 single-handed thirty-hour longcase clock by Robert Watts (Philip Snowden)

Tebbutt (1975) records four other longcase clocks by Robert Watts, three of which are illustrated. Two of these have eight-day movements and square brass dials, with both minute divisions round the outside of the chapter ring and quarter hour divisions round the inside. Another is a thirty-hour clock which has been converted to two hands. A further clock by Robert Watts, now in a private collection, is illustrated here.

From 1719 until 1759 Robert Watts was engaged by the warden of Browne's Hospital, Stamford, to maintain and repair the turret clock thought to have been installed by his father, **John Watts (senior)**, in 1684. All his vouchers for this work are illustrated in *A Clock At Browne's Hospital Stamford* (Lee 2001). Robert Watts also maintained and repaired the clocks at Burghley House, Stamford, between 1757 and 1759 (Tebbutt 1975, 67).

WATTS, William, of Uppingham [*circa* 1750]
A *circa* 1750 thirty-hour longcase movement and dial signed 'Wm Watts Uppingham' was reported to Newarke Houses Museum, Leicester, in 1988. The case was missing. The 575mm (11in) square brass dial had cut-outs behind the separate silvered chapter ring and a semi-circular calendar aperture below the centre. It also had false winding holes and cast brass urn and scroll spandrels. It was signed above the centre of the dial. The chapter ring had inner hour markings but the single hour hand was missing. No other information on this maker has come to light.

The movement of the Robert Watts clock of circa 1750 (Philip Snowden)

Robert Watts' signature on a thirty-hour clock in private ownership at Braunston in Rutland

WHITEHURST, John I, FRS, of Derby (1713-88)
John Whitehurst I was a Fellow of the Royal Society, scientist and eminent maker of turret and domestic clocks. He was the son of John Whitehurst, a clockmaker of Congleton, Cheshire. John Whitehurst I FRS worked in Derby, but after 1775 he moved to London to pursue

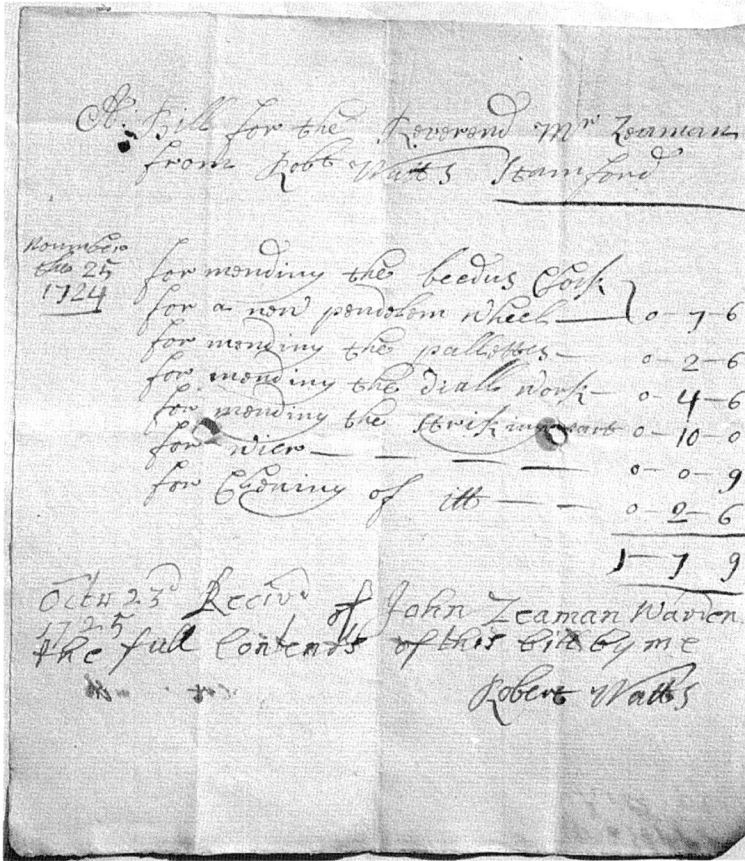

Robert Watts' voucher dated 25 November 1724 for repairing and maintaining the turret clock at Browne's Hospital, Stamford. It includes for a new 'pendolom' wheel, mending the 'pallettes', mending the 'diall work', and mending the 'striking part', as well as 'for Clening of itt.' (Michael Lee/Trustees of Browne's Hospital)

his scientific interests. In 1785 the Derby clockmaking business was taken over by his nephew, John Whitehurst II (1761-1834). His eldest son, John Whitehurst III (1788-1855), became a partner in the company on completion of his apprenticeship in 1809. At this time the business, known then as Whitehurst & Son, was based in Iron Gate, Derby. In 1826 the business was 'burgeoning mightily' and was moved to a new factory at 1 Cherry Street. Whitehurst & Son were one of two companies invited to design and tender for the new clock at the Palace of Westminster ('Big Ben') in 1846 but the contract was eventually awarded to **Edward Dent**. John Whitehurst III died in 1855 and the business carried on for two years under Thomas Woodward. In 1857 it was acquired by William Roskell of Liverpool, but had ceased trading by 1862 (Craven 1996, 169-74). Uppingham Churchwardens' Accounts record that John Whitehurst I supplied a new clock for the church in 1776 at a cost of £65. It was replaced in 1898.

WILKINS, John, of Oakham [1741-51]
John Wilkins was working in Oakham in the middle of the eighteenth century. Hambleton Churchwardens'

Accounts record that he maintained the church clock from 1741 until 1755 at a cost of 5s per year. A *circa* 1740 eight-day longcase clock by this maker was sold by the Heycock family of East Norton in November 1972. It had a brass dial, the centre of which was engraved with birds and a fruit basket, and a well-made oak case with step-moulded hood (Daniell 1975, 42).

WILKINS, Ralph, of Stamford, Lincolnshire [1775]
Ralph Wilkins is recorded in *Old Clocks & Watches & Their Makers* (Britten 1932, 860) as a clockmaker in Stamford in 1775.

WILKINSON, Samuel, of Stamford [1770]
Loomes (1976b, 253) records that Samuel Wilkinson was apprenticed to **Thomas Rayment** in 1770.

WILSON, James, of Stamford, Lincolnshire [1787-1803]
The Wilsons were a large family of clock and watchmakers working in the Stamford area from the late eighteenth century and a number of their clocks are known to be in private ownership. James Wilson worked in All

Saints' Place, Stamford, as a clock and watchmaker, gold and silversmith, and engraver. In the *Lincoln, Rutland & Stamford Mercury* of 26 April 1799 he announced '... that he has taken into partnership his nephew THOMAS HAYNES, who jointly solicits a continuance of their [his 'friends' and 'the public'] further favours and support'. In the same issue he itemised his 'extensive stock in trade' which he wished to dispose of by 'private contract'. This included '... a curious assortment of watches in gold, silver and metal cases [and] a great number of curious clocks, such as spring clocks, spring timepieces in solid brass frames; a year clock in fine mahogany case, a quarter clock, and several 8-day clocks in fine mahogany cases; and a great variety of clocks in fine wainscot cases' (Tebbutt 1975, 68). Edmund Monck, brother of **Thomas Monck**, was a journeyman clockmaker to James Wilson until 1790.

James Wilson maintained and repaired the clock at Clipsham Church from 1787 until at least 1795, and also Tinwell church clock from 1795 until he retired in 1803. His nephew **Thomas Haynes** then took over the business.

WILSON, Joseph T, of Stamford, Lincolnshire [1818-68]

Joseph Wilson was probably the installer of the Lyndon Hall stables clock. Although the name on the setting dial is Wilson, Stamford, the clock movement was actually made by **John Smith of Clerkenwell**. There is no date on the clock, but it is probably *circa* 1860, when the stables were erected. Joseph T Wilson advertised himself as a chronometer, watch and clockmaker, as well as silversmith, jeweller, engraver and optician. An advertisement in the *Lincoln, Rutland and Stamford Mercury* of 22 May 1818 records that Joseph had 'commenced business upon the premises so long occupied by his late uncle, Mr James Wilson, facing All Saints' Church near the Butter Market'. This advertisement also includes: 'J.W. particularly calls the attention of his friends to his Horizontal and Duplex watches, Pocket Chronometers, and Time-keepers, on improved principles, which will not vary in time by a transition from heat to cold, like watches with common vertical escapements.' He also advertised in *Newcombe's Compendium, Sharp's Compendium, Stamford & Lincolnshire Compendium* and *Dolby's Almanack & Stamford Compendium & Directory* on numerous occasions between 1818 and 1868. One of his watches with

A circa 1860 two train mahogany bracket clock by Joseph Wilson of Stamford (Philip Snowden)

an English lever movement and numbered 7399 is in Lincoln City Museum (Tebbutt 1975, 70 & 71).

WILSON, Ralph, of Stamford, Lincolnshire (1729-1829) [1767-1801]

Edith Weston Churchwardens' Accounts record that a Mr Wilson was paid £9 8s 6d in 1767 for repairing the church clock, a duty he performed for the next six years. The Mr Wilson mentioned here is almost certainly Ralph Wilson, clockmaker, watchmaker and silversmith of Stamford, whose shop was opposite the George Inn in St Martins. At the end of the eighteenth and the beginning of the nineteenth century there were watch and clockmakers of the same surname with businesses in Peterborough, Bourne and Spalding and it is thought that they were all related. Ralph Wilson died at Peterborough in 1829, aged 100 years (Tebbutt 1975, 69 & 73).

Chapter 4

Gazetteer of the
Bells, Scratch Dials, Sundials & Clocks of Rutland

This section contains available historical and current details of the bells, scratch dials, sundials and clocks for every parish in Rutland. It is inevitable that there are relevant documents, particularly those in private ownership, which have not been searched, and there may be clocks and sundials which are not included in the Gazetteer. The authors will be pleased to hear of any omissions, via Rutland County Museum. The churchwardens' accounts were a major primary source for this study, particularly concerning the county's church bells and clocks. The survival of these important documents ranges from complete to nothing, the overall average being in the order of twenty-five per cent. It appears, therefore, that a great deal of information has been lost forever.

Arrangement of Gazetteer Entries

- **Parish Name**.
- **Plan of the Church**: This shows the location of features referred to in the text, such as bells, scratch dials, sundials and clocks. These drawings are only approximately to scale.
- **Church Dedication** and grid reference.
- **Parish Records**: These are the parish documents held at the Record Office for Leicestershire, Leicester and Rutland, Long Street, Wigston, Leicestershire. These documents have a reference beginning DE. The only exceptions are Glebe Terriers on microfilm (MF 450), and the Henton photograph series (see below). Only the parish documents listed have been searched in detail.
- **Bell History**: This is based on information contained in the parish records and any other relevant primary or secondary source. Sources other than those at the Record Office for Leicestershire, Leicester and Rutland are identified within the text. Full details of secondary sources are given in the Bibliography.

The Ven E A Irons, Archdeacon of Oakham, was the incumbent at North Luffenham from 1900 to 1923. During this period he made notes on many aspects of the ecclesiastical history of Rutland parishes. These manuscript notes are held in Leicester University Library and are generally referred to here as 'Irons' Notes', followed by the Library accession reference, for example: (MS 80/1/3).

- **Bell Details**: This section includes a plan of the current bellframe, showing the orientation of the frame, the disposition of the bells, the ringing circle, the bellfounders and the dates when the current bells were cast. These drawings are not to scale.

Details of the bells are given in ascending numerical

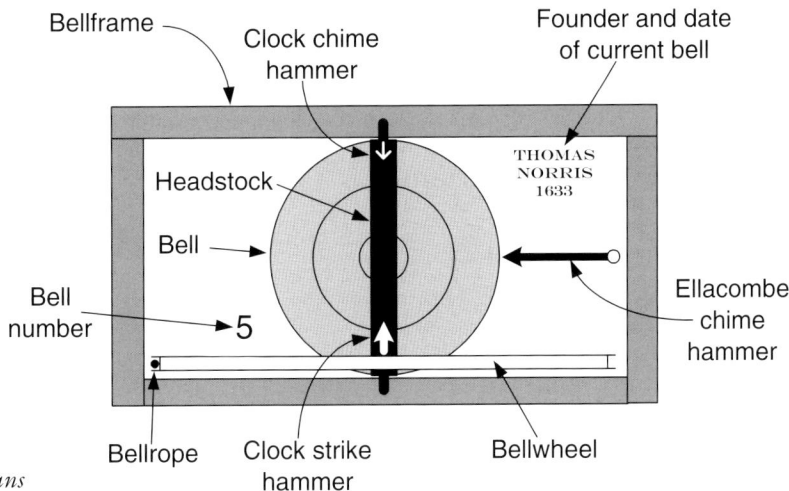

THOMAS NORRIS 1633

Labels: Bellframe, Clock chime hammer, Founder and date of current bell, Headstock, Bell, Bell number, 5, Ellacombe chime hammer, Bellrope, Clock strike hammer, Bellwheel

Key to bellframe plans

order. Note that the smallest bell, apart from any Angelus, Priest's or Sanctus Bell, is referred to as the treble, and the largest bell is referred to as the tenor.

Details include the date when the bell was cast, its note if known, its diameter across the mouth, its weight, whether or not it has canons, and the headstock details. Bell diameters and weights are traditionally given in imperial units and these are shown together with approximate metric equivalents. The founder's name and location is also given.

'Cast by' indicates that the founder is definite. 'Ascribed to' indicates that the founder has been allocated as a result of considering bell features and styles. Where there is no reference in parentheses after the ascription the bell has been ascribed by the authors.

Bell inscriptions are shown in bold capitals followed by a translation if the inscription is in Latin. For example:

[26] OMⴼIA [5] FIAⴼT [5] AD [5] ~ GLORIAM [5] DEI [5]

(Let all things be done to the Glory of God)

Inscriptions in gothic style are as shown in the following example:

𝔐𝔢𝔩𝔬𝔡𝔦𝔠 𝔊𝔢𝔯𝔢𝔱 𝔇𝔬𝔪𝔢𝔫 𝔆𝔞𝔪𝔭𝔞𝔫𝔞

Where an inscription is in one line on the bell, but flows on to two lines in the text, this is indicated by ~, for example:

J. TAYLOR & CO. FOUNDERS ~ LOUGHBOROUGH 1877

Numbers in brackets, such as [26], indicate the locations and identities of bellfounder's marks, devices and decorations. Details of these are given in Chapter 2 — Bellfounders, and in Appendix 1 — Bellfounders' Marks, Devices, Decoration and Lettering on Rutland Bells. Where coins are represented, this is shown by **O**.

General notes on the bell then follow. If details of an earlier bell have survived these are included under the sub-heading **Former bell**. These are mainly from North's *Church Bells of Rutland*.

• **Bellringing Customs**: This section includes all those customs recorded by Thomas North in 1880, together with any found in the parish records, and any which are followed today. A full explanation of all known customs is given in Appendix 3 — Rutland Ringing Customs.

Other headings and details, including **Handbells**, **Ringers' Rules**, **Hourglass** and **Dedication Service**, have been inserted if appropriate to that particular parish.

• **Scratch Dials**: The introductory paragraph indicates the dates of walls on which scratch dials have been incised. In most cases two drawings for each scratch dial are included, one showing the location, the other showing the layout together with a table of dimensions. All linear dimensions shown are in millimetres. All angles are

measured clockwise from a vertical line above the centre of the scratch dial. These drawings are not to scale.

• **Sundials**: This section includes brief details of any church sundial noted. A comprehensive list, with details, of all ecclesiastical and secular sundials is included as Appendix 4 — Rutland Sundials.

• **Clock History**: The notes on **Bell History** above are also relevant here.

• **Clock Details**: Known details of the present and any previous clocks are recorded here in tabular form.

• **Other Ecclesiastical and Secular Buildings**: All known bells, scratch dials, sundials and clocks on, attached to or near to other ecclesiastical and secular buildings within the parish are generally recorded under the name of the building or area (*see also* Appendix 4 — Rutland Sundials).

GENERAL COMMENTS CONCERNING THE GAZETTEER

• **Quotations**

Extracts from primary sources, churchwardens' accounts in particular, are verbatim, except that dates and monetary amounts have been standardised.

• **Technical Terms**

A brief explanation of technical terms used is given in the Select Glossary.

• **Illustrations**

Unless otherwise credited, all photographs and drawings are by the authors.

The Henton Photographic Collection in the Record Office for Leicestershire, Leicester and Rutland is a series of some 2000 dated photographs taken by George Henton in the 1890s and early 1900s. Those of Rutland are generally from the Great War period and are referenced as, for example, (Henton 994).

Two complete series of drawings of Rutland churches exist. A *circa* 1793 collection is held by Rutland County Museum under references F10/1984/1 to 67. These have been listed and described, and some illustrated, by Geoffrey K Brandwood (1989). A *circa* 1839 collection is held by Uppingham School Archives and these have been reproduced in *Rutland Churches before Restoration*, edited by Gillian Dickinson (1983).

• **Abbreviations**

The following abbreviations are used:

AHS: Antiquarian Horological Society
BHI: British Horological Institute
MSS: Manuscripts
NRO: Northamptonshire Record Office
OS: Ordnance Survey
PCC: Parochial Church Council
RCM: Rutland County Museum
SD: Location of table sundial on OS maps

ASHWELL

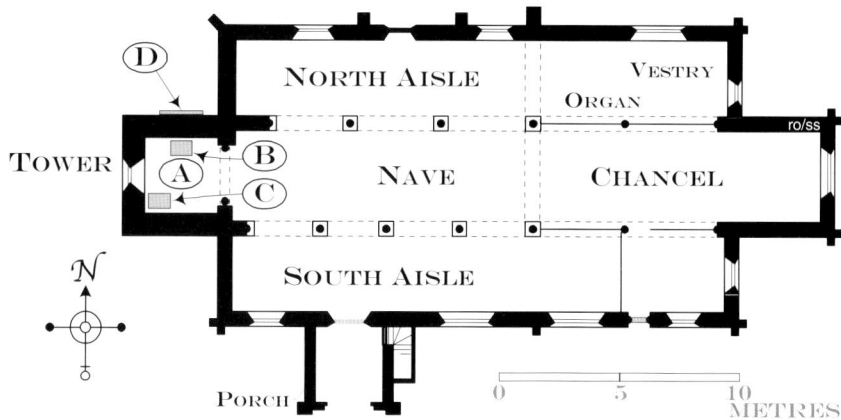

Plan of St Mary's Church
A: Six bells in the belfry.
Ringing chamber at the base
of the tower
B: Clock on the first floor of
the tower
C: Former clock retained in
the clockroom
D: Clock dial
Access to the clockroom and
belfry is by ladders

ST MARY

Grid Ref: SK 866137

PARISH RECORDS

Information concerning the church bells and clock can be found in the Churchwardens' and Overseers of the Poor Accounts. These survive from 1690 to 1808 (DE 5199/6).

BELL HISTORY

The following extracts appear to indicate that Ashwell had its own watchman in the late seventeenth century:

1690

Item given to the watchman when the watch was set forward	2d
Spent in Seting out the watch and going to pay their expenc	8d
charges about the watch men wee alowd them 4 pence a night once about	

Ashwell Church almost certainly had five bells in 1708. Three of these, the present third, fourth and sixth, were cast or recast in that year and still hang in the tower today. There is no record of payment in the accounts, implying that one or more benefactors paid for these new bells. It seems that they were a great attraction to ringers from nearby parishes, as indicated by the following extract:

1708 Dec 23

spent of oakham Ringers	7s 0d
spent of Langham Ringers 1 time	7s 0d
spent of Langham Ringers two times	5s 0d
spent of the Towne Ringers	10s 0d
spent of some out town Ringers	3s 6d
spent of Stabelford [Stapleford] Ringers	6s 4d
spent of some Cockmore [Cottesmore] Ringers	3s 0d

The churchwardens considered that their bells were of great importance but the cost of maintaining them was a large drain on their resources as seen in the following examples:

1711		
May 29	pd Tho Beridge for hanging ye bells £5 10s 0d	
1714	Guy Wilbon for shooting the Bel ropes	2s
1752		
May 17	pd when ye Bell yokes was taken down	1s 3d
	pd for Ale when ye Bell wheels was bargained for	1s 0d
	pd for Ale when ye Bells was done	1s 0d
June 13	Paid Blackburn & Bell for Repairing the Bells	£9 3s 7d
	pd for oyl for ditto	8d
	pd for putting in ye Great Bell Clapper	8d
1758		
April 8	pd for Oyl for the Bells	8d
Sept 9	pd John Meredith for a Bell Wheel	£1 11s 0d
	pd for Iron Work for Ditto	1s
Sept 28	pd Jno Meredith for Wood and Work in Mending a Bell Frame	£1 0s 0d
	pd for Help to take up the Bell	2s 0d
	pd Jno Hutchings for his pulleys to take up ditto	6d
Sept 30	pd for 12lb of Iron Dogs to hold the Bell Frame	4s 0d

In 1760 there was a problem with the treble, the current second bell. It had to be taken down and sent to Thomas Hedderly of Nottingham to be recast:

1760	pd for Ale for 6 Men when the Bell was took down	1s 0d
	pd for Ale when the Bell was Weighd	1s 0d
	pd for Carrying ye Bell to Nottingham	4s 0d
	pd for 2 Letters from Nottingham	8d
Sept 22	pd to the Bellfounder	£8 18s 0d
Oct 4	pd Jno Edgson for a New Bell Wheel, and taking down, & putting up the Little Bell and other Work, and the Blacksmiths Work	£2 6s 0d
Oct 14	pd for the Carriage of ye Bell from Nottingham	10s 0d

In 1786 it became necessary to recast the present fifth bell, the bellfounder being Edward Arnold of Leicester.

It is interesting to note that in the 1780s Edward Arnold also supplied bells to other churches in the vicinity: the then second at Brooke (1780), the then treble at Whissendine (1785) and the Priest's Bell at Barrowden (1786). He also installed the six-tune chime barrel at nearby Stapleford in 1785. Details are again recorded:

1785-86

pd for Ale for taking the Bell Down and Loading 3s 8½d
pd for the Carrage of the Bell to Witsondine
[Whissendine]   2s 6d
pd for bringing ye Ropes Gibs and Tools
from Tigh [Teigh]   2s 6d
pd for Carrying the Gibs home and Bring planks back 2s 6d
pd for ye Carrage of the Bell from Witsondine to
Leicster   10s 6d
pd Mr Arnold his Bill for the Bell   £23 16s 9d
pd for Carrage of the Bell from Leicester to
Uppingham   £1 0s 0d
pd Mr Webstee Bringing the Bell from Uppingham
and Loding   6s 6d

The ring was increased to six bells in 1850 by the addition of a new treble. This was hung over the former treble in an extension to the existing timber frame. The following extract is from the Sales Day Book of the Whitechapel Bell Foundry:

1850 Nov 29   Rev^d Mr Yard, Ashwell
To A Bell 3cwt 3qr 25lb @ £6 6s 9d per cwt   £25 0s 8d
Clapper   £1 0s 0d
To mans time journey and expenses, taking
pitch of Bells. Making pitch fork, tuning the
whole of the Bell putting in Crown Staples
and adjusting the Clappers   £6 10s 0d

Church accounts, which might provide further details of the bells during the nineteenth century, have not survived. No further information concerning the history of the bells has been found.

Bell Details

1 1850. Treble. Diameter 692mm (27¼in). Weight 202kg (3cwt 3qr 25lb). Canons retained. Timber headstock. Cast by C & G Mears, Whitechapel Bell

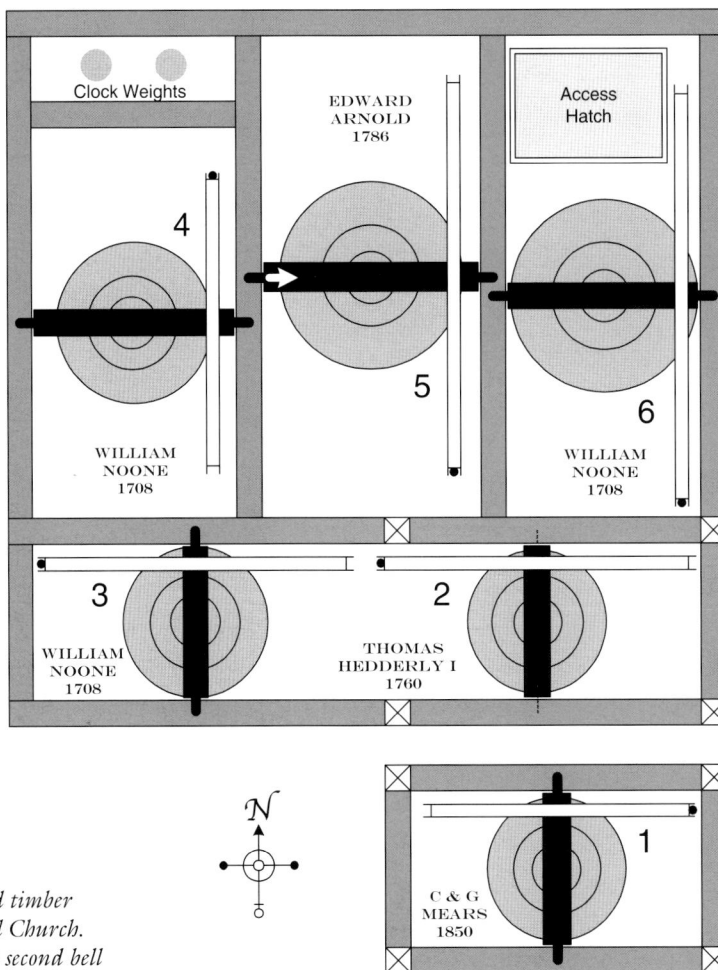

Plan of the low-sided timber bellframe at Ashwell Church. The treble is over the second bell

Foundry, London. On the waist:

[44] MISERICORDIAS DOMINI IN
~ ETERNUM CANTABO B. B. 1850

(I will sing the mercies of the Lord for ever)

On the soundbow:

C ET G MEARS LONDINI FECERUNT

(Made by C & G Mears of London)

B. B. was Beckford Bevan, the brother-in-law of Canon Thomas Yard, rector of Ashwell from 1850 to 1875 (North 1880, 118).

C ET G Mears Londini Fecerunt

The inscription above the soundbow of the treble at Ashwell

2 1760. Diameter 686mm (27in). Weight approximately 229kg (4cwt 2qr). Canons retained. Timber headstock. Recast by Thomas Hedderly I of Nottingham.

EX : DONO BARTHOLEMEI BVRTON
~ ARMIGERI ANNO DOM 1760

(Donated by Bartholomew Burton Esquire)

IOHN CHAMBERLAIN C:W
~ THO HEDDERLY FOUNDER

There are six heart-shaped arrows [43] between the canons. Decoration [42] is above the inscription band and also on the soundbow. There is a further decoration around the lip of this bell but it is too indistinct to illustrate. Bartholomew Burton was a former owner of Ashwell Hall (North 1880, 118).

Noone's eagle and rabbit stamp [41] on the third bell at Ashwell

Former bell

No details of this bell have survived.

3 1708. Diameter 737mm (29in). Weight approximately 254kg (5cwt). Canons retained. Timber headstock. Cast by William Noone of Nottingham.

GOD [40] SAVE [40] QVEEN [40] ANNE [40]
~ ANNO [40] DOMINI [40] 1708 [41]

4 1708. Diameter 800mm (31½in). Weight approximately 305kg (6cwt). Canons retained. Timber headstock. Cast by William Noone of Nottingham.

TEMPORE [40] IOHANNI [40] BVLL [40]
~ ET [40] NICHOLAI [40] COALE [40]
~ WARDENS [40] 1708

(In the time of John Bull and Nicholas Coale
Churchwardens)

Decoration [42] is below the inscription band with [41] at each quarter.

5 1786. Diameter 927mm (36½in). Weight approximately 432kg (8cwt 2qr). Canons retained. Timber headstock. Recast by Edward Arnold of Leicester.

[45] EDWᴰ ARNOLD LEICESTER FECIT 1786 O O

There is decoration [46] below the inscription band. Both [45] and [46] are of inferior quality. The image of the obverse and reverse of a George III golden guinea, minted in 1776, follow the date on the inscription band, shown as **O O** above. The coin would have been pressed into the cope to obtain the impressions. The clock strikes the hours on this bell.

Former bell

No details of this bell have survived.

Coins on the inscription band of the fifth bell at Ashwell

6 **1708**. Tenor. Note E. 946mm (37¼in). Weight approximately 495kg (9cwt 3qr). Canons retained. Timber headstock. Cast by William Noone of Nottingham.

[40] D̄c̄c̄ [40] C̄ampana [40] S̄acra [40] F̄iat [40]
~ T̄rinitatc̄ [40] B̄eata [40] 1708

(Let this bell be blessed by the Holy Trinity)

There is decoration [42] below the inscription band.

CLOCK BELL

A clock bell is believed to have been housed in a late seventeenth-century cupola built on the tower. For further details see under Clock History.

BELLRINGING CUSTOMS

On Sundays the treble was rung at 8am and both the treble and second at 9am. All the bells were chimed for Divine Service. The Pancake Bell was rung at noon on Shrove Tuesday and the Gleaning Bell at 8am and 6pm during the harvest (North 1880, 118).

The Churchwardens' Accounts give details of many reasons for ringing the bells, including the following examples:

1691	**Nov 4**	given to the ringers for ringing for King William's birth day and for his arriving on the English shore at his first coming for our deliverance from popery one shilling	1s 0d
1692	**May 26**	given to the ringers for ringing for the good news of our victtory over the french fleet	2s 0d
	April 11	To ye Ringers being ye Kings Coronation day	2s 6d
1704	**Sept 7**	Given to the Ringers upon ye day of thanksgiving for ye Victory over the Bavarian Army	1s 0d
1714		Gave the Ringers on the 5th November	2s 6d
		Gave the ringers on the Thanksgiving Day	2s 6d
		gave the Ringers when the Archdekon was heer	1s 0d

Bells were rung when a parishioner died and at the subsequent funeral service. The following examples are taken from the disbursements of the Overseers of the Poor:

1704	**March 13**	pd Guy Wilburn for Making Ann Masons Grave and Ringing her Bell	1s 0d
1760	**April 17**	pd Jno Holland for ye Bell, Grave, & Register	1s 10d
		pd for Ale for the Chimers, & Bearers	2s 4d
1784	**May 3**	pd William Willburn Sarah Castertons bell and grave etc	1s 10d
		pd the [clerk?] for chiming at Dito	1s 0d

The bells at Ashwell Church have not been used for full-circle ringing since the 1970s owing to the condition of the frame and a defect in the tower structure.

An example of the lettering on the Ashwell tenor

Today the third bell is chimed for five minutes before every service.

CANDLE HOLDER

An early stone candleholder projects from a wall in the ringing chamber. A candle was the only source of light for the ringers in the early days of ringing.

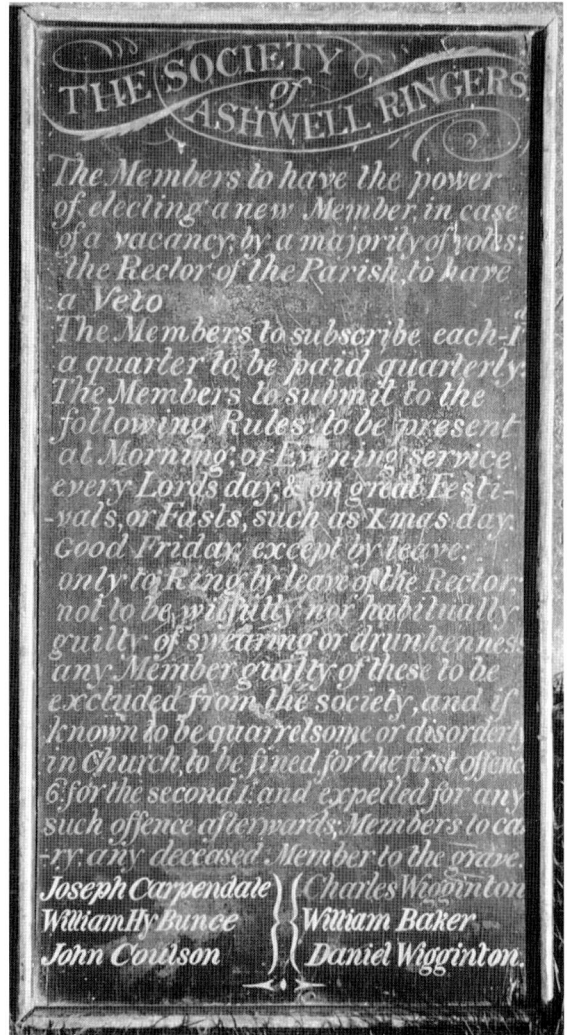

THE SOCIETY of ASHWELL RINGERS

The Members to have the power of electing a new Member, in case of a vacancy, by a majority of votes; the Rector of the Parish, to have a Veto

The Members to subscribe each-1 a quarter to be paid quarterly. The Members to submit to the following Rules: to be present at Morning, or Evening service every Lords day, & on great Festi-vals, or Fasts, such as Xmas day, Good Friday, except by leave; only to Ring by leave of the Rector, not to be wilfully nor habitually guilty of swearing or drunkenness any Member guilty of these to be excluded from the society, and if known to be quarrelsome or disorderly in Church, to be fined for the first offence 6 for the second 1 and expelled for any such offence afterwards. Members to carry any deceased Member to the grave.

Joseph Carpendale, Charles Wigginton, William Hy Bunce, William Baker, John Coulson, Daniel Wigginton.

Ashwell Ringers' Rules

RINGERS' RULES

Ringers' Rules, possibly of nineteenth-century origin, are set out on a board at the base of the tower.

SUNDIAL

There is no evidence of a sundial but one would certainly have been required to regulate the church clock. The entry for 1692 under 'Clock History' may refer to a sundial.

CLOCK HISTORY

The Churchwardens' Accounts include:

1690	for the clock keeping	5s 0d
1691 April 15	Item to Nicholas Day ... for mending the Clock hammer	1s 2d
1692 May 17	to Mr Pepper for drawing the dial upon the steeple [tower]	8s 0d
	Item for a clock rope for the great plumet [clock weight]	1s 3d

The above is confirmation that there was an early clock here but little is known of it. It probably had a timber frame and would have been very similar to other clocks installed about this time by John Watts of Stamford. Later accounts however do give some clues.

The entry of 1697 below probably indicates that the clock had been here for some years, and that of 1708 confirms that it was a single-handed clock. The dial would therefore have had four divisions between the hours. The clock was enclosed in a wooden case in 1709. It is possible that a new pendulum clock was installed at this time to replace an old foliot clock. Although it required regular attention the clock continued to work throughout the eighteenth century. Two Oakham clockmakers, Blackburn and Simpson, are named in the accounts. John Simpson regularly attended the clock, making repairs until 1792:

1697 May 29	for mending ye clock frame	2d
1698 April 26	Paid to Rich'd Phillips for coming three times to mend ye clock	£1 5s 0d
1708	spent of the man that help with the Clock hand	1s 0d
	pd for a wate [weight] for the Church Clock	2s 4d
1709 June 10	pd for makeing the Clock Case	4s 4d
	pd for nales for the Clock Case	1s 3d
1711 May 29	pd ye Clock mender	£1 1s 6d
1729	Charges Mr blackburn when he came about the Clock	1s 6d
1748 April 18	pd for a Dial board painting and a spindle for Ditto and a new Hand for ye Church [clock]	£1 1s 0d
	for ye Carpenter at Ditto	1s 2d
	pd for nails for Ditto	1d
1783 Dec	John Simpson for mending the Clock	5s 0d

The clock winder from 1735 to 1778 was John Holland and for the next ten years William Willburn. William

From a drawing of circa 1793 (RCM F10/1984/1) of Ashwell Church showing the cupola which probably housed the clock bell. The date of its removal is unknown but a drawing of circa 1839 in Uppingham School Archives shows that it had gone by this date. It may have been removed when the new clock was installed in 1805, as this would have struck the hours on the tenor bell, making the bell in the cupola redundant

Capendale then took over this responsibility for at least eighteen years, during which time a new clock was installed in 1805. They were all paid £2 a year for this duty.

In the accounts of *circa* 1690 there are many payments made for work done with regard to 'the over loft in the steeple'. This may point to the date for the erection of the cupola. It was painted in 1749, and it required repairs in 1755:

1749	pd for painting ye Cupaloe	£1 3s 6d
	pd Jno Holland for Cleaning ye Cupaloe and Clock House	1s 8d
	pd for a wether cock	5s 0d
	pd for use of the ladders	6d
1755	pd Jno Edgson for mending ye Cupaloe	10s 6d
1757 March 14	pd Jno Thurlby for slating ye Cupaloe	7s 6d
March 22	pd for ale when the Cupola was mended	1s 4d

Rear view of the old clock movement stored in the clockroom at Ashwell Church

An old clock movement on its original wooden stand is stored in the clockroom. There is no maker's name on the frame, and the setting dial, the usual location of the clockmaker's signature, is missing. This is almost certainly the clock installed in 1805 and the following extract from the accounts indicates its source:

1804 Oct 16   Going to Melton about the Church
                          Clock   5s 0d
1805 June 29 Pd half the Church Clock   £20 0s 0d

From the cost it seems quite likely that this was a second-hand clock and expert opinion (members of the AHS Turret Clock Group) suggests that it is a marriage of two clocks. The extended barrels are of a later date than the rest of the movement and would have been added to convert the clock from thirty-hour to eight-day duration.

White (1877, 664) records that at Ashwell 'It was proposed to place a new clock in the tower in 1887'. However, the replacement clock was not commissioned until three years later. Although it has no maker's name, the two-train flatbed turret clock was supplied by Thwaites & Reed of Clerkenwell, London, and probably installed

by a local clockmaker.

According to Dick Tidd, a former clock winder, lightning has struck this church several times and on one such occasion it travelled down the clock weight lines setting fire to the timber weight duct, evidence of which can still be seen in the base of the tower.

CLOCK DETAILS

Details of the 1805 clock:

Maker:	Unknown, but supplied by a Melton Mowbray clockmaker
Signed:	Setting dial missing
Installed:	1805
Cost:	£40
Frame:	Narrow birdcage with extended barrels
Trains:	Going and striking
Escapement:	Anchor
Pendulum:	Missing
Striking:	Countwheel hour-striking
Weights:	Missing
Winding:	Probably hand wound weekly
Dial:	Diamond-shaped wooden dial
Note:	Setting dial and fly also missing

The present movement at Ashwell Church by Thwaites and Reed of London

The clock dial on the north face of the tower at Ashwell Church in May 2000. It is based on the original early eighteenth century dial, although this would have been marked out for a single hour hand with quarter-hour divisions inside the chapter ring

Details of the present clock:

Maker:	Identified as Thwaites & Reed of Clerkenwell, London
Signed:	No name or date on the clock
Installed:	1890
Frame:	Cast-iron flatbed
Trains:	Going and striking
Escapement:	Deadbeat
Pendulum:	Wooden rod and cast-iron lenticular bob
Rate:	50 beats per minute
Striking:	Countwheel hour-striking
Weights:	Cast-iron
Winding:	Hand wound weekly
Dial:	Diamond-shaped wooden dial. New replica dial in 2000
Location:	North face of the tower

The setting dial of the clock installed by Thomas Large at Cottesmore Hunt Kennels

The clock turret at Cottesmore Hunt Kennels in 2001. Note that the dials are showing different times as the dial works had been disconnected when this photograph was taken

COTTESMORE HUNT KENNELS

A clock turret with three dials was included in the design when Cottesmore Hunt Kennels were built in 1890. The small two-train flatbed clock movement by William Evans of Handsworth, Birmingham, was supplied and installed by Thomas Large of Melton Mowbray.

CLOCK DETAILS

Details of the present clock:

Maker:	William Evans of Handsworth, Birmingham
Installed by:	Thomas Large of King Street, Melton Mowbray in 1890
Signed:	'Thos. Large. Melton Mowbray' on the setting dial
Frame:	Cast-iron flatbed
Trains:	Going and striking
Escapement:	Deadbeat
Pendulum:	Wooden rod and cast-iron cylindrical bob
Rate:	60 beats per minute
Striking:	Countwheel hour-striking
Weights:	Cast-iron
Winding:	Weekly
Dials:	Three convex copper dials on a turret. Gilded numerals and hands on a red background
Dial erected:	When clock installed
Clock bell:	In the cupola above the dials. No details available

ASHWELL HALL

Ashwell Hall, half a mile to the south of the village, was built in 1879. The octagonal timber clock turret on the converted stable block has one small round dial facing west. The original mechanical clock has been replaced by a synchronous electric movement.

AYSTON

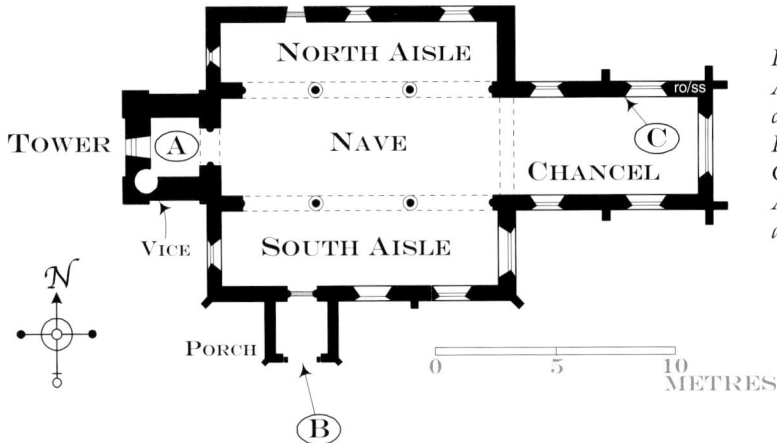

Plan of St Mary the Virgin's Church
A: Four bells in the belfry. Ringing
chamber at the base of the tower
B: Sundial
C: Scratch dial
Access to the belfry is the vice and then
a ladder

ST MARY THE VIRGIN   Grid Ref: SK 859009

PARISH RECORDS

The Churchwardens' Accounts Book 1833-1958 (DE 5476/6) is the only source in the parish records containing information on the church bells. The only relevant items are isolated references to paying the bell ringers, buying and 'shooting' bell ropes and 'Cleaning the Bell Chamber'. There is no mention of a clock.

BELL HISTORY

There is a ring of four bells in the tower. The second bell, ascribed to John Rufford of Toddington, was cast *circa* 1365 and is considered to be the oldest bell in Rutland. The three older bells were tuned and rehung without canons by John Taylor in 1877 when a new fourth was added to the ring. The present oak frame and fittings were installed at the same time, the bells being attached to timber headstocks by bolts through the crown. The total cost of the new installation was £207 2s 0d (North 1880, 119). The belfry was refurbished in the 1980s.

BELL DETAILS

1 *Circa* 1550. Treble. Diameter 756mm (29¾in). Weight approximately 279kg (5cwt 2qr). Canons removed. Timber headstock. Ascribed to Robert Newcombe I of Leicester.

[78] [11]

AMBROSE [24]

[48]

(Divine)

Marks [48] and [78] are indistinct. Letters are like [108]. Scheduled for preservation (Council for the Care of Churches).

2 *Circa* 1365. Diameter 800mm (31½in). Weight approximately 305kg (6cwt). Canons removed.

Timber headstock. Ascribed to John Rufford of Toddington (Pearson 1989).

[4] AVE [50] REX [50] GENTIS [50] ANGLORVM

(Hail King of the English Nation [Edward III])

Letters are like [112]. North suggests that the inscription is the first line of a hymn (North 1880, 120). Scheduled for preservation (Council for the Care of Churches).

The initial cross, first word and stop on the John Rufford bell at Ayston Church

3 1626. Diameter 876mm (34½in). Weight approximately 356kg (7cwt). Canons removed. Timber headstock. Cast by Toby Norris I of Stamford.

[26] NON [90] VERBO [90] SED [90] VOCE [90] ~ RESONABO [90] DOMINE [90] LAVDEM 1626.

(Not by word but by voice will I resound thy praise O Lord)

Below the inscription band:

R [53] HILL [53] R [53] ROYCE [53] Ch [53] WH ~ [53] TOBIE [53] NORRIS [53] CAST [53] ME [53]

North (1880, 119) has B HILL and R BOYCE.

4 1877. Tenor. Note G. Diameter 1060mm (41¾in). Weight 638kg (12cwt 2qr 7lb). Cast without canons. Timber headstock. Cast by John Taylor & Co of Loughborough.

J. TAYLOR & CO. FOUNDERS
~ LOUGHBOROUGH 1877

On the waist:

LAUS DEO.

(Praise be to God)

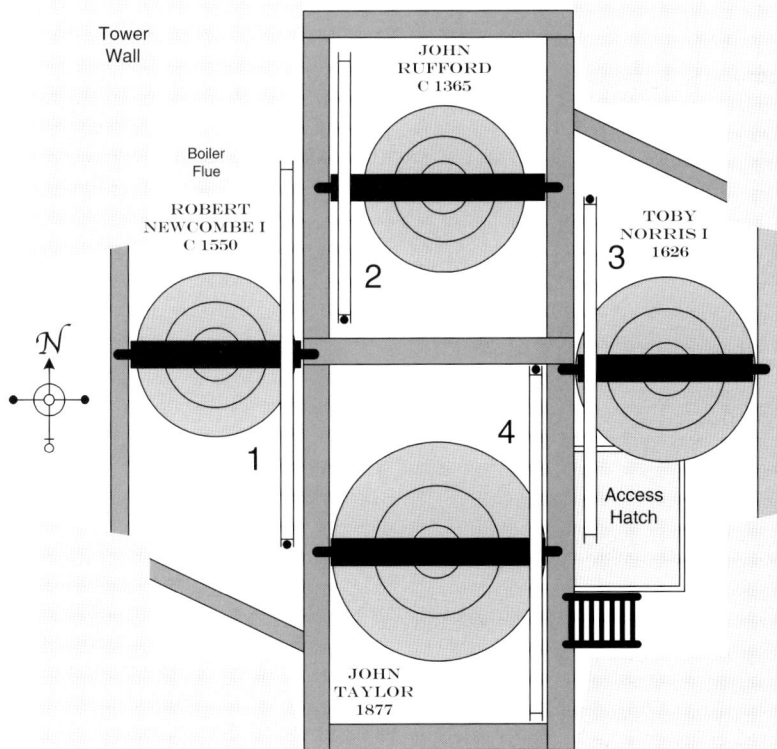

Plan of the bellframe at Ayston Church

Donated by the Rector, the Rev Sir J Henry Fludyer, Bart, to commemorate his fiftieth year as Curate and Rector of Ayston. The bell cost £96 14s 7½d (North 1880, 119).

Bellringing Customs

On Sundays the treble was rung at 9am. The bells were chimed for Divine Service and if there was to be a sermon, the tenor was rung. After the morning service the treble and second bells were chimed. The Pancake Bell was rung at noon on Shrove Tuesday. Before and after the Death Knell there were three tolls on the tenor for a male and two for a female (North 1880, 119).

A bill dated 17 March 1773 and made out by John Rudkin, Constable for the Parish of Ayston, gives the charges incurred for the funeral of a pauper, who died whilst passing through the parish from 'Knell to Lester' (Phillips 1911-12, 94). Amongst these charges was 1s 4d 'for the Bell'. There are various payments made to ringers in the Churchwardens' Accounts and itemised payments for Christmas occur between 1899 and 1918.

Today Ayston Church has three services a month and a bell is chimed for two minutes prior to the start. If bells are required at a wedding ceremony, arrangements are made with a visiting band as there are no local ringers.

Scratch Dial

There is a scratch dial inside the church on the west jamb of the north-east window of the chancel. This wall dates from the fifteenth century and the scratch dial is believed to be on a relocated stone. In its present position it is ineffective.

NORTH WALL OF THE CHANCEL

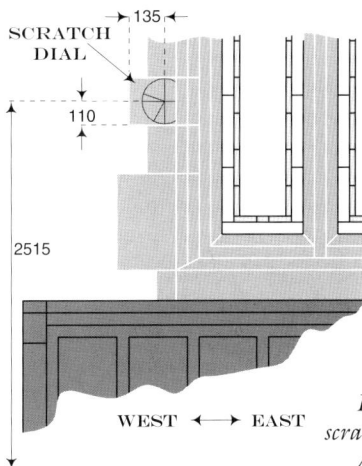

Location of the scratch dial inside Ayston Church

Scratch dial details

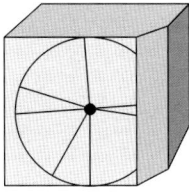

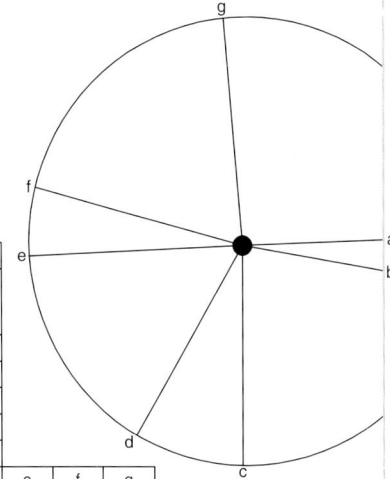

AYSTON							
Location	Inside the church - on the west jamb of the north-east chancel window						
Condition	Good						
Gnomon Hole Diameter	10						
Gnomon Hole Depth	45						
Height above ground level	2515						
Circle Diameter	210						
Line Ref	a	b	c	d	e	f	g
Length	70	74	105	105	105	105	105
Angle (°)	88	100	180	210	267	285	355

The sundial in the gable of Ayston Church porch. The arabic numerals are quite weathered, but the gnomon is in good condition

SUNDIAL

The sundial in the gable of the church porch was originally on a cottage, now demolished, which was standing on the north side of the road to Ridlington in 1910. It was known as Sundial House.

Sundial House circa 1910 (Betty Finch)

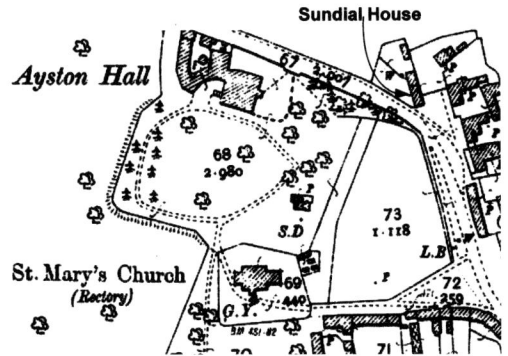

AYSTON HALL

The early 1900s OS Second Edition 25 inch map shows that there used to be a table sundial in the grounds of Ayston Hall.

The early 1900s OS Second Edition 25 inch map showing a table sundial (SD) in the grounds of Ayston Hall. The location of the former Sundial House is also shown

BARLEYTHORPE

This parish does not have a church.

BARLEYTHORPE HALL

When the Earl of Lonsdale built new stables to Barleythorpe Hall in 1870 a clock and dial were included in a central gable. The bellcote for the small clock bell above the dial is incorporated between twin chimney stacks. The dial has translucent glazing and was originally illuminated from behind by gas lighting. The gas was supplied from the Earl's own gas works in Manor Lane. To control the lighting the clock had a 'gas wheel'. This wheel, which turned once every twenty-four hours, had pins which opened and closed a gas valve via a lever. Electricity is now used to illuminate the dial, the only one of its kind in Rutland. The clock movement was located immediately behind the dial but it has now been replaced by a synchronous electric movement.

CLOCK DETAILS

Details of the original clock as installed:

Maker:	Dent
Signed:	'Dent, 61 Strand, London. 1871' on the minute setting dial
Installed:	1871
Frame:	Small cast-iron flatbed. Serial number 239 on the pendulum cock bar
Trains:	Going and striking
Escapement:	Pinwheel
Pendulum:	Temperature compensating bimetallic rod and cast-iron cylindrical bob
Rate:	60 beats per minute
Striking:	Countwheel hour-striking
Bell:	250mm (10in) bell in a bellcote between twin ashlar chimney stacks
Weights:	Lead
Winding:	Hand wound weekly
Dial:	Fabricated copper skeleton dial with white translucent glazing. Located in a gable over the stables
Illumination:	Originally gas illumination. 24 hour gas wheel on clock
Notes:	1. Separate silvered seconds and minutes setting dials
	2. Silvered countwheel marked in hours with roman numerals
	3. The fork of the crutch is two horizontal pins at the ends of flat springs
	4. Described in Dent catalogue as a 'No 1 clock'

When the stables at Barleythorpe Hall were converted into housing in the year 2000 a table sundial was incorporated as a garden feature. The dial is engraved on a block of limestone from Greetham Quarry. The development is now known as Clock House Court.

A table sundial that used to be in the grounds of Barleythorpe Hall was relocated at an unknown date to the garden of Westbourne House, Belton in Rutland. It was removed from there in 1997, but its new location is unknown.

The clock dial and chimney bellcote at Barleythorpe stables, now Clock House Court

The Barleythorpe Hall stables clock by Dent of London prior to restoration

(right) The restored Dent movement of the Barleythorpe Hall stables clock seen from the dial side

(below) The minute setting dial. The gas wheel revolves once in twenty-four hours and in the original installation its pins would lift a lever to keep a gas valve closed. During the night, a gap in the pins allowed the lever to drop, thus opening the valve to provide illumination. The threaded holes are at half-hour intervals and pins could be added or removed to adjust the illumination period. The lever and valve were removed when the dial was converted to electric illumination, which is now controlled by a photocell

The numbered countwheel

The pinwheel escapement and seconds dial

The pendulum crutch is unusual in that it has two horizontal pins at the ends of flat springs. The springs ensure that the crutch is always in contact with the pendulum rod when it is being impelled. The energy absorbed by each spring during the overswing is given back to the pendulum during the impulse. Stops disengage each spring once the pendulum has swung back past the central position (Robey 2000, 403)

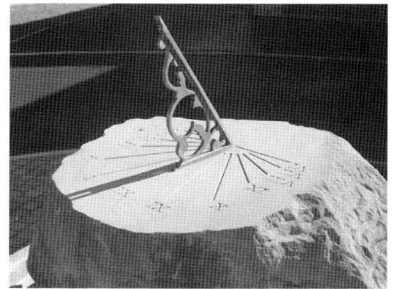

The new table sundial at Clock House Court, Barleythorpe

BARROW

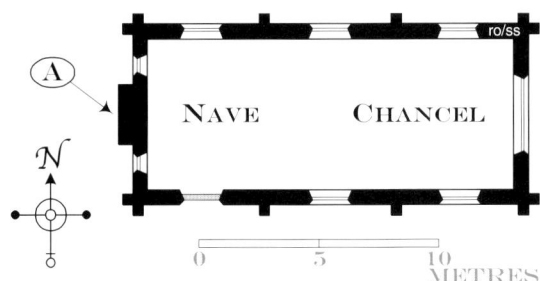

Conjectural plan of the Chapel of St Mary Magdalene
A: One bell in a double bellcote

ST MARY MAGDALENE   Grid Ref: SK 890153

This chapel was demolished in 1974.

CHAPEL HISTORY

A thirteenth-century chapel existed at Barrow on land to the east of the village but it is presumed to have closed around 1560 and was certainly in ruins by 1660. The site was excavated and walls exposed by A R Traylen in 1969. The building of a new chapel on a site within the village was completed in 1830, the Consecration Service being held in 1831. This second chapel fell into disrepair and was finally demolished in 1974 (Traylen 1988, 11).

PARISH RECORDS

The few surviving parish records relating to Barrow Chapel are included with those of Cottesmore.

BELL HISTORY

When the second chapel was built in 1830 a gabled double bellcote was erected over the west end. A *circa*

1839 drawing (Uppingham School Archives) shows just a single bell in the bellcote, possibly indicating that only one bell was ever installed.

There was still only one bell in 1880 when it was described as being 'modern' (North 1880, 120). It was re-hung in 1903 and some details of the work carried out at that time are contained within the following church receipts:

Dr to Emerson & Co Cottesmore re Barrow Chapel.
1903 March 27 to April 17
Bricklayers & Labourers time erecting Scaffolding, taking down Bell Frame & Bell, making new Bell Frame, getting up & fixing Bell, repairing apex stone of Arch & refixing it in proper position   £6 6s 7d
(DE 1920/81/164)

Dr to J T Hollis
1903 April 16
New ironwork & fittings complete for Bell Frame ... new clapper & clasp & fitting to Bell, 2 new spindles, 2 hooks on Beam ends ... journey, & time at same   £1 18s 0d
(DE 1920/81/132)

BELL DETAILS

1 *Circa* **1830**. Size and weight details not available. Ascribed to Thomas Mears II of the Whitechapel Bell Foundry, London (George Dawson). No inscription. The bell was removed when the chapel was demolished in 1974. Its fate is not known.

BARROW HOUSE

A table sundial is shown in the grounds of Barrow House on the early 1900s OS Second Edition 25 inch map. Barrow House is at SK 894152.

Barrow Chapel from a drawing of circa 1839 (Uppingham School Archives)

BARROWDEN

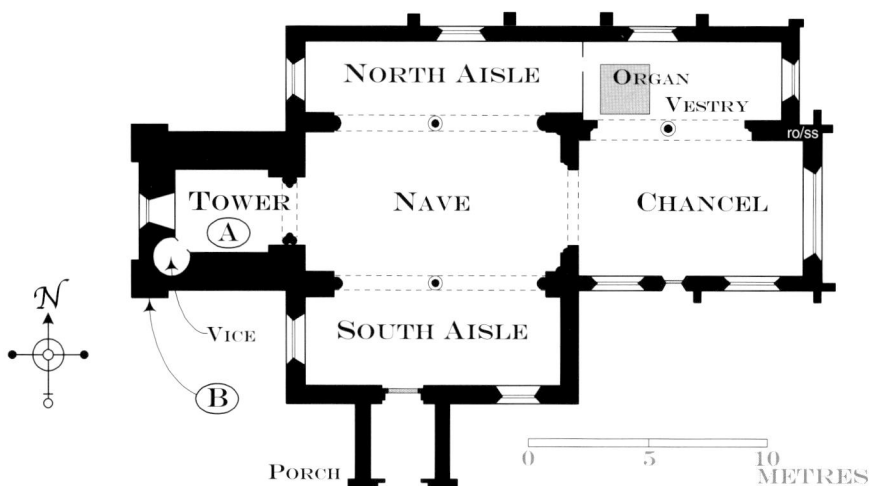

Plan of St Peter's Church
A: Six bells and a Priest's
Bell in the belfry
Ringing chamber at the
base of the tower
B: Sundial
The vice provides access to
the belfry

ST PETER
Grid Ref: SP 945999

PARISH RECORDS

The only sources of information concerning the bells in the parish records are an estimate for the restoration of the bells and spire in 1915-16 (DE 2021/43) and a report on the bells dated 12 August 1926 (DE 2021/44), both by John Taylor of Loughborough.

BELL HISTORY

There were five bells in the tower *circa* 1550 but the tenor may have been sold before 1605 (Irons' Notes, MS 80/1/3). No information is available regarding the history of the early bells but of the four present in 1605, three had been cast at Leicester: one by Thomas New-combe II *circa* 1570 and two by a member of the Watts family, possibly Francis, in 1595. The ring was returned to five in 1706 by the addition of a new bell cast by Alexander Rigby. Thomas North (1880, 120) records

Stone tablet on the west wall of the nave at Barrowden Church

that the then treble was also cast by Rigby in 1706. This was almost certainly a recasting of an earlier bell. A Priest's Bell was added in 1786.

All the bells were rehung by the Whitechapel Bell Foundry in 1857 'at a cost of £102 8s, raised by subscription' (North 1880, 121). Engraving on stones between the two north-facing windows in the belfry records this work:

THESE BELLS WERE
REHUNG BY
J. HURRY, FOR
MESSRS. MEARS,
Wᵀ CHAPEL LONDON
A.D. 1857
REVᴰ. C. ATLAY
RECTOR
I. JOHNSON, CLERK
H. MASON, AND
T. TAYLOR,
CHURCHWARDENS
T. SWANN

It is interesting to note that a bell at Charwelton in Northamptonshire is inscribed 'J. TAYLOR AND SON FOUNDERS LOUGHBOROUGH MDCCCXLIIII [1844] J. HURRY AGENT NORWICH' thus showing that J Hurry was an agent for both Mears and Taylors (North 1878, 220).

In 1915 the then treble, second and tenor were recast and the ring of five rehung in a metal high-sided frame for six bells by John Taylor. An analysis of the expenditure incurred shows that the total cost was £187. A separate note indicates that the cost of a new treble would have been £40.

A report and estimate of 1926 supplied by John

Plan of the bellframe at Barrowden Church

Priest's Bell
EDWARD
ARNOLD
1786

5

6

WATTS
1595

4

THOMAS
NEWCOMBE II
C 1570

JOHN
TAYLOR
1915

3

2

1

JOHN
TAYLOR
1915

JOHN
TAYLOR
1915

JOHN
TAYLOR
1990

Access from
vice

N

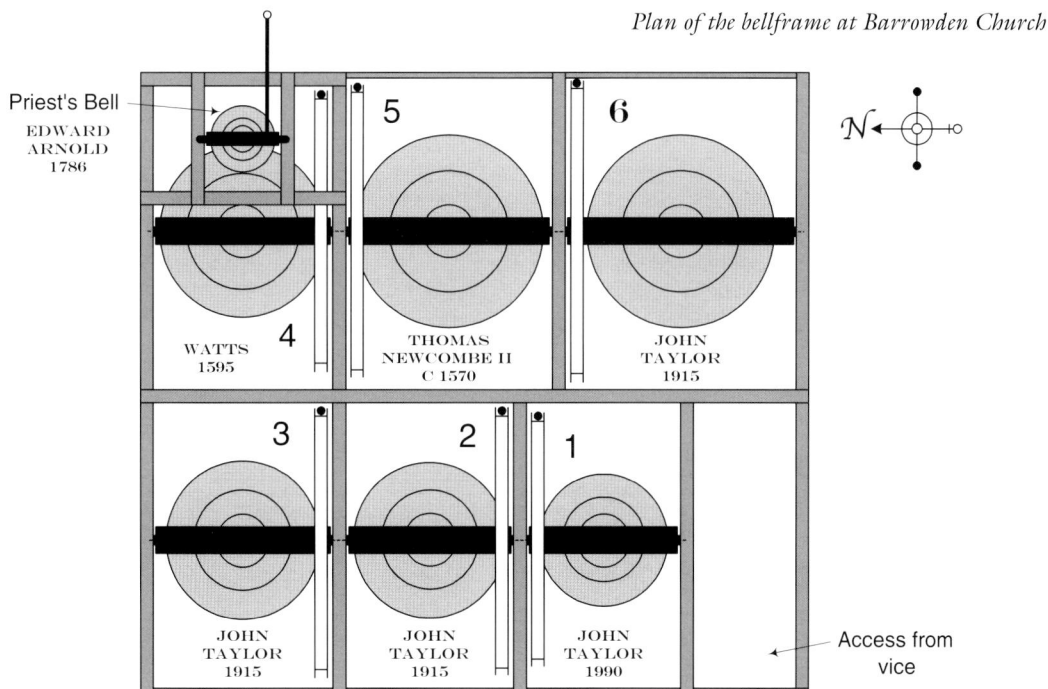

Taylor refers to the installation of a tolling hammer for the tenor (£9), refurbishing the bearings and painting the frame (£32), and supplying and hanging a new 'smaller bell' (£78). The additional cost of an inscription on this bell was 6d a letter. However, none of this work was carried out and it was not until 1980 that Taylors refurbished the frame. A new treble was added to the ring in 1990.

BELL DETAILS

1 1990. Treble. Diameter 635mm (25in). Weight approximately 152kg (3cwt). Cast without canons. Cast-iron headstock. Cast by John Taylor & Co of Loughborough.

[54]

On the waist:

[55] SAINT PETER

On the waist, opposite:

19 [56] 90

The lettering is like [130] and [131]. The inscription

The inscription on the treble at Barrowden Church using fifteenth-century style lettering probably owned by John Smith of Louth, Lincolnshire

band is filled with decoration [54]. This and the repaired Priest's Bell were dedicated on Sunday 27 May 1990 by the Bishop of Peterborough, the Right Rev Bill Westwood. Both bells were installed in the tower on 30 May 1990.

2 1915. Diameter 692mm (27¼in). Weight 203kg (3cwt 3qr 27lb). Cast without canons. Cast-iron headstock. Recast by John Taylor & Co of Loughborough.

[14] ALEXANDER [57] RIGBY [57]
~ MADE [57] ME [57] 1706

On the waist:

RECAST 1915
ARTHUR HUTCHINGS:
RECTOR:

On the waist, opposite:

[56]

Former bell

Cast in 1706 by Alexander Rigby of Stamford. Diameter 737mm (29in).

[14] ALEXANDER RIGBY MADE ME 1706

In 1880 the canons on this bell were broken and it was fastened to the headstock by bolts (North 1880, 120).

3 1915. Diameter 746mm (29⅜in). Weight 251kg (4cwt 3qr 21lb). Cast without canons. Cast-iron headstock. Recast by John Taylor & Co of Loughborough.

com [13] com [13] and [13] pray 1595 [3]

On the waist:

The new treble is hoisted into the belfry at Barrowden in 1990 (Nicholas Meadwell)

RECAST 1915
ARTHUR HUTCHINGS:
RECTOR:
On the waist, opposite:
[56]

Former bell
Cast in 1595 by Watts [possibly Francis] of Leicester (Pearson 1989). Diameter 775mm (30½in).
com [13] com [13] and [13] pray 1595 [3]

4 1595. Diameter 814mm (32¹⁄₁₆in). Weight 324kg (6cwt 1qr 14lb). Canons removed. Cast-iron headstock. Cast by Watts [possibly Francis] of Leicester (Pearson 1989).
[3] god [13] save [13] the [13] queene 1595
Scheduled for preservation (Council for the Care of Churches).

god save the queene 1595

The black-letter inscription on the fourth bell at Barrowden

5 *Circa* 1570. Diameter 841mm (33⅛in). Weight 312kg (6cwt 0qr 15lb). Canons removed. Cast-iron headstock. Ascribed to Thomas Newcombe II of Leicester.
[2] A B C D E F G H I [22]
The founder's marks and letters are spread evenly around the inscription band. The letters are like [108]. Scheduled for preservation (Council for the Care of Churches).

6 1915. Tenor. Note A flat. Diameter 968mm (38⅛in). Weight 503kg (9cwt 3qr 17lb). Cast without canons. Cast-iron headstock. Recast by John Taylor & Co of Loughborough.
[14] ALEXANDER [57] RIGBY [57]
~ MADE [57] ME [57] 1706
On the waist:

RECAST 1915
ARTHUR HUTCHINGS:
RECTOR:
On the waist, opposite: [56]

Former bell
Cast in 1706 by Alexander Rigby of Stamford. Tenor. Diameter (36in).
ALEXANDER RIGBY MADE ME 1706
North (1880, 120) recorded the date on this bell as 1704.

Barrowden Priest's Bell awaits its new headstock prior to being rehung in 1990 (Nicholas Meadwell)

PRIEST'S BELL

1786. Diameter 311mm (12¼in). Weight approximately 25kg (2qr). Canons retained. Timber headstock. Cast by Edward Arnold of Leicester.

EDW^D ARNOLD FECIT 178ð [49] [49]

Probably a memorial to the Rev Joseph Digby who was rector for thirty-two years from 1754.

A framed history of the church written by the church-wardens in 1862 states:

> A Saint's, or Sanctus bell which until quite recently was used for prayers but is now unfortunately out of repair, is suspended in the belfry from a frame in the east window.... This bell was rehung in 1889, when the upper portion of the spire, struck by lightning a short time previously, was removed and rebuilt (Kelly 1925, 720).

It was rehung by Nicholas Meadwell with new fittings and a new frame in 1990 and is now referred to as a Priest's Bell. The Re-dedication Service was on 27 May 1990.

BELLRINGING CUSTOMS

Prior to the Priest's Bell being damaged and becoming unusable *circa* 1840, it was rung after morning service when there was to be no sermon after Evensong. At the Death Knell there were three tolls for a male and two tolls for a female. Peals were also rung after the services held on important festival days (North 1880, 120). The Pancake Bell was rung on Shrove Tuesday (Traylen 1988, 35).

At a death the tenor used to be chimed once for each year of a person's life but this custom ceased in the 1970s. When Father Brian Scott moved to Barrowden in September 1983 he began to ring the Angelus on the

Barrowden Church sundial

treble bell, thrice three times, every day at noon. When the Priest's Bell was rehung in 1990, he chimed that instead, continuing to do so until he left the village in October 1998.

Today there is an active band of bellringers. The bells are rung for the Sunday services held at 8am and 11am and for all services at Easter and Christmas. They are also rung for weddings and if requested, for funerals.

SUNDIAL

A sundial of unknown date is located at the south-west corner of the tower, about six metres (20ft) above ground level.

BELMESTHORPE

Belmesthorpe is a small hamlet about half a mile south-east of Ryhall. It does not have a church although it did have a chapel, dedicated to St Mary the Virgin, in medieval times. This chapel still existed in 1636 when it is recorded that the chancel was out of repair. The site still retained the name of Chapel Yard in 1811 but it had completely disappeared by 1935 (*VCH* II, 269). There is no record of a bell or clock at this chapel.

BLUE BELL PUBLIC HOUSE

The Blue Bell public house is thought to be the oldest property in Belmesthorpe. Its name does not appear to have any connection with the name of the village or a possible bell at the chapel.

The sign of the Blue Bell public house at Belmesthorpe

BELTON IN RUTLAND

Plan of St Peter's Church
A: Six bells in the belfry.
Ringing chamber at the base
of the tower
B: Clock on the first floor of
the tower
C: Clock dial
D: Sundial
E: Scratch dial
The vice provides access to
both the clockroom and the
belfry

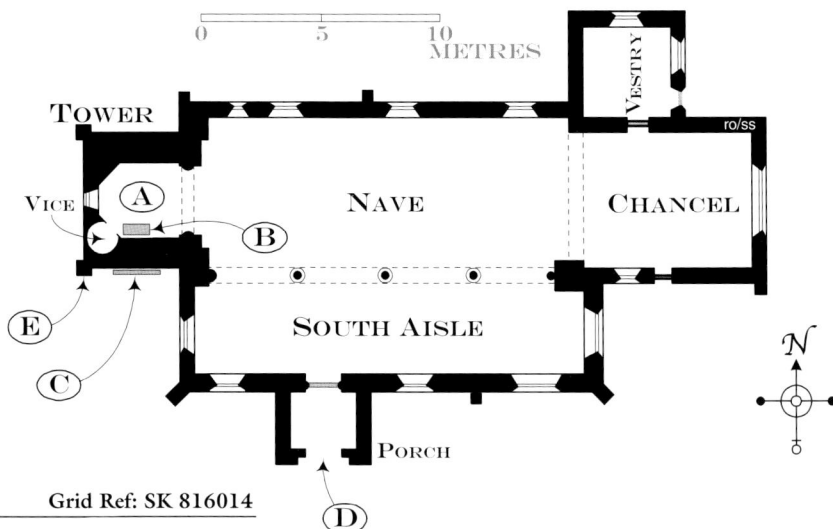

ST PETER   Grid Ref: SK 816014

PARISH RECORDS

The parish records include the faculty of 1911 for recasting the four old bells and adding two new bells (DE 1815/5), and a cutting from the *Grantham Journal* describing the dedication ceremony for the bells and lich-gate in 1911 (DE 1815/25/5). The Churchwardens' Accounts (DE 1785/12) which survive from 1826 to 1908 include occasional references to expenditure on bellropes and paying the ringers, but maintaining the clock in running order seems to have been the major cost. The Vestry Minute Book 1869-92 (DE 1815/15) provides some interesting snippets of information concerning the old clock and bellringing. Glebe Terriers are available on microfilm (MF 495).

BELL HISTORY

A Glebe Terrier for Belton, dated 13 December 1633, states that the church had 'a low broad steeple without a spire, within are three Bels, and a clock' (MF 495).

In 1681 an Archdeacon's Visitation ordered that the bells had to be recast and rehung (*VCH* II, 31). An analysis of the inscriptions on the bells as they were in 1880 suggests that only one of these bells was actually recast in 1681. By the end of the seventeenth century there were definitely four bells in the tower and three of these, if not all of them, were cast at the Norris foundry in Stamford. The then treble was recast in 1730.

A faculty dated 6 May 1911 reveals that the bells were in an unsafe condition and out of tune. As a consequence two new bells and four recast bells were installed on 30 August of that year in a new high-sided steel and iron frame. All six bells were cast without canons and mounted in cast-iron headstocks. The cost was £250 and the

installation was timed to commemorate the Coronation of King George V. The work was carried out by Gillett & Johnston of Croydon.

BELL DETAILS

1 1911. Treble. Diameter 638mm (25⅛in). Weight 190kg (3cwt 3qr 0lb). Cast without canons. Cast-iron headstock. Cast by Gillett & Johnston of Croydon.
CAST BY GILLETT & JOHNSTON CROYDON
On the waist:

> **ELIZABETH WARD GAVE ME**
> **~ TO THE GLORY OF GOD, 1911**

There is a band of decoration [58] below the inscription band. Donated by Elizabeth Ward, a grazier and resident of Belton. She was the widow of George Godfrey Ward, also a grazier.

2 1911. Diameter 711mm (28in). Weight 220kg (4cwt 1qr 8lb). Cast without canons. Cast-iron headstock. Cast by Gillett & Johnston of Croydon.
CAST BY GILLETT & JOHNSTON CROYDON
On the waist:

> **MY VOICE WILL IN CONCERT RING**
> **IN HONOUR OF BOTH GOD AND KING**
> **PRESENTED BY F. & C. GOUGH,**
> **~ BELTON HOUSE 1911**

There is a band of decoration [58] below the inscription.

3 1911. Diameter 775mm (30½in). Weight 275kg (5cwt 1qr 19lb). Cast without canons. Cast-iron headstock. Recast by Gillett & Johnston of Croydon.
GILLETT & JOHNSTON CROYDON
On the waist:

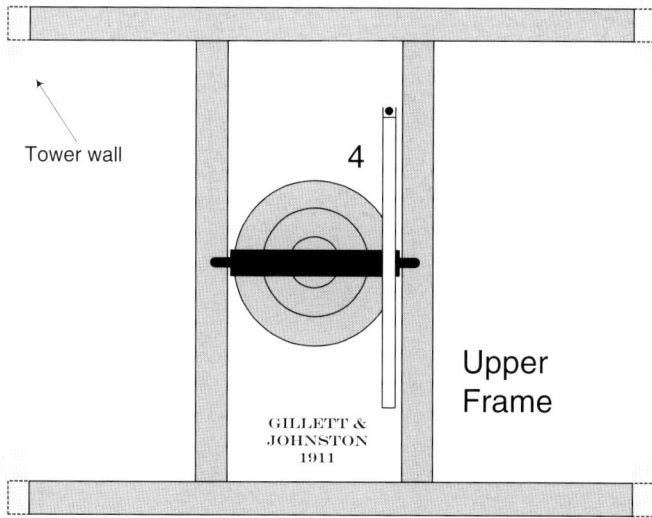

Plan of the bellframe at Belton Church

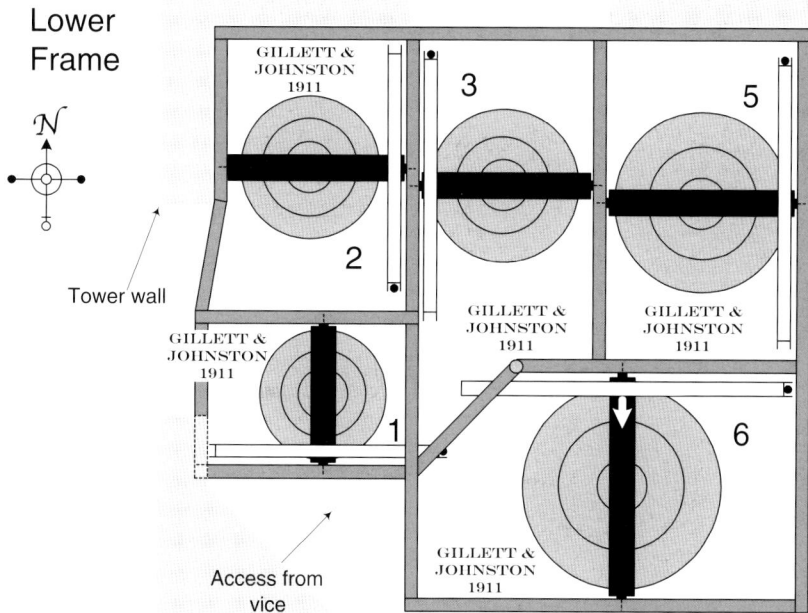

GLORIA PATRI FILIO & SPIRITUI SANCTO. 1730.
(Glory be to the Father, and to the Son,
and to the Holy Ghost)
**THOMAS PILKINTON. GENT. GAVE ME 1664.
RECAST. 1911.**
There is a band of decoration [58] below the inscription band. Note the missing G in 'Pilkinton'.

Former bells
The original treble. Cast in 1664, probably by Thomas Norris of Stamford.
THOMAS PILKINGTON GENT. GAVE ME 1664
Thomas Pilkington of Belton, the donor of this bell, was married in 1663 at Lyndon to a Mrs Thomazin Collins of Hambleton. His second wife was Mary Smith, half

An illuminated list of the bells at Belton prepared by Gillett & Johnston in 1911

sister to Sir Isaac Newton (North 1880, 121). Thomas Pilkington sold the manor at Belton to the Verney family *circa* 1673 (*VCH* II, 29).

Recast in 1730 by Thomas Eayre II of Kettering. Diameter 737mm (29in).
GLORIA PATRI FILIO & SPIRITUI SANCTO 1730 ~ THOMAS PILKINGTON GENT. GAVE ME 1664

4 1911. Diameter 813mm (32in). Weight 308kg (6cwt 0qr 8lb). Cast without canons. Cast-iron headstock. Recast by Gillett & Johnston of Croydon.

GILLETT & JOHNSTON CROYDON

On the waist:

1681. RECAST. 1911.

There is a band of decoration [58] below the inscription.

Former bell
Cast in 1681 by Toby Norris III of Stamford. Diameter 787mm (31in).
[10] W WORTH TOBIE ИORRIƧ CAƧT ME 1681

5 1911. Diameter 889mm (35in) diameter. Weight 402kg (7cwt 3qr 19lb). Cast without canons. Cast-iron headstock. Recast by Gillett & Johnston of Croydon.

GILLETT & JOHNSTON CROYDON

On the waist:

1695. RECAST. 1911.

There is a band of decoration [58] below the inscription band.

Former bell
Cast in 1695 by Toby Norris III of Stamford. Diameter 826mm (32½in).

[10] TOBY NORRIS CAST ME 1695

6 1911. Tenor. Note G. Diameter 991mm (39in). Weight 519kg (10cwt 0qr 24lb). Cast without canons. Cast-iron headstock. Recast by Gillett & Johnston of Croydon.

GILLETT & JOHNSTON CROYDON

On the waist:

1660. RECAST. 1911.

On the waist, opposite:

G. J. PATTISON	VICAR
F. GOUGH	
J. GROCOCK	CHURCHWARDENS

There is a band of decoration [58] below the inscription band. The clock strikes the hours on this bell.

Former bell
Cast in 1660 by Thomas Norris of Stamford. Tenor. Diameter 914mm (36in).

[26] THOMAS NORRIS MADE MEE 1660

It had decoration [52] on the inscription band.

Belton's third bell, the former treble, before being recast in 1911 (David Griffiths)

John Grocock, the village carpenter at Belton, was also churchwarden and a bellringer (David Griffiths)

DEDICATION SERVICE

A service to dedicate the new lich-gate and bells was held on the afternoon of Wednesday 30 August 1911. The *Grantham Journal* dated 9 September 1911 gives a full account of the service and the celebrations: 'The parishioners are delighted with their new acquisition ....' The service, at which there was a large congregation, was led by the Ven E M Moore, Archdeacon of Oakham. When the bells were dedicated it must have been a proud moment for the local ringers - Messrs C Reeve (treble), W Allen, C Marlow, T Atkin, A Jarman and F Bindley (tenor) - when they raised the six bells and commenced ringing. Peals were rung during the evening by visiting ringers including several 'Oakham campanologists' who 'cycled over and took part in the various peals'. Peals were also rung throughout the following weekend. On the Saturday night, as on the Wednesday night, handbell ringers helped to provide entertainment in the Village Hall.

BELTON BELLRINGERS

There is a memorial in the ringing chamber to members of St Peter's Guild of Change Ringers who served in the Great War. It was unveiled on 6 December 1919 by Col C H Jones CMG, of Uppingham. Three of the men listed rang the bells at the Dedication Service in 1911. Unfortunately two were killed in action (Phillips 1920, 24).

HANDBELLS

A handbell was used at Belton Church of England School, until it closed in 1971, to announce the start of lessons. Today it is still in regular use at Leighfield Primary School, Uppingham. As in many Rutland villages there was a set of handbells in Belton at the turn of the twentieth century. They were used to mark the celebrations at the Coronation of George V in the Village Hall on 22 June 1911. There are no handbells in the village now.

BELLRINGING CUSTOMS

On Sundays the treble was rung at 7am and the treble and second at 9am or midday depending on whether the Divine Service was to be in the morning or afternoon. For this service three bells were chimed and then the tenor bell, known as the Sermon Bell, was rung for five minutes. The Pancake Bell was rung at 11am on Shrove Tuesday and a Daily Bell used to be rung at 5am and at 1pm (North 1880, 121).

The late Edith Sleath, a lifelong resident of Belton in Rutland, remembered her grandmother Elizabeth Marlow, who was born in 1859, talking about the church bell being rung to inform the workers in the fields that it was 'dinner-time'. This is also referred to in a note from the incumbent (*Leicestershire & Rutland Notes & Queries* I, 1891, 275):

117. Belton Bells

The Rev. C. H. Newmarch sends the following notes about the bells of Belton, Rutland:-

A 'dinner bell' is rung daily at one o'clock to let people in the fields know the time. Origin unknown: it has been rung from time immemorial. The payment for it is included in the Sexton's salary, it being part of his duty.

There *used* to be a bell rung at 5a.m., summer and winter, and an old servant of mine, who was parish clerk, rang it; but it has been discontinued many years. The older parishioners remember it being rung.

We have also had, until within the last twenty years or so, a 'pancake bell' rung at noon on Shrove Tuesday. This has been discontinued only within my own incumbency.

At the Death Knell three tolls were rung for a male and two tolls for a female, both before and after the knell. These tolls were repeated three times. At funerals prior to 1880 all the bells used to be chimed, but by this date only the tenor was tolled (North 1880, 121).

As a result of the Burials Act of 1880 a Vestry Meeting on 19 April 1881 agreed was 'that if the use of the bell be desired by Nonconformists, and is asked for, it will not be refused'.

Sheila Sleath remembers hearing the Death Bell up until the mid 1950s, but by the early 1970s it was only rung when specifically requested. This bell was last rung in November 1992 to announce the death of a regular bellringer.

Unfortunately, Belton has not had its own band of bellringers since before 1992. Today the bells are rung

SCRATCH
DIAL

TOWER

200

2100

Location of the scratch dial on the south-west buttress of the tower of Belton Church

GROUND
LEVEL

WEST ⟷ EAST

Scratch dial details

e

d

Mortar line

c

b

a

BELTON IN RUTLAND					
Location	SW buttress of tower				
Condition	Faint				
Gnomon Hole Diameter			24		
Gnomon Hole Depth			20		
Height above ground level			2100		
Line Ref	a	b	c	d	e
Length	330	178	160	180	100
Angle (°)	163	179	193	225	343

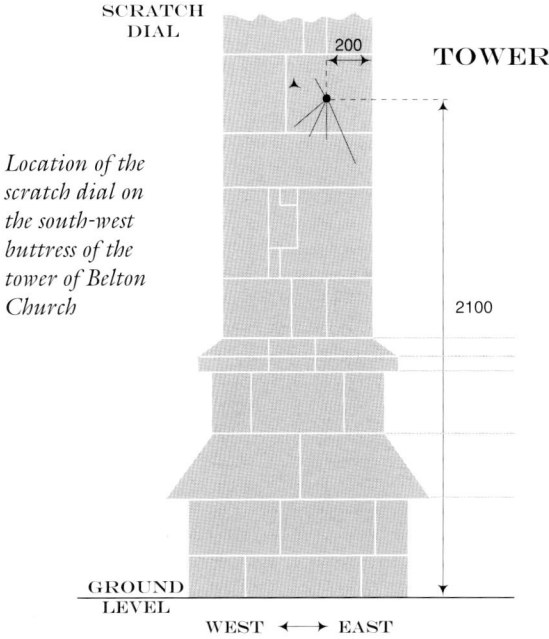

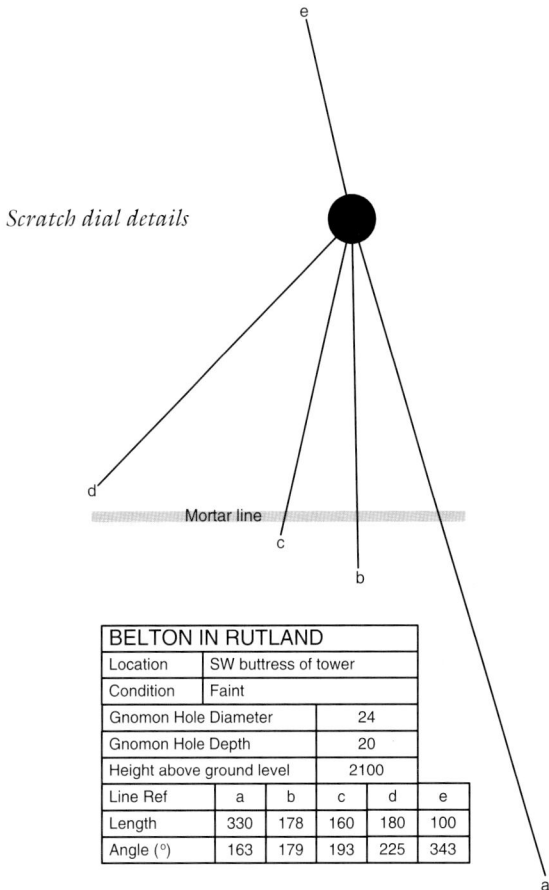

by visiting ringers, some from as far away as the United States of America.

Two of the smaller bells are chimed for fifteen minutes, then the Priest's Bell for three minutes prior to the start of a Sunday Service. The bells are chimed before the Midnight Service at Christmas and for the service which precedes the War Memorial Service. If requested, ringers from outside the village are available for weddings.

The ringing of church bells was prohibited during the early years of the Second World War, so it was a very special occasion when a quarter peal of Cambridge Surprise Minor was rung for the Golden Wedding Anniversary of Edith and Harold Sleath on 31 October 1992.

HOURGLASS

There used to be an iron bracket for an hourglass projecting from a pillar near the pulpit (White 1863, 822) and this would have been used for timing sermons. This is the only record of a church hourglass in Rutland although it is quite likely that there were many more.

SCRATCH DIAL

There is a scratch dial on the fifteenth-century southwest buttress of the tower. The lines are very shallow and can only be seen in any detail when the sun is in the west.

SUNDIAL

A vertical-declining sundial is set into the gable of the south porch which was rebuilt in 1841. This may indicate the date of the dial. The lines and numerals were re-cut as a Millennium project.

CLOCK HISTORY

The present clock, installed in 1887, may be the third

The scratch dial on the south-west buttress of the tower of Belton Church

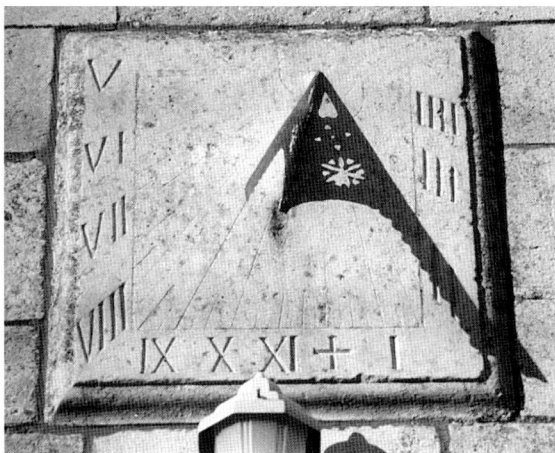

Belton Church sundial after restoration

clock at Belton Church. A Glebe Terrier (MF 495) confirms that there was a clock here as early as 1633. It would have had a verge and foliot escapement, but it is unlikely to have had a dial. As the Churchwardens' Accounts prior to 1826 are missing there is no record of this early clock or its replacement.

Evidence of an earlier clock remains in the form of old pulley hooks, and wooden blocks on the floor which may have been used to locate a clock stand. Clock weights found at the home of a former churchwarden may have belonged to an eighteenth-century clock with pendulum control. An 1839 drawing of Belton Church (Uppingham School Archives) clearly shows a dial with a stone bezel, and a photograph of *circa* 1880 (private collection) confirms that the dial had one hand. Accord-

Stone clock weights found when a wall was demolished at Corby Cottage in Nether Street, Belton. These could have belonged to the eighteenth-century clock

ing to the Vestry Minute Book it was the sexton's duty, *circa* 1869, to wind the clock every day after he had rung the Dinner Bell, confirming that it was a thirty-hour clock.

The Churchwardens' Accounts show that the clock was serviced and repaired by several Uppingham clock-makers. William Aris II worked on it in 1826 but John Houghton carried out the annual maintenance from this date until 1848:

1835-36   Pd Houghton for cleaning & repairing
the Church Clock 2 years   £1 4s 0d

After 1848 it seems that little work was done on the clock until 1855 when Robinson and Flint were paid £1 9s 0d for cleaning and repairing it.

Thomas Aris, the son of William II, repaired and maintained the clock regularly from 1859 until 1876. The minutes of the Vestry on Tuesday 31 May 1870 record that the Parish Council was considering the best means of repairing the church clock. Evidently a decision was made and Thomas Aris was employed to do this. There are no further entries after 1876 relating to re-pairs. Following the installation of a new clock in 1887 there are several references to paying amounts to Pinney, which were probably for minor repairs rather than regu-lar maintenance.

A plaque in the porch records that the present clock was installed as a memorial to Robert Baines, who lived at Old Hall, next to the church:

> The clock in the church tower was
> erected by Mary and Eliza Baines
> in loving memory of their dear
> brother Robert White Baines
> who died may 31ˢᵗ 1887.

Clock Details

Details of the present clock:

Maker:	John Smith & Sons of Derby
Signed:	'Pinney & Son Uppingham 1887' on the setting dial
Installed:	1887
Memorial:	Robert White Baines of Old Hall, Belton
Frame:	Cast-iron flatbed
Trains:	Going and striking
Escapement:	Pinwheel
Pendulum:	Wooden rod and cast-iron cylindrical bob
Rate:	54 beats per minute
Striking:	Countwheel hour-striking
Weights:	Cast-iron
Winding:	Hand wound weekly
Dial:	Skeleton dial with blue background and gilded roman numerals and hands
Dial erected:	Probably 1887
Location:	South face of the tower
Notes:	1. Smith's plaque has been removed from the frame
	2. Replaced an earlier clock and dial but few details are available

BELTON HOUSE

Belton House was built sometime between 1780 and 1830. There is a small house bell hung outside near to the kitchen. When the house was sold *circa* 1970 the stables were converted into private accommodation. The stable clock turret was retained but the movement removed.

BELTON GARAGE

A birdcage movement with anchor escapement, now in Rutland County Museum, was the stable clock at Loddington Hall, Leicestershire, until about 1900. The late Mr A Ringrose, the proprietor of Belton Garage, subsequently acquired it, together with its hemispherical clock bell by Thomas Eayre II, dated 1753. He installed the clock in a barn to the rear of his garage, and on his retirement he donated it to the museum where it is now on display (1969.428). The bell has not survived (*see* Chapter 3 — Clockmakers, Thomas Eayre II). The single-handed dial which he made for this clock has now been placed on the west facing gable of Mill View in Back Lane, Belton, which occupies the site of the former Belton Garage. It now has a synchronous electric movement.

BELTON VICARAGE & OTHER BUILDINGS

A table sundial used to be in the garden of the former Vicarage at SK 815011 and is shown on the early 1900s OS Second Edition 25 inch map. It is also shown on a

The turret clock over the former stables to Belton House

postcard of the same period. Until 1997 there was a sundial in the gardens of Westbourne House, to the west of Back Lane. It is understood that it was originally in the grounds of Barleythorpe Hall. There are other early sundials in the village including one above the door of Southview Cottage in Main Street. The numerals and lines are missing, but the gnomon holes were evident before it was covered over by a modern dial. This house was occupied by John Grocock, churchwarden and bellringer in the early 1900s. There are remains of another sundial on the rear wall of Gorse View in Nether Street.

The birdcage clock movement by Thomas Eayre II from Loddington Hall stables. It is now on display in Rutland County Museum (1969.428)

BISBROOKE

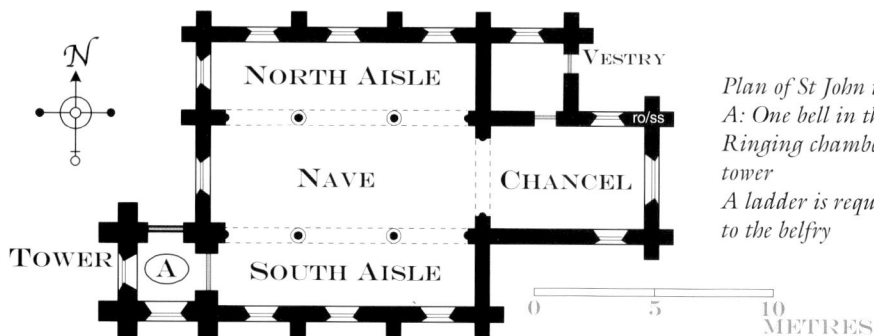

Plan of St John the Baptist's Church
A: One bell in the belfry
Ringing chamber at the base of the tower
A ladder is required to gain access to the belfry

ST JOHN THE BAPTIST   Grid Ref: SK 887996

PARISH RECORDS
There is a continuous run of Churchwardens' Accounts for 1798-1869 and 1869-1981 (DE 2902/19 & DE 5048/2).

BELL HISTORY
The present church was built on the site of the former medieval church which was demolished in 1871. A late eighteenth-century drawing of the church (RCM F10/1984/6) shows that it then had a double bellcote with a single bell over the west gable. A photograph of 1870 shows only a single bellcote in this position. This photograph, which was taken just before the church was demolished, also shows that the bell had a wheel and that weather hoods protected the bell and its fittings.

The parish records contain very little information

A late eighteenth-century drawing of Bisbrooke Church showing a double bellcote with one bell (RCM F10/1984/6)

An 1870 photograph of Bisbrooke Church from the north-west showing that it then had a single bellcote with weather hoods (Robert Boyle)

about the bell, apart from frequent expenditure on replacing the bellrope. In 1803 a Mr Wade was paid a bill for the bell amounting to £7 3s 2d. This probably refers to the reconstruction of the bellcote and the rehanging of the bell.

This earlier bell was larger than the present one and by the middle of the 1800s it was cracked. Apparently it was carefully repaired with putty and then painted. Understandably, this was not an effectual repair and the bell was subsequently recast in 1871 for the new church, the founder being paid £17 together with the metal from the old bell (North 1880, 122). This bell was fitted with a new headstock by Nicholas Meadwell in 1991 when it was rehung with new fittings.

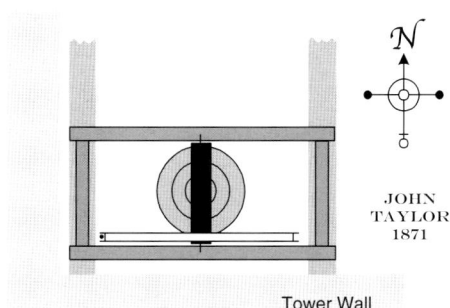

Plan of the timber bellframe at Bisbrooke Church. It rests on two timber joists in the south-west corner of the belfry

Bell Details

1 **1871**. Diameter 438mm (17¼in). Weight about 63kg (1cwt 1qr). Cast with canons. Timber headstock. Recast by John Taylor & Co of Loughborough.

J. TAYLOR & CO LOUGHBOROUGH 1871

Former bell
No details of this bell have survived.

Bellringing Customs
A bell was rung to call people to Vestry Meetings as well as for the annual 'Duke's Court', and the Gleaning Bell was rung during harvest until 1871. Before the Death Knell there were three tolls for a male and two tolls for a female (North 1880, 122).

Today the single bell is rung for five minutes before each service.

The restored Bisbrooke church bell in its frame in 1991 (Nicholas Meadwell)

BRAUNSTON IN RUTLAND

ALL SAINTS   Grid Ref: SK 832066

Parish Records
Churchwardens' Accounts, receipts and other documents survive from 1793. The periods covered are 1793-1821 (DE 2249/17), 1857-92 (DE 2249/18), and 1892-1936 (DE 2249/19). There is also a bundle of receipts from 1803 to 1842 (DE 2249/45-76). These documents provide an excellent record of expenditure in connection with the bells. There are no references to a clock prior to the installation of the present movement and dial in 1879.

Bell History
There were four bells at Braunston Church in 1880 and this number has since been increased to six. The oldest bell is the present third ascribed to Thomas Newcombe II of Leicester, *circa* 1570.

The earliest references to the bells are included in a

letter and attached notes to the churchwardens from the Ven E A Irons dated 18 April 1904 (loose papers in DE 2249/17). In each case the repair may refer to the bell fittings rather than to a bell requiring recasting:

1613 May 15 Robert Lewin and Kellum ffowks, wardens, are reminded to certifie by St. John Baptist's day of the repair of the bell that is broken.
1614 June 3 The wardens are warned to repaire the churche and to amend the bell ther broken.

At this time there were at least two bells in the belfry, the present third cast *circa* 1570 and the present fifth cast *circa* 1601. Thomas Norris and William Noone added the present tenor and fourth bells in 1660 and 1710 respectively. These may be recastings of earlier bells.

In 1914 the then ring of four was rehung in a new steel and iron low-sided frame for six by John Taylor. All four early bells have lost their canons and are now bolted through the crown to cast-iron headstocks A new treble

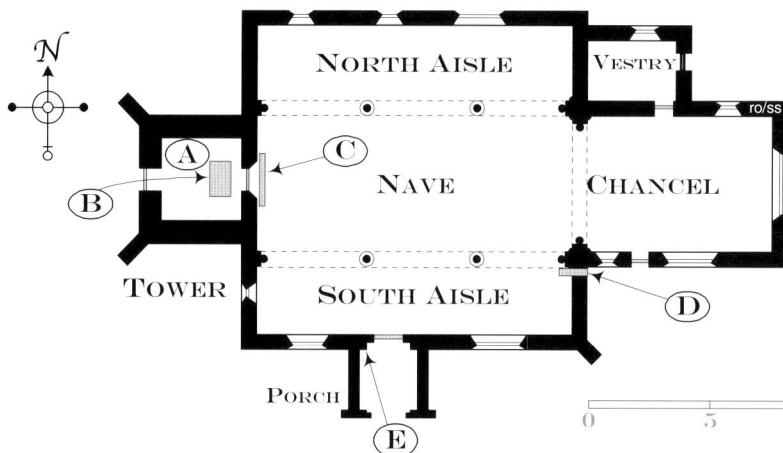

Plan of All Saints' Church
A: Six bells in the belfry
Ringing chamber at the base of the tower
B: Clock on the first floor of the tower
C: Clock dial over tower parapets
D: Location of former sundial
E: Scratch dial
Access to the clockroom and belfry is by ladders

and second were cast in 1967 by the Whitechapel Bell Foundry to make the present ring of six. The parishioners raised over £700 to pay for the new bells and the Dedication Service, held on 5 October 1968, was conducted by the Rev Ernest Orland, the then president of the Peterborough Guild of Church Bellringers.

Payments to ringers, repairs to the bells and the provision of new bellropes, as in most other parishes, were a constant drain on the financial resources of the churchwardens. However, in Braunston, help may have been available from one of its charities:

By an indenture dated 19 April 1636 a piece of land called the Wisp was conveyed to certain persons, the rent to be applied towards the maintenance of a preacher to preach in the chapel at Braunston, or in default thereof for and towards the repairs of the parish church of Braunston and the bells therein ... (*VCH* **II**, 37).

BELL DETAILS

1 1967. Treble. Note E. Diameter 679mm (26¾in). Weight 201kg (3cwt 3qr 24lb). Cast without canons. Cast-iron headstock. Cast by Mears, Whitechapel Bell Foundry, London. Low on the waist:

MEARS
19 [59] 67
LONDON

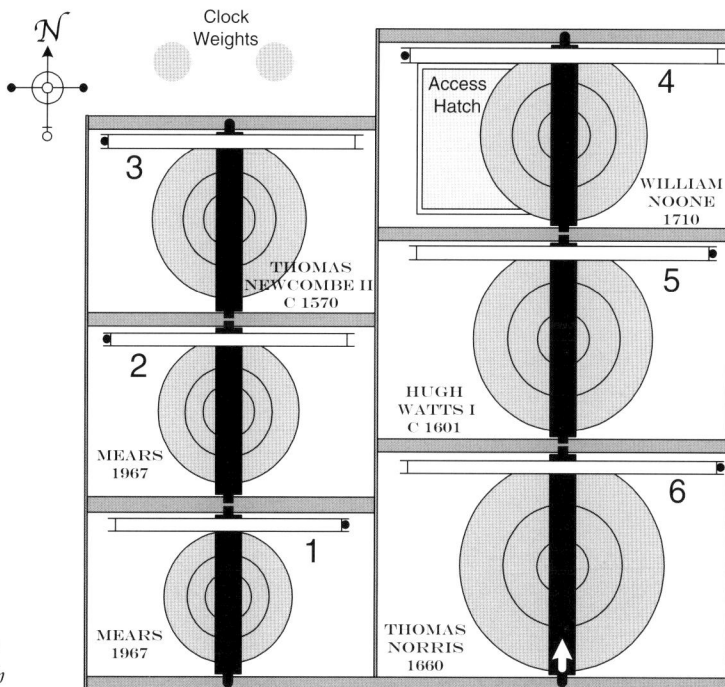

Plan of the bellframe at Braunston Church

Whitechapel Bell Foundry delivers one of the new bells to Braunston Church in 1967 (John Beadman)

2 1967. Note D. Diameter 724mm (28½in). Weight 237kg (4cwt 2qr 19lb). Cast without canons. Cast-iron headstock. Cast by Mears, Whitechapel Bell Foundry, London. Low on the waist:

MEARS

19 [59] 67

LONDON

3 *Circa* 1570. Note C. Diameter 791mm (31⅛in). Weight 279kg (5cwt 2qr). Canons removed. Cast-iron headstock. Ascribed to Thomas Newcombe II of Leicester.

[2] S T ɦ O Μ ₳ [27]

(St Thomas)

Letters and marks are spaced evenly around the inscription band. All letters are like [108]. Scheduled for preservation (Council for the Care of Churches).

The lettering and decoration on the fourth bell at Braunston

4 1710. Note B. Diameter 838mm (33in). Weight 306kg (6cwt 0qr 3lb). Canons removed. Cast-iron headstock. Cast by William Noone of Nottingham.

W RAWLINGS [40] T BRYON [40]

~ WARDENS [40] 1710 [40]

Below the inscription is a band of decoration [42].

5 *Circa* 1601. Note A. Diameter 902mm (35 ½in). Weight 406kg (8cwt). Canons removed. Cast-iron headstock. Cast by Hugh Watts I of Leicester.

 PℝₐISℲ Tɦϑ JOℝDϑ [3]

All letters are like [16] and [17].

6 1660. Tenor. Note G. Diameter 1003mm (39½in). Weight approximately 508kg (10cwt). Canons removed. Cast-iron headstock. Cast by Thomas Norris of Stamford.

[26] [52] THOMAS [52] NORRIS [52]

~ MADE [52] MEE [52] 1660 [52]

The clock strikes the hours on this bell.

BELLRINGING CUSTOMS

On Sundays a bell was rung at 8am if there was to be Divine Service, after which the day of the month was tolled. If there was no morning service the day of the month was tolled at noon. For Divine Service the bells were chimed for twenty minutes, followed by the Sermon Bell. After the service, another bell was rung if Evensong was to be said. A Gleaning Bell was rung during harvest and the Pancake Bell was rung on Shrove Tuesday until about 1875. At the Death Knell three times three tolls were given for a male and three times two for a female, both before and after the knell. The relatives could also request that the age of the deceased be tolled, but this custom had ended before 1880. There was an ancient custom here to have three tollings before a funeral. The first was to give warning, the second an hour later was to call the bearers together, and the third, again an hour later, was for the funeral (North 1880, 123).

The following are examples from the Churchwardens' Accounts stating when the bells were rung in celebration:

1799 May 30	Paid Amb: Ruddell 5s for Nelson & 5s for Warrens Victorey	10s 0d
1808 Sept 12	pd the Ringers Ale for good News [Peninsular War — defeat of the French by Wellington 27 Aug 1808] Ringers Ale	5s 0d 12s 6d
1814 June 10	Paid Ringers & Singers for Peace £1 5s 0d	
1857 Nov	Ringers on celebration of Gunpowder plot	5s 0d
1862	Ringers in celebration of Xmas & New yrs day	10s 0d

The accounts include regular payments to the ringers from 1793 to the 1840s. Receipts reveal that the ringers were supplied with ale, tobacco and cheese for their services. The last record in the Churchwardens' Accounts for ringing the bell 'on celebration of Gunpowder plot' was in 1862. The current church leaflet states that the benches in the porch seated those who were entitled to bread and dole at Christmas when a bell was rung at noon on St Thomas' Day (21 December).

Today Braunston Church has a band of bellringers and there is a regular practice on Wednesdays, in rotation with other local churches. The bells are rung for all Sunday services and for the Easter Sunday and Christmas Day services. They are also rung during New Year's Day and at midnight on every third year. On Remembrance Sunday the bells are rung half muffled for the 10am service.

They are rung for the Dedication Service on the occasion of the village's annual event known as 'The Happening'. If requested they are rung for weddings and funerals.

Scratch Dial

There is a well preserved early scratch dial inside the south porch, to the west of the main door. It is on the capital stone to the column. The south doorway was probably built in the thirteenth century.

Sundial

A *circa* 1839 drawing of the church shows a sundial set high on the east end of the nave wall (Uppingham School Archives). All traces of this dial have disappeared.

Clock History

The present clock, installed in 1879, is almost certainly the first clock in the church. It was installed at the cost of Mr Evan Hanbury, a local landowner and later High

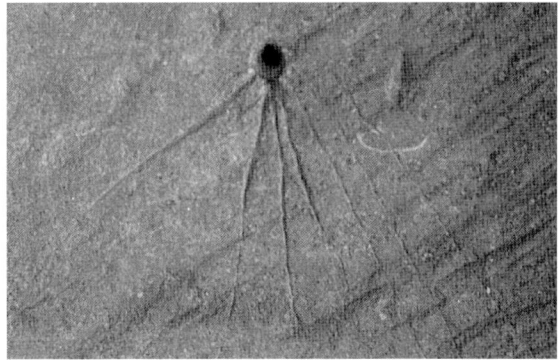

The scratch dial near the south door at Braunston Church

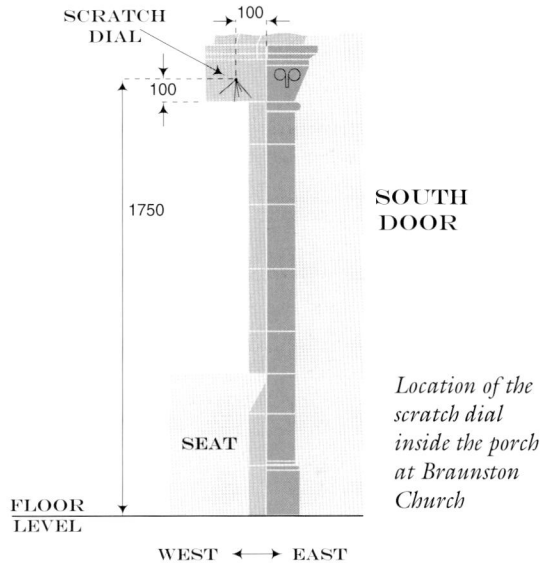

Location of the scratch dial inside the porch at Braunston Church

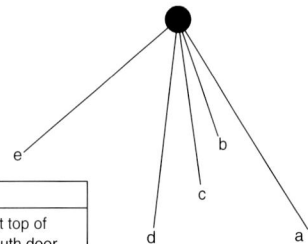

BRAUNSTON					
Location	Inside porch - at top of west jamb of south door				
Condition	Good				
Gnomon Hole Diameter			12		
Gnomon Hole Depth			15		
Height above ground level			1750		
Line Ref	a	b	c	d	e
Length	120	60	80	100	100
Angle (°)	147	160	172	187	230

Scratch dial details

From a circa 1839 drawing of Braunston Church showing a sundial on the south wall of the nave (Uppingham School Archives)

Sheriff of Rutland, who stipulated that the clock should face east so that his employees could see the time from the Brooke road. The Churchwardens' Accounts show that the parish also made a contribution:

1879 Towards the Church Clock as arranged by
Parish meeting   £10 0s 0d

Richard Buckby was the village carpenter and the following payment was probably for the pine clock case:

1880 Dec 11   Mr Buckby, for work belonging to
the Church Clock   £8 11s 0d

From 1882 Charles Payne of Market Square, Oakham, was responsible for maintaining and repairing the clock, and an item in 1899 may refer to painting the clock dial:

1895 April 18 Payne, attending & correcting
church clock   £1 10s 0d
1899 March 28 Munton, Bill for Painting
Church Clock   12s 9d

From 1901 Robert Corney, 'Manufacturing Jeweller, Silversmith, Watch and Clock Maker' of The Clock House, High Street, Oakham (*Matkin's Almanack* 1902, 75), and his successors, carried out repairs on the clock at frequent intervals up to 1940. For example:

1901 April 15 Corney, Repairing & cleaning
Church Clock   15s 0d

In 1999 the clock and dial were refurbished as a Millennium project. Prior to this the clock had not struck the hours for many years.

The Gillett & Bland clock movement at Braunston Church

CLOCK DETAILS

Details of the present clock:

Maker:	Gillett & Bland of Croydon
Signed:	'Gillett & Bland, Croydon' on the setting dial and on the frame
Installed:	1879
Donated by:	Mr Evan Hanbury with a donation of £10 by the churchwardens
Frame:	Cast-iron flatbed
Trains:	Going and striking
Escapement:	Deadbeat
Pendulum:	Wooden rod and cast-iron cylindrical bob
Rate:	48 beats per minute
Striking:	Countwheel hour-striking
Weights:	Cast-iron
Winding:	Hand wound weekly. Both trains have geared winding and bolt and shutter maintaining power
Dial:	Solid round cast-iron dial with gilded hands and roman numerals on a blue background
Location:	On the tower parapets facing east, looking over the body of the church
Note:	Until 1999 the dial was white with black hands and numerals

Braunston Church in 2000 showing the unusual location of the clock dial

BROOKE

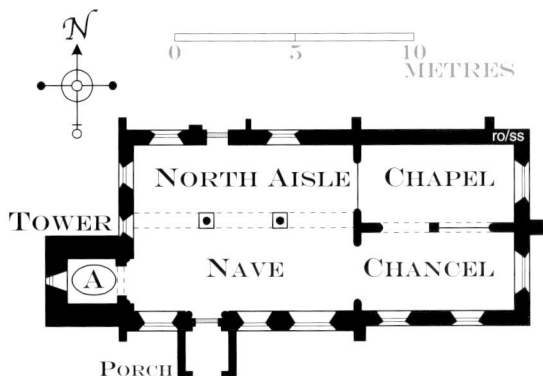

Plan of St Peter's Church
A: Six bells in the belfry. Ringing chamber at the base of the tower. Access to the belfry is by ladders

ST PETER
Grid Ref: SK 850057

PARISH RECORDS

Relevant surviving documents include the Churchward-ens' Accounts 1869-1921 (DE 2250/8), Vestry Minute Books 1787-1869 (DE 2250/9) and 1871-1905 (DE 5022/3), and the PCC Minute Book 1921-74 (DE 5022/4). A faculty dated 11 November 1991 (DE 5039/6) relates to the new bellframe installed in 1992. A copy of *The Book of Bells* (DE 5390), a celebration in words, photographs and watercolours of the appeal and fund raising for the restoration of the bells in 1992, can be inspected in the church.

BELL HISTORY

The earliest bell is the present second by Toby Norris I of Stamford cast in 1610. Prior to 1992 there were four bells in an oak frame which was thought to date from the sixteenth century (Pearson 1989). The frame had been untouched since 1811 and the bells had become unringable before 1938 (Powell 1938). However, although the early documents are almost devoid of references to the bells it seems that they were being used until at least 1891 when '3 New Bell Ropes' were purchased for £1 7s 0d.

The PCC Meeting of 24 January 1922 discussed an architect's report in which it was suggested that a new timber frame should be installed. Although the meeting agreed to this no action was taken. The bells were again discussed at the PCC Meeting on 27 April 1959. Here it was indicated that a 'considerable sum' was needed for belfry repairs. An architect's report considered at the May meeting of that year stated that the belfry was in better condition than had been anticipated. Again, no action was taken.

In 1991 the Beadman family of Braunston offered the gift of a new tenor subject to the restoration of the existing four bells. Nicholas Meadwell, then Steward of the Rutland Guild of Bellringers, offered his services free of charge to refurbish the belfry. This latter gift saved £10,000 thus making the venture possible. As a result of these offers, the PCC Meeting of 1 May 1991 decided to start a fund-raising programme, the target being £15,000. £3,000 was expected from trusts and the remainder was to be raised by the parish. The prospect of having the church bells in ringing order again after almost a century of silence was an inspiration to many.

Initially it was proposed to replace the timber frame and install the four original re-tuned bells and the new tenor in a steel frame for six. The sixth bell, a new treble,

was to be added at a later date as funds allowed. However, fund raising was so successful that the PCC were able to order both bells at the same time.

The old frame had to be fully recorded and carefully removed so that it could be preserved and, if necessary, re-assembled elsewhere. It was discovered during removal that the whole structure of the frame was very close to collapse. The new bells were cast at Taylor's Bell Foundry in Loughborough on 3 April 1992. They were delivered to the church on 1 May by engineers from the Aerial Erector Flight of the Mechanical Engineering

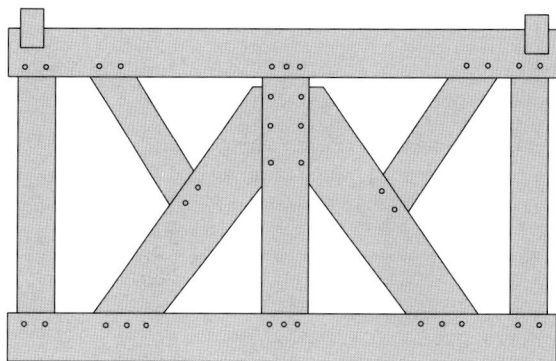

Drawing of a truss from the old oak bellframe at Brooke Church

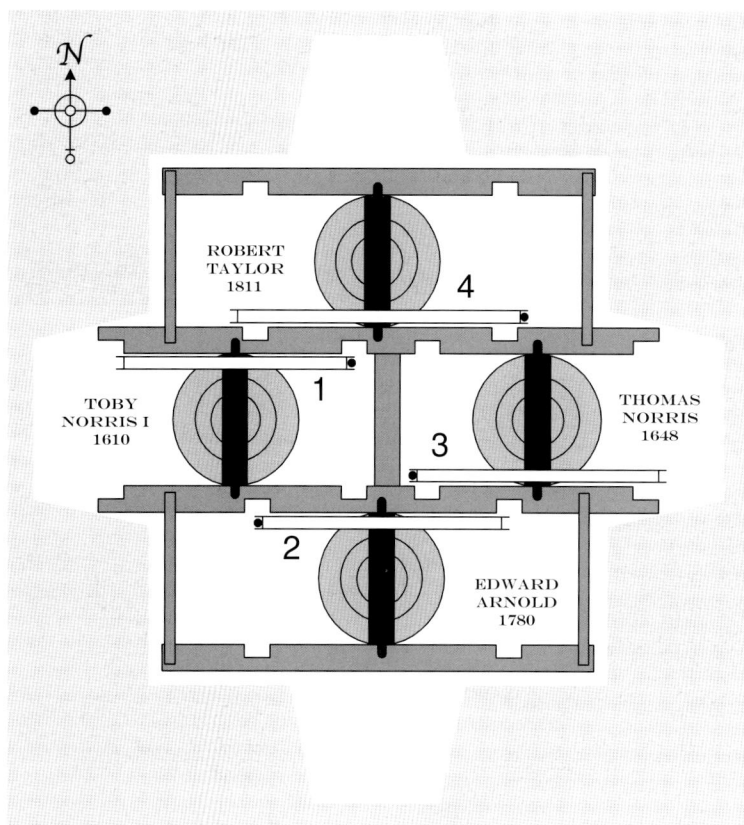

Plan of the old timber bellframe at Brooke Church. Note the cut-outs in the side members to allow the bells to ring full circle

One of the Brooke bells starts its journey to Loughborough (Nicholas Meadwell)

Casting the new treble for Brooke Church (John Beadman)

The new treble arrives in Brooke (Harold Killingback)

(above, left) John Beadman helps to manœuvre the bell that his family donated into Brooke Church (Harold Killingback)

(above, right) The new tenor being hoisted into the tower of Brooke Church (Harold Killingback)

(left) The new tenor and treble at Brooke Church ready for the first peal (Harold Killingback)

Squadron based at RAF North Luffenham, who provided all the heavy transport free of charge. Work on installing the new frame and hanging the refurbished and new bells was completed by the end of May. Archdeacon Bernard Fernyhough dedicated them on Friday 5 June 1992.

Bell Details

1 **1992**. Treble. Diameter 597mm (23 ½in). Weight 143kg (2cwt 3qr 8lb). Cast without canons. Cast-iron headstock. Cast by John Taylor & Co of Loughborough.

The canon-retaining headstock now fitted to the second bell at Brooke

THOU ART PETER [54]

On the waist:

CHARLES MAYHEW VICAR
JOAN NORTON HAROLD KILLINGBACK
~ CHURCHWARDENS

On the waist, opposite:

19 [56] 92

Decoration [66] is below the inscription band.

2 1610. Diameter 635mm (25in). Weight 147kg (2cwt 3qr 16lb). Canons retained. Cast-iron canon-retaining headstock. Cast by Toby Norris I of Stamford.

[14] IESVS [39] SPEDE [62] ME [62] CVM [62]
~ VOCO [62] VEИITE [62] 1610

(Jesus speed me, come when I call)

North (1880, 123-4) had this bell and bell 3 reversed [the then treble and second].

3 1780. Diameter 686mm (27in). Weight 171kg (3cwt 1qr 14lb). Canons removed. Cast-iron headstock. Cast by Edward Arnold of St Neots.

EDWD. ARNOLD ST. NEOTS FECIT
~ 1780 [45] O O

O O are the obverse and reverse of an unidentified coin from the reign of George III. When this bell was removed during the 1992 restoration the canons were found to be broken and the top of the bell was cracked. The crack was repaired by welding.

4 1648. Diameter 737mm (29in). Weight 218kg (4cwt 1qr 5lb). Canons removed. Cast-iron headstock. Cast by Thomas Norris of Stamford.

1648 [63]

5 1811. Diameter 806mm (31¾in). Weight 274kg (5cwt 1qr 16lb). Canons removed. Cast-iron headstock. Cast by Robert Taylor of St Neots.

R: TAYLOR FOUNDER. ST. NEOTS. 1811 [67]
~ H: ORTON. CHURCHWARDEN [67]

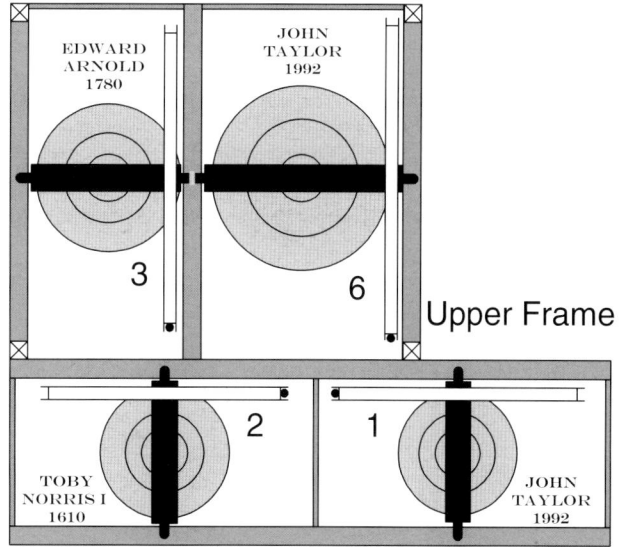

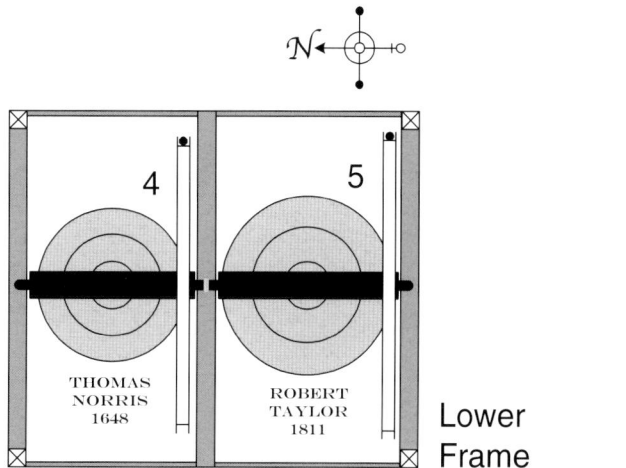

Plan of the 1992 bellframe at Brooke Church

6 1992. Tenor. Diameter 902mm (35½in). Weight 420kg (8cwt 1qr 2lb). Cast without canons. Cast-iron headstock. Cast by John Taylor & Co of Loughborough.

HE THAT HATH EARS TO HEAR
~ LET HIM HEAR [65]

On the waist:

A GIFT OF
THE BEADMAN FAMILY
1992

On the waist, opposite:

19 [56] 92

Below the inscription band is decoration [60]. Above the soundbow is decoration [64].

MEMORIAL

The following is from a framed notice in the tower:

ST PETER
BROOKE
THE STORY OF HOW A TINY VILLAGE CAME TO RAISE £25,000 TO RESTORE ITS ANCIENT BELLS AND TO HANG TWO NEW ONES, MAKING A PEAL OF SIX, IS A MOVING AND EXCITING TALE OF HARD WORK, LOVING CARE AND DETERMINATION. THIS TALE IS RECORDED IN THE "BOOK OF BELLS" WHICH IS HOUSED IN THE CHANCEL. SOME OF THE GIFTS WERE GIVEN IN MEMORY OF LOVED ONES. WE WISH TO PLACE ON RECORD THEIR NAMES.

IN MEMORIAM
JAMES GLENN ELLIS
ALFRED & LUCY JONES
FRANK & KATHLEEN JONES
SIDNEY & EDITH JONES
BASIL JUDD
MAMIE KENNEDY
BETTY KILLINGBACK
BESSIE WHITELAW
JULIA ANNE WRIGHT

CANDLE HOLDER

An egg-shaped face carved in stone is built into the inside north wall of the tower and was used by the bellringers as a candleholder.

Bellringers' candleholder at the base of the tower at Brooke Church

BELLRINGING CUSTOMS

At the Death Knell there used to be three tolls for a male and two tolls for a female. The bells were chimed and the Sermon Bell [tenor] rung for Divine Service (North 1880, 124).

Today the bells are rung for all church services except for the monthly 8am service. They are also rung at Christmas, Easter, New Year and for weddings. On Armistice Day they are rung half-muffled, and they are tolled for funerals if requested.

BURLEY

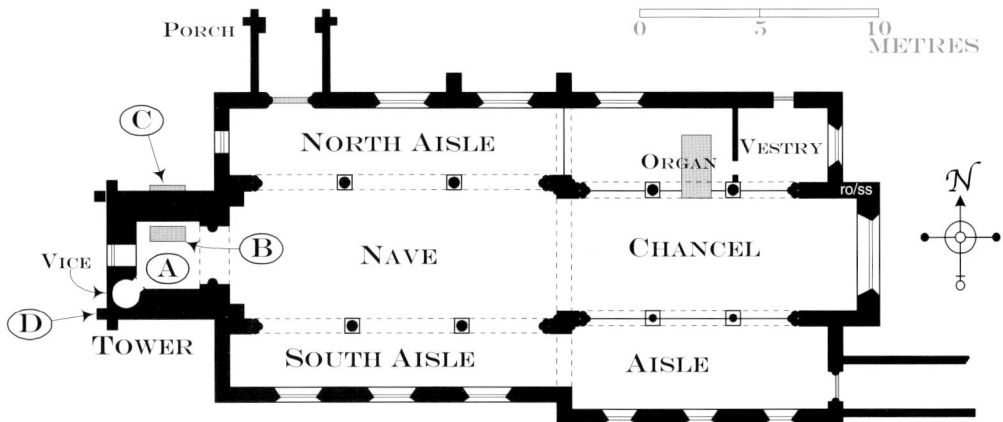

Plan of Holy Cross Church
A: One bell in the belfry. Ringing chamber at the base of the tower
B: Clock on the first floor of the tower
C: Clock dial
D: Scratch dial
Access to the clockroom is by the vice
Access to the belfry is by a ladder from the clockroom

HOLY CROSS

Grid Ref: SK 883102

A redundant church in the care of the Churches Conservation Trust.

PARISH RECORDS

There are no surviving Churchwardens' Accounts, although some accounts relating to church restorations are quoted in *History of Burley-on-the-Hill* (Finch 1900, 16-17). The only relevant parish documents are the Vestry Minute Book 1894-1967 (DE 2756/11), PCC Minute Books 1927-70 and 1971-80 (DE 2756/12-13), a bundle of correspondence dated 1975 and 1976 concerning the restoration of the turret clock (DE 2756/31/1-21), and correspondence from John Taylor & Co regarding the refurbishment of the bell fittings (DE 2756/32/1-3).

BELL HISTORY

The present bell was cast in 1705 by Alexander Rigby who by then had taken over the Norris bell foundry at Stamford. There is a tradition that it was cast from a former ring of smaller bells although there is no documentary evidence to support this. This new bell was probably installed as part of the restoration programme for the church begun in 1700.

Originally the present bell was hung by its canons from a timber headstock with a wheel for full-circle ringing. In 1978, a survey report by John Taylor stated that the bell fittings were in poor condition, the headstock bearings were worn out and the bellwheel was broken. However the bellframe was found to be in good condition. It was proposed to rehang the bell in the existing frame but with a new cast-iron headstock and a new wheel. In the event, the bell was hung dead from a steel joist placed across the frame and a chime hammer installed. During this work the canons and the iron clapper-staple were removed. The old timber headstock is preserved at the base of the tower.

BELL DETAILS

1 **1705.** Diameter 1099mm (43¼in). Weight 813kg (16cwt 0qr 1lb). Canons, clapper and staple removed. Chime hammer. Bell suspended from steel beam. Cast by Alexander Rigby of Stamford.

> [14] [57] ALEXANDER [57] RIGBY [57]
> ~ MADE [57] ME [57] 1705 [57]
> [14] [57] BVRLEY [57] IN [57] RVTLAND [57]

The largest surviving bell by this founder and his only surviving bell in Rutland. Scheduled for preservation (Council for the Care of Churches).

CLOCK BELL

There was a clock bell here sometime before 1880 but it is said that it was taken down and sent to the rectory at

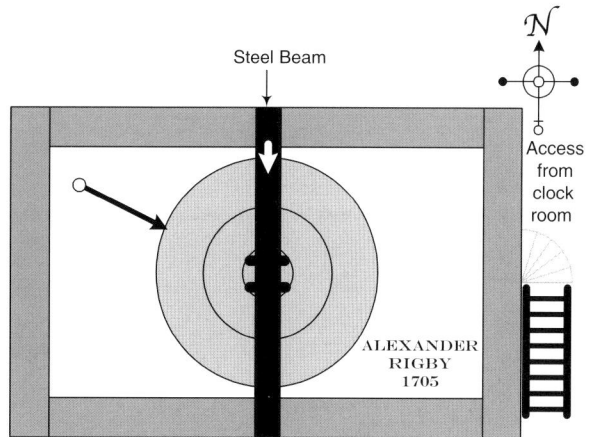

Plan of the timber bellframe at Burley Church

Oakham (North 1880, 124). Beyond this, its fate is not recorded. A *circa* 1793 drawing of the church shows an open framed cupola on the tower carrying a small bell (RCM F10/1984/9). This is almost certainly the clock bell referred to here and the unusual position may have been chosen so that it could be heard from the mansion and by estate workers. A later drawing of the church (Uppingham School Archives) shows that the cupola and bell had been removed by 1838. Clock bells were also placed outside the tower at Ashwell, Glaston and Morcott.

A conjectural drawing of Holy Cross as it might have appeared circa 1703, soon after the cupola had been installed (Canon John R H Prophet). This is based on a circa 1793 drawing of the church (RCM F10/1984/9)

BELLRINGING CUSTOMS

At the Death Knell there used to be thrice three tolls for a male and thrice two tolls for a female, both before and after the knell. The same number of tolls were given before and after a funeral (North 1880, 124).

Although the church is now in the care of the Churches Conservation Trust, services are occasionally held here and the Rigby bell is chimed for a few minutes prior to the start of them. It is also chimed at a funeral if requested by the bereaved family.

SCRATCH DIAL

There is a crudely constructed scratch dial on the west-facing buttress, at the south-west corner of the four-teenth-century tower. It has been relocated from a south-facing position.

SUNDIAL

The following item in the accounts of the 1700 restoration may refer to a sundial (Finch 1900, 17). A sundial would have been required to regulate the church clock:

1703 The Church Dyall whitewashing   6d

CLOCK HISTORY

Parliamentary Forces destroyed Burley on the Hill in 1645. Daniel Finch, the second Earl of Nottingham, purchased the ruined mansion and the estate from the Duke of Buckingham in 1694. He immediately began preparations to build another house, the main body of

which was completed by 1700. It took several more years to complete the internal work, to landscape the gardens and to restore the church.

The following extracts from the Church Restoration Accounts (Finch 1900, 16-17) probably indicate when the clock was installed. Note that 'finger board' is an old term for clock dial and that the hands were often referred to as 'fingers'. The 'frame for ye couples' was the cupola shown in the *circa* 1703 conjectural drawing:

1699 Mr Norman, ye carpenter for work at ye church
         clock & finger board   9s 0d
1703 Mr. Norman ye Carpenter for fixing ye Floor of
         ye clock in ye steeple   12s 0d
         For sawing & workmanship above the frame for
         ye couples [cupola] on ye Steeple   £3 15s 6d

The clock is signed 'Joseph Knibb London 1678' and is by far the most important church clock in Rutland, being one of only five known turret clocks by this eminent maker. Knibb was involved with the invention and development of the anchor escapement and consequently the clock at Burley has attracted a great deal of attention in horological circles.

Knibb enjoyed royal patronage and one of his turret clocks was placed over the State Entrance to the quadrangle at Windsor Castle. On it was his signature 'Joseph Knibb Londini 1677'. This clock was removed in 1829 and there is no trace of it now. It has been suggested that the Burley clock was the twin of the Windsor Castle clock but there is no evidence to support this (Jagger 1983, 27-9).

The following entry recorded in the Secret Service Payment Book (Jagger 1983, 27-9) may indicate the cost of the clock and its installation at Windsor Castle:

1682 to Mr. Knibb by His Said Majesty's Command
         upon a bill for clock work   £141

The Earl of Nottingham held several high ranking positions, including that of First Lord of the Admiralty from 1679/80 to 1684 and Secretary of State for War from 1688 to 1693. It was probably through these con-

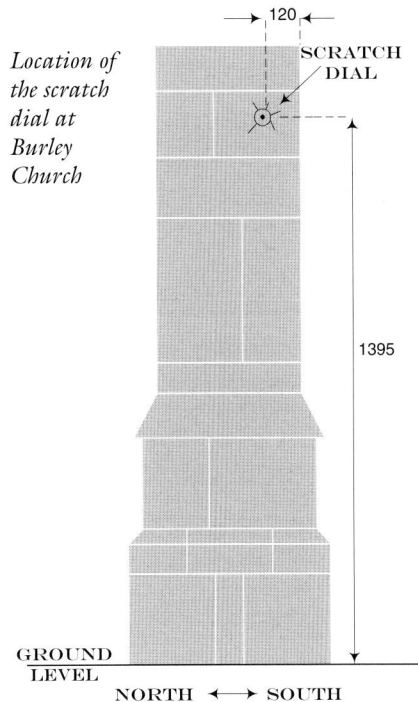

Location of the scratch dial at Burley Church

Scratch dial details

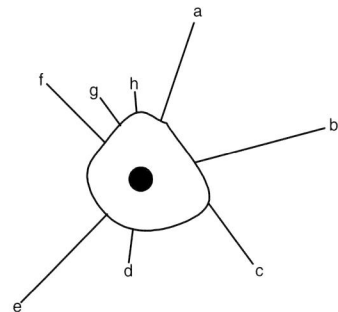

BURLEY								
Location	West facing buttress at south-west corner of the tower							
Condition	Good							
Gnomon Hole Diameter			12					
Gnomon Hole Depth			13					
Height above ground level			1395					
Line Ref	a	b	c	d	e	f	g	h
Length	50	65	36	15	60	40	15	10
Angle (°)	20	75	150	190	225	315	340	355

nections that he came to acquire a clock made by Knibb.

There are no early references to the clock in the surviving parish records. By 1975 it had long ceased to be in working order and it was in a sad and neglected state. A letter to the editor in *Antiquarian Horology* (Pearson 1975, 580) records the 'discovery' of the clock. The following is an edited version:

Dear Sir,

Now that full publicity has been given to the Knibb clock at Burley, Rutland, it may be an opportune time to provide details of the original discovery of this rarity ....

Being stationed at RAF Cottesmore I had noticed Burley church hidden away in its clump of trees overlooking Oakham, on many occasions. After several months the urge to discover if the church possessed a clock became too strong to ignore.

I arrived outside the churchyard gates one Saturday in September, 1969 and enquired at the little cottage adjacent to the gate as to whether I could investigate the tower. I was given the "go ahead" with a caution that the tower and flooring might be unsafe! Undeterred I took the keys and crossed the churchyard, pausing to study the exterior of the tower. Only the bare ring and a couple of iron stays remained to show there had once been a dial on the North face.

Once inside I found the doorway to the staircase and ventured up. At the first level I encountered a stout floor - stout it must have been as two enormous lead weights and a couple of equally sturdy pulleys were resting in a corner having (apparently) broken their ropes and fallen from higher up. I figured any floor taking that kind of punishment would also carry me — so I examined the weights. They were too heavy to move and I continued up to the clock floor. There I found a wooden clock case or box housing an open frame movement of obvious antiquity. With mounting excitement I noticed that the escapement was equipped with a curious sliding linkage to a central pendulum, then I noticed the nut under the pendulum bob was fashioned with exaggerated rounded "wings" like some Knibb long case bell nuts.

Further examination showed the clock to have been of exceptional quality compared with the average tower clock, the wheels were finely finished under the dirt and corrosion. On looking round the frame bars you can imagine how I felt on seeing an inscription on the left hand end of the facing upper flat bar. Shining my hand lantern on this area and rubbing with my palm revealed the now well known inscription in the form:

Joseph Knibb
London 1678

On inquiring about the church authorities (I couldn't believe no-one was aware of this treasure) I was put in touch with a Mr. Hoare (then churchwarden) ....

During my first meeting with Mr. Hoare I commenced in enthusiastic terms to describe my find - he responded in a classic manner by saying "You will be telling me we have a Knibb or something in our church in a minute"! From this you will gather he has some knowledge of clocks, but he had no idea what was contained in the Burley church!

I returned to photograph and sketch the clock and layout ....

After some thought I wrote to Dr. Beeson and asked if he could arrange a survey and some official approach to the Church, with a view to saving the clock. This, I presume, resulted in the current appeal.

The dial of this clock is curiously positioned. I have found that, usually, rich patrons caused "their" clock dial to face either their own residence, or the most obvious, ostentatious direction possible. This is not so here ....

Yours faithfully

BRIAN PEARSON

In 1975 a decision was made to restore the clock and an appeal was launched to cover the cost of restoration. David Nettle of the AHS acted as adviser and co-ordinator for the project. The Clocks Sub-Committee of the Council for Places of Worship offered a grant of £900 from the Hayward Foundation and the remainder of the total cost of £1,585 was raised by donations and fund raising events. John Smith & Sons of Derby carried out the restoration work. A report on a visit to Smith's works in January 1976 by the Turret Clock Group of the AHS gives some details of the work required:

A number of mechanical turret clocks were in for overhaul and the centre piece of the whole visit was the clock by Joseph Knibb dated 1678, from Burley, Rutland. This

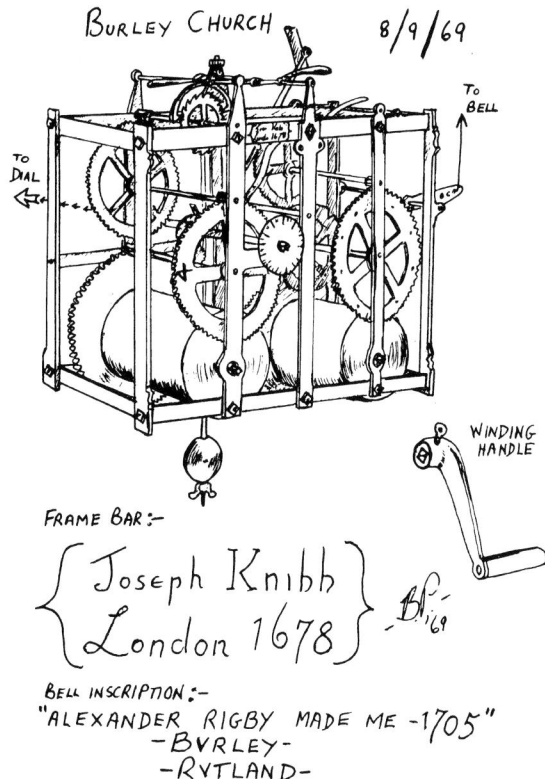

Brian Pearson's sketch of the Knibb clock at Burley Church (Pearson 1975, 581)

Knibb's signature on the clock frame at Burley Church
(Chris McKay)

*The counterpoised sliding rods which link the pallet arbor
to the crutch on the Knibb clock at Burley* (Chris McKay)

clock had been overhauled with great skill and feeling and
was on its final tests before returning to Burley. The works
had finished the clock in black, but had painted green those
parts which were not original. Some of these, for example
the fly, were renewed during the overhaul; others have been
provided in the past and clearly were not by Knibb's hand.
The clock now has a pair of automatic winders and a glass
fibre dial. An interesting discovery during the overhaul was
the presence of a blacksmith's or iron master's mark inside
the front cross bar. Unfortunately it is very indistinct and
cannot be identified (Nettle 1976, 701).

The clock was returned to the church on Monday 26
January 1976 and a Service of Thanksgiving for its restoration was conducted by the Rev J Marshall on Sunday
29 June.

CLOCK DETAILS

Details of the present clock:

Maker:	Joseph Knibb, London
Signed:	On the frame: Joseph Knibb London 1678
Installed:	*Circa* 1700
Cost:	Not recorded
Frame:	Wrought-iron birdcage frame, 686mm (27in) wide, 610mm (24in) high and 419mm (16½ in) deep
Trains:	Going and striking
Escapement:	Anchor
Pendulum:	Iron rod and lead bob suspended inside the frame

*The pendulum suspension bracket
and rating nut on the Burley clock*
(Chris McKay)

*Inspecting the clock on the day of its return: Mr and Mrs Joss Hanbury, then
owners of Burley on the Hill, Jack Gale, carpenter, and Raymond Hill the
churchwarden at Holy Cross, Burley* (Leicester Mercury)

Rate:	60 beats per minute
Striking:	Locking plate hour striking
Weights:	Originally cylindrical lead weights. Sold for scrap on conversion to automatic winding
Winding:	Smith's epicyclic automatic winding units
Dial:	Round glass-fibre dial with gilded hands and roman numerals on a black background
Location:	On the north face of the tower
Note:	An earlier dial was 25mm (1in) thick plate glass. The original dial may have been wooden and diamond shaped with a single hour hand

The clock has some unusual features. The pallet arbor communicates with the crutch arbor by means of two sliding rods linked together in the middle and counterpoised at the opposite ends. The crutch arbor is pivoted between an extra pair of frame bars and both the crutch and the pendulum hang inside the frame. The suspension bracket on one of these bars carries the pendulum rating nut which has a brass ring engraved from 1 to 12. The bracket is decorated with a fleur-de-lys which acts as an index. The lifting and locking levers are also pivoted between these frame bars. These interact through short cam-shaped stubs. The second wheel in the going train bears the numbers 1 to 4 for quarters and circles for eighths.

When the restored clock was returned to Burley it was fitted with two new automatic winding units by Smith of Derby (Chris McKay)

CALDECOTT

ST JOHN THE EVANGELIST   Grid Ref: SK 868937

PARISH RECORDS
The Churchwardens' Accounts 1807-1922 (DE 4195/14), the Annual Vestry Minute Book 1905-75 (DE 5272/12) and the PCC Minute Book 1934-82 (DE 5272/13) are the only relevant sources in the parish records. All three include useful historical information concerning the bells and the church clock.

BELL HISTORY
Until 1951 there were five bells in an old oak frame, all cast in 1696 by Toby Norris III of Stamford. As reported by J Tailby (*Gentleman's Magazine Library* IX, 266) the bells were very lucky to have escaped serious damage during a thunderstorm in the early hours of Sunday 30 July 1797:

... about six the lightning was remarkably frequent and vivid. In the short intervals between each flash the dark-ness appeared as great as at midnight, and torrents of rain poured down. The claps of thunder, particularly three, were tremendous and awful, causing the bold as well as the timid to be astonished and tremble.

At about this time the top of the spire came down causing considerable damage to the tower:

The frames and wheels of the bells are also shattered, so as none at present can be used except the fourth, which can only be tolled. Whether the bells are damaged is not at present known, it being very dangerous to go up to them; for when I took this account (the wind being high), the stones and mortar kept frequently falling upon the leads and in the churchyard.

The top of the old spire now stands near the porch.

Church records which may have given details of repairs as a consequence of this catastrophe have not survived, but the frame was obviously repaired or renewed and the bells rehung. The Churchwardens' Accounts imply that the bells were used for the whole of the

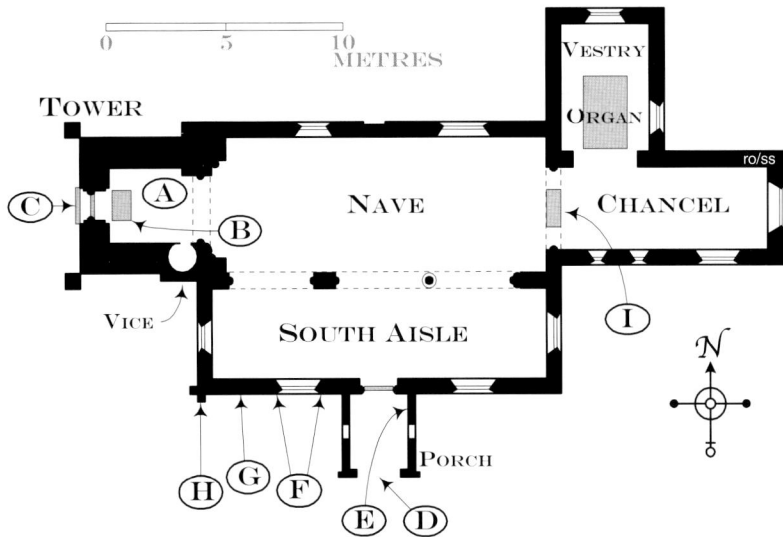

Plan of St John the Evangelist's
Church
A: Six bells in the belfry
Ringing chamber at the base of
the tower
B: Clock on the first floor of the
tower
C: Clock dial
D: Sundial
E: Sundial dedication plaque
F: Scratch dials 1 & 2
G: Scratch dial 3
H: Scratch dials 4 & 5
I: Location of bellcote removed
in 1976
The vice provides access to both
the clockroom and the belfry

nineteenth century, particularly during the first half, when the bellropes were repaired or replaced every few years. The following extracts also indicate some of the work carried out on the bells during this same period:

1810 March 26	pd Mr Wright for set of new Beellropes	
		£1 11s 6d
1820 Feb 8	pd Jeffs repairing the bell well [wheel] & work at church	£1 12s 0d
1823 Jan 20	pd Blacksmith bill for Mending the Grate [Great] bell	18s 11d
1842 Oct 28	Mr Geesons bile [bill] for Brasses and Grodgsons [gudgeons] for the Beles [bells]	£1 2s 6d
	Mr Christian for Beleropes	£1 5s 0d
	F Stevenson's Bill for Reparg Bells	£5 9s 4½d
	H Jeffs' Bill for do	£7 4s 6d

Thirteen shillings worth of ale was the standard payment made to the singers and ringers at Christmas. Henry Jeffs, noted in the above accounts, was the village carpenter and a beer retailer (White 1846, 662).

In 1887, John Taylor surveyed the belfry with a view to rehanging the bells in a new frame. Evidently there were insufficient funds for the churchwardens to accept the quotation.

Amounts paid for new and repaired bellropes from 1900 to 1916 indicate that the bells continued to be rung during this period. However in 1934, it was recorded at a PCC Meeting that 'serious attention would have to be paid in the near future to the condition of the Bells and bell-chamber'. Taylors of Loughborough subsequently inspected the bells and although their report was considered at the July meeting no action appears to have been taken.

It was not until 1951 that sufficient funds were avail-

able to refurbish the belfry, when the five old bells were tuned and rehung in a new high-sided metal frame for six. This work, carried out by Taylors, had been completed by April 1952. In 1980 a hole was found in the then third bell. This was recast in 1985 at a cost of £912. A new treble, costing £1,649, was added to the ring at the same time. The work also included painting the frame and refurbishment of the headstock bearings.

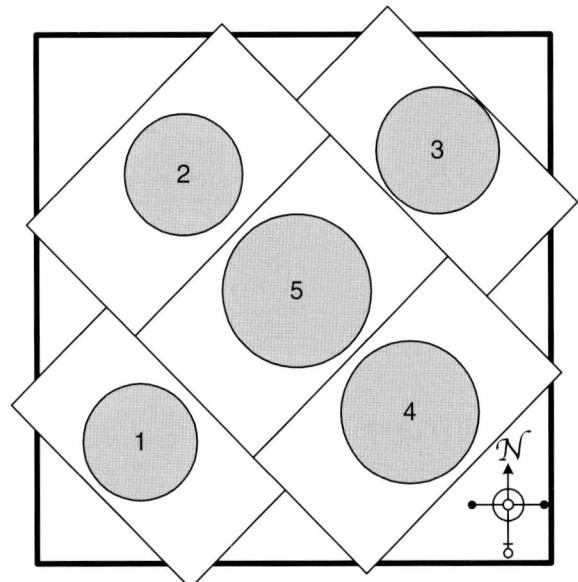

From a sketch of the timber bellframe prepared when John Taylor surveyed the belfry at Caldecott Church on 21 February 1887. Notes taken at the time include 'Canons broken on the 4th', 'Frame in a bad state' and '4th wheel 1 piece' (closed archives of John Taylor)

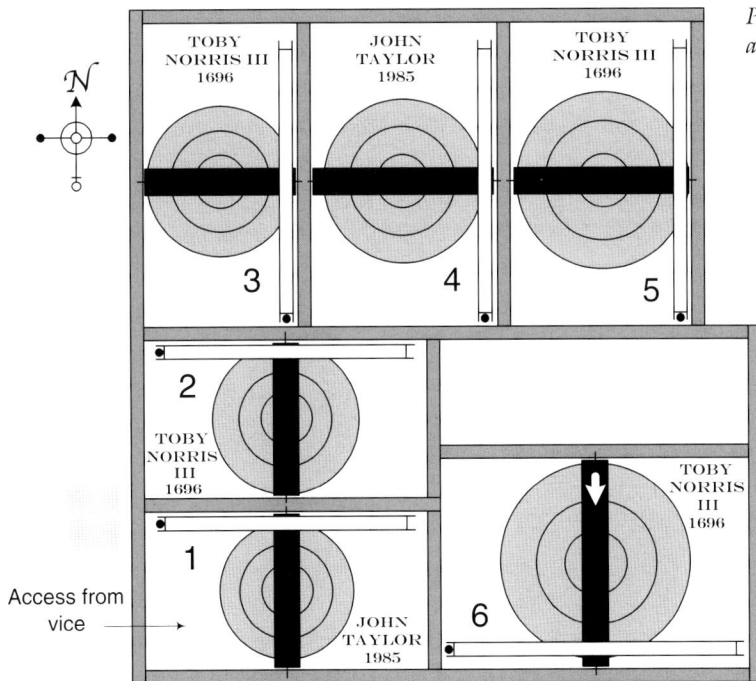

*Plan of the bellframe
at Caldecott Church*

BELL DETAILS

1 1985. Treble. Diameter 673mm (26½in). Weight 211kg (4cwt 0qr 17lb). Cast without canons. Cast-iron headstock. Cast by John Taylor & Co of Loughborough.

On the waist:

[56]
1985
M. A. C.   F. W. C.   C. R. S.   D. V. S.

This bell was donated in memory of Charles Robert Shortt. His initials are placed upon the bell, alongside those of his wife Dorothy Vera and his mother and father-in-law Mary Ann and Frederic William Cox.

2 1696. Diameter 718mm (28¼in). Weight 208kg (4cwt 0qr 11lb). Canons retained. Timber headstock. Cast by Toby Norris III of Stamford.

[10] PETER BROWNE GAVE ME
~ TO THIS TOWNE 1696

There are traces of decoration [70] on the inscription band. Peter Browne is believed to have been related to the Rev Robert Browne who was born at Tolethorpe Hall, Rutland, and who later founded the Brownists. The altar tomb of Peter Browne used to be in the churchyard near to the south side of the tower. On the west end of it there was a brass plate bearing the inscription 'Vnder this ston was buried ye body of Peter Browne May 16th, 1711, aged near 59 years' (*Gentleman's Maga-*

zine Library **IX**, 265). This plate was stolen in 1840 and the tomb was taken away by a mason with permission from the then vicar, Parson Gilham (Phillips 1911-12, 70).

3 1696. Diameter 737mm (29in). Weight 200kg (3cwt 3qr 20lb). Canons retained. Timber headstock. Cast by Toby Norris III of Stamford.

[10] 66

The founder's initial cross [10] and partial date **66** are on opposite sides of the bell. The rest of the inscription is missing except for a short length of decoration [70]. There is a 'blow-hole' on the shoulder above the date. As the present fifth bell records that Toby Norris CAST VS ALL FIVE in 1696, the 66 must be a founder's error.

4 1985. Diameter 775mm (30½in). Weight 274kg (5cwt 1qr 17lb). Cast without canons. Cast-iron headstock. Recast by John Taylor & Co of Loughborough.

TOBY NORRIS 1696

On the waist:

[56]
RECAST 1985

Former bell
Cast in 1696 by Toby Norris III of Stamford. Diameter 800mm (31½in). Weight 229kg (4cwt 2qr 1lb).
1696

5 1696. Diameter 838mm (33in). Weight 270kg (5cwt 1qr 8lb). Canons removed. Cast-iron headstock. Cast by Toby Norris III of Stamford.
[10] [70] **TOBY** [70] **NORRIS** [70] **CAƺT** [70] ~ **VS** [70] **ALL** [70] **FIVE** [70] **1696** [70]

6 1696. Tenor. Note G. Diameter 933mm (36¾in). Weight 413kg (8cwt 0qr 15lb). Canons retained. Timber headstock. Cast by Toby Norris III of Stamford.
[10] **JOHN** [70] **CHAPMAN** [70] **ROBART** [70] ~ **COLLWELL** [70] **C** [70] **W** [70] **1696**
The clock strikes the hours on this bell.

SANCTUS BELL

There was a Sanctus Bell here at some time but there are no references to it in the surviving records. The fifteenth-century Sanctus bellcote was over the east gable of the nave and it is shown without a bell on a *circa* 1839 drawing of the church (Uppingham School Archives). The bellcote was rebuilt when the pitch of the nave roof was raised in 1863, but was removed in 1976 after being damaged (Dickinson 1983, 37).

BELLRINGING CUSTOMS

On Sundays the treble was always rung at 8am. Prior to Sunday services all the bells were chimed and then the Sermon Bell [tenor] was rung. If there was a morning service a bell was rung at the end to indicate that there would be an afternoon service. This bell was rung at noon if there was no morning service. At the Death Knell there used to be three tolls for a male and two tolls for a female, both before and after the knell. The Pancake Bell was rung on Shrove Tuesday and prior to 1880, the Lenten Bell was rung daily at 11am throughout Lent. Peals were rung on Christmas Eve (North 1880, 125). The Church-wardens' Accounts reveal that singers as well as ringers were paid annually at Christmas from 1807 until 1849.

Up to the 1960s, Mr Stanger the churchwarden always rang the Pancake Bell on Shrove Tuesday and tolled the Passing Bell. A different sequence of tolls was also used to denote whether the deceased was a man, woman or child.

Caldecott still has its own band of ringers and this occasionally joins with another village band for practice sessions. If sufficient ringers are available, the bells are rung for the monthly 11am and 6pm Sunday services. Otherwise they are chimed. The bells are usually rung for the Crib Service on Christmas Eve and for the service held on Easter Day. They are sometimes rung on Christmas Day and New Year's Eve. They are rung for visiting preachers. A special effort is made to ring for special services, one being the Harvest Festival. Peals are rung for weddings upon request. On Armistice Day they are rung half-muffled. The bells are rung at funerals as 'special tributes'.

The Caldecott tenor with its timber headstock and original canons

OLD FATHER TIME

A report of 1797 gives the following information: 'The chancel is separated from the nave by a pointed arch. A corresponding one separates the nave from the steeple, which is now walled up. On the upper part (which is the back part of the neat deal gallery) is painted Time with his scythe and an hour-glass, and Death' (*Gentleman's Magazine Library* IX, 266). This painting no longer exists.

SCRATCH DIAL

There are five scratch dials here, all on the late thirteenth-century south aisle wall to the west of the porch.

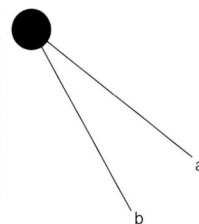

CALDECOTT 1		
Location	East jamb of south aisle window west of the porch	
Condition	Poor	
Gnomon Hole Diameter		20
Gnomon Hole Depth		20
Height above ground level		1700
Line Ref	a	b
Length	100	100
Angle (°)	127	150

Prior to weathering this scratch dial may have been similar to No. 2

Scratch dial 1 details

SOUTH AISLE

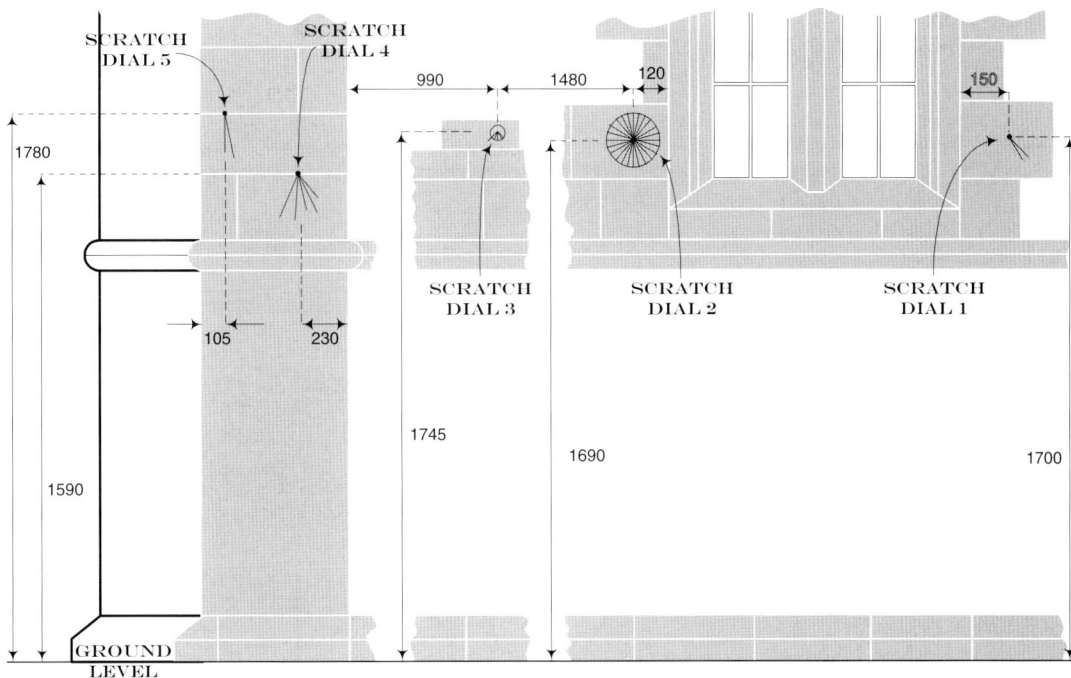

Location of the scratch dials at Caldecott Church

WEST ⟷ EAST

Scratch dial 2 details

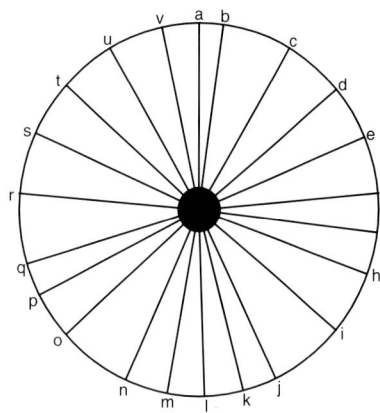

CALDECOTT 2											
Location	West jamb of south aisle window west of the porch										
Condition	Good										
Gnomon Hole Diameter			22								
Gnomon Hole Depth			30								
Height above ground level			1690								
Circle diameter			180								
Line Ref	a	b	c	d	e	f	g	h	i	j	k
Length	90	90	90	90	90	90	90	90	90	90	90
Angle (°)	0	8	30	50	67	85	97	110	130	155	166

Line Ref	l	m	n	o	p	q	r	s	t	u	v
Length	90	90	90	90	90	90	90	90	90	90	90
Angle (°)	178	190	204	228	243	253	275	293	312	330	348

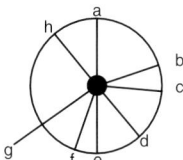

CALDECOTT 3								
Location	South aisle wall west of porch							
Condition	Average							
Gnomon Hole Diameter			8					
Gnomon Hole Depth			10					
Height above ground level			1745					
Circle Diameter			64					
Line Ref	a	b	c	d	e	f	g	h
Length	32	32	32	32	32	32	50	32
Angle (°)	0	72	95	140	180	200	234	320

Scratch dial 3 details

Scratch dial 4 details

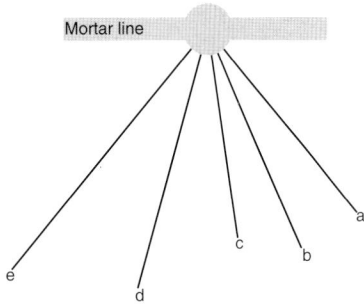

CALDECOTT 4					
Location	South aisle buttress west of porch				
Condition	Average				
Gnomon Hole Diameter		25			
Gnomon Hole Depth		Filled in			
Height above ground level		1590			
Line Ref	a	b	c	d	e
Length	115	120	100	130	150
Angle (°)	140	156	171	190	220

CALDECOTT 5		
Location	South aisle buttress west of porch	
Condition	Average	
Gnomon Hole Diameter		15
Gnomon Hole Depth		Filled in
Height above ground level		1780
Line Ref	a	b
Length	160	130
Angle (°)	170	185

Scratch dial 5 details

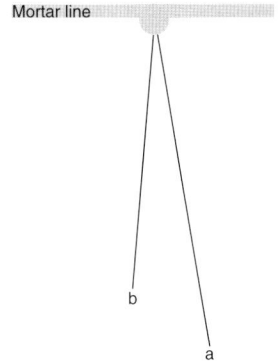

SUNDIAL

The sundial on the porch gable was erected in 1935. The motto is 'Your sunny hours alone I tell' and a brass plaque inside the porch records the gift as follows:

> THIS TABLET IS PLACED HERE BY THE
> INHABITANTS OF CALDECOTE
> TO THE GLORY OF GOD.
> THE SUNDIAL WAS ERECTED OVER THIS CHURCH
> PORCH TO COMMEMORATE THE SILVER JUBILEE
> OF THEIR GRACIOUS MAJESTIES KING GEORGE V
> AND QUEEN MARY, ON 6TH MAY 1935
> AND WAS THE GENEROUS GIFT OF THE
> REV. O. R. PLANT. RECTOR OF THE PARISH.

CLOCK HISTORY

The clock is an early wrought-iron birdcage movement with side-by-side trains. The movement is unsigned but it has similarities to early clocks by Thomas Eayre II, in particular the rating butterfly at the top of the pendulum. 'George Woodcock 1724' is engraved on a wooden plaque in the clockroom, and he may have been the donor. Originally it was a single-handed clock, but it was converted to two hands in 1983.

From the start of the available records in 1807 until 1833 a Mr Broom was paid to clean the clock annually for twenty-six years. He also carried out some repairs. From 1834 to 1865 John Fox, a clockmaker who lived at Great Easton, was paid the same amount for attending to the clock. Uppingham clockmaker Thomas Aris took over the responsibility for two years, although his bills were probably for specific repairs:

1851 Feb 21   Pd to J Fox for Cleaning Clock   5s 0d
Easter 1864 to Easter 1865
                              Mr Aris's bill   £1 6s 8d

The porch sundial erected at Caldecott Church in 1935

From 1869 John Holman looked after the clock and the last entry referring to it is:

Easter 1878 to Easter 1879
                              Holman for Clock   5s 0d

From the absence of any reference to the clock in later accounts it seems that it may have been out of action for the next forty or so years until it was restored in 1922

The rating butterfly at the top of the pendulum of Caldecott church clock is similar to those seen on Eayre clocks

From a photograph of Caldecott Church by George Henton, taken 27 May 1913, showing the original single-handed dial (Henton 946)

(*VCH* **II**, 182). At a meeting of the PCC in 1934 the rector included 'the complete overhauling of the Church Clock' amongst his list of things that needed to be done. In 1959 the question of renovating the clock dial or providing an electric clock was considered, but it was agreed at a later meeting that a better heating system would be installed instead. The matter of the clock was again raised in 1970:

> Following enquiries regarding the Church Clock, a visit had been made by the director of a firm of clock makers who subsequently sent a report regarding the present

clock. This seems to be completely worn out, & his estimate for the new non-striking clock is £335, plus extra for a striking unit, & for disconnecting the striking at night. It was felt by the meeting that it was impossible to embark on a project of this sort at the present time.

Caldecott church clock, possibly an early movement made by Thomas Eayre of Kettering. It was converted to automatic winding in 1983

The anchor escapement of Caldecott church clock

In 1980 it was recorded in the minutes that a 'Mr Allsop was anxious for the clock, which is of historic merit, to be restored to working order'. In 1983 the clock was restored in memory of Mrs J Singlehurst. The work also included the conversion to two hands and the installation of automatic winding. The original diamond-shaped timber dial, which had deteriorated beyond repair, was replaced by a replica at the same time.

CLOCK DETAILS
Details of the present clock:

Maker:	Almost certainly by Thomas Eayre II of Kettering
Signed:	Not signed
Installed:	Probably in 1724
Frame:	Birdcage
Trains:	Going and striking
Escapement:	Anchor
Pendulum:	Wrought-iron pendulum with lead bob
Rate:	60 beats per minute
Striking:	Locking plate hour-striking
Weights:	Original cylindrical lead weights removed from the clock, but retained in the church
Winding:	Huygens automatic winding system to both trains
Dial:	Diamond-shaped timber dial with a blue background, gilded roman numerals and hands. Replica dial for two hands, installed in 1983. Original dial was marked out for one hand with four divisions between the hours
Location:	West face of the tower

THE OLD SUN INN

An early 1900s photograph shows that the former Old Sun Inn, next to the old water mill, had a sundial over the front door (RCM 1972.65.K2). The building is now part of Mill Garage but the dial stone with two gnomon fixing holes still exists. Apart from the church, this is believed to be the only remaining sundial in the village, but the MSS of a Mr Barnett stated that in Caldecott 'Sundials were formerly rather numerous ...' (Phillips 1911-12, 72).

An early 1900s photograph of the former Old Sun Inn, Caldecott, showing the sundial over the front door (RCM 1972.65.K2)

CLIPSHAM

ST MARY   Grid Ref: SK 971164

PARISH RECORDS
The only relevant sources are the Churchwardens' Accounts 1779-1847 (DE 3066/3) and a Glebe Terrier of 1705 (MF 495).

BELL HISTORY
The earliest reference to the bells at Clipsham is in Irons' Notes of 1605:

> it was said there had been three 'tunable' bells, but 5 or 6 years ago one bell was taken by Mr. Geo. Butler of Lile Lodge [Leigh Lodge] Ridlington, by the consent of some of the parish, but not of all (*VCH* II, 44).

All three of the present bells were cast in the Norris foundry at Stamford. The earliest, dated 1657 is by Thomas Norris.

The accounts show frequent expenditure for bellropes during the period 1780-1835. In 1827 the price of three new ropes was 15s. Allowances of 3s 6d and 3s for the

old ropes were made by the ropemaker in 1830 and 1834. New bellwheels were provided in 1780 and on two occasions one of the bells had to be lowered for repairs:

1787 Jan 6	Belropes Cloklines	15s 0d
	The Bel Reparring	15s 9d
	Paid to John Pritty	1s 2d
	for Helping Pritty to low the Bell	6d
1793 Feb 5	Paid for Bellrops & Clocklines	15s 0d
	For Helping to low the Bell & Ale	2s 6d
March 2	paid to the Clark for Cleaning the Church after Bells work	2s 0d

Documents dated 20 March 1851 record that the bells were rehung in a new oak frame on Easter Monday of that year (Parsons 1983).

Eayre & Smith rehung the bells in 1981. In the new installation they hang dead by their canons from a steel frame mounted above the old oak frame. Each bell has a chime hammer operated from an Ellacombe chiming frame on the west wall at the base of the tower.

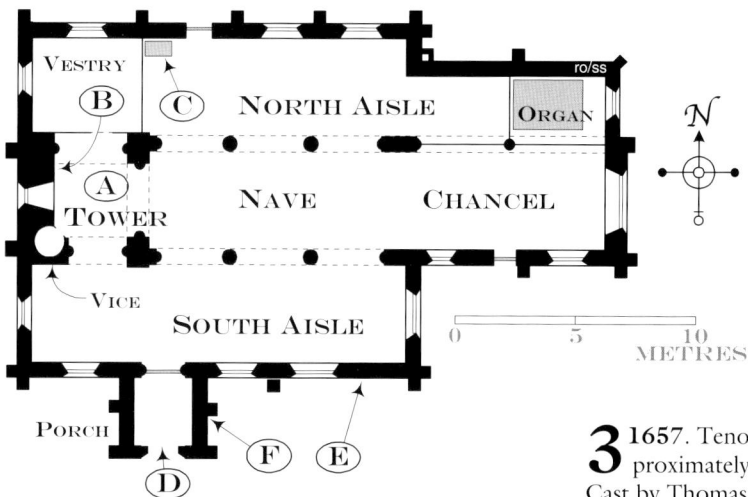

Plan of St Mary's Church
A: Three bells in the belfry. Ringing chamber at the base of the tower
B: Ellacombe chiming frame in the ringing chamber
C: Former clock movement
D: Location of a former sundial
E: Scratch dial 1
F: Scratch dial 2
The vice provides access to the former clockroom and the belfry

BELL DETAILS

1 1671. Treble. Diameter 660mm (26in). Weight approximately 203kg (4cwt). Canons retained. Cast by Thomas Norris of Stamford.

[10] [52] 1671 [52]

2 1675. Diameter 737mm (29in). Weight approximately 254kg (5cwt). Canons retained. Cast by Toby Norris III of Stamford.

[26] [69] W [69] WIⱭG [69] TOBIEAƧ [69]
~ ⱯORRIƧ CAƧT [69] ME 1675

W Wing was probably the donor.

3 1657. Tenor. Diameter 813mm (32in). Weight approximately 700kg (6cwt 1qr). Canons retained. Cast by Thomas Norris of Stamford.

[26] **THOMAS NORRIS MADE MEE 1657**
O

O is the reverse of a James I silver shilling on the soundbow. It is in mirror image and is directly below the S of Norris. Unlike other Rutland bells with coins, there is no obverse showing the King's head and this is possibly because the bell was cast during the Commonwealth period. The bell has noticeable indentations on the soundbow as a result of being struck by the clock hammer.

Plan of the bellframes at Clipsham Church. The old oak bellframe has been retained below the new steel frame

The reverse of a James I silver shilling on the soundbow of the third bell at Clipsham. It is in mirror image

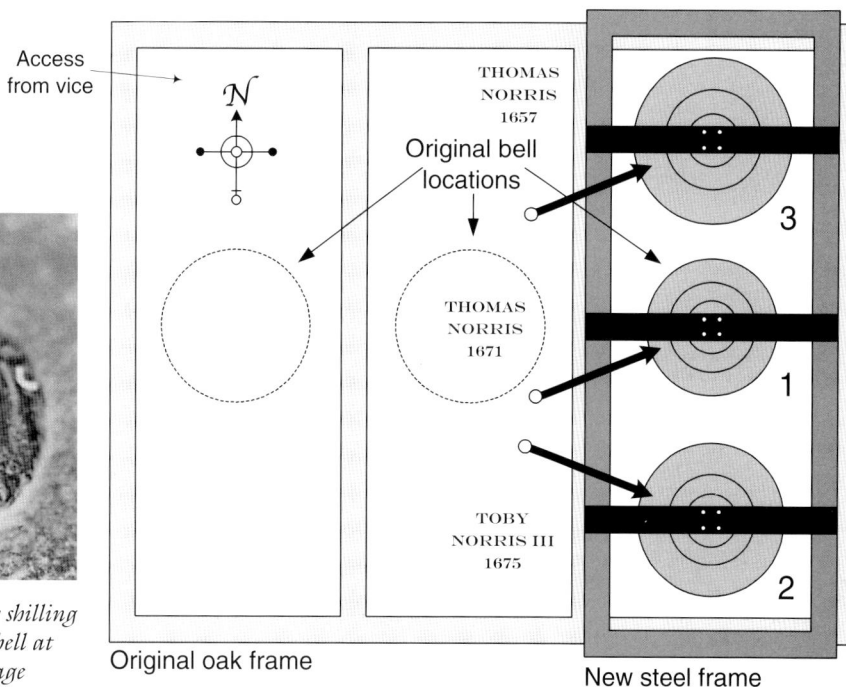

BELLRINGING CUSTOMS

There are no details in the Churchwardens' Accounts of bells being rung for special occasions.

On Sundays a bell was rung at 8am if Morning Prayer was to be said and again after morning service if Evensong was to be said. For Divine Service the bells were chimed and a Sermon Bell was rung. The Gleaning Bell was rung at 8am and 6pm during harvest. At the Death Knell there were thrice three tolls for a male and thrice two tolls for a female, both before and after the knell (North 1880, 127).

Today the bells are chimed before the Sunday Service normally held at 11.15am, and for all additional services.

SCRATCH DIALS

There are two scratch dials, one on the south wall of the south aisle and the other on a relocated stone on the east-facing wall of the porch. Both are on fourteenth-century stonework.

SUNDIAL

A sundial is shown on the porch of the church in a drawing of *circa* 1839 (Uppingham School Archives). There is no longer a sundial in this position.

CLOCK HISTORY

The timber framed clock made by John Watts of Stamford in 1688 is probably the only clock to be have been installed in the church at Clipsham. It is mentioned in a Glebe Terrier of 1705 (MF 495): 'It [Item] in the steeple a Clock and three Bells'.

Following removal from the tower in 1982 and subsequent restoration by Brian Sparkes, it was placed on display in the north aisle of the church. Although now 'retired' after over two hundred and fifty years of faithful service, it is still in good working order. This early timber framed turret clock with a pendulum and anchor escapement is one of only about thirty or so known, all of which are found in the Midlands. Of these, only the clocks made

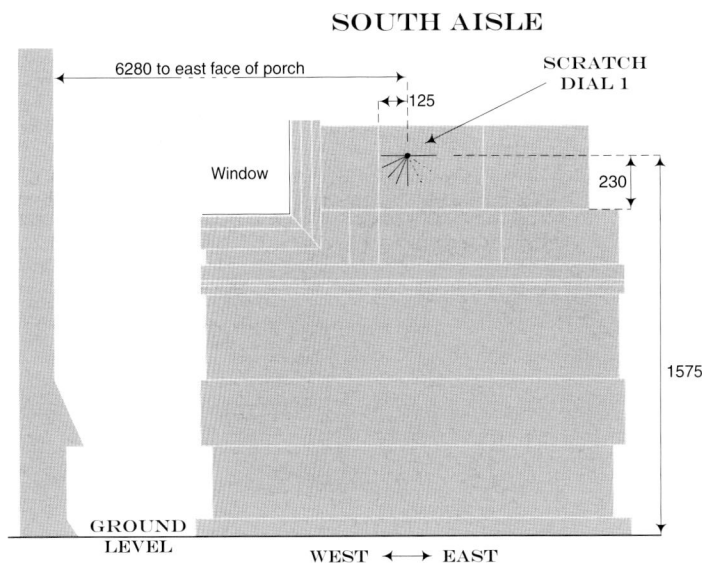

SOUTH AISLE

6280 to east face of porch

125

SCRATCH DIAL 1

Window

230

Location of scratch dial 1 at Clipsham Church

1575

GROUND LEVEL

WEST ←→ EAST

CLIPSHAM 1								
Location	South wall of south aisle, east of the east window							
Condition	Excellent							
Gnomon Hole Diameter	15							
Gnomon Hole Depth	25							
Height above ground level	1575							
Line Ref	a	b	c	d	e	f	g	h
Length	150	150	150	170	150	160	160	150
Angle (°)	90	120	150	180	195	217	238	270
Pock radius	150	-	-	150	150	150	150	150

Note that lines b and c are of a different form and may have been added later

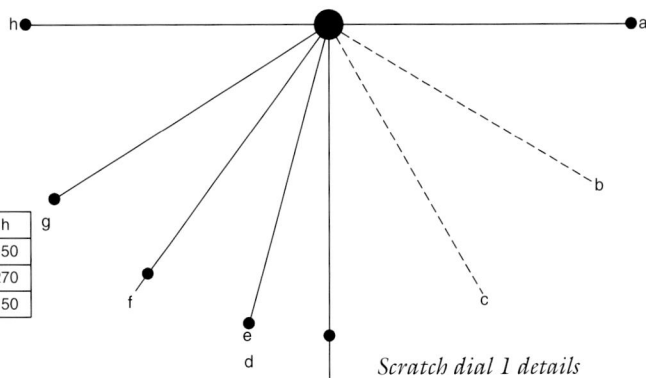

Scratch dial 1 details

EAST WALL OF PORCH

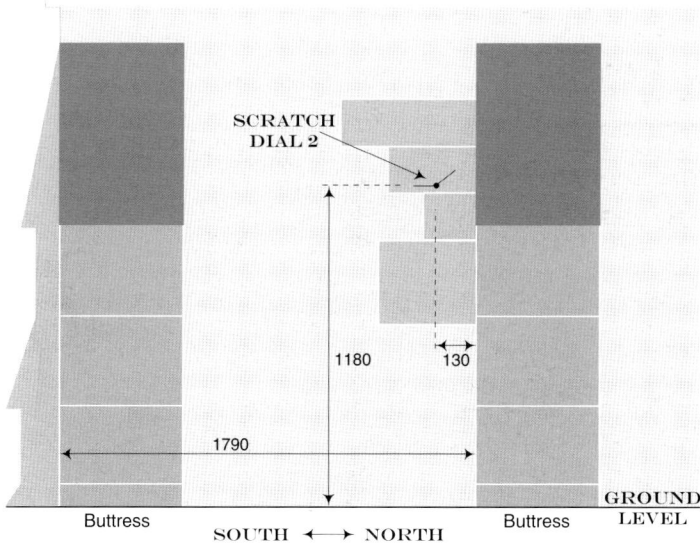

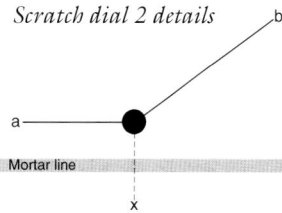

SCRATCH
DIAL 2

1180   130

1790

Buttress   Buttress

SOUTH ←→ NORTH

GROUND
LEVEL

Location of scratch dial 2 at Clipsham Church

Scratch dial 2 details

a   b

Mortar line

x

CLIPSHAM 2		
Location	Between the buttresses on the east facing porch wall	
Condition	Average	
Gnomon Hole Diameter		12
Gnomon Hole Depth		15
Height above ground level		1180
Line Ref	a	b
Length	55	85
Angle (°)	90	235

Note that this scratch dial is on a relocated stone and is inverted. Angles are therefore measured from datum line x.

From a circa 1839 drawing of Clipsham Church showing a sundial in the gable of the south porch. It appears to be covering the lower half of a niche (Uppingham School Archives)

by John Watts have the maker's initials and date carved on the frame, which is very fortunate for the researcher.

John Watts installed his clocks in a number of Rutland and north Northamptonshire churches and, as well as Clipsham, those from Empingham, Edith Weston, Greet-

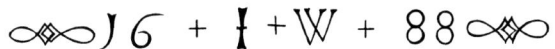

ham, Nassington and Apethorpe still survive. The Clipsham clock did not have an exterior dial but this was not uncommon with early clocks. It announced the time by striking the hours on the tenor bell.

The available accounts include many references to the clock. 'Clocklines' and 'oile' were purchased almost every year:

1786	**April 18**	paid for Oile for ye Clock	6d
	July 22	pd for a Wire for ye Clock	1s 4d
	Dec 31	Bellropes & Clocklines	15s 0d

Three Stamford clockmakers, James Wilson, John Brumhead and Thomas Haynes, attended the clock from 1787 until 1847.

James Wilson was paid for maintenance from 1787 until 1795 and probably did so until he retired in 1803. Except for the years 1807 and 1808, when John Brumhead looked after the clock, Thomas Haynes, nephew of James Wilson, followed by his son Robert, were responsible for the clock each year until 1847:

1787		pd Jas Willson's Bill for reparing the Church Clock	18s 6d
		pd for Oile for ye Clock	6d
1807	March 23	Mr Brumhead Bill to the Clock	18s 0d
		Bellropes & Clock Lines	17s 6d
1809	April 3	Paid Mr Haines for clening Clock	10s 6d

It appears that the clock was heavily restored at some

J 6 + I + W + 88

John Watts' signature on the clock frame at Clipsham. He signed most of his clocks with I W [Iohannes Watts] and the date

(above) Clipsham Church. The John Watts' clock in 1982 before being removed for restoration (Brian Sparkes)

(right) The restored clock, now in the north aisle at St Mary's Church, Clipsham

time during the 1800s (Lee 2000, 33), including the replacement of a slotted countwheel by a pin countwheel. This is also the view of Brian Sparkes who was able to inspect the movement in great detail when he restored it 1982-3. These are his comments on the repairs and alterations that he found:

Brass was scarce and expensive in early times and in its original form the clock doubtless had iron wheels throughout. When after many years wear and tear necessitated some of them being replaced the replacement wheels were a slightly different size from the iron originals. The holes in the upright iron bars had to be blocked and redrilled a little lower down and the number of teeth on these new wheels also differed from the originals. The pendulum would have had to be lengthened as suggested by the extension piece which has been grafted onto the legs of the clock at some stage. The setting looks later still and it incorporates a friction clutch to facilitate setting the clock to time. In its original form the lifting pins were mounted in the rim of the great wheel - the holes are still there and the escapement had to be disengaged to let the wheel train

run quickly when setting to time. The brass bushes in which the axles [arbors] are pivoted would have worn and been replaced or relined several times during the life of the clock. One of the strike levers has been repaired by gas or arc welding at some fairly recent date.

CLOCK DETAILS

Details of the present clock:

Maker:	John Watts
Signed:	16 + IW + 88 on the frame
Installed:	1688
Cost:	Not recorded, but probably about £10
Frame:	Posted wooden frame with wrought-iron pivot bars
Trains:	Going and striking, side by side. Going train great wheel turns exactly once every 2 hours
Escapement:	Anchor
Pendulum:	Wrought-iron pendulum rod with lead bob
Rate:	Originally 60 beats per minute but later lengthened

Striking:	Countwheel hour striking
Weights:	Stone weights
Winding:	Wound daily when in commission
Dial:	No dial

CLIPSHAM HALL

A two-train flatbed turret clock was installed in the stable block at Clipsham Hall in 1904. The stables were converted into dwellings in the early 1990s and that part of the east wing which includes the clock is now known as Clock House.

CLOCK DETAILS

Details of the present clock:

Maker:	G & F Cope & Co of Nottingham
Signed:	On the setting dial and on the frame
Installed:	1904
Frame:	Two part cast-iron flatbed
Trains:	Going and striking
Escapement:	Pinwheel. Escapement pallets on the pendulum rod
Pendulum:	Wooden rod and cast-iron bob hanging inside the movement
Rate:	48 beats per minute
Striking:	Countwheel hour-striking
Weights:	Cast-iron
Winding:	Weekly
Dials:	Two limestone dials with incised numerals and divisions. One facing west over the stables yard and the other facing east
Clock bell:	In the tower above the dials. Dated 1904
Note:	Restored in 1985 when the stables were converted into dwellings

A table sundial is shown in the grounds of Clipsham Hall, to the east of the house, on the early 1900s OS Second Edition 25 inch map. Today, this sundial still exists, but it has a new dial plate following the theft of the original in 2000. Another sundial, with an identical pedestal, is located in the garden to the west of the house. The brass dial is signed W Deane and incorporates points of the compass and the Equation of Time.

An overhead view of the flatbed turret clock by Cope of Nottingham at Clock House, formerly part of the stable block at Clipsham Hall (John Ablott). Note the pendulum hanging inside the movement, suspended from a bridge over the going train by a very wide suspension spring. The escapement pallets are on the pendulum rod and this spring minimises any twist in the pendulum swing, thus ensuring the correct action of the escapement

One of a matching pair of table sundials in the private gardens of Clipsham Hall. The dial is 175mm (7in) square

The west-facing clock dial at Clock House, Clipsham Hall stables

COTTESMORE

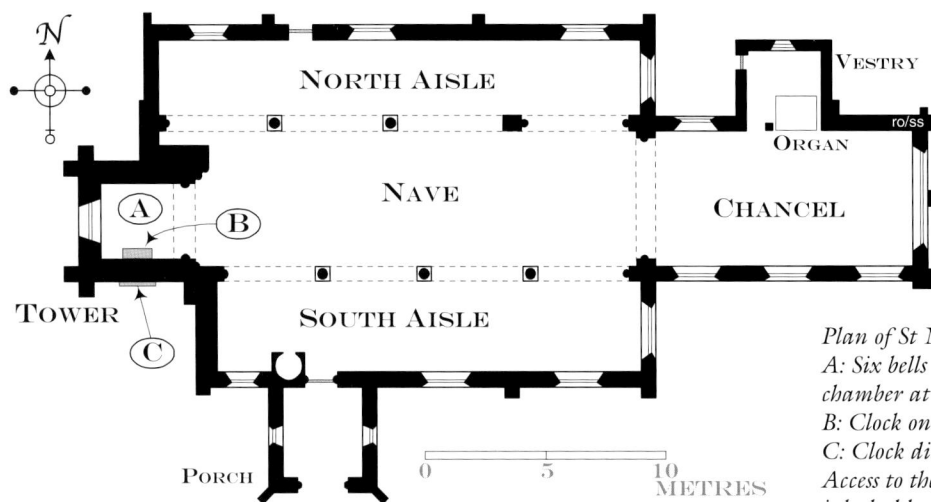

Plan of St Nicholas's Church
A: Six bells in the belfry. Ringing chamber at the base of the tower
B: Clock on the first floor of tower
C: Clock dial
Access to the clockroom and belfry is by ladders

ST NICHOLAS   Grid Ref: SK 902136

PARISH RECORDS

There is an excellent archive of 'church receipts' covering the periods 1782-84, 1830, and 1841-47 (DE 1920/75/1-91), 1848-1945 (DE 1920/76 & 85) and 1946-52 (DE 1920/87). This collection, which consists of more than 1400 individual documents, provides a great deal of detailed information on the work carried out to both the clock and the bells. The following are also available:

A report of August 1899 on the condition of the bells, with a specification and an estimate (DE 1920/30/1-4).
Accounts for the new clock in 1909 including bills and a list of subscribers (DE 1920/32/1-7).
A faculty dated 8 December 1933 for restoring and rehanging the bells (DE 1920/19).
A booklet (DE 1920/36) and Order of Service (DE 1920/35) for the re-dedication of the bells on 8 February 1934.
Church Account Book 1933-47 (DE 1920/86).
A faculty dated 7 September 1935 for adding a new treble (DE 1920/20).
Specification, estimate and correspondence of 1935 regarding the new treble (DE 1920/37/1-20).
The Order of Service for the dedication of the new treble on 11 October 1935 (DE 1920/38).
A booklet of 1935 entitled *The Bells of St Nicholas Cottesmore* (DE 1920/39), together with newspaper cuttings and draft letters of 1936 regarding the bells (DE 1920/40/1-7).

BELL HISTORY

The earliest bell at Cottesmore is the present third by Henry Oldfield II of Nottingham. It is dated 1598 and is one of only three bells in Rutland to carry the Royal Arms. In 1660 Thomas Norris of Stamford supplied three new or recast bells. His son Toby Norris III later cast a new tenor and this was duly recorded on the occasion of an Archdeacon's Visitation (Irons' Notes, MS 80/5/24):

1699 April 7
To certify the casting & hanging the Great Bell

After 1791, the bells were in such poor condition that they were no longer rung (DE 1920/75/91), and it was not until 1844 that the belfry was completely refurbished. The work included a new oak frame, and the bells were rehung with new fittings. Richard Coverley the village carpenter carried out the majority of this work, and two other carpenters assisted him for eleven days and three labourers for four days each. Labour charges were £3 11s 0d. The oak used for the bellframe, and the boarding for twelve windows in the steeple plus painting cost £31 15s 0d. It appears that Richard Coverley supplied and refitted all new parts to the bells with the exception of two bell clappers, and the total charge for the materials was £74 8s 2d. He charged £1 5s 0d for five new bellropes (DE 1920/75/36).

The Wymondham Ringers rang the first peal on the refurbished bells. Their names are recorded as follows:

Treble	James Woolman
Second	Edward Price
Third	Philip Lee
Fourth	James Lee
Tenor	George Lee (DE 1920/75/91)

Most of the repairs to the bells and many of the clock repairs from the 1840s right through to the end of the

Prior to the installation of the new frame, John Laxton installed two new floors in the tower of Cottesmore Church as shown in this bill (DE 1920/75/26)

One of the headstocks removed from the belfry of Cottesmore Church in 1899 remains at the base of the tower

nineteenth century were entrusted to the Hollis family of Cottesmore who were blacksmiths, wheelwrights and builders. The following are typical :

1840	May 1	Bell Claper md & new Boult	1s 6d
1854	Oct 5	28 New Washers Wrench making & Irons repairing to Bells	4s 6d
1864	Jan 11	New Bell Stay & fixing	2s 6d
		Block making & fixing in crown of bell	4s 6d
1887	June 13	4 Keys to Bell	1s 0d
	Sept 30	2 new roller spindle to bell blocks	2s 0d

Taylors of Loughborough inspected the bells in November 1885 and they reported that the tenor was cracked and needed recasting. They also advised that the fourth bell required turning and rehanging with new fittings (DE 1920/79/68). Taylors were keen to obtain the work, as shown in their letter of 10 December 1885 to Mr J P Hollis, churchwarden:

We assure you that we are doing the work at the lowest possible figure compatible with the use of the best materials and workmanship — indeed the present state of trade prompts one to offer prices not fairly remunerative — this we have done in your case in order to obtain work (DE 1920/79/70).

Taylors did secure the work and when the bells reached the foundry it was found that the fourth bell was also cracked and needed recasting.

The final bill from Taylors, dated 8 April 1886, outlines the work done in the February of that year. The total was £88 19s 3d for recasting two bells with new fittings and this included an allowance of £75 13s 9d for the metal from the old bells. Carriage between Ashwell Station and the foundry was charged at £3 (DE 1920/79/43).

In 1899 the fittings on the three smallest bells were renewed and all five bells were tuned and turned. Taylors

carried out this work at a cost of £62 4s 10d (DE 1920/80/15).

A faculty dated 8 December 1933 reveals that the old oak bellframe had become unsafe 'owing to weaknesses in its construction & the presence of the Death Watch Beetle in the beams, for which reason the bells could not be rung'. The faculty stated that at a meeting held on 25 October 1933, it was agreed to accept an offer made by Mr John Hollis JP of Hornsea, Yorkshire, an old resident of the village, to have the bells put in working order as a memorial to his wife who had recently died. The faculty allowed for the rehanging of the bells with new fittings in a cast-iron frame, the frame to be supported by steel girders grouted into the walls of the tower.

The following year a new treble was donated by Mrs

Annie Barnes who donated the new treble at Cottesmore in 1935 (R Lees)

Annie Barnes in memory of her husband Henry, who had been connected with the church for fifty years. In order to accommodate this new bell and maintain the ringing circle, the old treble dated 1660 was hung in a new frame over the third and fourth bells. Canon E Guilford dedicated the bell at a joint Dedication and Thanksgiving Service on 11 October 1935.

A notice at the base of the tower provides details about more recent work on the bells:

> Bell clappers refurbished and wooden pulley wheels renewed. Work carried out by John Taylor Ltd of Loughborough. Bell room and bell frame cleaned and frame, headstocks and wheels repainted by Dickie Lees. Work carried out between February and May 1994.

At this time, Mr Lees was the Bellringing Secretary.

BELL DETAILS

1 **1935**. Treble. Diameter 702mm (27⅝in). Weight 214kg (4cwt 0qr 24lb). Cast without canons. Cast-iron headstock. Cast by John Taylor & Co of Loughborough.

On a divided inscription band and continuing on the waist:

The Royal insignia on the treble at Cottesmore

[141]
G R
V
1935
TO THE GLORY OF GOD
AND IN MEMORY OF
HENRY BARNES,
1858 - 1932.
"O GIVE THANKS UNTO THE GOD OF HEAVEN.
FOR HIS MERCY ENDURETH FOR EVER."

On the waist, opposite:

[47]

Decoration [71] is around the inscription band. [141] is a crown. This bell was dedicated in the Silver Jubilee Year of George V's reign.

2 **1660**. Diameter 756mm (29¾in). Weight 247kg (4cwt 3qr 13lb). Canons removed. Cast-iron headstock. Cast by Thomas Norris of Stamford.

[26] [52] THOMAS [52] NORRIS [52]
~ MADE [52] MEE [52] 1660 [52]

This bell is hung in a low-sided frame over the third and fourth bells.

3 **1598**. Diameter 838mm (33in). Weight 308kg (6cwt 0qr 8lb). Canons removed. Cast-iron headstock. Cast by Henry Oldfield II of Nottingham.

[72] [73] GOD [73] SAVE [73] hIS [73]
~ ChVRCh [73] 1598 [73]

Decoration [7] is below the AV of SAVE.

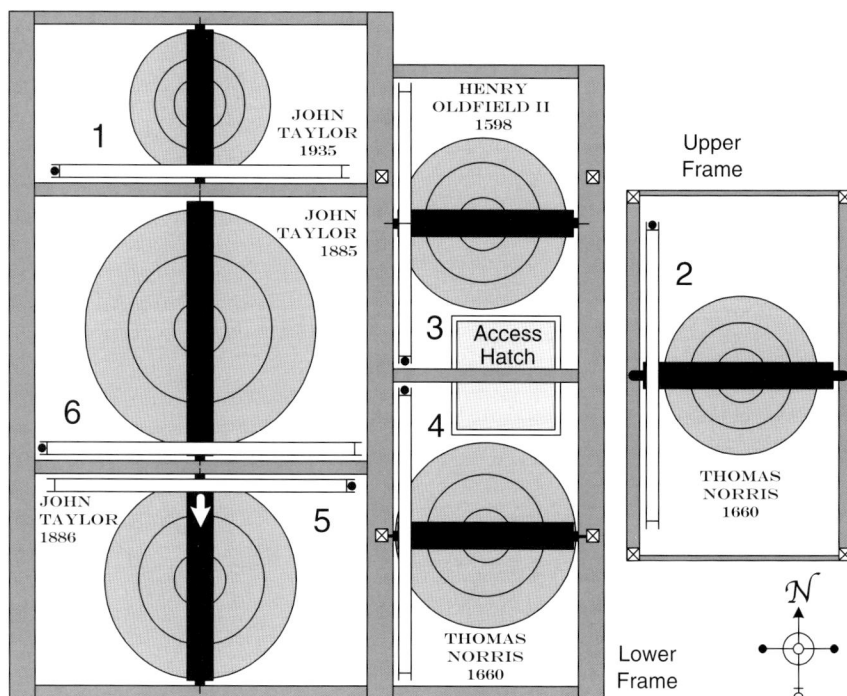

Plan of the bellframe at Cottesmore Church. The second bell is over the third and fourth bells

Cottesmore Church. Henry Oldfield's decoration [73] on the third bell

(right) The Royal Arms [68] on the third bell at Cottesmore

Below the date and high on the waist:
Royal Arms [68] surmounted by **ER**.
Letters are like [116]. Scheduled for preservation (Council for the Care of Churches).

4 1660. Diameter 879mm (34⅝in). Weight 342kg (6cwt 2qr 26lb). Canons removed. Cast-iron headstock. Cast by Thomas Norris of Stamford.

**[10] [52] THOMAS [52] NORRIS [52]
~ MADE [52] MEE [52] 1660 [52]**

5 1886. Diameter 984mm (38 ¾in). Weight 496kg (9cwt 3qr 1lb). Cast without canons. Cast-iron headstock. Recast by John Taylor & Co of Loughborough.

**RECAST BY JOHN TAYLOR AND CO
~ LOUGHBOROUGH 1886.**

The clock strikes the hours on this bell.

Former bell
Cast in 1660 by Thomas Norris of Stamford. Diameter 997mm (39¼in).

[10] THOMAS NORRIS MADE MEE 1660

6 1885. Tenor. Note F. Diameter 1108mm (43⅝in). Weight 712kg (14cwt 0qt 3lb). Cast without canons. Recast by John Taylor & Co Loughborough.

**J: TAYLOR AND CO BELLFOUNDERS
~ LOUGHBOROUGH 1885.**

Former bell
Cast in 1699 by Toby Norris III of Stamford. Tenor. Diameter 1105mm (43½in).

**THO CHRISTIAN JOHN HARDY
~ TOBY NORRIS CAST ME 1699**

BELLRINGING CUSTOMS
On Sundays the treble was rung at 8am, followed by the treble and second at 9am. The bells were chimed for Divine Service and at the close of morning service a bell was rung. The Gleaning Bell was rung during harvest. Peals were also rung at the Great Festivals. At the Death

Knell there were thrice three tolls for a male and thrice two tolls for a female both before and after the knell (North 1880, 127).

Today Cottesmore has a band of bellringers. Each Sunday the third bell is chimed for Communion at 8.30am and the bells rung for the 9.45am service. Traditionally the bells are also rung for all Christmas Services, and on Armistice Day they are rung half-muffled for the Remembrance Service. If requested they are rung for weddings, special occasions and funerals. At the latter, the tenor is tolled and occasionally rung half-muffled.

CLOCK HISTORY
The earliest reference to a clock at Cottesmore is:

1784 To John Hutchings

July 30	For looking after the clock	18s 0d
	For oil for the Clock	6d

Little is known of this early clock, but a drawing of the church *circa* 1839 (Uppingham School Archives) shows that it had a dial on the south face of the tower. The costs of its repair and maintenance were a constant drain on the churchwardens' funds as shown in the surviving bills. When Richard Coverley, the village carpenter, submitted a bill dated 1844 for making a new bellframe and re-hanging the bells, he also included for 'assisting Mr Sutton with clock'. This Mr Sutton was Augustus Sutton of Cottesmore Hall who had a particular interest in

Extract from Richard Coverley's bill of 1844 for work on the bells and clock at Cottesmore (DE 1920/75/36). Mr Sutton was Augustus Sutton of Cottesmore Hall who was also involved in the repair of the clock at Greetham Church

From a voucher of 1854 presented by Stephen Rodely, clockmaker of Oakham. A note below this extract states that he agrees to 'repair the chimes & keep them going for 12 months for 30/-' and to 'repair & clean clock for 10/-'. The churchwardens paid this bill on 13 April 1855 (DE 1920/76/5)

church clocks (*see* Greetham — Clock History & Chapter 3 — Clockmakers).

Interestingly, the accounts reveal that the clock had a chime barrel which played tunes on the church bells. Specific details of this have not survived, but normally tunes would be played every three hours, with a different tune for each day of the week. The only other chime barrel recorded in Rutland was at Oakham Church prior to 1858 but very little is known of it. Fortunately, one of Stephen Simpson's bills (DE 1920/76/106) provides a little more information about the Cottesmore clock:

1849 March 15
> to Watch part Bushd the holes, repd the lifting peice repd the stoping peice — and plated the pendlum   15s 0d
> Striking part A new click and spring to the roler for the line   7s 6d
> A new plate to the Locking Wheel & new pins & screws to let off the chimes new wire and hooks to the hammer and cleand   15s 0d
> to the Chimes, All new copper wire to the hammers the Cranks at the Bells rep — new Cotters to shift the Tune, Examined all the Work and made wright and Cleand £1 5s 0d
> [Total]   £3 2s 6d
>
> Settled August 30th S Simpson

The last mention of the 'chimes' in the parish documents is in a bill submitted by W Hollis, and this probably indicates their demise:

1867 Recd old Chimes Do moving [the chimes] in Belfry & getting down — Time Man 15s 0d
         (DE 1920/78/146)

The surviving accounts show that the clock and chimes at Cottesmore were repaired by a succession of Oakham clockmakers including William Payne (1847 to 1849), Stephen Rodely (1854 and 1855), Stephen Simpson (1849 to 1857), John Cooke (1861) and Charles Payne (1867 to 1874). The Hollis family of Cottesmore also carried out many repairs to the clock from 1840 until the new clock was installed in 1909. The following examples are typical:

From Fountain's bill of 1909 for the Cottesmore clock case (DE 1920/32/3)

1844 **To J Hollis & Son**
     Rod lengthened to clock hands 3lbs 1s 5½d
     Clock Hammer repaird   1s 0d
     New leavers & standards to Hammer 12½lbs
              5s 2½d
     Dial plaiting 8d & 1 large staple at 5d 1s 1d
     2 New eyes & 2 purls to Clock 1s 2d
Oct 14 1 Days work assisting to Fix Clock Up 3s 6d
Nov 21 Clock Hammer mend 1 new end to leaver 3½ etc   2s 6d
         (DE 1920/75/23)

1859 **To W Hollis**
April 29 Clock Repairing. Barrel & Spindle Mendg 1 New Copper Hand & 1 Hand Mendg New Screws & Wire & Hammer Repairing 6s 6d
         (DE 1920/77/95)

1862-63 **To Mr Hollis**
May 3 Church Clock Mendg   9d
Feb 7 Clock Mendg New Spring & Pendalum Repaird   1s 6d
         (DE 1920/77/28)

It is interesting to note that in 1874 he supplied an iron hand to the clock:

1877 **To Wm Hollis**
Aug 21 Mans time repairing clock 2s 0d
Sept 4 Clock frame taking to pieces spindle getting out & straighten New steel Catch Wheel & fitting Mans Time putting on 5s 3d
Oct 31 Clock Hammer mendg new piece & welding in Mans Time at same & fixing in 4s 6d
         (DE 1920/78/16)

In 1909 a decision was made to replace the old clock and £105 10s was raised by public subscription. A list of the subscribers and the amounts donated is included in the Parish Records. The new clock and a hand-chiming apparatus for five bells were supplied and installed by John Smith & Sons of Derby at a cost of £82. Mr R Fountain, a local carpenter, built a new clock case and new casing for the weights.

*The turret clock by John Smith &
Sons at Cottesmore Church, installed
in 1909* (DE 1920/32/3)

In 1784 John Hutchings was paid 18s a year for winding the clock and this is the earliest reference to this task. In 1839 Thomas Bloodworth is recorded as being the clock winder and was responsible for the clock until 1863. George Broom took over this task in 1864, followed by Richard Broom, presumably his son, in 1908. Mrs Sarah Broom was the clock winder during the First World War. From 1839 until the 1940s, the annual salary for this duty remained constant at £1 10s. As well as winding, it was the duty of the winder to carry out day-to-day maintenance on the clock and the cost of oil was often included in the accounts.

In February 1934 the clock was cleaned and overhauled by John Smith & Sons of Derby at a cost of £5 15s 0d.

CLOCK DETAILS

Details of the present clock:

Maker:	John Smith & Sons of Derby
Signed:	On the setting dial and front of the frame
Installed:	1909
Cost:	£70
Frame:	Cast-iron flatbed
Trains:	Going and striking
Escapement:	Pinwheel
Pendulum:	Wooden rod and cast-iron cylindrical bob
Rate:	53 beats per minute
Striking:	Locking plate hour striking
Weights:	Cast-iron
Winding:	Hand wound weekly
Dial:	Skeleton dial with blue backing plate. Gilded roman numerals and hands
Location:	South face of the tower

EDITH WESTON

ST MARY   Grid Ref: SK 927054

PARISH RECORDS

The following records were searched for information concerning the bells and clocks:

Churchwardens' Accounts 1757-1847 (DE 1937/18) and 1903-30 (DE 1937/19).
A file of correspondence regarding the church bells 1920-77 (DE 5032/38).
The faculty for restoring the church bells dated 1951 (DE 5032/37).
A file of correspondence regarding the church clocks from 1920 to 1976 (DE 5032/39).
A set of photographs of the restored former clock taken by RAF Wittering in 1966 (DE 5032/73-79).
Parish Surplus and Parish Fund — Minutes & Accounts 1897-1934 (DE 1937/21).

Church Expenses Account Book 1892-1931 (DE 1937/20).
Vestry Minute Book 1820-45 (DE 1937/24) & Vestry and PCC Minute Book 1886-1937 (DE 1937/25).

BELL HISTORY

The oldest bell at St Mary's is the present fourth dated 1597 and it may be a recasting of a much earlier bell. If so, the ascribed founder, Matthew Norris, probably retained the old inscription. The fifth was cast in 1621 by Toby Norris I, and the tenor by Henry Penn in 1723. Other than these details the early history of the bells is unrecorded.

Frequent entries for new bellropes in the Church-wardens' Accounts show that the bells were well used up to the mid nineteenth century. By contrast very little

0   5   10
METRES

TOWER   NORTH AISLE   VESTRY
                                    ORGAN
                                                      ro/ss

VICE   Ⓐ   NAVE   CHANCEL
              Ⓑ

Ⓒ   SOUTH AISLE   TRANSEPT

PORCH

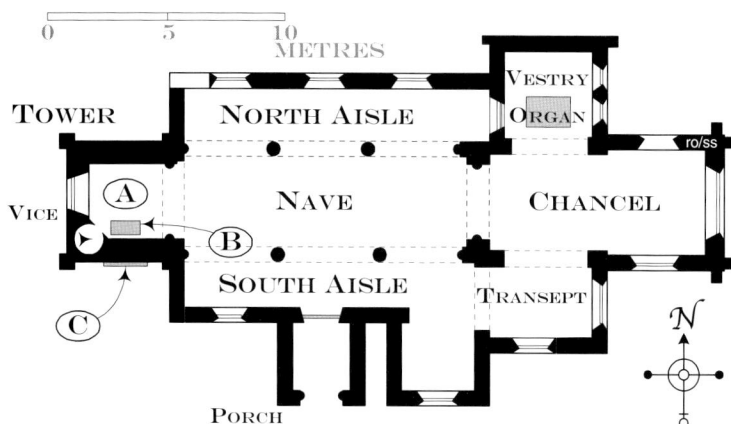

Plan of St Mary's Church
A: Six bells in the belfry. Ringing
chamber on the first floor of the tower
B: Clock on the second floor of the
tower
C: Clock dial
The vice provides access to both the
clockroom and the belfry

appears to have been spent on actual repairs to the bells. Some examples of both are given below:

1759	pd John peet for a bellclaper mending and two new Stays and nails	6s 2d
1769	pd Charles Willmott for mending the Church Bells	6s 0d
1788	Mending bell wheels	7s 6d
1807 May	3 New Bell Ropes & putting up too	£1 1s 0d
1842 March 30	Mr Darnell's Bill Ropes paid	12s 0d

By 1920 the bells and bellframe were causing concern. A quotation was obtained that same year from John Taylor for rehanging the bells in a new metal frame and it included for recasting the tenor. This quotation provides information about the old frame:

The framework is very old and decayed and is quite unfit to stand the strain of the bells ringing in full swing. It is noticed that this frame has been wedged up to the walls of the tower in various places. This undoubtedly steadies the frame somewhat but is a very bad practice as undue strain is thereby thrown on to the tower walls which is likely to do serious mischief to the structure if the ringing of the bells is continued. If these wedges were removed the frame would probably collapse altogether under the ringing of the bells, and taking everything into consideration we strongly advise that the bells should not again be rung in full swing in their present condition.

Taylors also stated that the tenor was 'of poor thin quality of tone' and that there was a possibility that it was cracked. They recommended recasting the bell with 'about 1cwt. of new metal to be added, to bring the bell to the requisite thickness and weight'. The cost of the new frame and for rehanging the bells was estimated at £275, with an extra £62 for recasting the tenor.

However, this quotation was not accepted and it was at a PCC Meeting held on 7 June 1923 that repairs to the belfry and bell framework were again considered. The minutes record that:

it was proposed that owing to the very bad & unsafe condition of the present wood structure on which the Bells were suspended, owing to the ravages of time and weather, it might be more economical and efficient to have a steel framework fitted if such could be procured — Messrs A Day and A Stevenson kindly consented to undertake the necessary work if such be considered satisfactory at the next Council Meeting [their offer was in fact to rehang the bells in a new oak frame].

New floors to the bell tower and clock chamber were essential, as both were rotten and unsafe. At this time there was an 'Edith Weston Restoration Fund', and it was decided to use this fund for the bell project.

Mears and Stainbank also provided a quotation for rehanging the bells in a new steel frame, but a decision was taken to accept the offer made by Arthur Day, the local blacksmith and Arthur Stevenson, landlord of the Wheatsheaf public house in Edith Weston (Kelly 1925, 724). The work was completed by the beginning of 1924.

On 10 March 1949 Taylors again quoted for refurbishing the belfry. They confirmed that since their previous inspection in 1920 a new oak frame had been installed, but that it was 'totally unfit to withstand the strain of the three bells being rung in full swing'. They continued to report that the

old fittings of the bells which were already worn out at the time [in 1920] were apparently used again when the bells were placed in the new frame, with the exception of the wheels which were renewed ... these fittings are in such a bad state that we advise that the bells should not be rung but should be chimed only.

It is interesting that this quotation does not regard the tenor to be cracked as previously reported in 1920. A badly fitted clapper affected the tone.

Taylor's quotation was for restoring the three existing bells and for adding three new bells to make a ring of six. All six bells were to be hung in a high-sided metal frame with new fittings, including elm headstocks for the fourth and fifth bells which were to retain their canons. The

West 2 Plan of the bellframe at Edith Weston Church

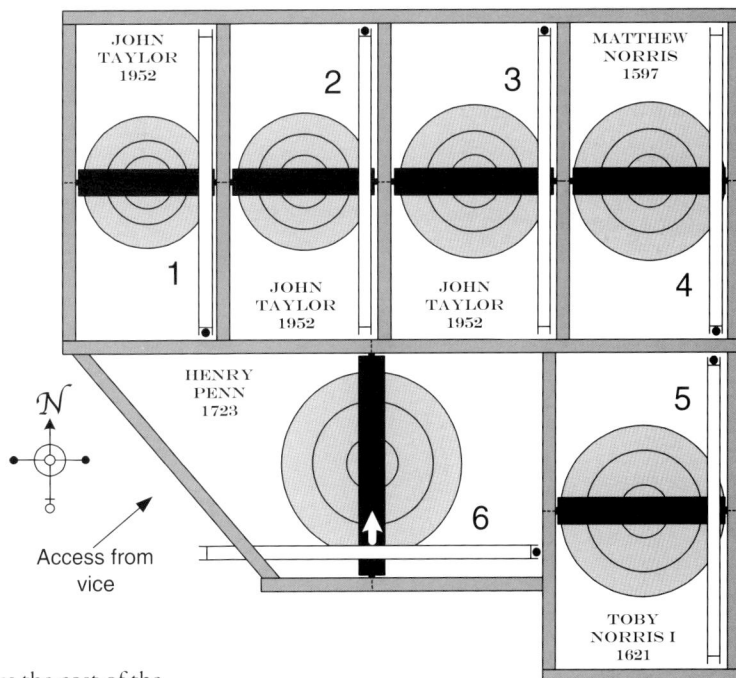

estimated cost of this work was £510 plus the cost of the new bells: treble £127, second £133, third £139. The inscriptions were charged at 1s per letter. The faculty for this proposal was granted on 23 November 1951 and the installation was completed in 1952. There is a memorial tablet in the ringing chamber recording the gifts of the parishioners and the individual donors.

BELL DETAILS

1 1952. Treble. Note F sharp. Diameter 648mm (25½in). Weight 194 kg (3cwt 3qr 8lb). Cast without canons. Cast-iron headstock. Cast by John Taylor & Co of Loughborough.

**JOHN TAYLOR & CO [65] FOUNDERS [65]
~ LOUGHBOROUGH [65] 1952 [65]**

On the waist:

**PERCY JAMES BEAUMONT
RECTOR 1952**

Louisa Beaumont, the Rector's wife, donated this bell.

2 1952. Note E. Diameter 673mm (26½in). Weight 206kg (4cwt 0qr 6lb). Cast without canons. Cast-iron headstock. Cast by John Taylor & Co of Loughborough.

**JOHN TAYLOR & CO [65] FOUNDERS [65]
~ LOUGHBOROUGH [65] 1952 [65]**

On the waist:

**ARTHUR DAY
RECTOR'S CHURCHWARDEN
1937 - 1949**

Arthur Day was the local blacksmith and this bell was donated to his memory by Catherine, his wife, who was

the sub-postmistress at Edith Weston. He and Arthur Stevenson installed the new bellframe in 1924.

3 1952. Note D. Diameter 718mm (28¼in). Weight 224kg (4cwt 1qr 18lb). Cast without canons. Cast-iron headstock. Cast by John Taylor & Co of Loughborough.

**JOHN TAYLOR & CO [65] FOUNDERS [65]
~ LOUGHBOROUGH [65] 1952 [65]**

On the waist:

**VERNON B. CROWTHER-BEYNON
RECTOR'S CHURCHWARDEN
1896 - 1912**

Vernon Bryan Crowther-Beynon FSA bequeathed this bell. He lived at Edith Weston Grange until he moved to Beckenham, Kent, in 1912. He was a founder member of the Rutland Archaeological and Natural History Society in 1902 and a frequent contributor to the *Rutland Magazine*. He died in 1941 at the age of 75 years (Waites 1987, 269).

4 1597. Note C sharp. Diameter 749mm (29½in). Weight 244kg (4cwt 3qr 5lb). Canons retained. Timber headstock. Ascribed to Matthew Norris of Leicester and possibly of Stamford (George Dawson).

**ƧVM ROƧA PVLSATA MVИDIA MARIA
~ VOCATA 1597**

(I being rung am called Mary the Rose of the World) North in 1880 lists the present fourth and fifth bells in reverse order to that shown here. It has been suggested that the inscription on this bell is an attempted copy of that on the ancient bell when it was recast (North 1880, 128). Scheduled for preservation (Council for the Care of Churches).

5 1621. Note B. Diameter 832mm (32¾in). Weight 303kg (5cwt 3qr 25lb). Canons retained. Timber headstock. Cast by Toby Norris I of Stamford.

[26] ИОИ [79] ƆLAMOR [79] SED [79] AMOR ~ [79] ƆAИTAT [79] IИ [79] AVRE [79] DEI ~ [79] 1621 [79] [5] T [5] P [5] [79] T [79] F [79]
(It is not noise but love that sings in the ear of God)

6 1723. Tenor. Note A. Diameter 889mm (35in). Weight 361kg (7cwt 0qr 13lb). Canons removed. Cast-iron headstock. Cast by Henry Penn of Peterborough.

JOHN [50] BULL [50] C.W [50] HENRY.
~ PENN [50] MADE . ME 1723 [80]

Above the inscription is a band of decoration [80]. In 1880 it was incorrectly recorded that this bell was cracked (North 1880, 128). The clock strikes the hours on this bell.

BELLRINGING CUSTOMS

On Sundays the treble was rung at 8am. For Divine Service the bells were chimed, after which the tenor was rung. At the Death Knell there were thrice three tolls for a male and thrice two tolls for a female (North 1880, 128).

During the first half of the nineteenth century, payments recorded in the Churchwardens' Accounts reveal that the bells were rung on 5 November and occasionally to celebrate other events:

1769		pd the 5th Novr for ringing	1s 0d
1805	Nov 7	paid the Ringers Nelson's victory	5s 0d
1809	Oct 25	paid the Ringers at the Jubilee	5s 0d
1821	July 19	Paid the Ringers on the King's [George IV] Coronation	10s 0d

At a Vestry Meeting 29 June 1829 it was 'Agreed that the men on the Road to Go Out at 7 O Clock in the morning & stop till 6 at night & from 12 O Clock till one for his [their] Dinner'. A bell was probably rung at these times to communicate the appropriate working hours.

Today Edith Weston has its own band of bellringers that practises regularly. The third bell is chimed before all of the Sunday services. If ringers are available the bells are rung for the 11am Easter Service, at the Midnight Service at Christmas, and at weddings if requested.

GREGORIAN CALENDAR

In 1752 Britain changed from the inaccurate Julian calendar to the Gregorian calendar. The eleven days difference between the two meant that 14 September followed 2 September. At the same time, the old custom of starting the year on Lady Day (25 March) was aban-

A note in Edith Weston Parish Records regarding the new Gregorian calendar

doned in favour of 1 January. The adoption of the new calendar is recorded on a loose scrap of paper in the Vestry Minutes Book (DE 1937/24).

CLOCK HISTORY

The old seventeenth-century wooden framed clock from Edith Weston Church is now on display in Rutland County Museum (L1975.11) as a working exhibit. Prior to its removal from the tower in 1928 it had served the community for nearly two hundred and fifty years.

Originally, the clock had no dial but indicated the time by striking the hours on one of the church bells. From the beginning of the surviving Churchwardens' Accounts, which start in 1757, there are frequent references to winding, servicing and repairing this clock. The following are the first three relevant entries:

1759	for Repairing the Church Clock	£1 18s 0d
1762	pd for 2 New Lines for ye Clock	16s 6d
1767	Pd Mr Wilson for Repairing The Church Clock	£9 8s 6d

The Mr Wilson noted here is Ralph Wilson, watchmaker and silversmith, whose shop was opposite the George Inn in St Martins, Stamford. He continued to be responsible for the Edith Weston clock until 1773. In 1775 a decision was made to provide the clock with a single hour hand and a stone dial. This is shown on a *circa* 1793 drawing of the church (RCM F10/1984/16). The work was entrusted to Richard Hackett, a clockmaker of Harringworth, and to Henry Stone, the local stonemason. A new clock case may have been supplied at the same time:

The dial installed by Henry Stone in 1775 at Edith Weston Church. The minute hand was added when the new clock was installed in 1928. The mark on the wall around the dial indicates that it originally had a stone bezel and this is confirmed by the circa 1793 drawing (RCM F10/1984/16)

1775

Spent on the Workmen at taking down the church
Clock   2s 0d
Paid for Meat and Drink for the Workmen at putting up
the Clock   £1 4s 2d
Mr Hacket's Bill for repairing the clock   £13 12s 6d
Henry Stone's Bill for the Clock Dial and putting it up
                                                        £2 17s 8d
A lock for the Clock case   9d
The Carpenter's Bill   £11 10s 6½d

The conversion work included fitting two new brass wheels in the going train and probably a longer pendulum. This would mean that a new escape wheel was also required. Although these wheels were removed when the clock was converted to verge and foliot control in the early 1960s, they were saved and are now displayed alongside the clock in Rutland County Museum. The hour wheel is engraved **R^D HACKETT HARRINGWORTH 1775**. The pendulum is missing.

Richard Hackett carried out more work on the clock in 1777. After this, other than maintenance provided by James Wilson of Stamford in the 1780s, it seems that little work was required on the movement until major repairs became necessary in 1809:

1809 Sept 10

Ale wen taken the clock down   1s 4d
Taken the clock to Stamford & bringin it Back again 10s 0d
Paid for ale   2s 0d
Mr Thos Monck Bill for Repairing the clock £15 16s 0d

Thomas Monck was a gunsmith, clockmaker and watchmaker and at this time had a shop in High Street, Stamford. He was in business with his brother Edmund who, until 1790, was a journeyman to James Wilson. Thomas serviced and repaired the clock until 1817. From 1818 William Aris II, who died in 1842, and subsequently his son, Thomas, both clockmakers of Uppingham, were responsible for the clock until at least 1847. After this date there is a gap in the accounts until 1903. The parish clerk, Edward Ablett, is named as the clock winder during the 1830s. The last recorded repair to this clock was in 1911:

1817 Dec 9	Paid Mr Aris, Uppingham, for Clening the Clock	10s 6d
1847 March 23	Aris's bill paid	£1 13s 0d
1911 Oct 6	Fromant (repairing Church Clock)	15s 0d

The Vestry and PCC Minute Book supplies the following information pertinent to the clock during the 1920s and 1930s:

In 1920 Mr George Naylor, sexton, was replaced by Mr E Russell and his annual salary was £10. His duties were 'to include the winding of the church clock'.

On the evening of 28 March 1928 a meeting was held in the village hall to consider the question of a new church clock. The Rector, churchwardens and about thirty parishioners attended the meeting. The Rev B P Payne explained that the old clock movement had been inspected by two clockmakers who had pronounced it 'quite worn out'. A decision was made to accept the quotation of G & F Cope of Nottingham to supply and install a clock with hour and half-hour striking at a cost of £60. An additional requirement was to preserve the character of the old clock dial, but if possible to put two hands in place of the one that then existed. Half of the required amount was promised by Colonel and Lady Cicily Hardy, providing the remainder could be raised by public subscription (DE 1937/25).

The clock had been installed by the end of June that year and a brass plaque on the north wall of the north aisle records the event:

The new clock was placed in the tower in the year of our Lord 1928 to replace the ancient 'One handed clock' believed to date from the time of the Commonwealth and has chimed away the hours for many generations in this place. The cost of the new clock was defrayed by generous contributions of patron, parishioners and old friends long associated with the parish of Edith Weston.

B. P. Payne   Rector
E. Makey   J. E. Andrew   Churchwardens

Although the present clock was supplied and installed by G & F Cope of Nottingham in 1928, it was manufactured in Germany by Phil Horz of Ulam-am-Donau (Hewitt 1994, 37). The going train has a remontoire. With this device, instead of directly turning the hands, the going train lifts a small weight. This weight is released every minute to turn the hands forward by exactly one minute and the result of this action can be seen by watching the dial. The purpose of the remontoire is to

The Phil Horz clock at Edith Weston which was supplied and installed by G & F Cope in 1928

The pendulum suspension and the going train remontoir fly on the Phil Horz clock at Edith Weston

G & F Cope's invoice for the Edith Weston clock (DE 5032/39)

improve timekeeping by isolating the escapement from variable forces caused by the effect of wind loading on the hands. It is the only clock in Rutland with this feature.

The successful installation of this clock was to lead to another clock by G & F Cope being installed at Tinwell Church the following year (DE 5032/39).

In 1932 Mr Bryan as Parish Clerk was made responsible for winding and regulating the church clock and later that year it was agreed that Mr Cope's employee, Mr Smith, who installed the clock, should annually inspect, oil and clean it for the sum of two guineas.

By April 1936 it was found that the clock was keeping very irregular time and it was thought that this might be due to excessive vibration of the supporting framework during bellringing. Copes solved this problem by replacing the wooden clock stand with two short steel girders set into the wall of the tower This provided a rigid mounting for the clock. As an additional precaution against variations in time-keeping it was suggested that a wooden case be fitted round the pendulum to protect it from draughts. Lady Hardy undertook to meet the cost of these alterations.

When the old clock was removed in 1928 it was stored in a partly dismantled state behind a bench at the west end of the church. In 1960 an offer to restore it was made by the Nottingham Branch of the Antiquarian Group of the BHI. The work, which was carried out to a very high standard by a group of eight experienced clockmakers, took five years to complete. The clock was eventually returned to the church in December 1965.

The presence of unused holes in the framework and

The parts removed during the restoration and conversion to verge and foliot escapement

the 'discovery of the date 1658 marked on the inner face of the annulus of each main wheel' (*BHI Journal* **109**, 1967) together with the fact that it was commonly referred to as the 'Commonwealth Clock', led the group to conclude that it was made before the anchor escapement and pendulum had been generally applied to turret clocks. [The first known turret clock with this new form of control was made by Joseph Knibb in 1669-70.] They decided that the unused holes were left over following the supposed conversion to anchor escapement by Richard Hackett in 1775. Their restoration therefore involved the removal of Hackett's conversion and the installation of a new verge and foliot escapement together with new wheels and pinions, all in wrought iron, to an appropriate style. Without the benefit of later research on similar clocks this was probably a reasonable conclusion at the time, but it is now considered to be inappropriate. The clock, however, is one of the few examples in a museum which demonstrates a working verge and foliot escapement.

Recent research by Michael Lee, who has restored or inspected all of the surviving clocks made by John Watts of Stamford, has led him to conclude that the Edith Weston clock is also by this maker (Lee 2000, 29). It was

The new verge and foliot escapement fitted to the Edith Weston clock

therefore probably made after 1675. If, like all other surviving John Watts' clocks, it had pendulum control, it would have started life with an anchor escapement. Hackett's conversion of 1775 was therefore to provide a hand for the clock. It is unlikely that the community would have accepted the unreliability and inaccuracy of a verge and foliot escapement when a far more accurate and reliable anchor escapement had been available for a hundred years.

Michael Lee dismantled the clock in 1991 with the intention of looking for the 1658 dates which were recorded, but not illustrated, by the Nottingham horologists. 'I inspected all the old metal wheels in great detail and was disappointed that none of the many rough marks looked like a date' (Lee 2000, 29). A further inspection was made in the year 2000 by co-author Robert Ovens and he was also unable to locate these dates. It is possible that they were originally very faint and are now not visible due to the cleaning and painting process. According to the church guide of 1976 the date was written 'in the old English scribe'. If so, it would be possible to misinterpret 88 for 58 making the date 1688, well into the time scale of when John Watts was supplying clocks in the area.

CLOCK DETAILS

Details of the 'Commonwealth Clock' now in the Rutland County Museum:

Maker:	Almost certainly John Watts of Stamford
Signed:	No name or date on the clock
Installed:	Possibly 1688
Cost:	Not recorded, but probably £10
Frame:	Posted wooden frame with wrought-iron pivot bars
Trains:	Going and striking
Escapement:	Converted to verge and foliot from anchor in the 1960s restoration
Rate:	21 beats per minute
Striking:	Countwheel hour-striking
Weights:	Stone weights
Winding:	Daily
Note:	Removed from the tower in 1928. Restored 1960 to 1965. Originally displayed in the church but on indefinite loan to the Rutland County Museum since 1975

Details of the present clock:

Maker:	Phil Horz, Ulam-am-Donau, Germany (Hewitt 1994, 37)
Signed:	No name or date on the clock
Installed:	Supplied and installed in 1928 by G & F Cope & Co of Nottingham
Cost:	£60
Frame:	Small cast-iron birdcage
Trains:	Going and striking
Escapement:	Deadbeat
Pendulum:	Wooden pendulum rod and cast-iron lenticular bob (approximately 250mm diameter)
Rate:	60 beats per minute
Striking:	Countwheel hour and half-hour striking
Weights:	Removed
Winding:	Smith's epicyclic automatic winder to each train, installed in 1987
Dial:	Circular limestone dial installed in 1775. Has hour and minute hands but marked out for a single hour hand
Location:	South face of the tower
Note:	The going train has a remontoire device which lifts a small weight to move the hands by one minute every minute

SUNDIAL HOUSE

The vertical sundial at Sundial House (SK 926055) has two sets of hour lines, one with arabic and the other with roman numerals. One possible explanation is that it is a relocated dial. A date stone above the dial records that the house was built in 1773.

The sundial at Sundial House, Edith Weston. The shadow is indicating approximately 2.25pm by the arabic numerals

EDITH WESTON HALL

A table sundial is shown in the garden of Edith Weston Hall on the early 1900s OS Second Edition 25 inch map.

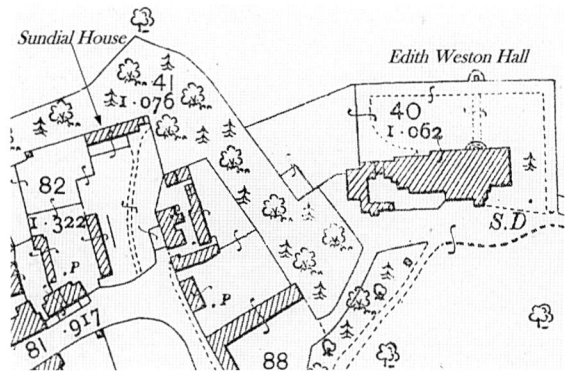

The locations of Sundial House and the table sundial (SD) in the garden of Edith Weston Hall are shown on this early 1900s OS Second Edition 25 inch map

EGLETON

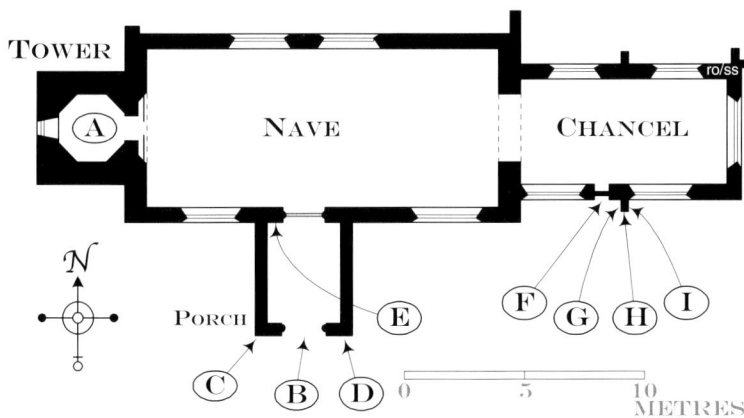

Plan of St Edmund's Church
A: One bell and a Priest's Bell in the belfry. Ringing chamber at the base of the tower
B: Sundial
C: Scratch dial 1
D: Scratch dials 2 & 3
E: Scratch dial 4
F: Scratch dial 5
G: Scratch dial 6
H: Scratch dials 7, 8, 9, 10 & 11
I: Scratch dial 12
Access to the belfry is by ladders

ST EDMUND
Grid Ref: SK 876075

PARISH RECORDS

There are no surviving records in the parish archive relevant to the history of the bells.

BELL HISTORY

The only early reference to the bells is a presentment recorded in Irons' Notes on an Archdeacon's Visitation:

1605 Mr Harbottle's father of Egleton about 50 or 60 years since sold the great bell (*VCH* II, 47).

There is one bell and a Priest's Bell in the tower as recorded by Thomas North in 1880, who stated that they were 'rung by a lever'. The absence of any inscription or founder's mark on either of the bells, coupled with the lack of any relevant parish records, means that the bells cannot be dated with any certainty. The former

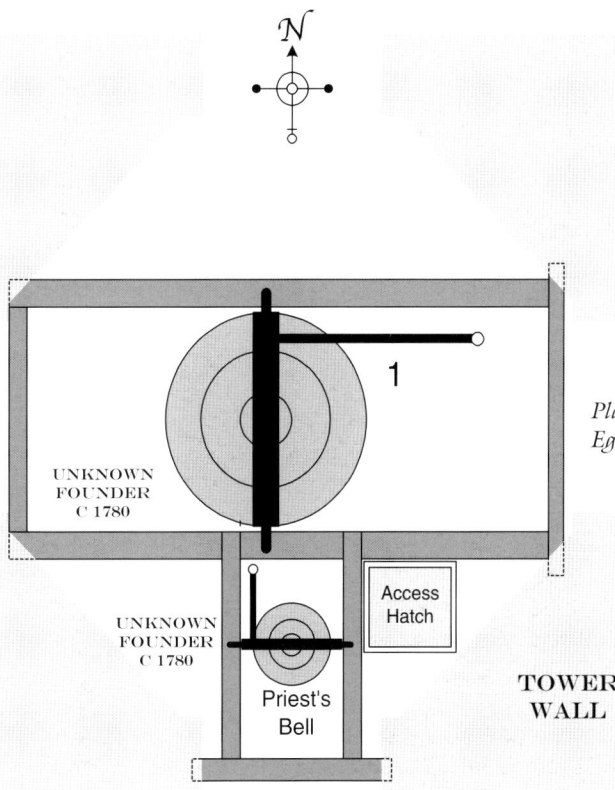

Plan of the bellframe at Egleton Church

tower is said to have been destroyed by fire (Phillips 1903-04, 175) and Brewer (1813, 59) described the tower as being 'quite modern'. A *circa* 1793 drawing (RCM F10/1984/17) confirms that the tower and the spire had been rebuilt by this date. The present bells may have been installed at the time of this rebuilding.

The Priest's Bell is hung in a side extension to the old oak frame that carries the larger bell. The frame was repaired and the bells rehung with new bearings and fabricated headstocks in 1991 at a cost of £600. Both bells have levers for chime ringing.

BELL DETAILS

1 **Probably *circa* 1780**. Note A flat. Diameter 991mm (39in). Weight approximately 559kg (11cwt). Canons removed. Fabricated steel headstock. Cast by an unknown founder. No inscription.

PRIEST'S BELL

Probably *circa* 1780. Diameter 387mm (15¼in). Weight approximately 47kg (3qr 20lb). Canons removed. Fabricated steel headstock. Cast by an unknown founder. No inscription. North (1880) recorded this bell as the treble.

BELLRINGING CUSTOMS

The Gleaning Bell was rung during harvest and at the Death Knell there were thrice three tolls for a male and

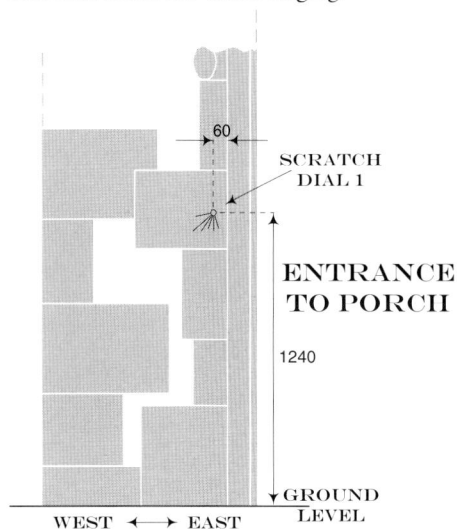

Location of scratch dial 1 on the south face of the porch, to the west of the entrance of Egleton Church

Scratch dial 1 details

EGLETON 1					
Location	West front of south porch				
Condition	Poor				
Gnomon Hole Diameter		25			
Gnomon Hole Depth		Filled in			
Height above ground level		1240			
Line Ref	a	b	c	d	e	f
Length	60	70	70	100	120	100
Angle (°)	168	180	196	215	225	260

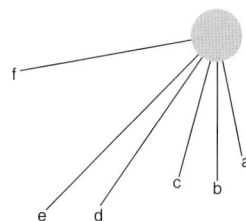

Scratch dial 2 details

EGLETON 2							
Location	East front of south porch						
Condition	Poor - lines faint						
Gnomon Hole Diameter		12					
Gnomon Hole Depth		Filled in					
Height above ground level		1560					
Line Ref	a	b	c	d	e	f	g
Length	102	100	90	110	100	110	88
Angle (°)	155	160	174	190	211	220	235

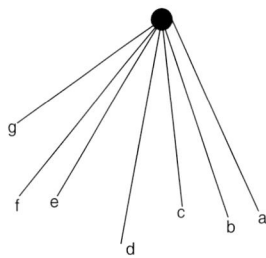

EGLETON 3							
Location	East front of south porch						
Condition	Poor - lines faint						
Gnomon Hole Diameter		25					
Gnomon Hole Depth		Filled in					
Height above ground level		990					
Line Ref	a	b	c	d	e	f	g
Length	144	102	110	120	124	122	100
Angle (°)	115	130	147	163	183	200	218

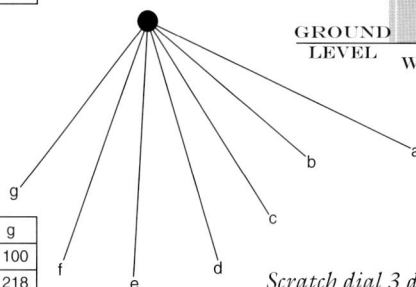

Location of scratch dials 2 and 3 on the south face of the porch, to the east of the entrance to Egleton Church

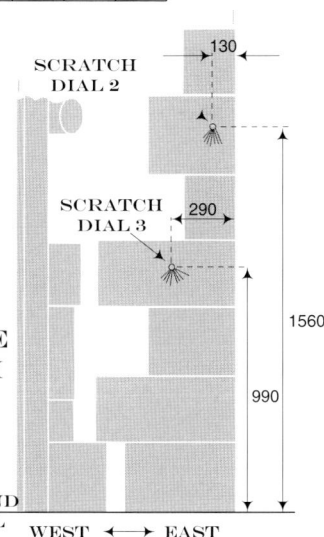

Scratch dial 3 details

SCRATCH DIAL 4

SOUTH DOOR

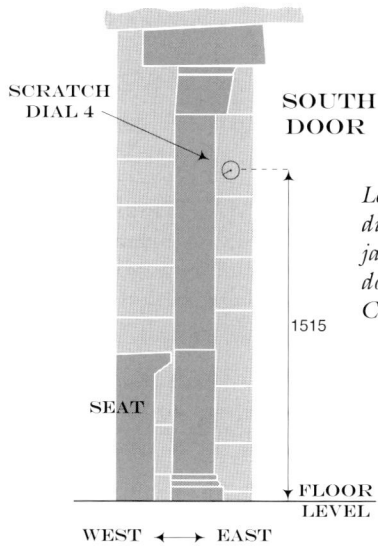

Location of scratch dial 4 on the west jamb of the south door at Egleton Church

1515

SEAT

FLOOR LEVEL

WEST ◄──► EAST

EGLETON 4		
Location	West jamb of south doorway	
Condition	Good	
Gnomon Hole Diameter	5	
Gnomon Hole Depth	Pock	
Height above ground level	1515	
Circle Diameter	82	
Line Ref	a	
Length	41	
Angle (°)	260	

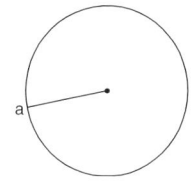

Scratch dial 4 details

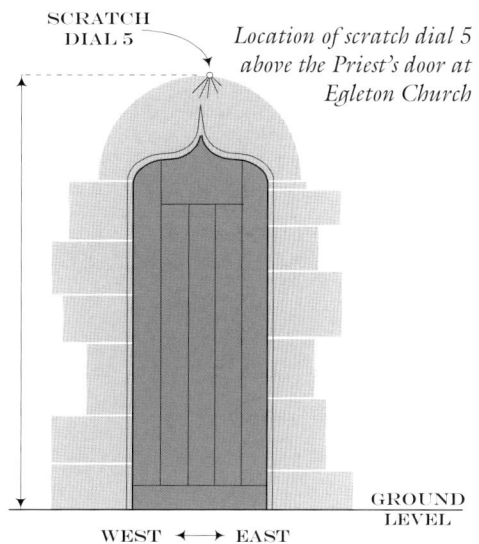

SCRATCH DIAL 5

Location of scratch dial 5 above the Priest's door at Egleton Church

1850

a

b

Scratch dial 5 details

e

c

d

EGLETON 5					
Location	Above Priest's door				
Condition	Poor - lines faint				
Gnomon Hole Diameter	Filled in				
Gnomon Hole Depth	Filled in				
Height above ground level	1850				
Line Ref	a	b	c	d	e
Length	122	128	178	202	202
Angle (°)	145	160	182	200	222

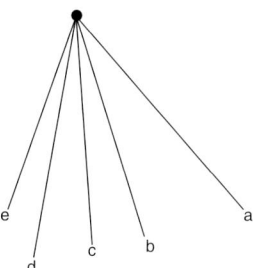

GROUND LEVEL

WEST ◄──► EAST

Location of scratch dial 6 on the west face of the chancel buttress at Egleton Church

260

SCRATCH DIAL 6

CHANCEL

1390

e

a

d

c

b

Scratch dial 6 details

EGLETON 6					
Location	West face of chancel buttress				
Condition	Poor				
Gnomon Hole Diameter	5				
Gnomon Hole Depth	Pock				
Height above ground level	1390				
Line Ref	a	b	c	d	e
Length	125	112	110	118	100
Angle (°)	138	162	176	190	199

Note that this scratch dial is on a relocated stone

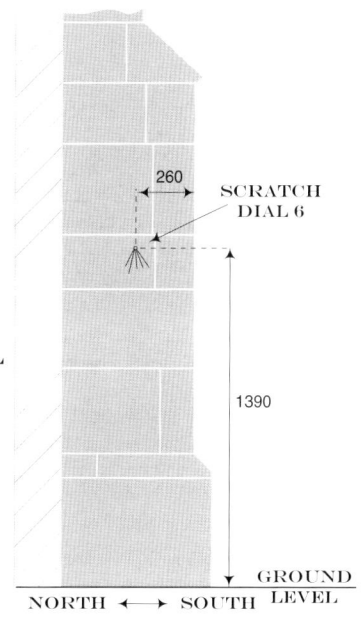

GROUND LEVEL

NORTH ◄──► SOUTH

Scratch dial 7 details

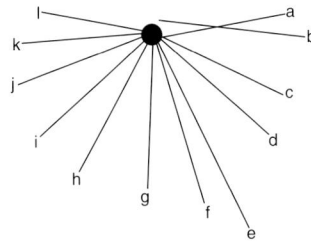

EGLETON 7												
Location	South face of chancel buttress											
Condition	Good											
Gnomon Hole Diameter	12											
Gnomon Hole Depth	10											
Height above ground level	1660											
Line Ref	a	b	c	d	e	f	g	h	i	j	k	l
Length	60	65	70	75	105	85	75	75	75	70	65	55
Angle (°)	80	95	115	130	154	163	183	208	228	249	265	280

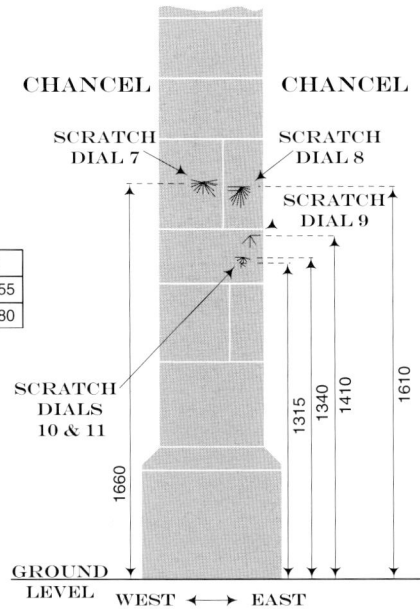

Scratch dial 8 details

EGLETON 8													
Location	South face of chancel buttress												
Condition	Good												
Gnomon Hole Diameter	7												
Gnomon Hole Depth	10												
Height above ground level	1610												
Line Ref	a	b	c	d	e	f	g	h	i	j	k	l	m
Length	57	67	85	77	90	108	114	110	110	94	50	50	50
Angle (°)	90	117	130	150	162	175	184	197	210	225	244	270	270

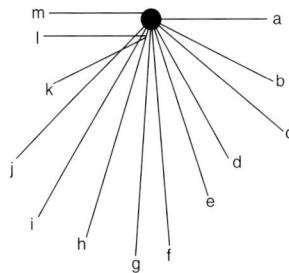

Location of scratch dials 7, 8, 9, 10 and 11 on the south buttress of the chancel at Egleton Church

Scratch dial 9 details

EGLETON 9				
Location	South face of chancel buttress			
Condition	Good			
Gnomon Hole Diameter	None			
Gnomon Hole Depth	None			
Height above ground level	1410			
Line Ref	a	b	c	d
Length	40	50	48	45
Angle (°)	90	130	190	220

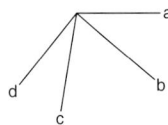

EGLETON 10 & 11				
Location	South face of chancel buttress			
Condition	Average			
Gnomon Hole Diameter	None			
Gnomon Hole Depth	None			
Height above ground level	1340 / 1315			
Line Ref	a	b	c	d
Length	50	43	40	40
Angle (°)	110	148	180	212

Line Ref	x	y	z
Radius	56	36	31
Angle (°)	90	137	180

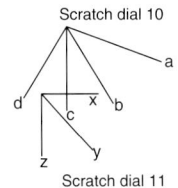

Scratch dial 10 and 11 details

thrice two tolls for a female (North 1880, 128).

The Sexton used to toll the church bell for a few minutes on Sunday morning at 8am to indicate that a service would be held at 11am, or at 11am if the service was to be held in the afternoon. The 11am bell also indicated to the villagers that it was time to take their Sunday dinners to the bakehouse, a custom held in many Rutland villages that died out just before the Second World War. The baker made a charge of 2d for cooking the roast and Yorkshire pudding (Traylen nd, Pt **1**).

Today the larger of the bells is chimed thirty-three times for each regular service at the church. It is also chimed at funerals, and for weddings if requested.

SCRATCH DIALS

There are twelve scratch dials on the south elevation of St Edmund's Church. Multiple scratch dials are not unusual, although it is rare to find so many on one church. Many of the dials at Egleton are very faint and can only be seen in the early morning or late afternoon when the sun is shining across the walls. The south door is twelfth century, the front face of the south porch is fourteenth century, and the south wall of the chancel is fifteenth century.

Location of scratch dial 12 on the east face of the chancel buttress at Egleton Church

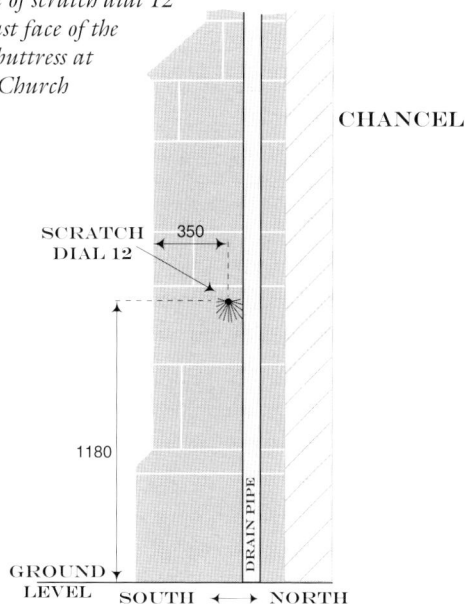

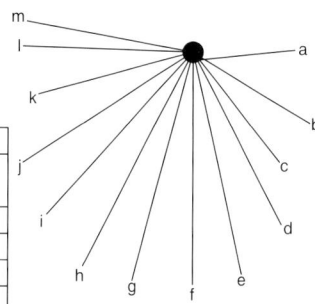

Egleton Church. Scratch dial 8 on the south face of the chancel buttress

Scratch dial 12 details

EGLETON 12								
Location	East face of chancel buttress							
Condition	Good							
Gnomon Hole Diameter			15					
Gnomon Hole Depth			10					
Height above ground level			1180					
Line Ref	a	b	c	d	e	f	g	h
Length	52	68	68	94	110	112	114	116
Angle (°)	85	120	140	152	167	180	195	208
Line Ref	i	j	k	l	m			
Length	108	100	80	84	84			
Angle (°)	223	238	255	272	290			

Note that this scratch dial is on a relocated stone

SUNDIAL

There is a sundial in the gable of the south porch which was probably installed when the porch was rebuilt during the restoration of 1872. In oblique sunlight the roman numerals, hour lines and decoration can be seen quite clearly. They are in slight relief and this may indicate that the dial was painted rather than carved, this effect being caused by the differential weathering of the painted and unpainted surfaces.

The porch sundial at Egleton Church seen in oblique sunlight. The hours are divided into quarters and there is a sunburst decoration round the top of the gnomon

EMPINGHAM

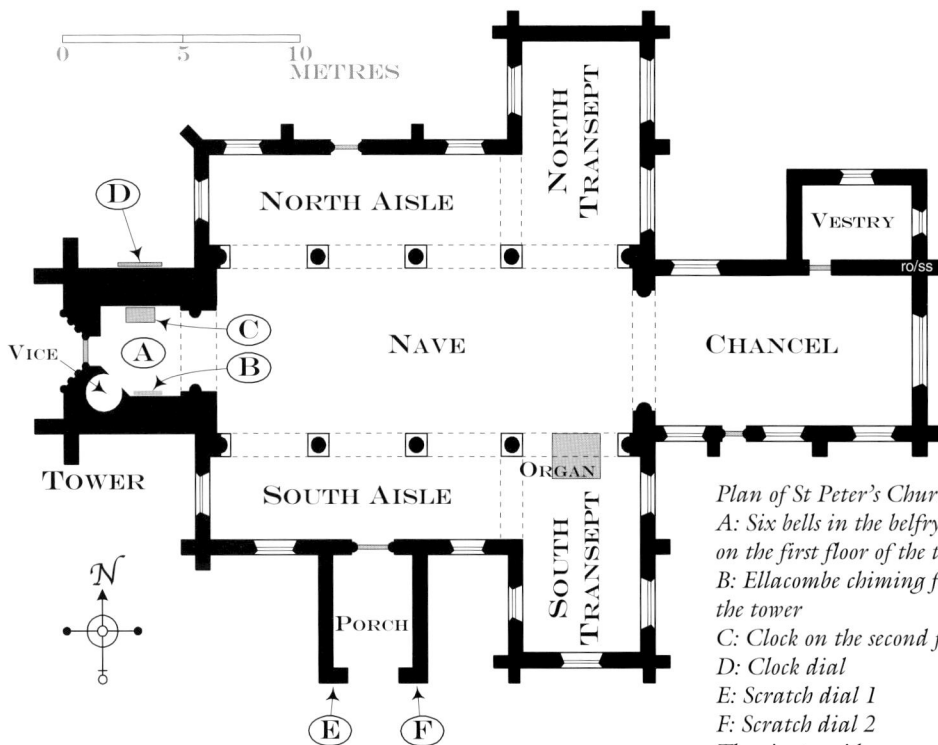

Plan of St Peter's Church
A: Six bells in the belfry. Ringing chamber
on the first floor of the tower
B: Ellacombe chiming frame at the base of
the tower
C: Clock on the second floor of the tower
D: Clock dial
E: Scratch dial 1
F: Scratch dial 2
The vice provides access to both the ringing
chamber and clockroom. Access to the belfry
is by ladder

ST PETER   Grid Ref: SK 951085

PARISH RECORDS

The only relevant surviving documents are the Church-wardens' Disbursements 1779-85 (DE 1617/1b) and the Vestry Minutes 1726-77 (DE 1617/1a), 1819-34 (DE 1617/2) and 1834-55 (DE 1617/3). They provide no historical information regarding the clocks and bells. However, a few details from the lost parish records are given in *A History of Empingham AD 500 - AD 1900* by J E Swaby.

BELL HISTORY

According to Thomas North the earliest recorded bell of five at Empingham in 1859 was dated 1548. This and two of the others dated 1611 and 1648 were cast by unknown founders. The other two bells dated 1661 and 1695 were cast in the Norris foundry at nearby Stamford and it is fair to assume that the other seventeenth-century bells were also cast in that same foundry. No doubt some of these bells were recasts of earlier bells. There is some evidence that bells were actually cast in the churchyard at Empingham by an itinerant founder, for during excavations in 1876 indications of a furnace and fused bell metal were discovered (North 1880, 11).

Details from the Whitechapel Sales Day Book of the new ring of five supplied to Empingham in 1859 (Whitechapel Archives)

George Mears & Co of the Whitechapel Bell Foundry recast all five bells in 1859 and this ring was used for the first time on 4 July 1859. Unfortunately, they were not well hung and by 1880 both the fourth and fifth bells were cracked (North 1880, 129). All five bells were replaced and a sixth bell added during the church restoration of 1895. These were all cast by John Taylor of Loughborough and hung in a new high-sided metal frame. The new clock was installed at the same time. When floorboards were removed during alterations in the tower the following words were revealed. They were written in an almost illiterate hand and were obviously from a set of Ringers' Rules (Ennis 1979, 16):

He that doth a bell overthrow,
Two pence shall pay before he go.

The bells were rehung by Eayre & Smith Ltd in August 1990. A chiming apparatus at the base of the tower was given in memory of Herbert Tibbert, 1900-71, in gratitude for his long service to the church as bellringer and verger.

BELL DETAILS

1 1895. Treble. Diameter 752mm (29⅝in). Weight 266kg (5cwt 0qr 26lb). Cast without canons. Cast-iron headstock. Recast by John Taylor & Co of Loughborough.

RING OUT THE FALSE : RING IN THE TRUE
~ [65] [65] [65]
On the waist: [81]
A. D. 1895
Below the inscription band is decoration [54]. The lettering is like [137].

Former bells
Cast in 1695 by Toby Norris III of Stamford.
TOBY NORRIS CAST ME T MITCHELL 1695
Recast in 1859 by George Mears & Co of the Whitechapel Bell Foundry, London. Diameter 762mm (30in).
G. MEARS FOUNDER LONDON 1859.

2 1895. Diameter 800mm (31½in). Weight 306kg (6cwt 0qr 3lb). Cast without canons. Cast-iron headstock. Recast by John Taylor & Co of Loughborough.
GLORY TO GOD IN THE HIGHEST
~ [65] [65] [65] [65] [65]
On the waist:
[81]
A. D. 1895
Below the inscription band is decoration [54]. The lettering is like [137].

Former bells
Cast in 1548. Founder unknown.
1548

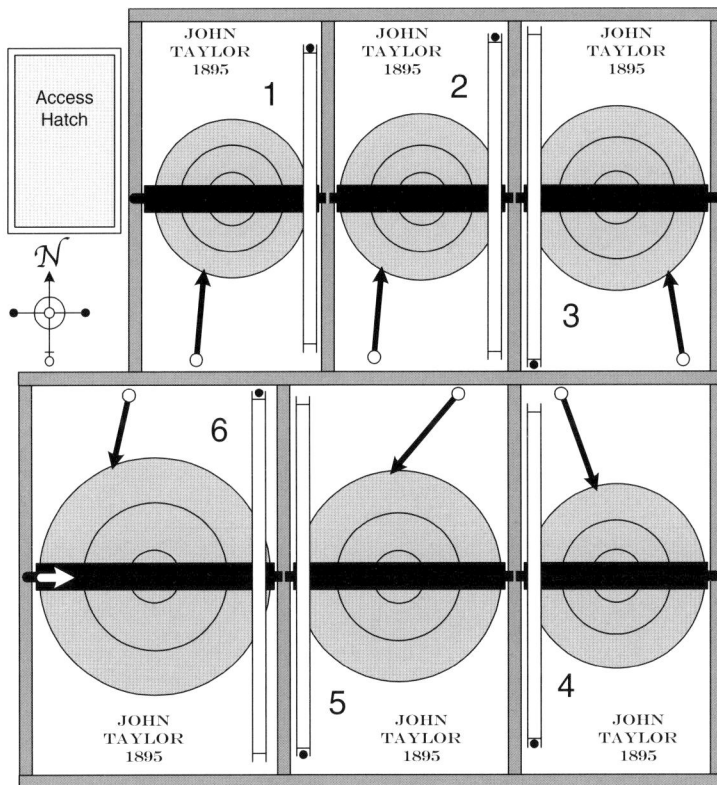

Plan of the bellframe at Empingham Church

Recast in 1859 by George Mears & Co of the Whitechapel Bell Foundry, London. Diameter 838mm (33in).

G. MEARS FOUNDER LONDON 1859.

3 1895. Diameter 867mm (34⅛in). Weight 379kg (7cwt 1qr 24lb). Cast without canons. Cast-iron headstock. Recast by John Taylor & Co of Loughborough.

ON EARTH PEACE : GOODWILL TOWARD
~ MEN [65] [65] [65] [65]

On the waist:

[81]
A. D. 1895

Below the inscription band is decoration **[54]**. The lettering is like **[137]**.

Former bells

Cast in 1648. Founder unknown but probably Thomas Norris of Stamford.

1648

Recast in 1859 by George Mears & Co of the Whitechapel Bell Foundry, London. Diameter 876mm (34½in).

G. MEARS FOUNDER LONDON 1859.

4 1895. Diameter 911mm (35⅞in). Weight 424kg (8cwt 1qr 11lb). Cast without canons. Cast-iron headstock. Recast by John Taylor & Co of Loughborough.

HOSANNA TO THE SON OF DAVID
~ [65] [65] [65] [65] [65] [65] [65] [65]

On the waist:

[81]
A. D. 1895

Below the inscription band is decoration **[54]**. The lettering is like **[137]**.

Former bells

Cast in 1661 by Thomas Norris of Stamford.

THOMAS NORRIS CAST ME 1661

Recast in 1859 by George Mears & Co of the Whitechapel Bell Foundry, London. Diameter 940mm (37in).

G. MEARS FOUNDER LONDON 1859.

This bell was cracked in 1880 (North 1880, 129).

5 1895. Diameter 1019mm (40⅛in). Weight 599kg (11cwt 3qr 5lb). Cast without canons. Cast-iron headstock. Recast by John Taylor & Co of Loughborough.

OMNIA FIANT AD GLORIAM DEI
~ [65] [65] [65] [65] [65] [65] [65] [65] [65]
(Let all things be done to the Glory of God)

On the waist:

GILBERTUS JOHANNES
DOMINUS DE AVELAND
(Gilbert John, Lord Aveland)
A. D. 1859

On the waist, opposite:

THOMAS LOVICK COOPER, M. A.　VICAR
WILLIAM FANCOURT ⎫
MARK CANNER ⎰**CHURCHWARDENS**

Below the inscription band is decoration **[54]**. The lettering is like **[137]**.

Gilbert John Heathcote, landowner, of Normanton Hall was created Baron Aveland in 1856 (Waites 1987, 278). Thomas Lovick Cooper, born 1801, was rector of the parish from 1831 to 1892. William Fancourt was a farmer in the village (White 1846, 622). Mark Canner was a grocer and draper (White 1877, 673).

Former bells

Cast in 1611. Founder unknown but probably Toby Norris I of Stamford. Weight 571kg (11cwt 1qr) (Whitechapel Archives).

OMNIA FIANT AD GLORIAM DEI
ANN MACKWORTH AND
~ THOMAS MACKWORTH ARMIGER 1611

Ann Mackworth and her son Thomas Mackworth were the donors of this bell. Ann was the second wife and widow of George Mackworth Esq of Mackworth in Derbyshire, and Empingham. She was buried at Empingham on 4 June 1612. Thomas was sheriff of the county in 1599 and created a baronet on 4 June 1619. He was buried at Empingham 22 March 1656 (North 1880, 129). Tenor bell until 1859.

Recast in 1859 by George Mears & Co of the Whitechapel Bell Foundry, London. Diameter 1067mm (42in).

G. MEARS FOUNDER LONDON.
~ OMNIA FIANT AD GLORIAM DEI.
GILBERTUS JOHANNES
~ DOMINUS DE AVELAND AD 1859
WILLIAM FANCOURT ⎫
MARK CANNER ⎰**CHURCHWARDENS**

The new donor of 1859 replaced the names of the donors on the 1611 bell. Thomas North did not record the name of Thomas Lovick Cooper as is included on the present bell. This bell was cracked in 1880 (North 1880, 129). Tenor bell until 1895.

6 1895. Tenor. Note F sharp. Diameter 1143mm (45in). Weight 791kg (15cwt 2qr 9lb). Cast without canons. Cast-iron headstock. Cast by John Taylor & Co of Loughborough.

BLESSED IS HE THAT COMETH IN THE
~ NAME OF THE LORD [65] [65] [65]

On the waist:

Ancaster
1895

On the waist, opposite:

THOMAS WILLIAM OWEN, M. A. RECTOR
ROBERT BARRAND ⎫**CHURCH**
WILLIAM THOMAS HUMPHREY ⎰**~WARDENS**

Below the inscription band is decoration **[54]**. The lettering on the inscription band is like **[137]**.

Ancaster

The donor's name on the waist of the tenor at Empingham

Gilbert Henry Heathcote-Drummond-Willoughby of Normanton Hall was created Earl of Ancaster in 1892 (*Burke's Peerage* 1967, 71). Thomas William Owen was rector of the parish from 1892 to 1921. Robert Barrand and William Thomas Humphrey were farmers in Empingham and the former was also the village blacksmith (White 1877, 673).

The clock strikes the hours on this bell.

Bellringing Customs

The Pancake Bell was rung on Shrove Tuesday but this custom had ceased by 1880. At the Death Knell there were three tolls for a male and two tolls for a female. The tenor was usually tolled for the Death Knell but at 1880 the third bell was used because both the fourth and tenor were cracked. The Gleaning Bell was rung during harvest at 8am and 5pm (North 1880, 129).

A History of Empingham provides the following information, presumably taken from Churchwardens' Accounts which were available at that time:

In 1760 the ringers were paid for ringing on 29 May (the Restoration of the Monarchy), 29 July, 1 August, 7 October and 22 October. The usual payment to the team was 2s 6d. In 1764 there were special rings on 29 May, the King's 'Bearthday', his 'Crownation' and 5 November.

In 1888 Thomas Bland, as clerk and sexton was paid 2s for tolling the bell for the dowager Lady Willoughby and 3s in 1892 for tolling the bell for H R H The Duke of Clarence, the eldest son of The Prince of Wales. In 1899 a Mrs Bland was paid 10s for ringing the Gleaning Bell (Swaby 1988, 45 & 47).

Ernest Mills recorded his memories from seventy years of living in the village in *Empingham Remembered*. The following extract from this work provides more details of bellringing customs in the village:

The church bells played a large part pre 1939. A full peal was rung at 6 a.m. on Christmas Day to herald in the birth of Christ, in addition to ringing before morning and evening service throughout the year. On the passing of the old year the tenor bell would be tolled mournfully for 10 minutes or so to midnight and then the full peal would ring in the new year.

Harry Sneath was verger and sexton of the church and as he worked locally

he was always available to ring the so-called Death Bell. This I think was the sixth and largest bell in the Church and he would be summoned as soon as the death was confirmed (except between 6 p.m. and 9 a.m.) to dong the bell. Three times for a male and twice for a female and

then once for each year of the deceased life. Pretty hard on him when an octogenarian passed on.

On the afternoon of a funeral

the death bell would toll about 1.30 to call the pall bearers, who suitably attired in deep black with bowler hats, would collect the bier from its storage place.

After the coffin had been collected the family mourners would follow the coffin to the north gate of the churchyard and

all the time this was happening the death bell would toll mournfully until the cortege was safely in the Church (Mills 1984, 14 & 19).

More information about gleaning customs in Empingham at the beginning of the twentieth century is given in *Leicestershire & Rutland within Living Memory* (Leicestershire & Rutland Federation of Women's Institutes 1994, 85). After the sheaves had been cleared from the cornfields a Gleaning Bell would be rung daily at 8am throughout the gleaning season. This was a signal to the village women that they could enter certain fields to pick up any spilled corn. Each gleaner had to pay three pence to the bellringer. The women, sometimes as many as forty, each holding her lunch and a large linen sheet, would meet at the church. The oldest woman was 'mistress of ceremonies' and she would lead the assembled group to the gleaning field waving a flag. The heads of gathered corn were heaped onto the sheets and when the day ended at 5pm each gleaner would tie her sheet by its corners and carry it home balanced on her head.

Today Empingham has a band of bellringers which practises for one hour each Tuesday evening. Three or four of the church bells are rung for thirty minutes, then the treble chimed for one minute before the 9.30am service on Sundays. In the recent past, the chime ropes would be pulled by a sidesman prior to the Sunday Service. The bells are rung at 9.30am on Christmas Day and although they are not rung during Holy Week, they are on Easter Sunday. As a mark of respect, the tenor is always chimed for one minute at 11am on Armistice Day. The bells are usually rung at weddings as the bride and groom leave the church, and this continues for thirty minutes.

Handbells

Ernest Mills in *Empingham Remembered* also includes the following:

There was also a set of handbells, the ringers of these toured the larger houses and played a selection of tunes on Christmas evening for which they received a donation and [were] regaled with beer and mince pies. Mr Charles Wilson, known to every one as Charlie, Miss Elsie Wilson's father, was owner of the bells and leader of this team and I believe the bells are still in existence in the Wilson family. Charlie was also a church bellringer and played the trumpet in Stamford Silver Band (Mills 1984, 14).

Oil Lamp

An oil lamp in the ringing chamber was used by the bellringers before electricity was installed.

Scratch Dials

There are two scratch dials on the mid fourteenth-century south porch. Both are on the south face. The dial on the east side of the entrance consists of a gnomon hole and vague remains of two lines and a full circle of pocks. It is in such poor condition that no meaningful dimensions can be taken.

Bellringers' oil lamp in the ringing chamber at Empingham Church

Location of the two scratch dials at Empingham Church

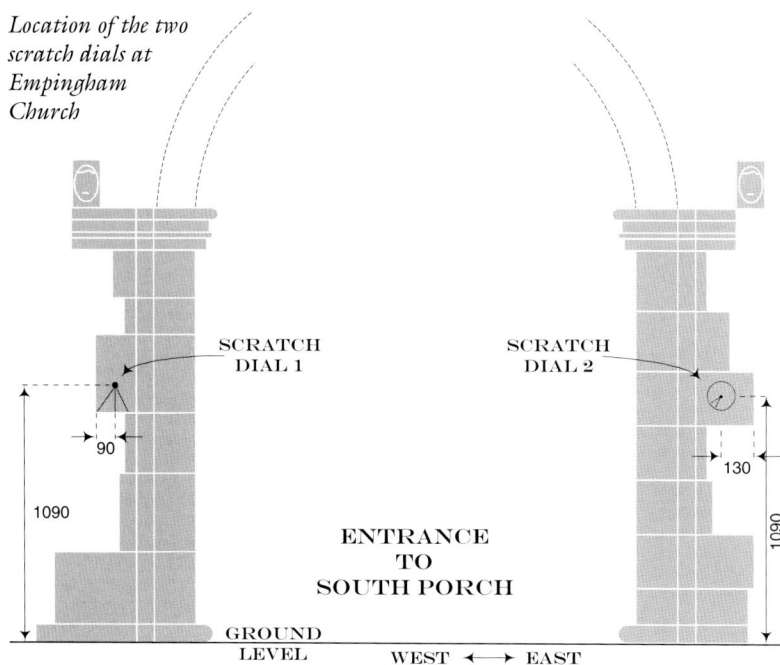

SCRATCH DIAL 1

SCRATCH DIAL 2

90

130

1090

1090

ENTRANCE TO SOUTH PORCH

GROUND LEVEL

WEST ←——→ EAST

Details of scratch dial 1

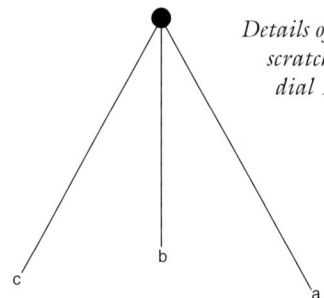

c   b   a

EMPINGHAM 1			
Location	West front of south porch		
Condition	Poor		
Gnomon Hole Diameter		10	
Gnomon Hole Depth		15	
Height above ground level		1090	
Line Ref	a	b	c
Length	130	110	140
Angle (°)	150	180	210

Clock History

The former clock at Empingham Church is now in Stamford Museum. It is signed **I W 1686** indicating that it was made in 1686 by John Watts of Stamford.

Few details of the clock's history have survived but it is known that it was refurbished in 1877 when a new wooden frame was provided. Fortunately, that part of the old frame with John Watts' signature was saved and fitted over the top rail of the replacement frame. Two names are carved on the new frame: **W. OGDEN. 1877** on the top rail and **T BLAND P. CLERK** on a side rail. William Ogden was the village carpenter, who was presumably the maker of the new frame, and Thomas Bland was the Parish Clerk.

John Watts' signature on the old Empingham church clock frame

One of the clerk's daily duties was to wind the clock:

In 1894 Thomas [Bland] had ... £1 for the clock
(Swaby 1988, 47)

After the new clock was installed as part of the church restoration in 1895, the Watts' clock was left in the clockroom. Here it remained until 1984, when a grant was obtained from the Council for the Care of Churches to pay for its restoration with the intention of displaying it in Stamford Museum. Michael Lee carried out the restoration work:

The mechanism had been pushed to one side, complete with all the working parts, when a new clock was installed. The only parts that had been disconnected were on the floor and leaning against the wall of the clock room. These parts were associated with the striking mechanism and had to be removed to make way for the new clock .... Apart from woodworm in the winding barrels and some worn

John Watts' Empingham clock before restoration
(Michael Lee)

Michael Lee with the dismantled Empingham clock
(Michael Lee)

The two stone weights belonging to Empingham church clock were found behind a table tomb in the churchyard. They are limestone with the weight loops held in by lead. The going train weight in the centre is 250mm (10in) high and weighs approximately 13.6kg (30lb). The striking train weight on the right is 325mm (13in) high and weighs approximately 31.7kg (70lb). The broken weight on the left was found when the clock was removed for restoration. It may indicate that one of the other weights is a replacement

bushes all that was needed was a good cleaning of all the parts (Lee 2000, 31-2).

The two stone clock weights were missing at this time, but they have since been found in the churchyard.

CLOCK DETAILS

Details of the clock replaced in 1895:

Maker:	John Watts of Stamford
Signed:	I W 1686
Installed:	1686
Cost:	Not recorded, but probably about £10
Frame:	Wooden frame
Trains:	Going and striking
Escapement:	Anchor
Rate:	60 beats per minute
Striking:	Countwheel hour-striking
Weights:	Original stone weights found in the churchyard in 2001
Winding:	Daily
Notes:	Replaced by the present clock in 1895. Remained in the clockroom until restored in 1984

Details of the present clock:

Maker:	John Smith & Sons of Derby
Signed:	On the setting dial and on the frame
Installed:	1895
Cost:	£88 (Swaby 1988, 45)
Frame:	Cast-iron flatbed
Trains:	Going and striking
Escapement:	Pinwheel
Pendulum:	Wooden rod and cast-iron cylindrical bob
Rate:	48 beats per minute
Striking:	Locking plate hour-striking
Weights:	Cast-iron
Winding:	Hand wound weekly
Dial:	Skeleton dial with gilded roman numerals, minute ring and hands
Location:	North face of the tower

The escapement of the Smith clock at Empingham

The present clock by John Smith & Sons which was installed in the church at Empingham in 1895

ESSENDINE

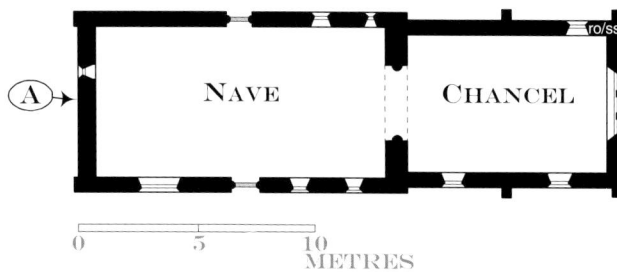

*Plan of St Mary the Virgin's Church
A: Two bells in a double bellcote
Bells chimed from the west end of the nave*

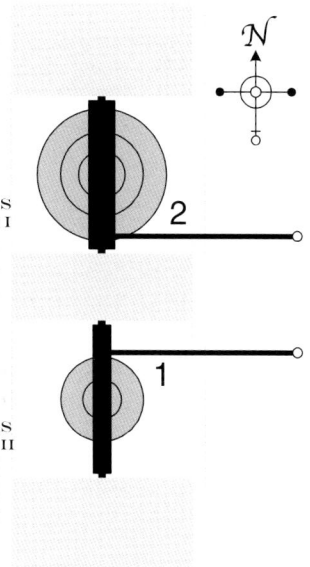

THOMAS
MEARS I
1805

*Plan of the
double bellcote
at Essendine
Church*

THOMAS
MEARS II
1823

ST MARY THE VIRGIN   Grid Ref: TF 049127

PARISH RECORDS

The surviving Churchwardens' Accounts (DE 2395/6-7) cover the period 1917-38. There are no references to the bells in these documents.

BELL HISTORY

The typical Rutland double bellcote was built in the thirteenth century and the early style was retained when it was rebuilt *circa* 1835. Nothing is known of the early bells. The present bells are dated 1805 and 1823, but neither has a founder's name or mark. They are hung for chime ringing. A lever on each headstock is connected to a bellrope inside the church by wires which pass through the roof.

One source (*VCH* II, 254) states that 'before 1888 the bells were rung from outside' and a drawing of *circa*

1839 (Uppingham School Archives) seems to support this. However, a drawing of the church *circa* 1793 by Nathan Fielding (RCM F10/1984/21) clearly shows that they were rung from inside the building. An anonymous drawing (RCM F10/1984/20) of approximately the same date confirms this and shows a structure which may well cover an opening in the roof for the bellropes. Blore's illustration of 1811 indicates that the ropes hung inside the church (Blore 1811, 26).

The implication is that the bells were rung from

From a drawing of Essendine Church circa 1793 showing the structure on the eastern side of the bellcote. Such a structure would have enabled access to the bells from inside the church (RCM F10/1984/20)

From a slightly later drawing of Essendine Church by Nathan Fielding showing that the structure had been removed from behind the bellcote and that the bells were rung from inside the church (RCM F10/1984/21)

inside the church until the west wall was rebuilt *circa* 1835. They were then rung from the outside until, at some time during the next fifty years, they reverted to being rung from the nave.

Bell Details

1 **1823**. Diameter 381mm (15 in). Weight approximately 51kg (1cwt). Canons retained. Timber headstock. Ascribed to Thomas Mears II of the Whitechapel Bell Foundry, London (George Dawson). On the lower waist:

> **THOMAS STEANS**
> **CHURCHWARDEN**
> **1823**

North (1880, 130) records the date of this bell as 1808 and has the bell order reversed.

2 **1805**. Diameter 457mm (18 in). Weight approximately 51kg (1cwt). Canons retained. Timber headstock. Ascribed to Thomas Mears I of the Whitechapel

Bell Foundry, London (George Dawson).

1805

Bellringing Customs
At the Death Knell there were thrice three tolls for a male and thrice two tolls for a female (North 1880, 130). Toady both bells are chimed for ten minutes before the Sunday Service held at 9am and for the service held at the same time on Christmas Day.

The bellcote at Essendine, looking east

EXTON

ST PETER & ST PAUL   Grid Ref: SK 920112

Parish Records
Churchwardens' Accounts exist from 1858 to 1870 (DE 3012/74) and from 1863 to 1924 (DE 3012/75). There are also three files of Church Accounts for the period 1902-24 and the year 1937 (DE 3012/76-8). Files of correspondence and other documents relating to the bells include:

Report on a survey of damage caused by a lightning strike in April 1843 (DE 3012/40).
Bills, receipts and correspondence of 1895 and 1896 regarding the restoration of the bells (DE 3012/68/1-14).
Printed order of service for the benediction of the bells on 29 January 1896 (DE 3012/69).
Bellringers' rules and a list of ringers of 1896 and 1897 (DE 3012/70).
Correspondence and specification for overhauling the bells by Taylors dated 1933 to 1938 (DE 3012/71/1-6).
Unpublished reminiscences: *Some Recollections of the Parish of Exton in the County of Rutland from AD 1875-1877* by Herbert Robert Wilton Hall, schoolmaster (DE 3012/130/1-3; Wilton Hall 1913).

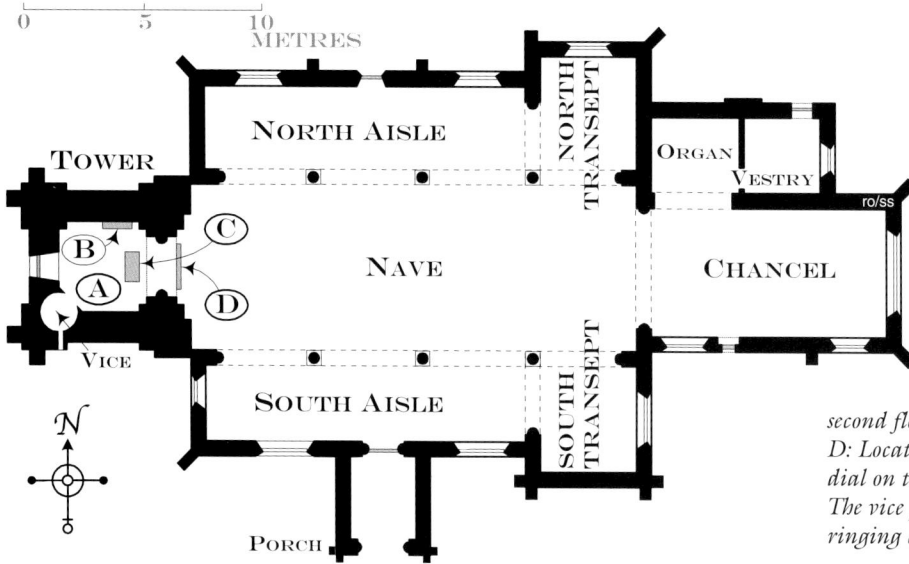

Plan of St Peter &
St Paul's Church
A: Six bells in the
belfry. Ringing
chamber on the first
floor of the tower.
The third bell can be
chimed from the base
of the tower
B: Ellacombe
chiming frame in
the ringing chamber
C: Location of the
former clock on the
second floor of the tower
D: Location of the former clock
dial on the east face of the tower
The vice provides access to the
ringing chamber and belfry

BELL HISTORY

The earliest bells for which some details are available are the ring of five cast by Toby Norris III of Stamford in 1675. Joseph Eayre of St Neots increased this to six by the addition of a new tenor in 1763. The tower and steeple were extensively damaged by a lightning strike on 25 April 1843 but it appears that the bells and frame survived intact.

An Ellacombe chiming frame was installed by John Taylor in the ringing chamber as recounted in the following extract from *Some Recollections of Exton*:

> The irregularities of the Bellringers had caused Mr Knox [Rector of Exton] much trouble. They usually departed from the belfry in a body on Sundays as soon as the bells were 'down', and never entered the Church. They seemed to think that they had a prescriptive right to ring the Bells for their own amusement when they chose, and to consume as much beer in the Belfry as they could convey thither. They would not be reformed, so Mr Knox introduced Ellacombe's Chiming Apparatus which could be manipulated by one man - Ezra Smith. Thenceforward the bells were very rarely rung (Wilton Hall 1913, 11).

Ezra Smith was paid regularly for chiming from 1874 right through to 1896 and his yearly salary was £2 10s 0d.

Examples from the accounts concerning the bells:

1870 Nov 7	Paid Mr Dennison for Bell Ropes	£2 8s 0d
1874 Oct 24	Tyers Bill for Chimeing	16s 0d
Jan 22	Littledyke's Bill for Ropes	£3 0s 0d
1882 Oct 25	Gleaning Bell	15s 0d
Nov 11	Chimeing Ropes	6s 0d

A new high-sided metal frame by John Taylor was installed in 1895 when both the treble and tenor bells were recast and the remaining Norris bells rehung without their canons. The total cost of recasting the two

The chiming
apparatus at
Exton Church
which was used
by Ezra Smith
and others

bells, including an allowance for the metal from the old bells was £272 16s 7d. The Bishop of Peterborough dedicated the new installation on 29 January 1896 and members from the St Martin's Society of Stamford were paid £2 7s 6d for ringing at this service.

Inspired by the work done within the belfry, a group of sixteen men immediately formed themselves into 'The Exton Ringers'. The available accounts show regular annual payments to 'H Green for Bellringers' up to 1921. H Green was one of the ringers to sign the original *Rules for Bell Ringers*.

Additional extracts from the Accounts:

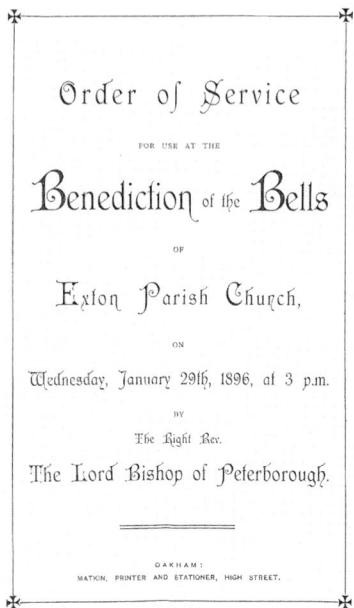

Exton Church. The Order of Service for the benediction of the bells

1908	Aug 25	Midland Railway Coy [Company] Carriage on Clapper from Taylors	8d
	Sept 4	Jno Wallace hanging Clapper	1s 0d
	Oct 13	Jno Taylor & Co repairs to Bell Clapper	16s 1d
1911	April 17	W A Ireland Painting Bell frames £2 10s repairs 12/6	£3 2s 6d
1915	March 11	Jno Taylor & Son 6 Bell Mufflers	£1 1s 4d

The bells were again rehung by John Taylor in 1934. The work included new gudgeons and ball bearings, and painting the frame. The total cost was £75.

BELL DETAILS

1 1895. Treble. Note E. Diameter 752mm (29⅝in). Weight 287kg (5cwt 2qr 17lb). Cast without canons. Cast-iron headstock. Recast by John Taylor & Co of Loughborough.

DONVM [52] DE [52] DOMINA CAMBDEN ~ [52] 169 [85]
(The gift of Lady Campden)

On the waist: [81]
RECAST 1895

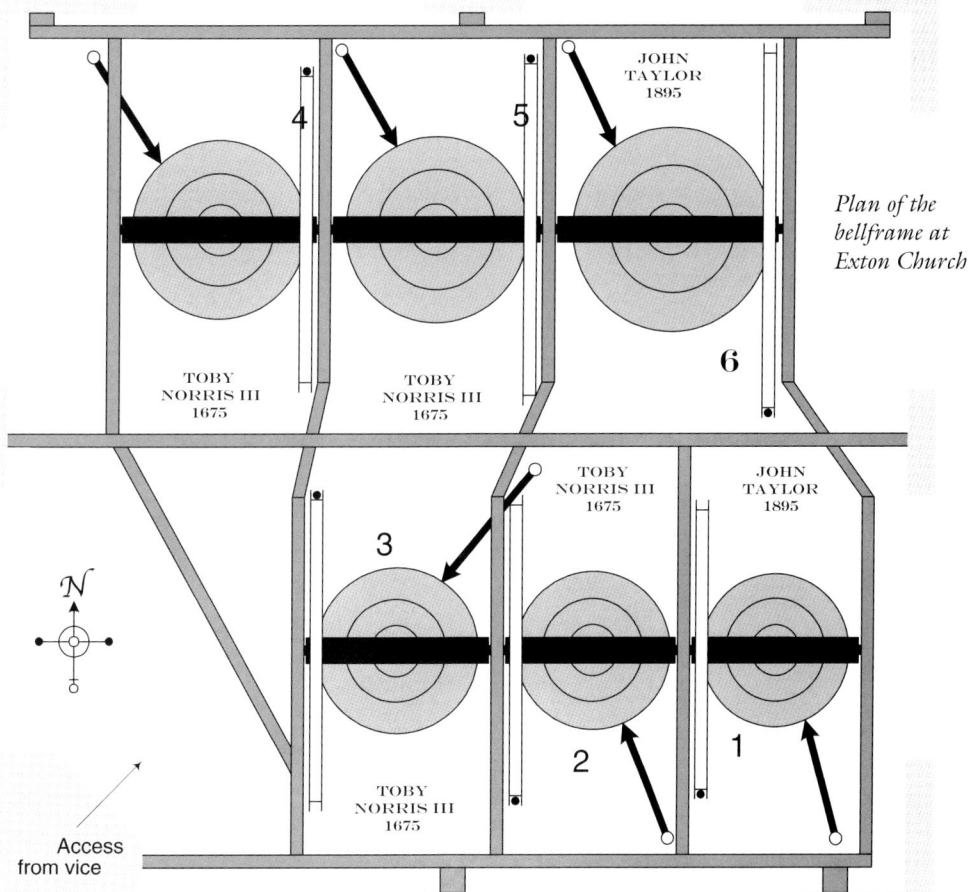

Plan of the bellframe at Exton Church

The marble figure of Lady Campden on the Grinling Gibbons monument in the north transept of Exton Church. She donated Exton treble in 1675

Decoration [54] is below the inscription band. The date should read 1675. Lady Campden was Lady Elizabeth Bertie, the fourth wife of Baptist Noel, the third Viscount Campden of Exton House. He died at Exton on 29 October 1683.

Former bell
Cast in 1675 by Toby Norris III of Stamford. Diameter 660mm (26in).

DOИVM DE DOMIИA CAMBDEИ 16Ƨ∠

2 1675. Note D. Diameter 787mm (31in). Weight 300kg (5cwt 3qr 18lb). Canons removed. Cast-iron headstock. Cast by Toby Norris III of Stamford.

[26] [52] GOD [52] SAVE [52] THE [52] KING ~ [52] 1675 [52]

3 1675. Note C. Diameter 832mm (32¾in). Weight 291kg (5cwt 2qr 25lb). Canons removed. Cast-iron headstock. Cast by Toby Norris III of Stamford.

[26] [52] GOD [52] SAVE [52] THE [52] KING ~ [52] TOBIEAƧ [52] ИORRIƧ [52] ~ CAƧT [52] ME [52] 1675 [52]

4 1675. Note B. Diameter 876mm (34½in). Weight 372kg (7cwt 1qr 8lb). Canons removed. Cast-iron headstock. Cast by Toby Norris III of Stamford.

[26] [52] GOD [52] SAVE [52] THE [52] KING ~ [52] TOBIEAƧ [52] ИORRIƧ [52]

~ CAƧT [52] ME [52] 1675 [52]

5 1675. Note A. Diameter 921mm (36¼in). Weight 386kg (7cwt 2qr 12lb). Canons removed. Cast-iron headstock. Cast by Toby Norris III of Stamford.

[14] [52] GOD [52] SAVE [52] THE [52] KING ~ [52] TOBIEAƧ [52] ИORRIƧ [52] ~ CAƧT [52] ME [52] 1675 [52]

6 1895. Tenor. Note G. Diameter 1029mm (40½in). Weight 639kg (12cwt 2qr 10lb). Cast without canons. Cast-iron headstock. Recast by John Taylor & Co of Loughborough.

JOSEPH EAYRE S^T NEOTS FECIT 1763 [65] ~ THOMAS HURST VICAR: W. SPRINGTHORP ~ CHARLES BROWN CHURCHWARDENS:.

On the waist:

[81]
RECAST 1895

Decoration [86] is below the inscription band.

Former bell
Cast in 1763 by Joseph Eayre of St Neots. Diameter 1016mm (40in). Weight 553kg (10cwt 3qr 16lb).

JOSEPH EAYRE ST NEOTS FECIT 1763 ~THOMAS HURST VICAR W SPRINGTHORP ~ CHARLES BROWN CHURCHWARDENS

This bell was recast because a crack was found in the crown. The weight was increased to 'produce a more fuller and richer tone' (DE 3012/68/8).

BELLRINGING CUSTOMS
The following is taken from *Some Recollections of Exton*. The Rev John Baker who had been the Assistant Curate of Exton from March 1874 to March 1878 supplied the information in 1914:

At funerals it was the custom for the Sexton to ring the 'carrying bell' for a few minutes, about an hour before the cortège was timed to arrive at the church: the passing bell was rung after a death; three times <u>three</u> tolls being given at the end, in case of a male, and twice two a female. The bell was also tolled at intervals of a minute for half an hour before a funeral, and was rung quickly for some 5 minutes after. In gleaning time a bell was rung some ten minutes at 9am for the people to begin their gleaning, and at 4pm for them to stop. For this the gleaners each made a small return to the Sexton (Wilton Hall 1913, Addenda).

The accounts of the 1860s show that the Clerk was paid 10s for ringing the bell during harvest.

The bells were chimed on Sunday for Divine Service using the Ellacombe chiming apparatus.

At Funerals a few tolls are given on the tenor bell as a warning for the bearers to assemble; the same bell is tolled (half a minute between each stroke) for half an hour before the funeral office is said (North 1880, 131).

At the beginning of the twentieth century, the following occasions for ringing the bells were recorded in the church accounts:

1903	Sept 4	H Green muffled Peal for late vicar on Sunday 30 Aug [19]03	3s 0d
1910	May 7	Bell Ringers for Muffling bells on the Death of King Edward VII	2s 0d
1919	Jan 11	Jas Bardwell Refreshments for Ringers on Armistice	5s 0d

Thatched Village (Buchan 1983, 121) gives an insight into the custom of gleaning at Exton in the 1930s. When carting of the sheaves was complete any remaining stalks of corn were raked up. During raking a stook was left in the field as a sign to the gleaners to keep out. A church bell was rung as the signal for the gleaners to start.

Today the bells at Exton are not rung or chimed for the regular services held throughout the year. They are however rung at weddings if requested by the bride and groom and if a local band of ringers can be found to do so.

Handbells

It is recorded in *Thatched Village* (Buchan 1983, 62) that on Christmas Day afternoons in the 1930s Exton handbell ringers walked round the village playing carols. When they had finished Mrs Buchan, the headmistress of the Roman Catholic School, gave them mincepies and beer.

Bellringers' Rules

The newly formed Exton Ringers of 1896 signed a printed agreement to observe the bellringers' rules and each member was given a copy. Amongst the rules was one that may have discouraged some from joining: 'That no Beer or Spirituous Liquors be brought into the Belfry.' As an afterthought an extra rule was added by hand: 'That 1d a week be paid by each member to form a fund for paying for broken stays' (DE 3012/70).

Clock History

Evidence of an early clock is given in Irons' Notes (MS 80/5/24):

1618 April 11 The clock there goeth not right but maketh a greate rumbling when it goeth

This early clock would have had a verge and foliot escapement and possibly a wooden frame. The more accurate and reliable anchor escapement was not available until after 1670. There is no further record of this clock and it is assumed that it was changed for a pendulum controlled clock, possibly in the early 1700s. A drawing of *circa* 1793 by Nathan Fielding shows a dial on the east face of the tower.

On 25 April 1843 the spire and tower were struck by lightning during a violent storm. A villager, William Webster, was a witness: 'The wall of the Tower opened like a pair of barn doors, and the Spire toppled over on to

From Fielding's 1793 drawing of Exton Church showing a clock dial on the east face of the tower (RCM F10/ 1984/23/2)

the Church' (Wilton Hall 1913, 3). The clock survived this disaster. The tower and spire had been rebuilt by 1846 but the dial was not replaced. The reason for this may have been due to lack of funds, or it was considered that a dial was not absolutely necessary, as it could not be seen from the village.

The Churchwardens' Accounts give three references concerning repairs to the clock:

1862-63	mending the Clock	£8 18s 10d
1867-68	repairing clock 7/6	7s 6d
1868-69	Repairing Clock 9/	9s 0d

The entry in 1868-69 is the last mention of the clock.

The following extract from *Some Recollections of Exton* throws some light upon its fate:

There had been a little trouble over the Church Clock. It had no dial, and struck the hours only. Some time before I went to Exton the Clock was out of order and 'wouldn't

The accounts for the year from Easter 1858 disclose that there was a working clock in Exton Church as shown in these details of the Clerk's salary (DE 3012/74)

go'. An enthusiastic Curate from Cottesmore who 'understood these things', having obtained Mr Knox's permission to overhaul the clock, took it to pieces. But he wasn't a Lord Grimthorpe: something went wrong and he never succeeded in putting it together again! Opportunely he had an effectual call to another sphere of usefulness. Whether he adventured reforming refractory Church Clocks, or forever after left them severely alone, this deponent knoweth not. But Mr Knox 'said things about it', and he had not left off saying them in my time (Wilton Hall 1913, 11).

CLOCK DETAILS
Details of the former clock:

Maker:	Unknown
Frame:	Probably posted wooden frame with wrought-iron pivot bars
Trains:	Going and striking
Escapement:	Probably anchor
Striking:	Probably countwheel hour-striking
Winding:	Daily
Dial:	Square wooden dial with hour hand only
Notes:	Dial not replaced when tower rebuilt in 1846. Out of commission by *circa* 1870

EXTON HALL

An engraving in Wright's *History and Antiquities of the County of Rutland* (1684, 49) of the south side of Exton House (known as Exton Old Hall) shows two sundials. The House and Chapel were virtually destroyed by a fire in 1810 and Exton Hall of today was built as a result. A table sundial is shown in the nursery garden on the early 1900s OS Second Edition 25 inch map.

ST THOMAS OF CANTERBURY'S ROMAN CATHOLIC CHAPEL

Thomas North (1880, 131) records that the old house-bell from Exton House was hung in the bellcote over the new chapel which was built in 1868. This bell still hangs in the bellcote but it has not been rung since the early 1970s.

St Thomas of Canterbury's Roman Catholic Chapel stands in the private grounds of Exton Hall. Access is allowed for services

BELL DETAILS
1771. Diameter approximately 635mm (25 in). Weight approximately 178kg (3cwt 2qr). Canons retained. Timber headstock. Cast by Joseph Eayre of St Neots.

　　　JOSEPH EAYRE ST NEOTS FECIT 1771

Hung for chime ringing. The bellrope is attached to a lever on the headstock.

EXTON SCHOOL

The former Roman Catholic School at Exton was built by the Earl of Gainsborough, of Exton Hall, in 1868. Although it no longer has a bell, the impressive bellcote over the south gable was retained when it was converted in to a private house.

The bellcote at the former Roman Catholic School in Exton

From an engraving of circa 1684 showing two sundials on the south elevation of Exton Old Hall (Wright 1684, 49)

GLASTON

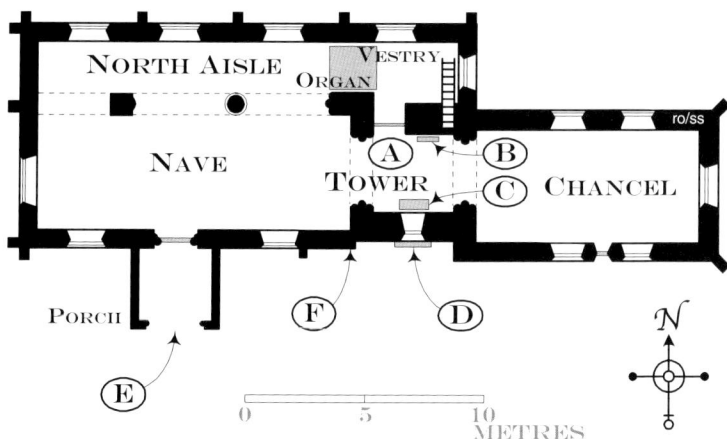

Plan of St Andrew's Church
A: Six bells in the belfry. Ringing
chamber at the base of the tower
B: Ellacombe chiming frame at the
base of the tower
C: Clock on first floor of the tower
D: Clock dial
E: Sundial
F: Scratch dial
Access to the clockroom is by stairs
from the Vestry. Access to the belfry is
by a ladder from the clockroom

ST ANDREW   Grid Ref: SK 896005

Parish Records

The most important source in the archive is a scrap-book
containing churchwardens' and other parish officers' disbursements and vouchers. This covers the period 1717-
1851 (DE 2575/46). There is also a bundle of Churchwardens' Accounts for 1866-68 (DE 5050/15-21) and
a Churchwardens' Account Book for 1889-1917 (DE
2575/47). Other relevant documents consulted include:

> Accounts dated 1867 for work on the church clock by
> James Sparkes (DE 5050/80).
> Parish of Glaston Select Vestry Book 1827-29 (DE
> 5050/231).
> An early twentieth-century inventory of church property
> including the bells (DE 5050/213).
> A document containing details of 'The Bell Fund' (DE
> 5050/26/70).
> A bundle of documents containing a plan, a report and
> the faculty for the restoration and rehanging of the bells
> in 1931, and a faculty for adding a sixth bell dated 31
> March 1937 (DE 2575/14/1-5).
> The faculty for installing an Ellacombe improved chiming
> apparatus dated 9 December 1965 (DE 5050/78).
> A report and quotation dated 24 January 1979 by John
> Smith & Sons, Derby, regarding the overhaul of the
> church clock (DE 5050/211).

Bell History

The earliest known documentary evidence regarding bells
in Rutland is an item dated February 1293 in *The Rolls
and Register of Bishop Sutton* (Hill 1958, 63-4). A letter
sent to the Dean of Rutland concerning the activities of
some men in the village clearly indicates that Glaston
Church then had more than one bell:

> It has recently been reported to us that certain unruly men
> spread a rumour that the rector of Glaston had died while
> he was away from the parish. They then behaved as though

they were enemies by entering the church and rectory of
Glaston, using armed violence, and are defending themselves there by force. They have removed the clappers of
the bells so that they cannot be rung to summon the
people to hear the word of God (translated from the Latin
by David Thomson and Philip Riley).

Until 1931 there were three bells in the tower, the
earliest being cast in 1598 by Newcombe and Watts of
Leicester, the other two being early Norris bells of 1616
and 1622. All three are preserved in the present ring
complete with their original canons.

Irons' Notes record that in 1605 Glaston Church
possessed just one bell of the three that it owned in 1600
(*VCH* 11,187). The following extracts from Irons' Notes
(MS 80/1/3) provide more information concerning
these early bells:

1610	**Nov 3**	Ward pres that our churche doth want bells as by lawe we ought to have
1611	**Nov 8**	Our churche is in decaye: the leades of the churche are oute of repaire. They have but one bell, they had 2 bells sent to Lester to cast & they were left
	Nov 26	To certify of the having againe of 2 bells that were lost
1612	**June 13**	Our churche to be oute of repaire & that it wanteth bels
1616	**June 14**	[reference to the repair of a bell]
	June 28	Rich Harris of Harringworth for not payeinge his levy towards the buyinge of a bell, having a ffarm in Glason.
1619	**May 11**	Anth Feild & Wm Wottse, wardens: warned to repair their steeple & their bells and to certify
	May 29	Anthony ffeld & Wm Witt monished to repayre their steeple & their bells
1620	**July 6**	They want a fore bell.

It was not until 1616 that the first of these two bells
[those that were 'lost' and then returned] was recast by

Toby Norris I at his Stamford foundry, and it is clear that the cost was covered by a levy. The other bell stated as being 'lost' was presumably the treble. The Church Survey of 1619 (Peterborough Diocesan Records, Church Survey 3) states 'they want A forebell, the second bell not hangd up' and this is partly confirmed by the note of 1620 above. The replacement bell was recast in 1622, also by Toby Norris I.

The Church Survey of 1681 reported that there was a need 'to rep [repair] the bells' (Peterborough Diocesan Records, Church Survey 7). This must be a reference to the state of the bell frame and fittings as the bells were then in good condition, as they are now.

The only references to the bells in the surviving Churchwardens' Accounts, which begin in 1717, are for the occasional replacement of bellropes and repairs to the clappers and bellwheels. There is no mention of repairing or renewing the frame. This old wooden frame, which was possibly of late seventeenth century origin, was to last for nearly two hundred and fifty years before being replaced.

The bells are now rung from the base of the tower, but in an earlier arrangement,

> to avoid ringing the bells from the church, a floor was built across under the tower at the springing of the arches and the upper part of the arches blocked, a ringing chamber being thus formed, which received its light by diamond leaded windows looking out under the arches into the nave and chancel. To reach the ringing chamber the East window of the aisle was blocked and a doorway formed in its South part, from which stone steps led down outside, and a wooden staircase inside passed up through the tower wall (Wordsworth 1889, Additions, 257).

According to Thomas North this ringing chamber had been removed before 1880 (North 1880, 132).

On 21 June 1887 there was a Royal Jubilee Thanksgiving Festival at Glaston. The first peal of the day was rung at 5am by Messrs King, G E Chapman, and C Stevens. A Jubilee Peal was then rung after a service of prayer and thanksgiving at 11am. 'It was agreed [by the Management Committee] that the Ringers be not allowed beer at 5am but at Luncheon time.' They were paid 2s 6d for their services (DE 5050/231).

In 1931 a decision was made to install a new frame, refurbish the existing bells and add two new bells to the ring:

> The bells were removed for cleaning, turning etc & were rehung in a new modern metal framework & two new bells were added. The wooden frame removed, partly oak & reputedly partly chestnut, was very decayed and in a dangerous condition. The work was carried out by Messrs. Taylor of Loughborough at a cost of just under £300. A further sum of about £25 was spent in mason's and joiner's work of preparation and completion to the tower walls and bell chamber flooring .... The peal was dedicated by the Very Rev. J. G. Simpson, Dean of Peterborough,

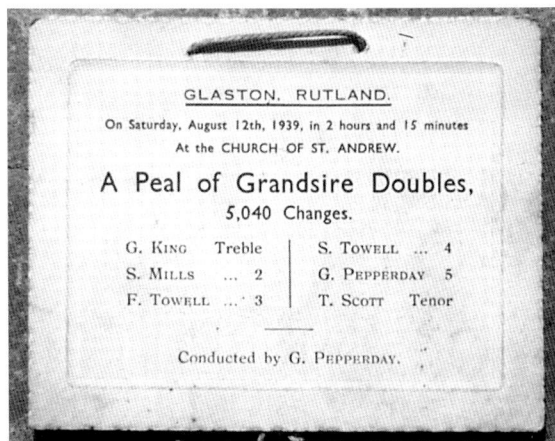

A number of notable peals have been rung at Glaston and the details of one are shown on this card hanging in the ringing chamber. Another notice in the tower records that on Wednesday 3 February 1965 a Quarter Peal of 1440 Plain Bob Minor, conducted by E A Jacques (tenor), was rung in forty-five minutes by six members of the Peterborough Diocesan Guild to the memory of 'Sir Winston S. Churchill, K.G., the Greatest Englishman of this Age'

on Dec 19th 1931 (Wordsworth 1889, Additions, 128).

A further faculty was obtained in 1937 to cast a new treble and this was dedicated by Bishop Lang on 26 June. In 1965 an Ellacombe 'improved chiming apparatus' was installed in memory of Eric Gore Browne.

The old clapper removed from the tenor in 1931 was acquired by J W Wallace and mounted on a wooden block. It is now in the Rutland County Museum (1985.20).

BELL DETAILS

1 1937. Treble. Diameter 584mm (23in). Weight 137kg (2cwt 2qr 23lb). Cast without canons. Cast-iron headstock. Cast by John Taylor & Co, Loughborough.
[95] JOHN TAYLOR & CO [95] FOUNDERS ~ [95] LOUGHBOROUGH
On the waist:

FROM
ERIC AND IMOGEN GORE BROWNE
IN THE YEAR OF THE CORONATION
OF KING GEORGE VI & ELIZABETH.
~ HIS WIFE.
1937
DEO DANTE DEDI
(God gave so I gave)

Lt Colonel Sir Eric Gore Browne DSO OBE TD MA (1885-1964) purchased Glaston House in 1932, following the death of Victoria, Lady Carbery. He died in 1964 and there is a memorial tablet to him on the south wall of the chancel.

Plan of the bellframe at Glaston Church

The second bell at Glaston showing the chime hammer

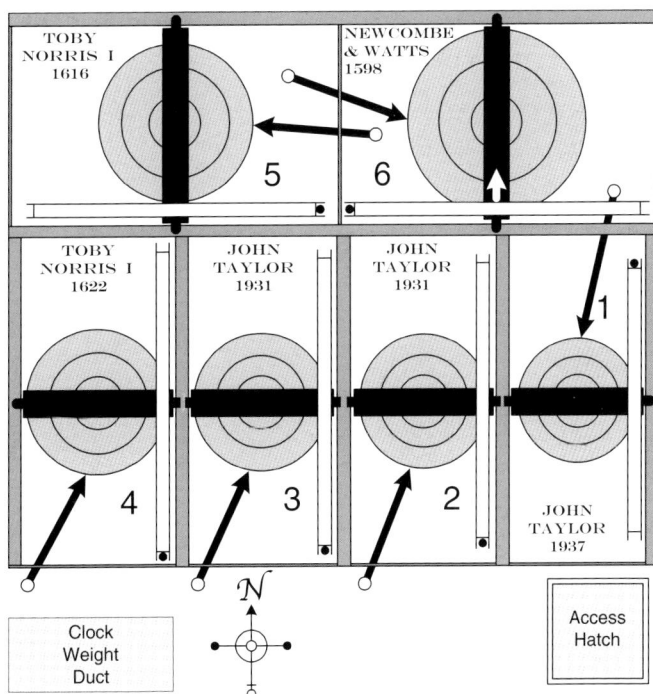

2 1931. Diameter 625mm (24⅝in). Weight 156kg (3cwt 0qr 8lb). Cast without canons. Cast-iron headstock. Cast by John Taylor & Co of Loughborough.

GLORIA IN EXCELSIS DEO [71]
(Glory to God in the highest)
On the waist:

VICTORIA CARBERY
On the waist, opposite:

[47]
1931

There is a memorial stone to Victoria, Lady Carbery (1843-1932) on the outside of the chancel wall to the west of the Priest's door. Her husband was the Hon W C Evans-Freke who became Baron Carbery in 1889. They originally lived at Bisbrooke Hall and then Laxton Hall. After his death in 1894 Lady Carbery moved to Glaston House.

3 1931. Diameter 664mm (26⅛in). Weight 184kg (3cwt 2qr 14lb). Cast without canons. Cast-iron headstock. Cast by John Taylor & Co, Loughborough.

IN TERRA PAX [71]
(Peace on earth)
On the waist:

R. WALTHAM
OLIM RECTOR

Richard Waltham was Rector of Glaston from 1890 to 1917. He retired to Torquay and died at Leamington in 1923. He is buried at Glaston. It is thought that his sister, Catherine, may have donated this bell.

4 1622. Diameter 695mm (27⅜in). Weight 175kg (3cwt 1qr 21lb). Canons retained. Cast-iron canon-retaining headstock. Cast by Toby Norris I of Stamford.

[26] OMИIA [5] FIAИT [5] AD [5] GLORIAM
~ [5] DEI [5] 1622 [5] E [5] C [5] T [5] A
(Let all things be done to the Glory of God)
Royal Arms [83] is high on the waist, directly below the initials and stops **E [5] C [5] T [5] A**.

5 1616. Diameter 781mm (30¾in). Weight 260kg (5cwt 0qr 14lb). Canons retained. Cast-iron canon-retaining headstock. Cast by Toby Norris I of Stamford.

[26] ИOИ [:] [:] CLAMOR [:] [:] SED [:]
~ [:] AMOR [:] [:] CAИTAT [:] [:] IИ [:]
~ [:] AVRE [:] [:] DEI [:] [:] 1616[:]
(It is not noise but love that sings in the ear of God)
[:] is stop [62].

6 1598. Tenor. Note B flat. Diameter 873mm (34⅜in). Weight 378kg (7cwt 1qr 22lb). Canons retained. Cast-iron canon-retaining headstock. Ascribed to Newcombe & Watts of Leicester.

[3]
[23] COELORUM CHRISTE PLACEAT
~ TIBI REX SONUS ISTE
(O Christ, King of Heaven, may this sound be pleasing to Thee)
Below the inscription band in two lines:

THOMAS BRUDNELL ANTHONU COLLy
~ WILLIAM HUTTON

The canon-retaining headstock on the tenor at Glaston

THOMAS BOWDELL 1598

All letters are like [16] and [17] with the exception of 'w' and 'y' which are like [110].

Anthony Colly was Lord of the Manor of Glaston and possibly the donor of this bell. He died in 1640. William Hutton was curate of the parish. He died in 1604. The marriages of Thomas Brudenell in 1584 and Thomas Bowdell in 1599 are recorded in the Parish Registers. Scheduled for preservation (Council for the Care of Churches). The clock strikes the hours on this bell.

CLOCK BELL

1793. Diameter (533mm) 21in. Weight 67kg (1cwt 1qr 8lb). Cast by Edward Arnold of Leicester.

E ARNOLD FECIT 1793

This hemispherical clock bell hung outside a south light of the spire until it was removed in 1931, but it may have been out of use since 1905 when the present clock was installed. After 1931 it was used as a plant container in the rectory garden until it was sold to John Taylor for scrap in October 1976. It is thought that this bell may have replaced a similar bell supplied by Thomas Eayre II of Kettering as part of the original clock installation in 1739 (Wordsworth 1889, Addition, 258).

BELLRINGING CUSTOMS

The Rev Christopher Wordsworth included details of ringing customs in *Glaston Parish Charities and Memorials* (Wordsworth 1889, Additions, 137):

> The following customs were I am told discontinued only a few years before my incumbency (1877) possibly on account of the old age of the last parish clerk, or the restoration of the church:
> The Pancake bell (no 2) rung on Shrove Tuesday (at noon).
> On Sunday 1st bell rung at 7am.
> 1st and 2nd bell rung at 9am.

Wordsworth also recorded those customs current at

The Edward Arnold clock bell hanging from a spire window at Glaston Church in 1915 (Henton 994)

the start of his incumbency:

> In 1877 I find only the 1st at 8am and I have accepted this as calling to the first service on 2nd, 4th (& 5th) Sundays in the month.
> At Death knell thrice three tolls are given for a male, thrice two for a female.
> If a person dies after sunset, this knell is deferred until the following morning.
> At funerals the tenor bell is tolled but in 1845 the bells were chimed.
> For Divine Service on Sundays and greater Festivals the bells are chimed for 10 minutes then the tenor is tolled.
> The bells are rung from the floor of the church (transcribed by Auriol Thompson).

Today Glaston has a small band of ringers. The tenor is chimed for five minutes before each service and if sufficient ringers are available the bells are rung for the Easter and Christmas Services. They are also rung for Benefice Services and Harvest Festivals.

HANDBELLS

The parish has a set of nine handbells and at the time of writing there was a proposal to have them restored. They

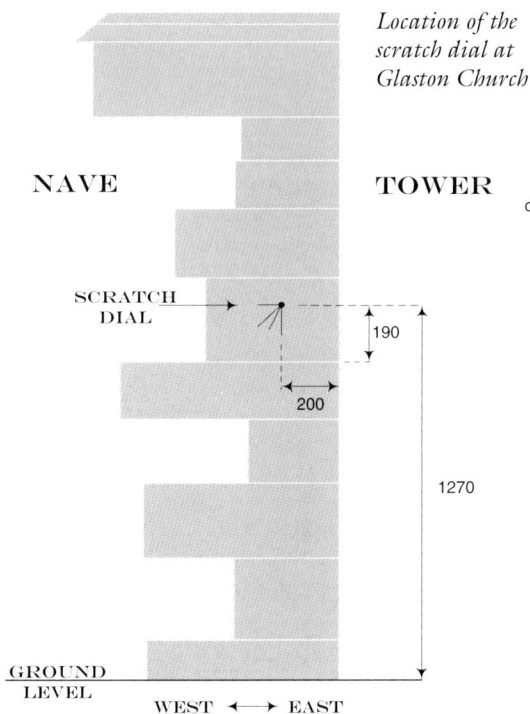

Location of the scratch dial at Glaston Church

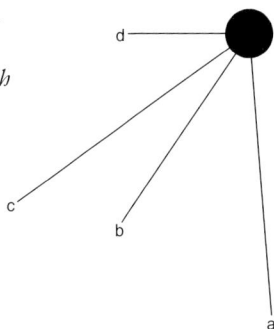

NAVE TOWER

SCRATCH DIAL

190

200

1270

GROUND LEVEL

WEST ◄──► EAST

GLASTON				
Location	South facing quoin at south-east corner of nave			
Condition	Poor			
Gnomon Hole Diameter		24		
Gnomon Hole Depth		60		
Height above ground level		1270		
Line Ref	a	b	c	d
Length	135	110	140	60
Angle (°)	175	215	235	270

d c b a

Scratch dial details

The vertical sundial in the gable of the south porch at Glaston Church

were rung many years ago as an accompaniment to carol singers when they toured the village.

SCRATCH DIAL

A very weathered scratch dial is located on a south-facing quoin at the south-east corner of the nave. The wall is twelfth century.

SUNDIAL

Fielding's *circa* 1793 drawing of Glaston Church (RCM F10/1984/25/2) shows a direct south vertical sundial in the gable of the south porch. The dial has survived together with its gnomon, but the numerals and lines have virtually weathered away.

CLOCK HISTORY

'1739. In this year the church was presented with a clock .... The inner dial bears the name of Thomas Eayre of Kettering and the stone brackets to support the face are dated 1739' (Wordsworth 1889, Additions, 258). Further details are given in the Churchwardens' Accounts:

1739	Horse & other Charges	4s 0d
	for going to Kettering to fetch ye Clock	10s 0d
	Expences there	1s 6d
	pd Fox for altering ye Clock	2s 6d
	pd for a Lock, hokes [hooks] and bands for ye clock case	1s 6d

The clock was almost certainly donated by a local benefactor, as the cost is not included in the accounts. It is also possible that the clock was supplied complete with dial, hand and bell, but there is no confirmation that Thomas Eayre actually installed it. In fact, this may have been carried out by Robert Fox, a clockmaker of Uppingham who was paid for 'altering ye clock'. Robert Fox continued to look after the clock until at least 1754:

1741	July 24	pd Mr Fox of Uppingham	7s 6d
1754		Pd for looking after the clock by Fox of Uppingham	5s 4d

A clock case mentioned in 1739, together with weight and pendulum cases were made by Robert Slater and the following are extracts from one of his bills:

1739	Dec 7	half a day Both [Slater and his helper] about the Weight Case	1s 0d
	Dec 13	About the Case half a day	7d
1740	Feb 11, 12, 13, 14	about the Case my Lad two [days]	6s 4d
	Feb 19, 20:	Stuff for the Bottom of the weight Case	8d

Nailes for the work	1s 2d
Board for Leges for the Case	8d
a plank for the Steple over the Clock Case	3s 0d
Board for the pendel case	4d

The fact that the clock required a pendulum case probably indicates that it had a long pendulum, a particular characteristic of some Thomas Eayre II clocks (*see* Chapter 3 — Clockmakers). His clock at Gaulby Church, Leicestershire, and the former clock from Loddington Hall stables, Leicestershire (see Belton in Rutland), now in Rutland County Museum (1969.428), both have long pendulums. All his clocks required daily winding.

From 1739 until 1776 the clock was wound by the Parish Clerk, John Fox:

1751 pd John Fox a smith for looking after the
Clock last yr   4s 0d
1754 For looking after ye Town Clock   4s 0d

It appears that the clock gave good service for the first twenty-five years and this was undoubtedly due to regular maintenance. The dial also received special attention:

1753 pd Tob Hippesley for painting Church Dial 5s 0d

However, in 1764 the surviving vouchers show that a considerable amount of work was carried out on the clock. It seems that this is when the clock was converted to two hands although the vouchers are not specific on this point. For this conversion it was necessary to remove the dial, paint and gild it with a minute ring outside the roman numerals, alter the 'dial works' and replace the dial with new lead flashing for weather protection. '2 trusses' were also provided to support the dial. These are probably the stone corbels that support the dial today. It is assumed that they were dated '1739' in order to record the original date of the clock installation. At the same time new weights and pulleys were provided, together with a new case for the weights. A new pulley system and a longer fall for the weights may have increased the duration between winding, probably from one to two days. Local craftsmen carried out all this work and extracts from their bills are given below:

1764
John Hand, plumber and glazier:
Feb 22   for Casting the Clock weights   5s 0d
March 30 for 0c 1qt 21 lb of New Lead for the
Clock dial   10s 2½d
for work putting it on   1s 0d
Francis Birch, carpenter:
Feb 25   To wood for Clock Pulles [pulleys] and
Putting up   1s 6d
For making Case for ye weights   1s 6d
For puting up Dial Board   1s 6d
John Fox, blacksmith:
March 20 a large bar and two Longue [long] Screws
and parls [poles?] and a Forearm [lever?] for
the pules [pulleys] and two new rolls [rollers
for the weight lines] for the Church Clock

and altring the dial worke	10s 0d

Edward Bingham:
31 March To painting & Gilding the Clock face £1 1s 0d
To working 2 trusses to suport do
[the clock dial]   2s 6d

The following items from Francis Birch's bill show that the roof of the clockcase was altered and later covered with lead sheet, probably in an attempt to provide better weather protection for the clock. The south light of the spire, which contained the clock bell, was immediately above the case and open to the elements:

1767 **Dec 19**   For altering the clock roof   8s 0d
1768 **Sept 28** 2cwt 2qtrs 0lb new lead for ye
clock   £2 18 4d
To a day's work laying it on   2s 6d

Richard Hackett, clockmaker of Harringworth, Northamptonshire, regularly maintained and repaired the clock from 1770 to 1782 :

1770 pd to Mr Hacket for reparing the Clock £1 19s 6d
1771 Paid Mr Hackets bill   15s 0d

William Aris I worked with Richard Hackett in Harringworth and he took over responsibility for the Glaston clock when Hackett died in January 1782. He continued until at least 1793, the last date of the available accounts.

William Aris' bill of 1785 for keeping the Glaston church clock in good order (DE 2575/46)

James Sparkes, clockmaker of Uppingham, carried out what appears to be a full restoration of the clock and dial in 1867. The following details are taken from his bill for this work (DE 5050/80):

1867 The Churchwardens of Glaston
Dr to Jas Sparkes

Painting & Gilding Clock dial & hand	£2 5s 0d
89 feet of New line	18s 0d
New collet & Nut to hands	2s 6d
New [illegible] Wood for fixing Clock	5s 0d
Iron work for fixing pulleys & Clock frame	18s 0d
New Tooth to Escape wheel reparing pallets and depth	10s 0d
New connecting rod with collets & studs	10s 0d
New adjusting peace	5s 0d
altering Crutch	2s 6d
3 New holes & reparing pivots	5s 0d
reparing Third wheel depth	2s 6d
filing up pinions as New	5s 0d
altering wheels to act on different part	5s 0d

Self and man 3 days fixing   £1 17s 0d
Carrage for Clock from Glaston & Back   7s 6d
Cleaning & painting Clock   <u>£1 10s 0d</u>
  £10 8s 0d

The next mention of the clock is in 1887 when on 30 June the Management and Provisions Committee, which had organised the Jubilee Festival held on 21 June, met to discuss 'the dispersal of the remainder of the fund in hand'. This totalled £2 7s 7½d and it was proposed by Mr Thurlby that 'the balance of the Jubilee Account be expended upon the repair of the face and hand of the Church clock ... Mr Thurlby's motion was put and carried (7 to 1)' (DE 5050/231).

At a public meeting to consider how the Coronation of King Edward VII was to be celebrated it was decided that the funds collected for the purpose should be spent in providing a feast for the village. It was agreed that any surplus should be put towards the repair or replacement of the church clock 'which was quite worn out'. In the event, sufficient additional funds were raised to enable the parish to order a new clock from John Smith & Sons of Derby. It was set going at 8pm on 13 September 1905 (Wordsworth 1889, Additions, 213). It appears that the clock was not working from the early 1930s until it was renovated in 1944. Automatic winding units were fitted to both trains in 1989 by Smiths of Derby.

Cost details for the restoration of the clock dial at Glaston Church. It is assumed that the rector paid the short-fall, as he is known to have done on another occasion (DE 5050/231)

CLOCK DETAILS

Details of the Thomas Eayre clock:

Maker:	Thomas Eayre II of Kettering
Signed:	Signed by Thomas Eayre on the setting dial
Installed:	1739
Cost:	Not recorded
Frame:	Wrought-iron birdcage
Trains:	Going and striking
Escapement:	Anchor
Rate:	Not known but thought to be a long pendulum as a special case was built for it
Striking:	Countwheel hour striking
Clock Bell:	Hemispherical clock bell hanging outside the south light of the spire
Winding:	Probably daily as other Eayre clocks of this period
Dial:	Probably as present dial. Dial corbels dated 1739

The clock dial at Glaston Church is supported by corbels dated 1739

The going train of Glaston church clock and its automatic winding unit

Details of the present clock:

Maker:	John Smith & Sons of Derby
Signed:	On the setting dial and on the frame
Installed:	1905
Cost:	£72
Frame:	Cast-iron flatbed
Trains:	Going and striking
Escapement:	Pinwheel
Pendulum:	Wooden rod and cast-iron cylindrical bob
Rate:	52 beats per minute
Striking:	Countwheel hour striking
Weights:	Old cast-iron weights removed
Winding:	Originally hand wound weekly. Smith's epicyclic automatic winding units fitted to both trains in 1989
Dial:	Hexagonal timber dial with a blue dial-plate. Gilded roman numerals and hands. Dial sits on corbels dated 1739
Location:	South face of the tower

RAILWAY MISSION CHAPEL

Glaston Railway Mission Chapel was one of three chapels erected along the Rutland section of the Manton to Kettering railway line when it was being constructed in the mid 1870s. Built in 1876, it was located amongst a settlement of navvies' huts in the field just above the southern entrance to Glaston tunnel (SK 904003). It was referred to by the navvies as 'The Cathedral'. The other chapels were at Seaton and Wing (see under Seaton and Wing). All three were closed on completion of the line in 1878.

> The exteriors of the chapels were soon to be distinguished from the huts which surrounded them by their little wooden steeples or turrets, and regularly every settler heard, or might have heard, the single bell chiming out its summons to prayer and praise. This was a simple thing, but it told on the people. The steeple made the building look a little different from its neighbours, and the bell it carried spoke with a voice every Sunday which all were obliged to hear, whether they heeded it or not (Barrett 1880, 103-4).

MAIN STREET, GLASTON

The only secular sundial found in Glaston is on the south wall of a cottage at the corner of Main Street and Church Lane. The gnomon has survived on this direct south dial but all the lines have weathered away. There are, however, faint traces of arabic numerals.

GREAT CASTERTON

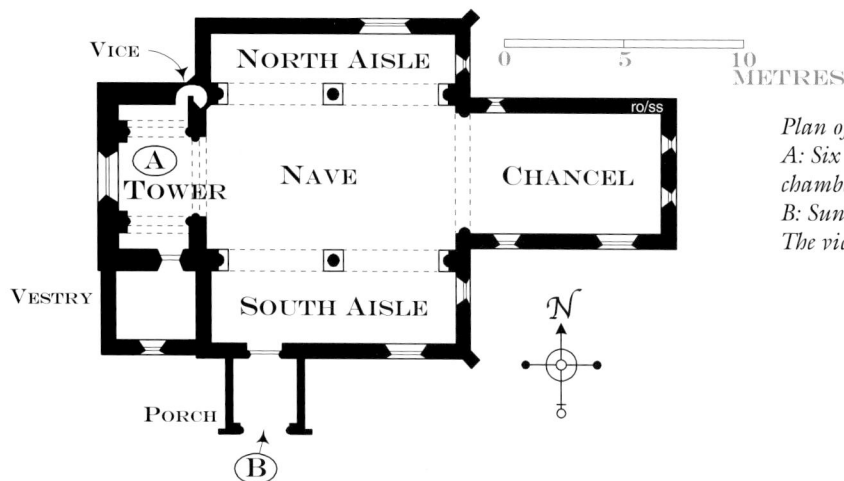

Plan of St Peter & St Paul's Church
A: Six bells in the belfry. Ringing chamber at the base of the tower
B: Sundial
The vice provides access to the belfry

ST PETER & ST PAUL   Grid Ref: TF 001088

PARISH RECORDS
Churchwardens' Accounts are available for 1871-1943 (DE 5198/42) and 1944-71 (DE 5198/43). Vestry Minute Books are also available for 1863-87 (DE 5198/44), 1888-1919 (DE 5198/45) and 1919-24 (DE 5198/45). Documents dated 1989 still held at the church include a report on the condition of the installation, as well as a quotation and a plan for the rehanging and augmentation of the bells, all by the Whitechapel Bell Foundry.

BELL HISTORY

At the end of the thirteenth century the church had a bellcote over the west gable but in the fifteenth century this was taken down and the present tower erected within the western end of the nave (*VCH* II, 233). Irons' Notes (MS 80/1/3) record 'Bells broken' in 1607, and this may refer to the state of the frame rather than to the bells themselves. There is no indication as to the number of bells in the tower. In 1718 a new ring of five by Henry Penn of Peterborough was installed in a new oak frame. This frame was to last until 1990, by which time it had deteriorated to such an extent that the bells were completely unringable.

The Churchwardens' Accounts show that after 1880 repair work on the frame and bellwheels was a fairly regular occurrence:

1885	Bill for repairing Bell wheels	15s 11d
1889	Crowson's Bill for repairs to Bell frames [and other work]	£4 8s 3d
1905	Mr Crowsons a/c (Bell hanging)	£14 18s 9d

William Crowson was the local carpenter and at some time a churchwarden.

At a Vestry Meeting held in 1916 'It was decided to have the bells looked at, as the tenor bell was reported to be in a dangerous condition'. On 26 December 1919 a Vestry Meeting again considered the state of the bells and the bellframe. A report from John Taylor & Co estimated that the cost of putting the bells in 'thorough good order' would be £297 10s. However, although it was not safe to ring the bells, it was agreed, following reassurances from Mr Crowson and the Parish Clerk, that they could continue to be chimed 'for the present'. Taylor's quotation was not accepted and it was to be another seventy-one years before the bells could be rung once again.

In 1989 a report on the condition of the old installation was prepared by the Whitechapel Bell Foundry and the following is a summary of their findings:

Although the bells were sound they were of poor tone and out of tune. They had been cast to a 'thin scale' which was probably the reason for the poor tonal qualities, but this could be improved by tuning. They also had cast-in iron clapper staples and it was recommended that these be cut out and independent staples used.

The ringing fittings dated from 1718, although the wheels and fourth bell headstock were later. The bells were hung from elm headstocks with plain gunmetal bearings let into the top timbers of the bellframe. The wrought-iron clappers were heavily worn and generally in a poor shape and the wheels were weak and falling apart. Overall, it was considered that the installation was generally in a derelict condition.

The old oak frame had five bells and an anti-clockwise rope circle. Bells three, four and five swung east to west, and bells one and two swung diagonally north-west to south-east. The under-frame consisted of four large timber beams running north to south. Immediately on top of these were four more large beams running east to west. Large wedges had been driven between the top timbers of the bellframe and the tower walls, to restrict the movement of the frame during ringing.

The belfry was open to the vault below which had a central trap for raising and lowering the bells. The bells had previously been rung from the ground floor but the rope holes had been filled in with concrete. The proposed installation would require new holes and because the drop from this point was some twenty-five feet, a rope guide was recommended.

The ring of five Henry Penn bells just after being removed from the tower at Great Casterton in 1990 (Michael Lee)

The old clappers are now displayed on a board at the base of the tower at Great Casterton Church

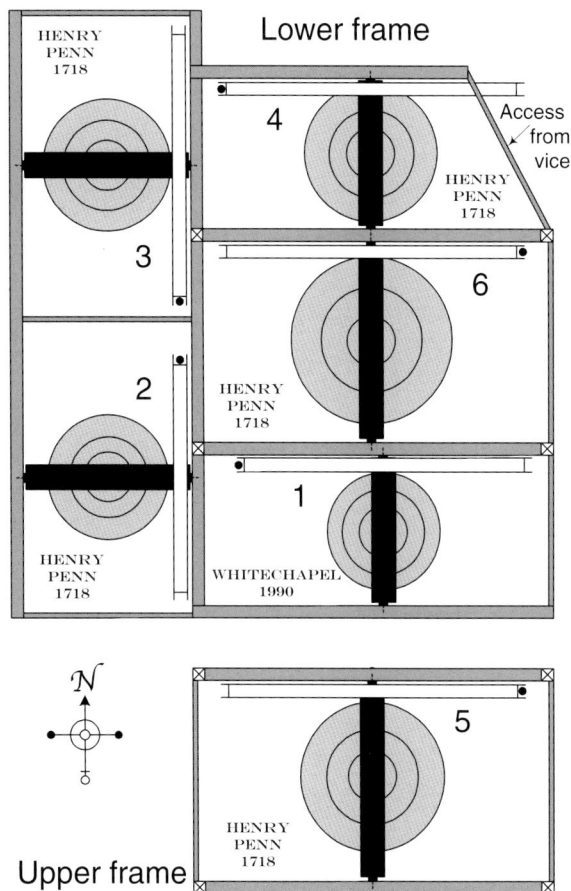

Plan of the bellframe at Great Casterton Church. The upper frame is over the tenor

BELL DETAILS

1 1990. Treble. Note F. Diameter 610mm (24in). Weight 161kg (3cwt 0qr 19lb). Cast without canons. Fabricated steel headstock. Cast by the Whitechapel Bell Foundry, London.

PETER [102]

On the waist:

J. BUTLER   RECTOR
MR. A. HALL } **CHURCHWARDENS**
MR. R. G. CADMAN }

On the waist, opposite:

19 [93] 90
WHITECHAPEL

2 1718. Note E flat. Diameter 648mm (25½in). Weight 154kg (3cwt 0qr 3lb). Canons retained. Fabricated steel canon-retaining headstock. Cast by Henry Penn of Peterborough. Former treble.

HENRY PENN FOVNDER 1718

3 1718. Note D flat. Diameter 673mm (26½in). Weight 147kg (2cwt 3qr 16lb). Canons retained. Fabricated steel canon-retaining headstock. Cast by Henry Penn of Peterborough. No inscription, decoration or bellfounder's marks. This and bells four and five may have been stock bells, cast during the winter months when there were few orders (Lee 1999, 81).

4 1718. Note C. Diameter 699mm (27½in). Weight 178kg (3cwt 2qr). Canons retained. Fabricated steel canon-retaining headstock. Cast by Henry Penn of Peterborough. No inscription, decoration or bellfounder's marks.

5 1718. Note B flat. Diameter 756mm (29¾in). Weight 204kg (4cwt 0qr 1lb). Canons retained. Fabricated steel canon-retaining headstock. Cast by Henry Penn of Peterborough. No inscription, decoration or bellfounder's marks.

6 1718. Note A flat. Tenor. Diameter 838mm (33in). Weight 275kg (5cwt 1qr 18lb). Canons retained. Fabricated steel canon-retaining headstock. Cast by Henry Penn of Peterborough.

GEORGE O MAXWELL   THOMAS
~ BROUGHTON CHURCHWARDENS 1718
North (1880, 126) has G MANCELL not GEORGE MAXWELL. **O** is an unidentified coin.

BELLRINGING CUSTOMS

The Gleaning Bell was rung at 8am and 6pm during harvest. At the Death Knell there were thrice three tolls for a male and thrice two for a female. For Divine Service the bells were chimed for a short time and then one bell was rung. This was followed by another short period of chiming and finally the tenor was rung (North 1880, 126).

It was also recommended that the new treble should be cast using an old nineteenth-century gauge so that its tonal qualities would match those of the old bells. This would result in an inferior bell by modern standards but it would more readily complement the existing bells.

The work of rehanging the bells was carried out by Nicholas Meadwell and Michael Lee. The new galvanised steel frame and fittings were supplied by the Whitechapel Bell Foundry who also tuned the bells and cast the new treble. The work was completed on 25 October 1990 and the bells were rededicated on Sunday 28 July 1991 by the Rt Rev W J Westwood, MA, Lord Bishop of Peterborough (Lee 1999, 79).

The church was able to pay for the renovation of the bells using interest received from the rental and eventual sale in 1979 of land that had been in its ownership from the reign of Queen Elizabeth I.

The unrestored south-east declining sundial in the porch gable at Great Casterton Church

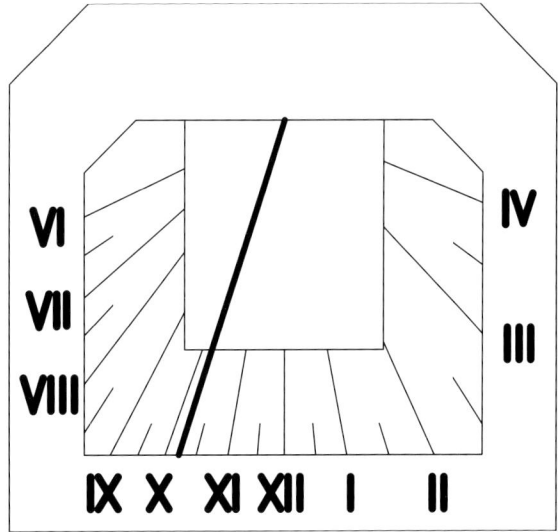

This drawing shows the proposed layout of the sundial at Great Casterton church just before restoration (Walter Wells). *The positions of the hour lines were calculated from the orientation of the dial face and the angles of the existing gnomon*

Today Great Casterton has its own band of bellringers and they practise weekly for one and a half hours on Thursday evenings. Generally all six bells are rung for half an hour and then chimed for three minutes before each Sunday Service. They are similarly rung for the Christingle and Carol Service and on Christmas Day. If requested for a wedding, the bells are rung for twenty minutes before and after the service. The bells are rung half-muffled for funerals at the request of the bereaved. Quarter peals are rung on special occasions and full peals by request for particular events.

SUNDIAL

A vertical sundial on the gable of the south porch is shown on a *circa* 1839 drawing of the church (Uppingham School Archives). There are no incised lines, indicating that this was a painted sundial. Only a few traces of roman numerals remain along the bottom edge.

GREAT CASTERTON PRIMARY SCHOOL

Great Casterton School was built in 1861 by the Marquis of Exeter. It cost £350 and was designed to accommodate 60 juniors and 24 infants. When it finally closed in 1962 it was demolished and a new primary school was built on the same site. The bellcote from the old school, complete with its 229mm (9in) bell, was saved and re-erected in the grounds of the new school. It is still used to call the pupils in from the playground.

The old school bellcote in the grounds of the new primary school at Great Casterton

GREETHAM

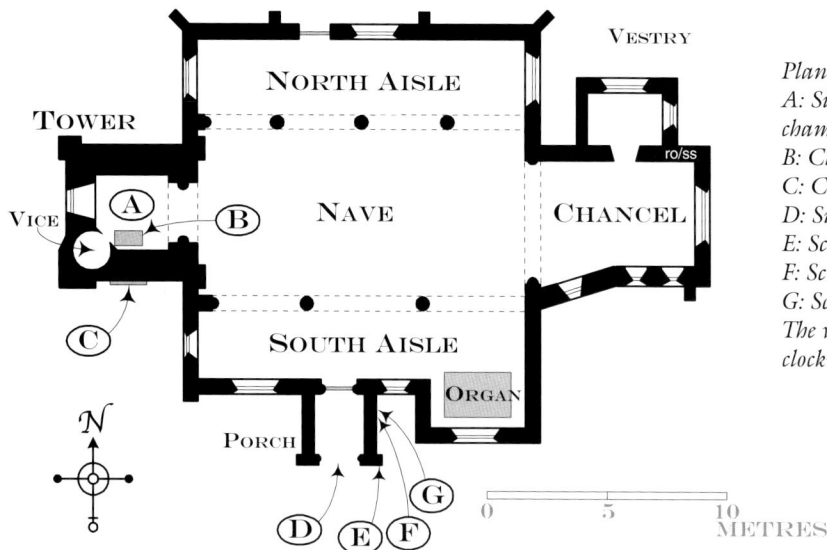

Plan of St Mary the Virgin's Church
A: Six bells in the belfry. Ringing
chamber at the base of the tower
B: Clock on the first floor of the tower
C: Clock dial
D: Sundial
E: Scratch dial 1
F: Scratch dial 2
G: Scratch dial 3
The vice provides access to both the
clockroom and the belfry

ST MARY THE VIRGIN   Grid Ref: SK 924147

PARISH RECORDS

Available records are the Church Accounts Book 1788-1873 (DE 2574/9) and the Churchwarden' Accounts 1875-1938 (DE 5187/8).

BELL HISTORY

A new oak bellframe was installed in 1787 and Thomas North (1880, 134) records the following as being cut into one of the members:

> T CHAR [illegible] Y   WM SHARMAN
> C Ward[n]   17 Fecit 87

William Sharman was a village carpenter and a church-warden in 1787.

White's *Directory* of 1846 records: 'In the tower are five bells, but all are cracked except one ... and efforts are now making to raise funds for re-casting the bells'. Evidently the fund-raising efforts were unsuccessful as 'the church in 1860 was "in a ruinous condition" with one of the bells "half out of the belfry window"' (Dickinson 1983, 58).

By 1880 there were only four bells and these 'were in a sad condition'. Only one, the treble, could be used and this had to be 'knocked' by means of a rope being attached to the clapper. The full extent of their condition is given under Bell Details below. The belfry must have been a depressing scene with broken bells lying haphazardly and bird droppings 'some inches thick on the bell frames' (North 1880,133). The dates on these bells were 1650, 1658, 1703 and 1741. The fate of the fifth bell recorded in 1846 is unknown.

The Church Accounts show that bellropes were purchased annually from 1789 to1793 but very few were acquired in the following period up to 1857. Two ropes were accounted for in 1818 indicating that at this date two of the bells may have been in commission. Only two more ropes were purchased before the bells were recast and rehung in 1923, one in 1883, and one in 1915.

John Taylor surveyed the belfry on 17 February 1914, presumably to provide a quotation for the rehanging of the bells in a new frame. Notes taken at the time record that there was a two-tier timber frame with two bells in each tier (closed archives of John Taylor).

1923 was a memorable year for the parishioners of Greetham, for the bells of St Mary's Church were rung as a peal again after almost one hundred and fifty years' silence. The church had been restored in 1897 but there were insufficient funds at this time to put the bells in order. It was later decided to have the four bells recast and rehung in a new steel and cast-iron frame as a memorial to those parishioners who had died in the First World War. A Memorial Appeal Committee was set up and the estimated cost of £600 (Phillips 1920, 242) was raised by subscription, fêtes, whist drives and other events. The four existing bells were recast by Gillett & Johnston and it was a creditable decision to retain the old inscriptions. The ring was increased to five as a result of the generosity of Mrs Sheldon who donated a new tenor in memory of her husband and her father. Two new floors were installed in the tower at the same time.

The Dedication Service took place at the beginning of July 1923, conducted by the Rev J H Charles, Vicar of Oakham. Amongst the congregation was a group of ex-

The new Greetham bells in their frame at the Croydon bell foundry in 1923 (Tony Traylen)

Upper frame over bell 2

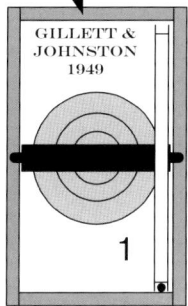

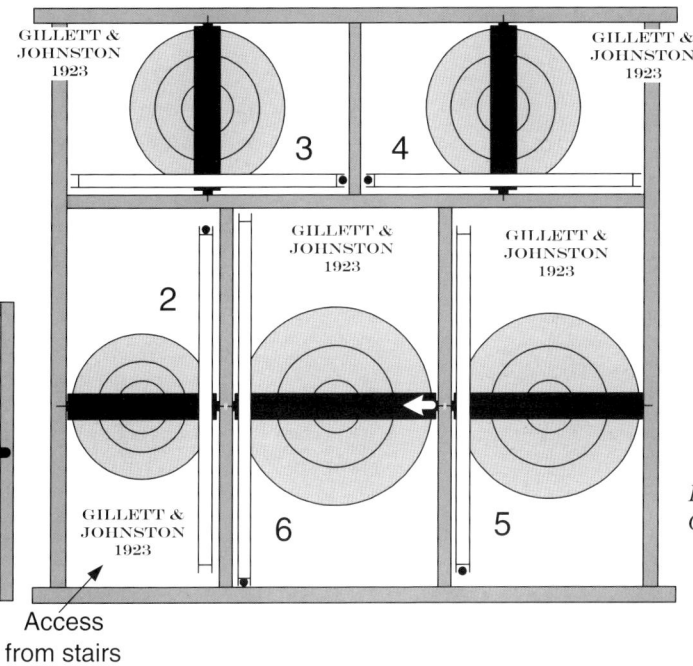

Access from stairs

Plan of the bellframe at Greetham Church

Servicemen who had paraded through the village accompanied by the Greetham Band. Mr Charles based his address on the uses of bells, and the concluding hymn was 'Lift them gently to the steeple'. The Last Post sounded before the first peal was rung on the new bells (*Grantham Journal* 7 July 1923).

When the new frame was installed in 1923 space was left for a sixth bell. This was donated by J W Kirk in memory of his wife, Annie Kirk, in 1949.

BELL DETAILS

1 **1949**. Treble. Note F. Diameter 641mm (25¼in). Weight 186kg (3cwt 2qr 19lb). Cast without canons. Cast-iron headstock. Cast by Gillett & Johnston of Croydon. On the waist:

IN MEMORIAM
ANNIE KIRK
OB. 3 SEP. 1944.

On the waist, opposite:

19 [74] 49
GILLETT & JOHNSTON
FOUNDERS, CROYDON

Donated by J W Kirk.

2 1923. Note E flat. Diameter 686mm (27in). Weight 208kg (4cwt 0qr 10lb). Cast without canons. Cast-iron headstock. Recast by Gillett & Johnston of Croydon.

RECAST BY GILLETT & JOHNSTON.
~ CROYDON. 1923.

On the waist:

LABOR IPSE VOLUPTAS
(The labour [of ringing] is itself pleasure)
THOMAS PARKER CHURCHWARDEN
J. EAYRE FECIT 1741.

On the waist, opposite:
[76]
Decoration [75] is below the inscription band.

Former bell

Cast in 1741 by Joseph Eayre of St Neots. Diameter 781mm (30¾in).

LABOR IPSE VOLUPTAS
THOMAS PARKER CHURCHWARDEN
J. EAYRE FECIT 1741

In 1880 this bell was struck by a rope attached to the clapper.

3 1923. Note D flat. Diameter 740mm (29⅛in). Weight 254kg (5cwt). Cast without canons. Cast-iron headstock. Recast by Gillett & Johnston of Croydon.

GILLETT & JOHNSTON. CROYDON. 1923 [76]

On the waist:

ALEX : RIGBY : MADE : ME : 1703 :
ROBERT : CV . REY : AND :
~ HENRY : CLARKE ⦂ W :

Decoration [75] is below the inscription band.

Former bell

Cast in 1703 by Alexander Rigby of Stamford. Diameter 851mm (33½in).

[14] ALEX [50] RIGBY [50] MADE [50] ME
~ [50] 1703 [50]
ROBERT [50] CV [piece missing] REY [50] AND
~ [50] HENRY [50] CLARKE [50] ⦂ W [50]

In 1880 this bell was standing with its mouth upwards in the north window of the belfry. Its crown was missing and there was a crack in its side.

4 1923. Note C. Diameter 775mm (30½in). Weight 270kg (5cwt 1qr 7lb). Cast without canons. Cast-iron headstock. Recast by Gillett & Johnston, Croydon.

RECAST BY GILLETT & JOHNSTON.
~ CROYDON. 1923 [76]

On the waist:

IS RC 1658

Decoration [75] is below the inscription band.

Former bell

Cast in 1658 by Thomas Norris of Stamford. Diameter 914mm (36in).

[26] IS RC 1658

In 1880 this bell was dismounted and resting on two planks in its pit. It was cracked and the canons broken.

5 1923. Note B flat. Diameter 867mm (34⅛in). Weight 388kg (7cwt 2qr 16lb). Cast without canons. Cast-iron headstock. Recast by Gillett & Johnston of Croydon.

RECAST BY GILLETT & JOHNSTON.
~ CROYDON. 1923 [76]

On the waist:

T. H THOMAS NORRIS MADE MEE 1650

Decoration [75] is below the inscription band.

Former bell

Cast in 1650 by Thomas Norris of Stamford. Diameter 1016mm (40in).

[10] TH [piece missing] THOMAS NORRIS
~ MADE MEE 1650

In 1880 this bell stood on a lower frame with its crown and other large portions missing.

6 1923. Tenor. Note A flat. Diameter 965mm (38in). Weight 521kg (10cwt 1qr). Cast without canons. Cast-iron headstock. Cast by Gillett & Johnston, Croydon.

GILLETT & JOHNSTON. CROYDON. 1923. [76]

On the waist:

IN MEMORY OF
WILLIAM BOSWORTH & JOHN SHELDON
~ CHURCHWARDENS
1862-1884:

Decoration [75] is below the inscription band. The clock strikes the hours on this bell.

BELLRINGING CUSTOMS

On Sundays the only available bell was 'knocked' first at 8am and then again before and after the service. The Gleaning Bell sounded at 8am and 6pm. Thrice 'three knocks' were given for a male and 'thrice two for a female, both before and after the knell' (North 1880, 134).

At the time of writing the village has its own band of bellringers which practises on Thursday evenings. Five or six bells are rung for fifteen minutes both before and after every service. All six bells are rung before a Carol Service and for the service held on Easter Day. One bell is rung muffled on Good Friday. On New Year's Day the bells are rung at noon for fifteen minutes, and for thirty minutes prior to the Harvest Festival. All six bells are

Location of scratch dial 1 at Greetham Church

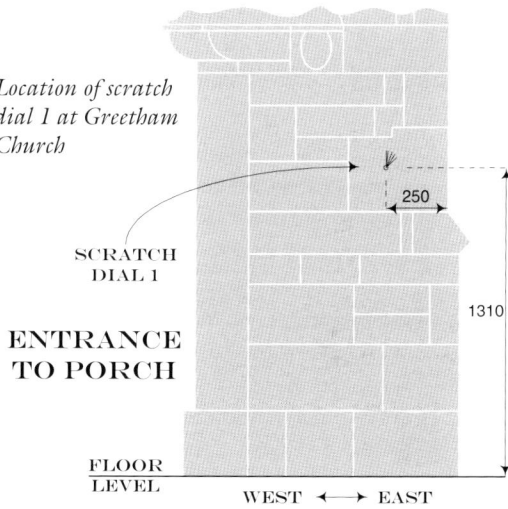

SCRATCH DIAL 1

ENTRANCE TO PORCH

250

1310

FLOOR LEVEL

WEST ←→ EAST

Scratch dial 1 details

GREETHAM 1					
Location	East front of the porch				
Condition	Poor				
Gnomon Hole Diameter	11				
Gnomon Hole Depth	25				
Height above ground level	1310				
Line Ref	a	b	c	d	e
Length	90	100	104	100	104
Angle (°)	180	186	198	215	236

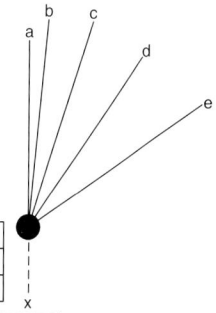

Note that this scratch dial is on a relocated stone and is inverted. Angles are therefore measured from datum line x.

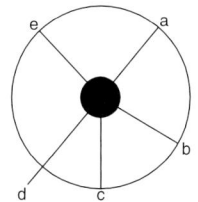

GREETHAM 2					
Location	East face of porch				
Condition	Poor				
Gnomon Hole Diameter	22				
Gnomon Hole Depth	18				
Height above ground level	1560				
Circle Diameter	88				
Line Ref	a	b	c	d	e
Length	44	44	44	55	50
Angle (°)	40	120	180	220	317

Scratch dial 2 details

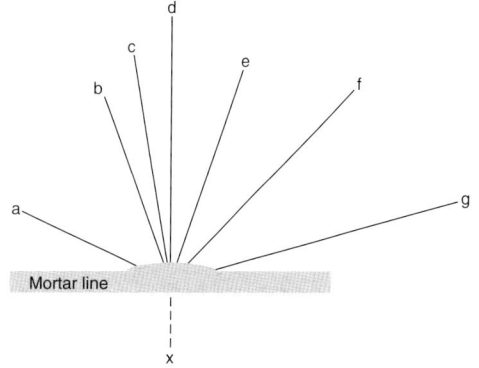

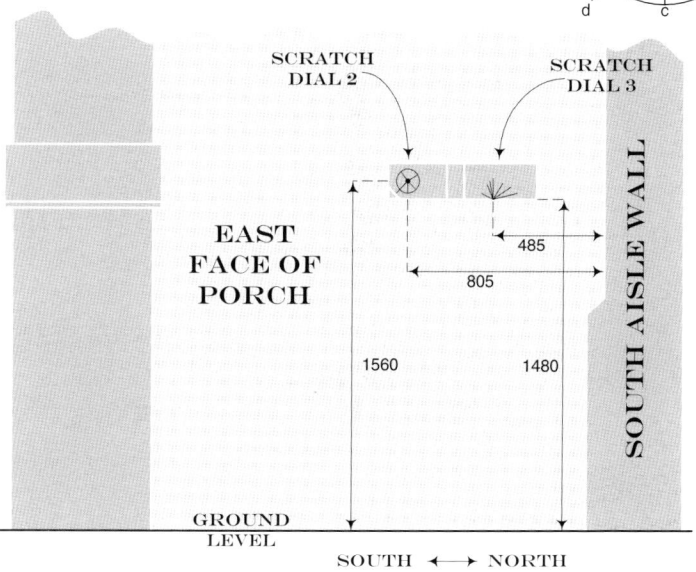

SCRATCH DIAL 2

SCRATCH DIAL 3

EAST FACE OF PORCH

485

805

1560

1480

SOUTH AISLE WALL

GROUND LEVEL

SOUTH ←→ NORTH

Mortar line

x

Location of scratch dials 2 and 3 at Greetham Church

usually rung for fifteen or twenty minutes after a wedding. The tenor is tolled at a funeral if requested by the bereaved family.

SCRATCH DIALS

There are three scratch dials, one on the front face of the porch and two on relocated stones set into the east wall of the porch. The porch was originally fourteenth century. It was partially rebuilt using existing stones in 1673

GREETHAM 3						
Location	East face of porch					
Condition	Average					
Gnomon Hole Diameter	Filled in					
Gnomon Hole Depth	Filled in					
Height above ground level	1480					
Line Ref	a	b	c	d	e	f	g
Length	80	95	105	128	108	130	150
Angle (°)	115	160	170	180	200	225	255

Note that this scratch dial is on a relocated stone and is inverted. Angles are therefore measured from datum line x.

Scratch dial 3 details

A drawing of the porch sundial at Greetham Church from an unpublished report on the church restoration (Finch 1897)

and the date of this restoration is shown on the apex stone of the gable coping.

SUNDIAL

There is a sundial in the gable of the south porch. It is badly weathered and the majority of the roman numerals have disappeared. Although the gnomon is missing, its slot in the face of the dial still remains. A *circa* 1839 drawing of Greetham Church shows a dial in the same position (Uppingham School Archives). It was probably installed when the porch was rebuilt.

CLOCK HISTORY

Apart from payments for winding and maintenance there is little mention of the old clock in the Churchwardens' Accounts. A framed newspaper cutting hanging in the base of the tower, dated 7 July 1923, states that the church clock had a single-handed dial and that it was to strike the hours on the new tenor. Bearing in mind the previous state of the bells, the clock probably had not struck the hours for a very long time.

In 1967 John Smith & Sons of Derby installed a new synchronous electric clock, bell strike unit and dial. The old clock was sold to Ralph Cox, a local antique dealer, for £30. Following full restoration it was exhibited at the Chelsea Antiques Fair from where it was sold to an American customer. It is now understood to be on display in the foyer of a hotel in New York. Fortunately, a photograph of the clock was taken at this fair. From this it can be seen that it has many features common to those made by John Watts of Stamford, and there is little doubt that it originated in the same workshop.

The only illustration found of the old dial is a photograph taken by Mr H Tempest of Nottingham in 1943 as part of a general survey of the church (NRO P4830/3). It shows that the dial had a white chapter ring with black roman numerals, and a single hour hand - the same hand as shown on the clock at the Chelsea Antiques Fair in 1967.

Most Watts' clocks have **I W** and a date engraved on the top cross-member of the frame. This detail is missing from the Greetham clock but it is quite possible that a new frame was provided when it was restored *circa* 1846: 'The clock, after being useless for half a century, has recently been repaired at the expense of A. Sutton, Esq' (White 1846, 613). This restoration is confirmed by an incomplete engraving on a board found in the church some years ago:

THIS CLOCK WAS REPAIRED AT THE [expense]
OF AUGUSTUS SUTTON ESQ[R] **OF**
~ COTTESMORE HALL A [D? date?]

The end of the board that carried the date is missing.

Augustus Sutton (*see* Chapter 3 — Clockmakers), born 13 January 1825, was the fifth son of Sir Richard Sutton, 2nd Baronet of Norwood Park and owner of Cottesmore Hall. Augustus later became Rector of West Tofts in Norfolk and it is recorded that he was 'a very clever mechanic, his speciality being the reconstruction and repairing of old disused church clocks' (Bird & Bird 1996, 166).

The first entry in the Churchwardens' Accounts with regard to the clock was in 1831 when a Mr Coverley was

The John Watts' clock movement from Greetham Church on display at the Chelsea Antiques Fair in 1967 (David Bland and Ralph Cox)

William Senescall, linen draper, tailor and grocer outside his shop in Main Street, Greetham. He wound the church clock until 1916 (David Bland)

paid £2 a year to wind it. It is believed that he was the same Richard Coverley, the Cottesmore carpenter, who had made the new frame and rehung the bells at Cottesmore five years earlier. His bill for this work shows that he also worked with Augustus Sutton on the clock at Cottesmore Church (see Cottesmore — Clock History). A member of the Coverley family was still carrying out the duty of clockwinder until 1907, after which it was taken over by William Senescall until 1916 and then by G Carrier through to 1922.

It was the clockwinders' duty to oil and clean the clock. They obviously maintained it very well as there is only one recorded occasion that the clock had to be repaired. This was carried out by the village wheelwright and carpenter:

1892 Jan 22 Paid T Stubbs, repairs Church
Clock £1 18s 4d

CLOCK DETAILS
Details of the former clock:

Maker:	Attributed to John Watts of Stamford
Signed:	Unsigned
Installed:	*Circa* 1700
Removed:	1967
Cost:	Not recorded, but probably about £10
Frame:	Wooden frame, trains side by side
Trains:	Going and striking
Escapement:	Originally anchor. Later converted to deadbeat
Pendulum:	Wrought-iron pendulum with lead bob
Rate:	Originally 60 beats per minute but later lengthened

Striking:	Countwheel hour striking
Weights:	Stone weights
Winding:	Daily
Dial:	White dial with black numerals and hand. Marked out for single hour hand
Hand:	Single hour hand
Restored:	*Circa* 1844 and 1967
Note:	Believed to be on display in the foyer of a hotel in New York

Details of the present clock:

Maker:	John Smith & Sons of Derby
Signed:	Engraved plate on housing: 'John Smith & Sons, Midland Clock Works, Derby, 1967, No 15036'
Installed:	1967
Cost:	Not recorded
Movement:	Synchronous electric motor
Trains:	Going and striking
Striking:	Electric striking unit with a countwheel
Dial:	Convex copper dial with a blue background and gilded numerals and hands, installed in 1967
Location:	South face of the tower

GREETHAM HOUSE

There used to be a bell at the rear of Greetham House (SK 925146) which was rung as a timekeeper for the estate workers. A table sundial is shown in the grounds on the early 1900s OS Second Edition 25 inch map.

VICARAGE

The old vicarage was built of local stone and was one of the finest small Georgian buildings in Rutland. When it was demolished some years ago the stone was disposed of in a local quarry. The sundial was lost in the demolition rubble but it was eventually rescued and positioned on the new vicarage (SK 926146) built on the same site (information from David Bland).

The table sundial on the lawn at Greetham House in 1954 (David Carlin)

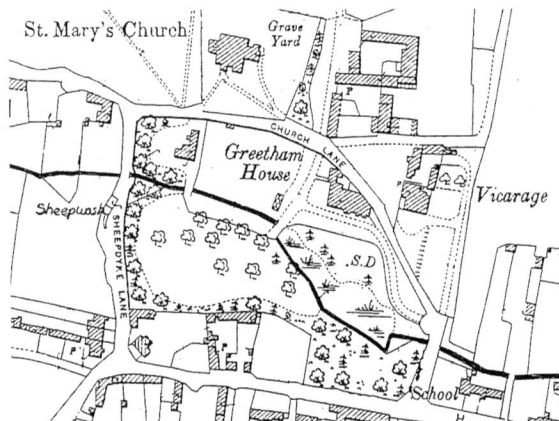

Greetham House together with the sundial (SD) in its grounds, the Vicarage and the School are shown on this early 1900s OS Second Edition 25 inch map

The dial of Greetham school clock was restored by Robert Ovens in 1998. It was originally white with black numerals and hands, and had 'Coronation 1911' in red across the centre

GREETHAM VILLAGE SCHOOL

A bell used to hang from a bracket at the rear of the village school, but it disappeared when the school closed in 1969 (information from David Bland). In 1911 a clock was installed in the south gable of the school to celebrate the Coronation of George V. Both clock and dial were supplied and installed by William Potts of Leeds. A synchronous electric motor now drives the hands, but the old movement has been preserved.

CHAPMAN'S COTTAGE

There was a sundial on the south gable of Chapman's Cottage at the east end of Main Street. The cottage was demolished in the 1970s. There was a timber yard here with logging pits for cutting timber, and a smithy.

RAM JAM INN

What has been described as a Saxon sundial on the east wall of the Ram Jam Inn (SK 945160) near Stretton was found in 1929 during site excavations (*VCH* II, 1935, 135) (*see* Chapter 1.1 — Historical Introduction).

Chapman's Cottage in Main Street, Greetham, before it was demolished in the 1970s. A sundial can be seen on the end gable

HAMBLETON

ST ANDREW   Grid Ref: SK 900076

PARISH RECORDS
The parish archive includes the Churchwardens' Accounts 1729-59 (DE 2209/51) and 1759-1879 (DE 2209/52). Documents concerning the bells include an estimate and bills of 1887 when the tenor was recast and a new treble added to the ring (DE 2209/43/1-2), and an analysis of costs of *circa* 1982 when the bells were rehung in a new metal frame (DE 5004 /32/1-2).

BELL HISTORY
A bell of *circa* 1510, possibly by Robert Mellour of Nottingham, hung in the belfry for 351 years. This and a Priest's Bell dated 1636 'were exchanged for the present 3rd in 1861' (North 1880, 164). 'In 1861 the tower and spire were repaired, the bells rehung, 1 bell re-cast, and 2 new buttresses built at the west end, at a cost of nearly £350' (White 1877, 678). John Taylor supplied the recast bell and in 1880 it hung alongside three bells by Toby Norris I, two dated 1610 and the other 1611.

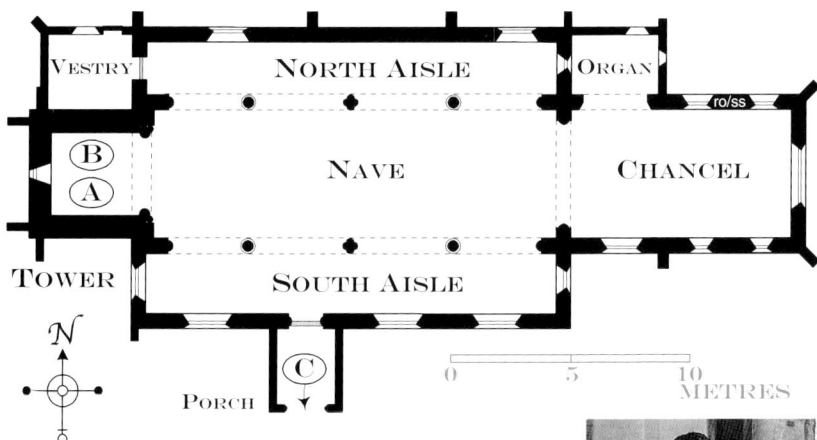

Plan of St Andrew's Church
A: Five bells in the belfry.
Ringing chamber at the base
of the tower
B: Former clockroom on the
first floor of the tower
C: Location of former sundial
Access to the belfry is by two
ladders

There is no mention of this work in the Churchwarden's Accounts.

Judging by the purchase of bellropes, the bells were well used in the 100 years from 1730. Some of the rope suppliers of this period are named in the accounts:

1731	Pd Ben Carter for Bellrops	9s 0d
1772 April 18	pd Smith of Oakham for Bell ropes	9s 0d
1802 April 10	Pd Mrs Toon for Bell Ropes and clockline	£1 2s 0d
1829 April 13	Darnell for Bellropes	£1 5s 0d

There are also frequent references to repairing the bells throughout the eighteenth century. The following are some examples:

1738	pd George Bains for a bell Boldrock [baldrick]	1s 0d
1760 April 4	Pd Jon Woodcock for mending ye Bell frames	3s 0d
1771 Dec 28	Pd Jon Pawlett a Bill for 2 Bell wheels	£4 12s 6d
1785 Oct 18	pd for a lather [ladder] to go in to the Bell Chamber	4s 0d

The Toby Norris I tenor of 1611 was recast in 1887 as it was cracked and the canons broken. The bells were again rehung at the same time, and the ring increased to five by the addition of a new treble to celebrate the Golden Jubilee of Queen Victoria. John Taylor's invoice of 31 August 1887 records that the cost to the parish was £127 18s 11d, less an allowance of £35 12s 3d for the old tenor.

The old frame was replaced in 1982 by Nicholas Meadwell, assisted by Charlie Hudson of Preston, Ray Bailey of Langham and Joseph Dickinson of Oakham. The new frame was made by Portal Fabrications to a design by John Taylor, who also supplied the new headstocks. New bellwheels were made in the church. The total cost of £6,599 was covered by loans, grants and fund raising events. The old clappers are now displayed in the base of the tower.

The Hambleton bells in 1982 with their old timber headstocks. The bells were rehung by (left to right) Nicholas Meadwell, Charlie Hudson, Ray Bailey and Joseph Dickinson (Brian Nichols)

BELL DETAILS

1 1887. Treble. Note D. Diameter 686mm (27in). Weight 206kg (4cwt 0qr 7lb). Cast without canons. Cast-iron headstock. Cast by John Taylor & Co of Loughborough.

VICTORIA JUBILEE 1887. A. D.
On the waist: [120]

2 1610. Note C. Diameter 781mm (30¾in). Weight 241kg (4cwt 2qr 27lb). Canons retained. Cast-iron canon-retaining headstock. Cast by Toby Norris I of Stamford.

[14] IESVS [39] SPEDE [39] ME [39] OMИIA ~ [39] FIAИT [39] AD [39] GLORIAM [39] ~ DEI [39] [62] 1610 [62]

(Jesus speed me. Let all things be done to the Glory of God)

3 1610. Note B. Diameter 826mm (32½in). Weight 307kg (6cwt 0qr 4lb). Canons retained. Cast-iron canon-retaining headstock. Cast by Toby Norris I of Stamford.

Plan of the bellframe at Hambleton Church

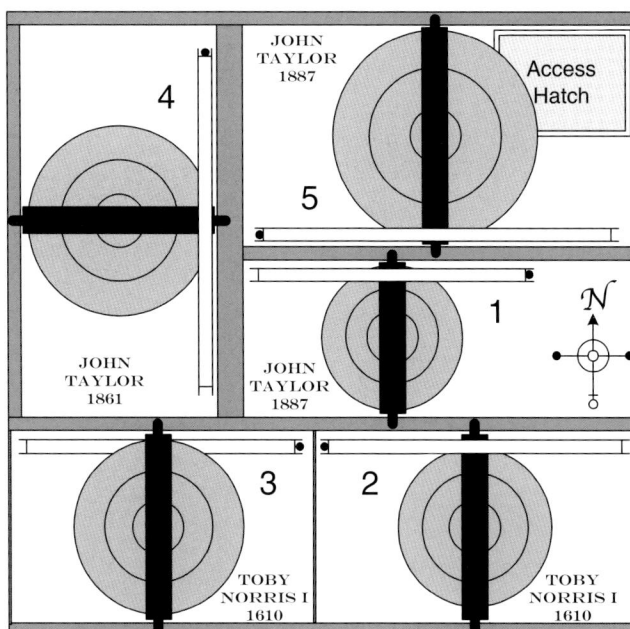

[26] ИOИ [39] CLAMOR [39] SED [39] AMOR ~ [39] ƆAИTAT [39] IИ [39] AVRE [39] DEI 1610
(It is not noise but love that sings in the ear of God)
North (1880, 134) dates this bell 1621 and omits AMOR.

4 1861. Note A. Diameter 921mm (36¼in). Weight 419kg (8cwt 1qr). Cast without canons. Cast-iron headstock. Recast by John Taylor & Co, Loughborough.
[91] **JOHN TAYLOR & Cᵒ FOUNDERS ~ LOUGHBOROUGH A:D 1861.** [92]

Former bell
Cast *circa* 1510, probably by Robert Mellour of Nottingham.

S. S. S.

The inscription indicates that it may originally have been a Sanctus Bell. This and the Priest's Bell were recast into the present fourth in 1861.

5 1887. Tenor. Note G. Diameter 1003mm (39½in). Weight 537kg (10cwt 2qr 8lb). Cast without canons. Cast-iron headstock. Recast by John Taylor & Co of Loughborough.
1611 A. D. LAUS DEO RECAST 1887.
(Praise be to God)
On the waist: [120]

Former bell
Cast in 1611 by Toby Norris I of Stamford. Diameter 965mm (38in). Weight 470kg (9cwt 1qr).
[10] ПOП . SOПO . АПIMАBƆS .
~ MORTƆORƆM . SED . АƆRIBƆS .
~ VIVEПTIƆM . 1611

(I sound not for the souls of the dead, but for the ears of the living)
In 1880 this bell was cracked and the canons broken (North 1880, 164).

PRIEST'S BELL
There was a Priest's Bell here dated 1636. It was sent to John Taylor's bell foundry when the fourth bell was recast in 1861 (North 1880, 164). No other details are known.

BELLRINGING CUSTOMS
In the Churchwardens' Accounts there are regular payments, presumably to parish clerks, for the 'clock and bell', but there are few details regarding specific bellringing duties or events. The following is an isolated example:

1853 June 3   Paid John Gregory Tolling the Bell for the Duke of Wellingtons Funeral on Nobr 18   1s 6d

The Gleaning Bell was rung during harvest (North 1880, 134), and amongst the Overseers of the Poor Accounts (DE2209/51) there are occasional payments for tolling a bell at a pauper's funeral:

1740   Pd for a Bell & Grave Wdo Bagley 1s 3½d
a Coffing 8s   9s 3½d
Pd for a Bell & Grave for George
English's child   1s 3½d

Today Hambleton has a band of bellringers and practice night is Wednesday from 7.30pm to 9pm. The band was formed in August 1999, the intention being to ring in the New Millennium. The bells are rung for the 9am services on the second and fourth Sunday of each month.

They are also rung at Christmas and for weddings.

SUNDIAL

Drawings of *circa* 1793 (RCM F10/1984/28) and *circa* 1839 (Uppingham School Archives) show a sundial on the porch gable, but no evidence of it remains. Its primary function would have been to regulate the church clock. There are three direct references to it in the Churchwardens' Accounts:

1751	Pd dor a new doing ye Dial	5s 0d
1763 May 2	Pd Mr Birridge for a new stile [gnomon]	4s 6d
1797 April 15	Pd Henry Stone a bill for a new dial & co taxed to £1 5s 0d	£1 15s 0d

The Henry Stone referred to here supplied a stone clock dial to Edith Weston Church in 1775.

CLOCK HISTORY

The Churchwardens' Accounts reveal that there was a clock at Hambleton Church in 1729 and they provide a detailed record of the winders and clockmakers involved with it until 1829, when it is assumed that the clock was abandoned. Whilst there are no specific details available, the clock would have been very similar to those installed by John Watts, and may well have been installed by him. It would have required daily winding and would have struck the hours on one of the available bells.

Sam Bagley is the first clock winder to be mentioned and when he died in 1748 his widow took over this duty. William Tyres was the next clock winder from 1754 until 1782, followed by Thomas Walker:

1729 March 28	paid Saml Bagley for Clock & Bell ½ a year	15s 0d
1749	pd Widw Bagley one year looking after ye Clock & Bell	£1 10s 0d
1773	pd Wm Tyres half a yr looking after Clock etc	16s 0d
1783 April 8	Pd Thos Walker for ringing the Bell and looking after the Clock	£1 12s 0d

Thomas Walker would have wound the clock over ten thousand times in the thirty years he was responsible for this duty. Edward Ward was the new winder in 1813, followed by Joseph Veasey from 1819 until 1829. After this date there is no further mention of the clock.

It appears that local craftsmen maintained and carried out repairs on the clock:

1748	Pd T Birridge for mending ye bell weel & a new spindle for ye Clock	1s 6d

[The spindle mentioned here probably confirms that the clock had a dial with a single hour hand]

1770 Nov	Pd John Line for repairing the Church clock	£2 10s 0d
1776 April 8	Pd John Pawlett for looking after the clock	5s 0d
1790 Dec 24	Pd Ed Wadking for Mending the Clock	1s 0d

From a drawing of circa 1793 by Nathan Fielding showing a sundial in the gable of the porch at Hambleton Church (RCM F10/1984/28)

1797 Sept 25	pd the Blacksmith for a new spring to the Clock hammer & other work	3s 0d
Oct 29	Pd John Vines Cleaning ye clock	2s 6d
1798 Jan 12	John Vines for putting ye clock to rights	1s 6d

The accounts also tell us which clockmakers attended the clock. Regular annual payments of 5s for maintenance were made to Stephen Blackburn of Oakham from 1731 until 1740 and then to John Wilkins also of Oakham until 1755. In 1759 there was obviously a serious problem with the clock as it had to be removed from the tower and taken to Stephen Blackburn's workshop. The exercise was repeated eight years later:

1759	pd Blackbourn for mending ye Church Clock	£1 12s 0d
	pd for carrying ye Clock to Oakham & back again	3s 0d
	pd for Ale wn ye Clock was set up	1s 4d
1767 Jan 3	pd Stefon Blackborn for Mending the Clock	£4 0s 0d
	pd for Caring the Clock to Oakham & fettching Back	3s 0d

John Simpson of Oakham repaired the clock in 1788 but it appears that another serious problem arose the following year for again the clock could not be repaired in situ:

1789 April 6	Pd Mr Furniss of Uppingham for repairing the Church Clock	£1 5s 0d
	Pd for carrying the Clock to Uppingham and bringing it back again	8s 0d

Joseph Furniss was paid 10s 6d for cleaning the clock in 1791, and two local men, Edward Wadkin and John Vines, repaired it many times over the next fifteen years.

The last clockmaker employed by the churchwardens was William Aris II of Uppingham. He was paid 12s a year for maintenance. His last visit was in 1822.

1810   April 20   pd Mr Ares Bill for church Clock
£6 0s 0d
pd William Tires for caring the Church
Clock to Uppingham   5s 0d
pd John Fryer for Bring the Church
Clock from Uppingham   2s 6d

HAMBLETON HALL

Walter Gore Marshall made his fortune in the brewery business. He built Hambleton Hall in 1881 when he came to Rutland to enjoy the fox hunting, and in particular the intensive social activities that went with it. He was a great benefactor of the village, having provided a new school, post office and cottages before he died in 1899. At the church he also paid for the rebuilding of the chancel and for the installation of the magnificent organ.

A four-dial clock turret was erected over Hambleton Hall stables to celebrate the sixtieth anniversary of the Coronation of Queen Victoria, and this fact is recorded on the dials. The turret may have been converted from, or replaced, an earlier turret as the clock movement is dated 1882.

Details of the former clock:

Maker:	John Smith & Sons of Derby
Signed:	On the setting dial and on the frame
Dated:	1882
Frame:	Cast-iron flatbed
Trains:	Going and striking
Escapement:	Pinwheel

Pendulum:	Wooden rod and cast-iron cylindrical bob
Rate:	40 beats per minute
Striking:	Countwheel hour striking
Clock Bell:	Single bell in cupola above the dials
Weights:	Cast-iron
Winding:	Originally hand wound weekly
Dials:	4 illuminated dials in a turret. Gilded copper hands and numerals
Notes:	Clock movement removed when stables converted to staff accommodation. Replaced in 2001 by a synchronous electric movement and electric striking unit

A vertical sundial on the front elevation of Hambleton Hall is in the same 'art nouveau' style as the dials of the stables clock and the Hambleton Post Office clock.

POST OFFICE

There is an 'art nouveau' clock on the front of the former Hambleton Post and Telegraph Office which was built for the village in 1898 by Walter Gore Marshall (*see* Chapter 1.7 — The Electric Telegraph and Standard Time). He originally wanted to donate a new church clock but there was no position where the villagers could easily see a dial. However, the central location of the Post Office made it ideal for the new village clock. It is of the same style as the clock dials and sundial at Hambleton Hall. The small brass weight driven movement, which has no maker's name, is located in the room behind the dial. It has been wound every week by George Bushell for nearly seventy years. His wife was Postmistress here for fifty years from 1932 for which she received the British Empire Medal. The Post Office closed in the 1980s but several of the fixtures and fittings were retained, including the old mahogany counter and an English dial wall clock by Robert Corney of Oakham.

The former two-train clock movement at Hambleton Hall stables by John Smith & Sons of Derby. The movement was removed from a room below the clock turret when the stables were converted. A synchronous electric movement now drives the hands

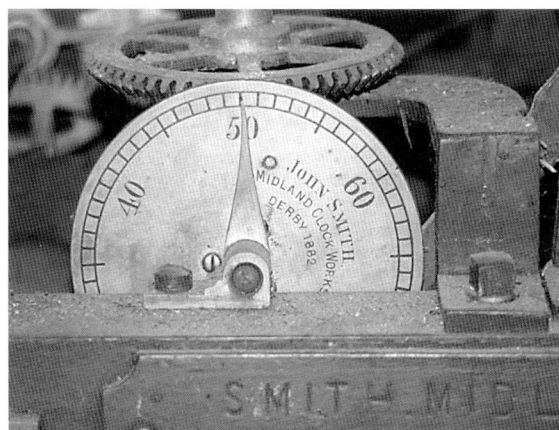

The setting dial of the former stables clock at Hambleton Hall. It is dated fifteen years earlier than the external dials

This four-dial clock turret with a hunting weather vane is over the entrance to the stables yard at Hambleton Hall. It has a clock bell in the cupola. The inscriptions on the dials are:

East and west facing dials:

Anno Sexagesimo Victoria Regina [In the sixtieth year of Queen Victoria]

North and south facing dials:

Anno Domini MDCCCXCVII [AD 1897]

The vertical sundial at Hambleton Hall. The inscription is translated as: 'It's drinking time now' and 'Time passes, friendship remains, it is the time to do good'

OTHER VILLAGE SUNDIALS

Two table sundials are shown on the early 1900s OS Second Edition 25 inch map. One was in the grounds of the Manor House in Ketton Road (SK 901076), the other in the garden of Old Hall (SK 899069). Neither dial exists today but the remains of old vertical sundials can be seen at Hilltop in Oakham Road (SK 900077) and on the south elevation of Home Farmhouse, in Ketton Road.

This early 1900s OS Second Edition 25 inch map of part of Upper Hambleton shows a table sundial (SD) in the grounds of the Manor House. The stables to Hambleton Hall are also shown to the east of the Hall

HORN

Horn no longer has a church and there are no surviving church records. The village of Horn had been depopulated by the 1500s by which time the church was in a ruinous state. Although there are no surviving records to confirm the fact, it seems likely that the church would have had at least one bell. By the early 1800s the church had completely disappeared but it is thought to have been in the western part of the parish, in Exton Park. As late as 1809, rectors were inducted by a thorn tree which marked the site (*VCH* II, 140). Today the exact site is unknown.

An annual open-air service is still held at Horn, usually on the third Sunday in June (information from Bernadette Wallace).

KETTON

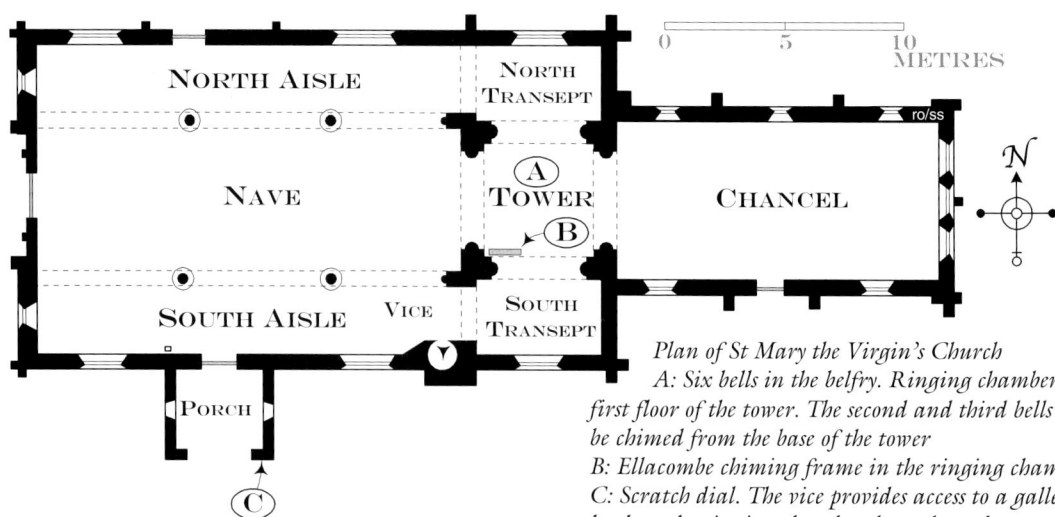

Plan of St Mary the Virgin's Church
A: Six bells in the belfry. Ringing chamber on the first floor of the tower. The second and third bells can also be chimed from the base of the tower
B: Ellacombe chiming frame in the ringing chamber
C: Scratch dial. The vice provides access to a gallery which leads to the ringing chamber door above the nave arch of the tower
Access to the belfry is by a ladder from the ringing chamber

ST MARY THE VIRGIN   Grid Ref: SK 982043

PARISH RECORDS

There are no surviving Churchwardens' Accounts. A Memoranda Book 1783-1881 (DE 1944/12) contains a varied selection of information about village history including lists of gleaners who paid for a bell to be rung during harvest. Also in the archive are two architects' reports concerning the tower, one by T Graham Jackson dated 26 Dec 1865 (DE 2995/28/1-4), the other by J C Traylen dated 9 Oct 1883 (DE 2995/45).

BELL HISTORY

The thirteenth-century tower, with its fourteenth-century spire, is considered to be the finest in Rutland. The spire of Leicester Cathedral was largely modelled on that of Ketton.

There would have been medieval bells here until the early 1600s, but between 1598 and 1640 five of the bells were cast or recast by five different founders. The present treble is a 1748 recast of one of these bells and the present third an 1897 recast of an early eighteenth-century bell.

In 1862 the architect T Graham Jackson MA inspected the fabric of the church. He found, amongst other things, that the tower and steeple were in urgent need of repair due to cracks in the structure. John Taylor also surveyed the belfry in the same year and his notebook contains a sketch of the layout as it was then. However, the report he prepared as a result of his visit has not survived.

It was not until 1865 that Graham Jackson was asked to consider what work was required to cure the problems

that he found. In his second report (DE 2995/28/1-4) he explained the cause:

A heavy peal of six bells was hung in the belfry stage of the tower, the massive framing that carried them was allowed to become so loose that the timbers swayed backwards and forwards with the motion of the bells and literally acted like battering rams against the slender piers that

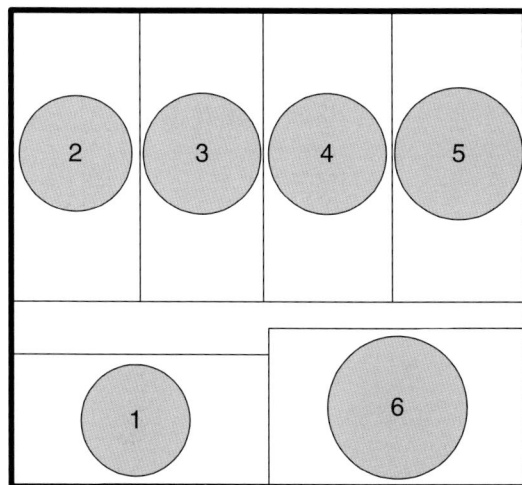

This drawing is based on a sketch in John Taylor's notebook and shows the layout of the bellframe at Ketton in 1862
(closed archives of John Taylor)

support the spire, and to make matters worse when this looseness was observed wedges were driven in between the framing and the piers so that none of the destructive force was wasted, or failed to reach the walls of the tower. Add to this that at various times in order to make more room for the bells the slender piers were still further robbed of their strength by having slices scooped out of them, and holes dug into them, and who then can wonder if the tower already overladen began to give way.

He stated that there appeared to be two options. The first was to preserve the tower as it stood by wrapping it with iron bands. This would mean that it could not be used for full circle ringing and he suggested that the bells should be removed and installed in a separate purpose-built tower on the north side of the church. The second option was to dismantle the tower stone by stone and rebuild it to the same design, but this was not favoured as 'the whole value of the building as an original work would be lost'. The estimated cost for each proposal was £150 for banding and £2,000 for rebuilding. In the event, the first option was chosen but the bells remained in the tower.

Thomas North in 1880 noted that the bells were not rung on account of the supposed danger to the steeple and also recorded that three of the bellwheels were broken (North 1880, 136).

This photograph shows the third and fourth bells at Ketton. The construction of the composite frame installed in 1897 can also be seen. The canons were removed from all five bells when they were rehung at this time

In 1897 the bells were again inspected by John Taylor and the report, published in the August 1897 issue of Ketton Church Magazine, revealed that the bells 'were in a very bad state indeed' and that the oak framework was 'most rotten and dilapidated'. The estimate for the recasting of the third bell was £39 4s 0d and the total

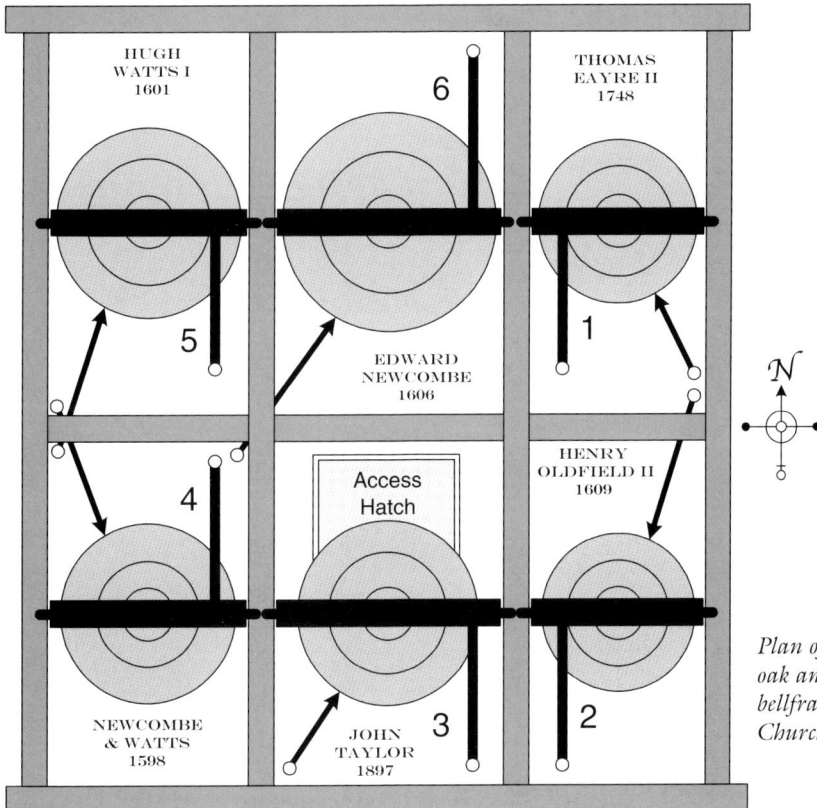

Plan of the composite oak and cast-iron bellframe at Ketton Church

estimate for all of the work to be done was approximately £115. As a result of this inspection the third bell was recast, the others retuned, and the whole peal rehung for chime ringing in a new composite oak and iron low-sided frame.

In 1890 there was an expenditure of £1 4s 6d on an apparatus for chiming three bells and in the late 1940s an Ellacombe chiming frame with ropes to hammers on all six bells was installed in the ringing chamber.

Bell Details

1 1748. Treble. Diameter 756mm (29¾in). Weight 263kg (5cwt 0qr 20lb). Canons removed. Timber headstock. Hung for chime ringing. Cast by Thomas Eayre II of Kettering.

NICHO : BULINGHAM : AB : ME : SVIS :
~ SUMTIBUS : HIC : COLLOCARI : CURAUIT.
~ 1640 : T : WOTTON W : ROWLATT :

(Nicholas Bullingham AB caused me to be placed here at his expense)

High on the waist:

C : W : 1748

Former bell
Probably cast in 1640 by Thomas Norris of Stamford. Diameter 749mm (29½in).

NICHO : BULINGHAM : AB : ME : SUIS :
~ SUMTIBUS : HIC : COLLOCARI : CVRAVIT:
~ 1640 : T WOTTON : W : ROWLATT C W : 1748.

Nicholas Bullingham, the donor of this bell, was baptised at Ketton on 26 October 1609. His father, also Nicholas, held the Prebendal Manor at Ketton under the Cathedral of Lincoln (North 1880, 135).

2 1609. Diameter 768mm (30¼in). Weight 252kg (4cwt 3qr 23lb). Canons removed. Timber headstock. Hung for chime ringing. Cast by Henry Oldfield II of Nottingham.

[34] I sweetly toling men do call to taste on meats that feeds
~ the soole 1609

Founder's mark [7] is immediately below initial cross [34]. Except for the ornate I [84] all letters are like [114].

The ornate I at the beginning of the inscription on the second bell at Ketton

Bell founder's marks, decoration and date on the fourth bell at Ketton

3 1897. Diameter 851mm (33½in). Weight 351kg (6cwt 3qr 17lb). Cast without canons. Timber headstock. Hung for chime ringing. Recast by John Taylor & Co of Loughborough.

[65] RECAST 1897 [65] MOSES [50] SISSON
~ [50] C. H. W. 1713.

On the waist:

A. SWIRE [65] VICAR.
G. FREESTONE & W. NUTT
C. WS.

On the waist opposite: [81]

Decoration [54] is below the inscription band.

Former bell
Cast in 1713 by Henry Penn of Peterborough. Diameter 813mm (32in).

MOSES [50] SISSON [50] CH [50] W [50]
~ HENRY [50] PENN [50] FVSORE [50] 1713.

A piece of the rim of this bell was reported to be missing in 1880 (North 1880, 136).

4 1598. Diameter 864mm (34in). Weight 329kg (6cwt 1qr 26lb). Canons removed. Timber headstock. Hung for chime ringing. Ascribed to Newcombe and Watts of Leicester.

[3]
[23] ME [40] ME [40] I [40] MEREL𝔶 [40]
~ 𝔴ILL [40] SING [40] 1598 [40]

Decoration [119] is below the inscription band. All letters are like [16] and [17] with the exception of **w** and **y** which are like [110].

Described in 1880 as a fine bell in perfect preservation (North 1880, 136). Scheduled for preservation (Council for the Care of Churches).

Part of the inscription on the fourth bell at Ketton

Some of the Ketton handbells

5 **1601**. Diameter 914mm (36in). Weight 400kg (7cwt 3qr 13lb). Canons removed. Timber headstock. Hung for chime ringing. Cast by Hugh Watts I of Leicester.

SARUE THE LORDE 1601 [3]

(Serve the Lord)

All letters are like [16] and [17].

6 **1606**. Tenor. Note F sharp. Diameter 1048mm (41¼in). Weight 517kg (10cwt 0qr 20lb). Canons removed. Timber headstock. Hung for chime ringing. Cast by Edward Newcombe of Leicester.

[31] BE · YT · KNOWNE · TO · ALL · THAT · ~ DOTH · ME · SEE · THAT · NEWCOMBE · OF · ~ LEICESTER · MADE · MEE · 1606

BELLRINGING CUSTOMS

Until 1914, the Gleaning Bell was rung daily at 9am and 5pm during harvest time and the collecting of corn was only allowed between these times in fields where all the stooks had been removed. School attendance figures were very low during this period, for children as well as adults were involved in gleaning. Each person was expected to pay 2d a week to the Parish Clerk for ringing the bell (Traylen nd, Pt 2).

The Ketton Memoranda Book (DE 1944/12) records the names of the gleaners from 1810 until 1837. In 1810 there are one hundred and four names, but in 1837 only eight. Interestingly, the date when the gleaning bell was first rung is also given. For example, in 1823 it was 1 September, in 1825 15 August, and in 1826 31 July.

Today, although Ketton does not have a regular band of bellringers, the bells are chimed for thirty minutes before each Sunday Service which begins at 10.30am. The second bell is chimed during the Eucharist, when this is part of the service. The bells are chimed for all Christmas Services, and on Armistice Sunday the tenor is chimed at 11am. The bells were chimed for the Millennium and the custom of ringing on New Year's Day may be rein-

stated in the future. If requested the bells are rung for weddings and for other special occasions.

HANDBELLS

Ketton has nineteen handbells which are rung at Christmas. Little is known of their history but they are thought to be earlier than 1900. They were found in a box in the library which was then a small room in Church House, which is to the west of the tower in the churchyard. The leather handles had disintegrated and they were generally in a poor state. They were subsequently refurbished by John Taylor of Loughborough and a band of bellringers used to tour the village playing them at Christmas. Today, they are rung at the Festival of Carols in the church and concerts are held at local retirement homes.

SCRATCH DIAL

There is a scratch dial in very poor condition on the east front of the early fourteenth-century south porch.

CLOCK HISTORY

The Ketton Memoranda Book includes many payments to a Mr Stanger for mending shoes, and at the end of

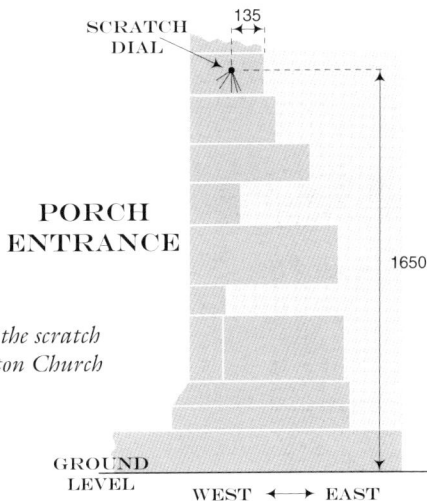

Location of the scratch dial at Ketton Church

KETTON					
Location	East front of the south porch				
Condition	Very poor				
Gnomon Hole Diameter		20			
Gnomon Hole Depth		20			
Height above ground level		1650			
Line Ref	a	b	c	d	e
Length	80	80	80	70	70
Angle (°)	148	159	180	216	250

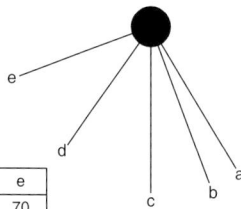

Scratch dial details

these are records of payments for winding the church clock. A similar payment is recorded in very sketchy accounts to a later parish clerk, John Joyce:

1824	Clerks wages & Clock	£1 3s 8d
1826	Clerks wages. Due at Michs last	2s 8d
	Clock 1 year	£1 1s. 0d
1869	Jn Joyce winding clock	£1 1s 0d

This is the only recorded evidence of a former clock. None of the early drawings of the church show a clock dial but it may not have had one. The tenor has an indentation on the soundbow due to being struck by a clock hammer.

ST MARY'S HOUSE

St Mary's House, High Street, Ketton (SK 980045), was originally set up as a 'penitentiary for reforming young women' and for training them in domestic service (Clough 1993, 143). When it was founded in 1893 under the authority of the Peterborough Diocesan Conference a chapel was incorporated as part of the establishment. The

Sisters of the Community of St Mary the Virgin, Wantage, ran the home until its closure in 1945 and it then became a church hall, known as Bishop Clayton Hall. When it was sold in 1994 and subsequently converted into private accommodation the single bell in the chapel bellcote was retained (information from Geoff Fox).

KETTON SCHOOL

Ketton National School had a bell dated 1833 which hung in a bellcote. When the school was demolished the bell was transferred to a modern bellcote over the entrance to the new primary school.

RAILWAY STATION

A *circa* 1910 postcard of Ketton railway crossing and station shows that the station building had a bellcote, but the bell was missing at the time of the photograph. It may have been used to warn of approaching trains. The station opened for passenger traffic on 1 May 1848 and closed on 6 June 1966 (Healy 1989, 102). The station building has since been demolished.

KETTON GRANGE

There is a vertical sundial on a south-facing gable of Ketton Grange and another over a door at Grange Cottage. A date stone on the west gable of Grange Cottage indicates that it was built in 1689.

A clock turret over the former stables at Grange Cottage has east and west-facing dials and a bell. The clock has been out of commission since before 1950.

Ketton railway crossing and station circa 1910 showing the bellcote on the station building (RCM 1972.65.2)

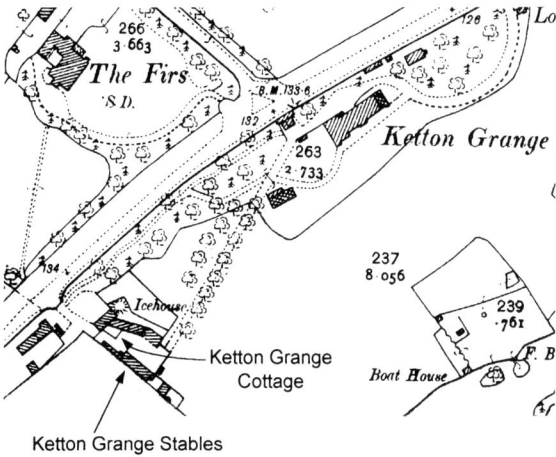

The early 1900s OS Second Edition 25 inch map shows the location of Ketton Grange (SK 985051) and Grange Cottage (SK 985051). Note also the table sundial (SD) in the garden of The Firs (SK 983051)

The posted frame turret clock by Alexander Simmons of Warwick in the former stables at Grange Cottage, Ketton

CLOCK DETAILS

Details of the clock at Grange Cottage stables:

Maker:	Alexander Simmons of Warwick
Signed:	Signed 'Alexander Simmons Warwick 1855' on the setting dial
Installed:	1855
Frame:	Posted frame
Trains:	Going and striking
Escapement:	Pinwheel
Pendulum:	Wooden rod and cast-iron lenticular bob
Rate:	33 beats per minute
Striking:	Countwheel hour striking
Weights:	Cast-iron, rectangular section
Winding:	Weekly
Dials:	East and west facing convex copper dials with gilded numerals
Clock bell:	In open cupola above dials
Note:	The pendulum hangs in a separate lath and plaster case

KETTON HALL & THE PRIORY

Table sundials are shown in the the grounds of Ketton Hall and The Priory on the early 1900s OS Second Edition 25 inch map.

The clock turret over the former stables at Grange Cottage, Ketton

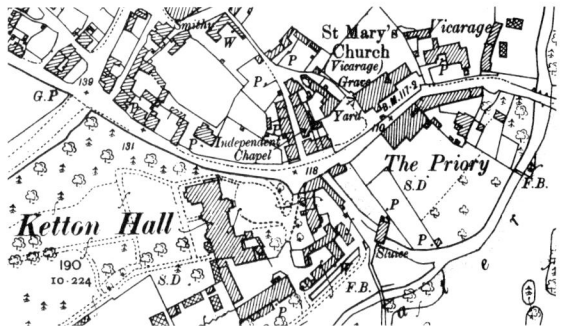

The early 1900s OS Second Edition 25 inch map showing the table sundials (SD) in the grounds of Ketton Hall (SK 980042) and The Priory (SK 982043). Elsewhere on this map another table sundial is shown in the garden of Geeston House (SK 986040)

LANGHAM

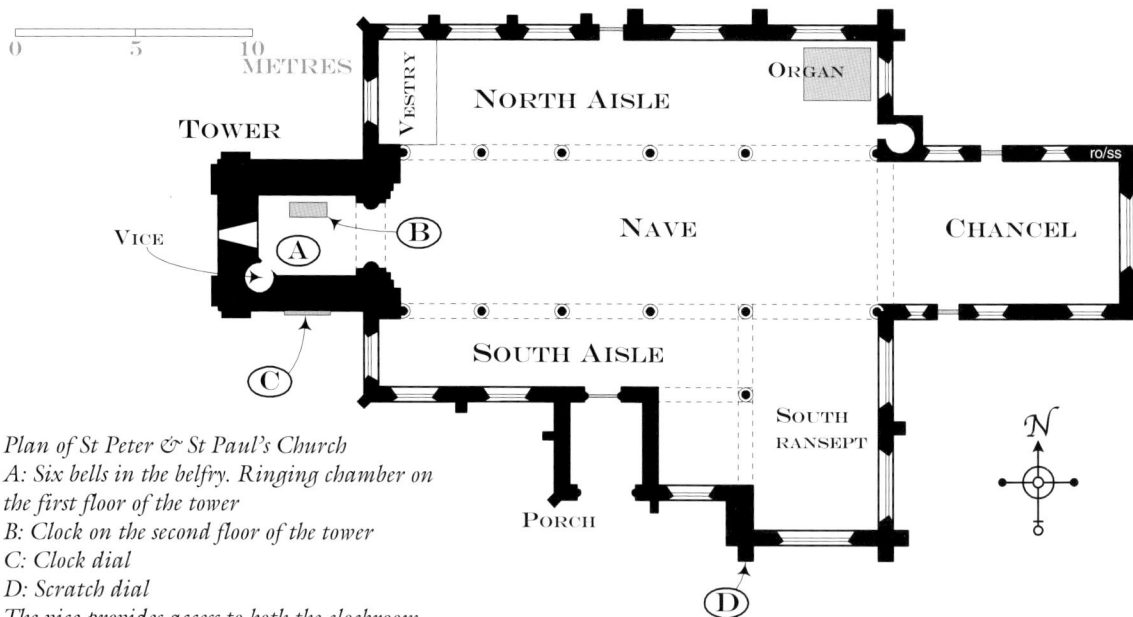

Plan of St Peter & St Paul's Church
A: Six bells in the belfry. Ringing chamber on
the first floor of the tower
B: Clock on the second floor of the tower
C: Clock dial
D: Scratch dial
The vice provides access to both the clockroom
and the belfry

ST PETER & ST PAUL   Grid Ref: SK 844112

PARISH RECORDS

The following records in the parish archive were searched:

Churchwardens' Accounts 1782-1846 and 1847-1949
(DE 2150/15 & 16).
PCC Minute Book from 1920-30 (DE 2150/17). This
includes church accounts 1899-1904.

BELL HISTORY

There have been six bells in the tower of Langham
Church since 1771 and the ring includes one of the
oldest bells in Rutland. It was cast by an unknown
fifteenth-century London founder and it is one of only
three bells in the county to be decorated with the Royal
Arms. The earliest references to the bells are recorded by
the Ven E A Irons in his Visitation Notes (MS 80/5/
20). One note records that in 1544 a Richard Hubbard
included a bequest in his will to the bells of the parish,
presumably for their maintenance. Another note refers
to the poor state of the belfry:

1621 April 21 The Bells & bellframes are in much
decay and the bells cannot be rung

It appears that as a result of this Visitation the churchwardens took some action as the bellframe was replaced
and the bells rehung in 1622.

A short length of oak beam inscribed **E C : H. H :
C.W : 1662**, probably part of the old bellframe, was

The old oak beam from Langham Church belfry

saved when repair work was being carried out in the
belfry *circa* 1896. It is now preserved in the church. At
one end is **RESTORED TO THE CHURCH BY T.
SWINGLER 1896**. The churchwardens whose initials
are recorded on the beam were probably Edward Cole
(senior or junior) and Henry Hubbert as listed in the
Hearth Tax return of 1665 (Bourn 1991, 25). Irons
recorded these names as Edward Coke and Henry
Hubbard.

The Archdeacon's Visitation Report of 1681 ordered
the Sanctus Bell to be repaired and rehung (MS 80/5/
20). This bell was not recorded by Thomas North in
1880, implying that it was lost, sold or used in recasting
one of the larger bells.

By the end of the seventeenth century there were at
least four bells in the tower, three of these having been
supplied by the Norris foundry at Stamford. Two further
bells were cast for the church, one in 1754 by Thomas
Eayre II of Kettering and the other in 1771 by Thomas
Hedderly I of Nottingham. Annual payments for bellropes
were made by the churchwardens in the last quarter of

the eighteenth century, implying that the bells were well used. Virtually nothing was paid for repairs to the bells during this period but in the first half of the next century it was found necessary to carry out work on the wheels and frame. Those responsible for these repairs were Mr Almond, Mr Riley and Joseph Faulks. In 1846 Mr Cort of Leicester supplied 'a brass bearing for the great Bell' costing 10s. 'Taylor and Son (Bellfounders)' were paid £11 8s 10d in January 1848 but no details are provided as to the nature of the work carried out.

When the second, a Norris bell of 1660, was recast in 1874 all the bells were rehung (North 1880, 136). This work was carried out by John Taylor and the Church Restoration Account of 1881 records that the total cost was £155.

The general accounts show that Fred Sewell, the local blacksmith, was paid for bell maintenance throughout the 1890s:

1896-97   Mr Fred Sewell repairing Bell frames etc   10s 0d
          Mr Edwn Mantle conveying bell hanger to
          and from Oakham Stn. two journeys   2s 0d

Repairs to the bellframe were numerous. Eventually a new frame was installed on the advice of John Taylor. Evidently, damage to the tower was being caused as a result of the poor state of the old wooden bellframe 'which was wedged between the walls' (Phillips 1903-04,146). This was a 'remedy' often adopted to overcome loose joints in timber bellframes. However, it allowed the vibrations from bellringing to be transmitted directly to the tower wall, which can eventually result in serious damage to the structure.

John Taylor installed a new high-sided metal frame, recast the treble and second, retuned the other bells and rehung the whole ring at a total cost of £239. This work was completed in 1900 and much of this was paid for through the generosity of Lt Col Sir Henry Clarke Jervoise. He is buried beside the south door of the church, his headstone emphasising 'his devotion to the restoration of the parish church'. A brass plaque in the church records his gift:

To the Glory of God
and in loving memory of
a dear Sister
these bells were restored
December 1900
by
H. Clarke Jervoise,
Janet Small of Dirnanean.
The two small bells were recast
and the whole rehung
on metal frames.

Refurbishment and improvement work between 1973 and 1975 included overhauling and rehanging the bells on ball bearings, and panelling the ringing chamber.

BELL DETAILS

1 1900. Treble. Diameter 759mm (29⅞in). Weight 297kg (5cwt 3qr 10lb). Cast without canons. Cast-iron headstock. Recast by John Taylor & Co of Loughborough.
[99] GRATA SIT ARGUTA RESONANS
~ CAMPANULA VOCE ∴ [49] [128]
~ THOˢ EAYRE FECIT, [104] 1754 [49]
(May this little bell be pleasant sounding with clear tone. Made by Thomas Eayre)
On the waist: [107]
                [65] RECAST 1900 [65]
Decoration [54] is below the inscription band.

Former bell
Cast in 1754 by Thomas Eayre II of Kettering. Diameter 787mm (31in).
GRATA SIT ARGUTA RESONANS
~ VOCE [49] THOˢ EAYRE FECIT O O 1754 [49]
O O was the impression of a coin or coins, the details of which have not been recorded.

2 1900. Diameter 813mm (32in). Weight 336kg (6cwt 2qr 13lb). Cast without canons. Cast-iron headstock. Recast by John Taylor & Co of Loughborough.
THOMAS NORRIS MADE MEE 1660
On the waist: [107]
                [65] RECAST 1900 [65]
Decoration [54] is below the inscription band.

Former bells
The original second. Cast in 1660 by Thomas Norris of Stamford. Diameter 838mm (33in).
THOMAS NORRIS MADE MEE 1660
Recast in 1874 by John Taylor & Co of Loughborough. Diameter 838mm (33in).
J. TAYLOR & CO FOUNDERS
~ LOUGHBOROUGH 1874.

3 1636. Diameter 876mm (34½in). Weight 361kg (7cwt 0qr 11lb). Canons removed. Cast-iron headstock. Cast by Thomas Norris of Stamford.
[26] [69] THOMAS [69] NORRIS [69]
~ MADE [69] ME [69] 1636 [69]
On the soundbow: O   O   O   O
O O O O are the images of four coins which are evenly spaced around the soundbow. They are the obverse and reverse (twice) of the same Charles I silver shilling. Although the coin is not dated it is believed to have been minted *circa* 1634-35.

4 *Circa* **1480.** Diameter 933mm (36¾in). Weight 426kg (8cwt 1qr 16lb). Canons removed. Cast-iron headstock. Cast by an unknown London founder (Council for the Care of Churches).
[1] Sit Nomen Domini Benedictum [36] [35]

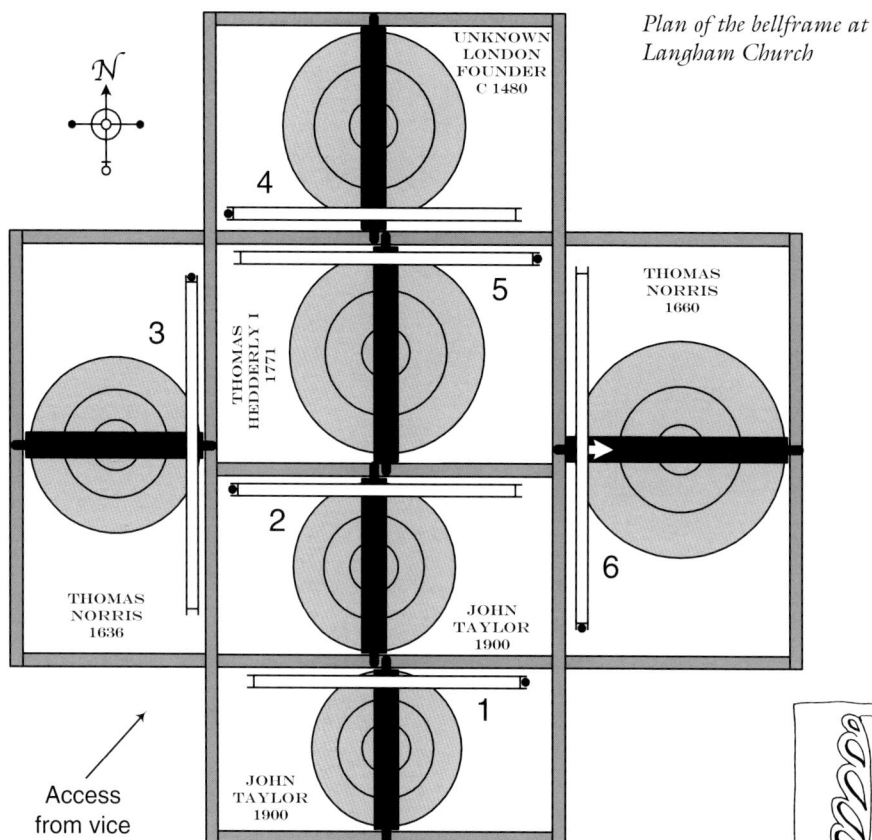

Plan of the bellframe at Langham Church

(Blessed be the name of the Lord)
Words and founder's marks are spaced evenly around the inscription band. A 'Brede mark' bell (*see* Chapter 3 — Bellfounders, London, John Daniel's successor). Scheduled for preservation (Council for the Care of Churches).

5 1771. Diameter 1022mm (40¼in). Weight 554kg (10cwt 3qr 17lb). Canons removed. Cast-iron headstock. Cast by Thomas Hedderly I of Nottingham.
THE CHURCHIS PRAiS i SOUND ALL ~ WAYS THOMAS HEDDERLY FOUNDER ~ NOTTING^M 1771
(The Church's praise I sound always)
Decoration [129] is below the inscription band. Bands of decoration [42] are placed around the collar and directly above the soundbow and the latter includes a king's head [133].

6 1660. Tenor. Note F. Diameter 1118mm (44in). Weight 688kg (13cwt 2qr 4lb). Canons removed. Cast-iron headstock. Cast by Thomas Norris of Stamford.
[10] [52] THOMAS [52] NORRIS [52] ~ MADE [52] MEE [52] 1660 [52] I ꜧ [52]
The clock strikes the hours on this bell.

These initials on the tenor at Langham may be those of a churchwarden or the donor

The king's head [133] *within the band decoration directly above the soundbow of the fifth bell at Langham*

BELLRINGING CUSTOMS
The Churchwardens' Accounts from 1782 to 1846 give an interesting insight into customs of that period. The bells were rung at Christmas, on New Year's Eve and on 'Feast Sunday'. Up until 1813 they were also rung regularly on 29 May 'at King Charles Restoration', and on 5 November for 'Gunpowder Plott'. The amount paid to the ringers on these and other occasions was generally 5s. A memorandum included in the accounts states that at a meeting 'of the inhabitants of the Parish of Langham' held on 9 April 1828, the ringers' fees would be 'discontinued on account of the depression of the times' (Phillips

1903-04, 148). This is reflected in the disbursements but payments resumed at the accession of Queen Victoria, an event which gave cause for further celebration:

1838 June 30 Paid Ringers for Coronation   15s 0d
1840 Feb 10   Pd ringers at Queens Wedding 10s 0d

In the next two years the ringers were also paid for the 'Birth of a Prince' [the future Edward VII] and his subsequent christening.

In 1786 the fee for tolling the funeral bell was 4d. The accounts record that the clerk was paid regularly for this duty and there were other occasions when he tolled the bell for the funerals of those of national importance. For example, in 1827 he was paid 1s for tolling the bell on the day of the Duke of York's funeral.

The Pancake Bell was rung on Shrove Tuesday and the Gleaning Bell at 9am and 5pm during the harvest. At the Death Knell three tolls were given for a male and two tolls for a female, both before and after the knell. During the winter months the Curfew Bell was rung at 8pm, after which the day of the month was tolled. Tradition says that a lady having lost her way one night was guided home by the sound of the Curfew Bell. In gratitude, she left an endowment ensuring that this bell continued to be rung. However, by 1880 there was no trace of this particular endowment (North 1880, 137), although this bell continued to be rung from Old Michaelmas Day to Old Lady Day until about 1912 (Traylen 1989, 34).

The accounts from 1847 show that it was the sexton's duty to ring the bells for the services, and there are a number of other entries for ringing the bell or bells on specific occasions:

1852 Nov 18   Pd C Sewel. Tollg Bell on the late
                   duke of Wellingtons Funeral   10d

1870 March 15 Pd Ed Dalby for Ringing Prayer
                   Bell   £1 0s 0d
1894-95   Tolling bell at Vicar's death &
                   funeral   2s 0d
1909 Dec 20   Ringing bells for confirmation   9s 0d
1912-13   Paid to ringers at the induction service
                   of the Rev J H Wood M A   17s 6d

The following information on customs remembered by inhabitants of Langham in the last century makes interesting reading.

On the day of a funeral the sexton would ring the treble for a few strokes, one hour before the funeral, to prepare the bearers. They, with the undertaker, would arrive at the home of the deceased, where they would be offered refreshments. Before a bier was used the coffin would be carried on the shoulders of the bearers, and when the cortège left the house the tenor rang one stroke every minute until it arrived at the church gate.

The schools closed during the harvest so that the children could join their mothers gleaning in the fields. Every woman wore an apron with a wide pocket in front. She would collect the heads of wheat and barley in this before transferring them into a sack, which at the end of the day was often carried home on her head.

The Rent Bell was traditionally rung in Langham but ceased in 1914. It announced the arrival of the steward to collect the half-yearly rents. He would receive this in an upstairs room of the Noel Arms. Rents of over twenty pounds for the year were paid on 'Big Rent Day' at Christmas and after the collection a dinner was held. The Steward would provide hot drink based on rum, songs were sung and a toast to the landlord followed. 'Little Rent Day' for cottagers was held on the previous day, but they received no refreshment.

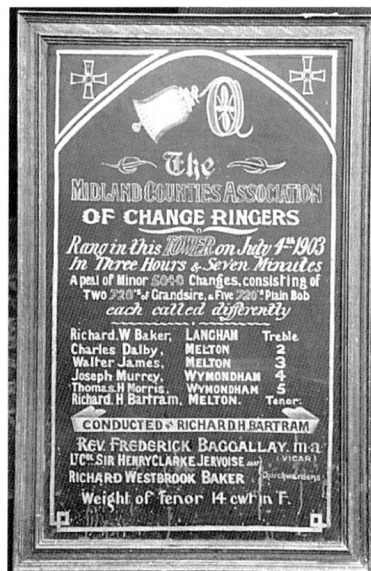

Two illuminated peal boards in Langham Church

A memorial in Langham Church to dedicated bellringers

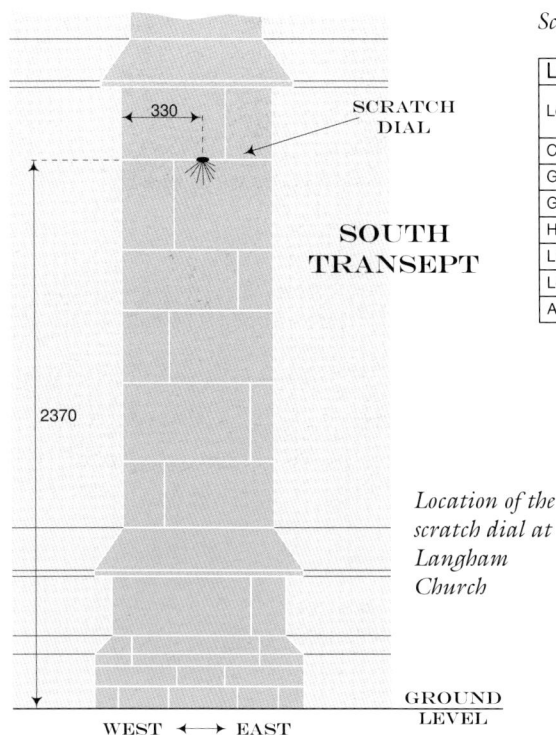

SCRATCH DIAL

330

SOUTH
TRANSEPT

2370

Location of the scratch dial at Langham Church

GROUND LEVEL

WEST ←→ EAST

Scratch dial details

LANGHAM								
Location	South face of south transept west buttress							
Condition	Poor							
Gnomon Hole Diameter			Filled in					
Gnomon Hole Depth			Filled in					
Height above ground level			2370					
Line Ref	a	b	c	d	e	f	g	h
Length	66	65	70	67	80	85	90	66
Angle (°)	114	126	156	163	178	191	217	232

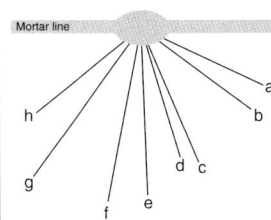

Langham's last town crier was heard in 1910. He was Mr Glenn, and according to tradition would ring his bell before shouting aloud news of forthcoming events. He also made private announcements which would cost between six pence and a shilling a time (Traylen nd, Pt 2).

The accounts show that Langham has a long history of bellringing and this continues with the present active band who ring for all services, and for funerals, birthdays and anniversaries when requested.

CANDLE HOLDER
An early stone bellringers' candle holder in the form of a cupped hand remains on the north wall of the ringing chamber.

SCRATCH DIAL
There is a scratch dial at an unusually high level on the west buttress of the thirteenth-century south transept.

CLOCK HISTORY
The absence of any accounts or other relevant records before 1782 means that there are no details of a clock at Langham prior to this date. However, the Churchwardens' Accounts reveal that there was certainly a church clock here in 1784, as a Mr Gibson was paid 6d 'for Wire for Clock'. No details are given as to its type or its maker, neither are there any clues given in the various entries

concerning work carried out on the clock. Later accounts reveal that very little was expended on it other than the purchase of clocklines and the occasional repair between 1826 and 1845. As might be expected it was the local Oakham clockmakers who were called in to carry out repair and maintenance work. In 1836 Thomas Cooke was one of the named repairers, and from 1848 until 1867 Stephen Simpson was paid for maintaining the clock.

During the general restoration of the church a new clock was donated by the Rev John Mould, then vicar of the parish. It was supplied and installed by Tucker of Gray's Inn, London, in September 1875, at a cost of £100.

It was the sexton's responsibility to see that the clock was wound and kept in good working order. From the 1870s to the 1890s Edward Dalby, George Sewell, Alfred Hubbard and Charles Burdett were paid for cleaning, repairing or winding it:

1877 July		George Sewell 2 months pay for the clock	1s 8d
	April 16	E Dalby half a years salary for chiming and c [clock]	10s 0d
1882 March 21		Alfred Hubbard for managing the Church Clock & winding the same	£1 6s 0d

The presentation plaque on the flatbed turret clock by Tucker at Langham Church

A tablet in Langham Church recording the restoration of the clock

Langham church clock escapement

| 1897-98 | C Burdett [sexton] for winding clock | £1 0s 0d |

Payments for repairs and maintenance are recorded to the end of the available accounts, and the following are some examples:

1900-01	Mr Payne, [of Oakham] attending to Church Clock	5s 0d
1925	Mr Holt for repairing Clock	£5 0s 0d
	Mr Crane for repairing Clock	£5 17s 6d
1944	Messrs Smith [of Derby] for Clock maintenance	£2 5s 0d

The following item was noted in the minutes of the PCC Meeting of 3 July 1925: 'That the clock be cleaned and if possible for future for it to be kept up to Greenwich Time'. In 1925 the source of 'Greenwich Time' would have been the local Post Office which received time signals via the electric telegraph.

In 1973 the clock was overhauled and fitted with Gillett & Johnston automatic winding units. A tablet in the church records that a full restoration was carried out in 1986 in memory of John Lowe Hassan. The work was carried out by John Smith & Sons of Derby who also fitted replacement automatic winding units.

CLOCK DETAILS

Details of the present clock:

Maker:	John Tucker of Gray's Inn, London
Signed:	On the setting dial 'Tucker, Theobalds Road, Gray's Inn, London. 1875'. Also signed on the frame
Installed:	1875
Cost:	£100
Frame:	Cast-iron flatbed
Trains:	Going and striking
Escapement:	Deadbeat
Pendulum:	Wooden rod & cast-iron cylindrical bob
Rate:	48 beats per minute
Striking:	Locking plate hour striking
Weights:	Cast-iron
Winding:	Smith's AW1 epicyclic automatic winding units fitted to the second arbor of each train in 1986
	Replaced automatic winding units installed in 1973, possibly by Gillett & Johnston
Dial:	Skeleton dial with black background and gilded roman numerals and hands
Location:	South face of the tower
Note:	Clock is housed in a pine case set on a platform and accessed by a ladder

From a circa 1890 photograph of the Manor House, Langham (RCM A.9243), showing the sundial and the family of Thomas Swingler & Son, the agents of the Earl of Lonsdale (information from Tim Clough)

Langham Manor House sundial in 2000

LANGHAM MANOR HOUSE

A vertical limestone sundial can be seen below a gable window on the seventeenth-century Manor House (SK 844114) in Church Lane. This sundial is engraved with roman numerals and has the words MORNING and EVENING above the gnomon.

LANGHAM HALL

A clock turret was placed over the entrance to Langham Hall (SK 845112), adjacent to the Hall stables, in 1929. It has four dials: two clock dials facing south and east, and two wind direction dials facing north and west, linked to the weather vane above. The north facing dial is opposite the church and it is said that the then owner of the Hall, Owen H Smith, stipulated this feature so that he could determine the wind direction for hunting purposes on his return from church services.

The small single-train clock movement is mounted in a bedroom cupboard below the turret.

Details of the clock at Langham Hall stables:

Maker:	John Smith & Sons of Derby
Signed:	Signed on the setting dial: 'John Smith & Sons of Derby 1929'
Installed:	1929
Frame:	Cast-iron flatbed
Trains:	Going train only
Escapement:	Pinwheel
Pendulum:	Wooden rod with cast-iron cylindrical bob
Rate:	52 beats per minute
Weight:	Cast-iron
Winding:	Weekly
Dials:	Two white clock dials with black numerals on a turret, facing east and south, and two matching wind direction dials
Notes:	The clock movement is mounted in a bedroom cupboard below the turret Repaired and overhauled by John Smith & Sons in 1940

RANKSBOROUGH HALL

A table sundial is shown in the garden of Ranksborough Hall on the early 1900s OS Second Edition 25 inch map.

The west-facing wind direction dial on the turret at Langham Hall

The clock movement at Langham Hall

The clock turret above the entrance to Langham Hall

LITTLE CASTERTON

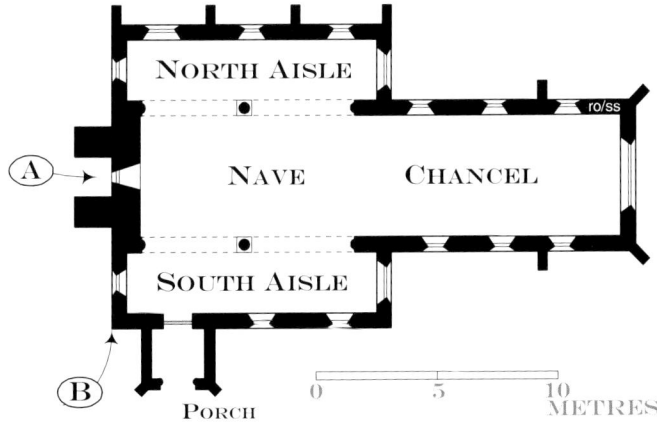

Plan of All Saints' Church
A: Two bells in a double bellcote
B: Scratch dial

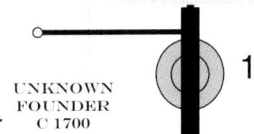

A plan of the bellcote at All Saints' Church, Little Casterton

ALL SAINTS   Grid Ref: TF 018099

PARISH RECORDS

The only relevant records in the parish archive are a Church Book 1832-75 (DE 2344/5) which contains Churchwardens' Accounts, Overseers' Accounts and Vestry Minutes, and a Vestry Minute Book 1876-1941 (DE 2344/6).

BELL HISTORY

The bellcote was added in the early thirteenth century and All Saints' Church had two bells from this date. Nothing is known of the small bell other than it was probably cast at the beginning of the eighteenth century. The tenor is the earliest bell in Rutland cast by Toby Norris I of Stamford.

The Churchwardens' Accounts and Vestry Minutes contain very little detail concerning the bells and only refer to the occasional purchase of bellropes and minor repairs carried out between 1911 and 1917.

This is the only church in Rutland where the bells are rung from outside. The bellropes are between the two buttresses on the west wall, and the space between has been roofed over for the comfort of the ringers. The bells are hung for chime ringing and were rehung on new timber headstocks by Nicholas Meadwell in 1990.

BELL DETAILS

1 **Circa 1700.** Diameter 343mm (13½in). Weight approximately 31kg (2qr 12lb). Canons retained.

The bellcote and bells at Little Casterton

From a circa 1793 drawing of Little Casterton Church showing that the southern bay of the double bellcote was then vacant (RCM F10/1984/13)

Timber headstock. Cast by an unknown founder. No inscription. The treble was missing in 1793 and the present bell may be a Priest's Bell from another church.

2 **1608**. Diameter 559mm (22in). Weight about (127kg) 2cwt 2qr. Canons retained. Timber headstock. Cast by Toby Norris I of Stamford.

CVM : VOCO : VENITE : D : B [14] A : 1608
(Come when I call)

BELLRINGING CUSTOMS
At the Death Knell there were thrice three tolls for a male and thrice two for a female (North 1880, 126). Today the treble is chimed for three and a half minutes and then the tenor for one and a half minutes before each service. They are sometimes chimed for weddings.

SCRATCH DIAL
There is a possible scratch dial in very poor condition on a quoin at the south-west corner of the south aisle, approximately 1.6m (63in) above ground level. The wall dates from the early thirteenth century.

LYDDINGTON

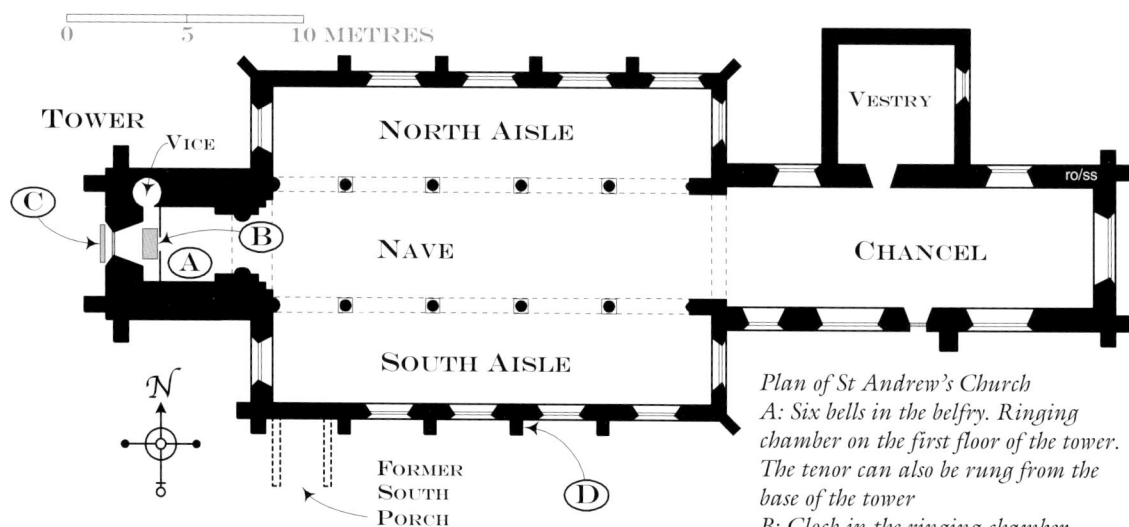

0 5 10 METRES

TOWER
VICE
NORTH AISLE
NAVE
SOUTH AISLE
CHANCEL
VESTRY
ro/ss

C
B
A
D

FORMER SOUTH PORCH

N

Plan of St Andrew's Church
A: Six bells in the belfry. Ringing chamber on the first floor of the tower. The tenor can also be rung from the base of the tower
B: Clock in the ringing chamber
C: Clock dial
D: Relocated scratch dial
The vice provides access to the clockroom, ringing chamber and belfry

ST ANDREW   Grid Ref: SP 876970

PARISH RECORDS
The archive includes a complete run of Churchwardens' Accounts from 1626 to 1959. These contain a great deal of information about the bells and clocks. The references are: 1626-1824 (DE 1881/41, with 1690-1721 at the rear of DE 1881/1a), 1824-1921 (DE 1881/42) and 1921-59 (DE 2989/1-2). In addition, some correspondence relating to the bells is held in the parish.

BELL HISTORY
Although no records of the early bells at Lyddington have survived, the Churchwardens' Accounts include details of work carried out on them. The following are some examples:

1628 Payed to Sharpe for mending ye bell claper and seven pounds of iron   4s 0d
      and for helping him about ye clapper we spent   7d
1630 Payed to John Lam for kepinge ye Bells ye first halfe yeare   10d

1649 Payed to Hough Sharpe for mending Seccond bell guging [gudgeon] & ye third Bell guging & two Rods of irone for ye third bell and for mending eight plates of iron for the third Bell   3s 6d
1666 Paid to Hugh Sharpe for iron and worke about ye great Bell [tenor]   6s 0d
1670 To Kenelme Waterfield mending Bell frames   2s 6d

A new oak bellframe was installed in 1677 as shown by the following which was engraved on one of the frame timbers (North 1880, 140):

JAMES HILLAND
ROBERT COLLWELL
CHIRCH WARDINS
IOHN BROWNE
FECIT
1677

It is interesting to note that **IOHN BROWNE FECIT** and the date **1701** were carved on the old frame at North Luffenham.

The four bells supplied by Toby Norris III of Stamford in 1694 are the earliest at Lyddington for which details are known. The largest of these, the present tenor, is inscribed **TOBYAS NORIS CAST VS ALL ORE IN 1694**, the **ORE** being a corruption of 'over', meaning 'cast us all over [again] in 1694'. Hence the date of the three undated bells, the present third, the fourth before it was recast, and the fifth, is known. The same founder supplied a fifth bell, the present second, in 1695. The bells were presumably donated by those, other than the churchwardens, named on the inscription bands.

John Browne carried out further work on the bells in 1692 and the following later item probably refers to his son:

1743 pd John Browne of Caldecott for work at the
Bells 4s 6d
Expended when he Came to see what wood
and Iorns [irons] was wanting 1s 0d

The following two references from the accounts imply that in addition to having five large bells, Lyddington Church also had a smaller one, presumably a Sanctus or Priest's Bell. Thomas North made no reference to this bell in 1880:

1727 May 22 paid for a rope for the little Bell 2s 6d
1758 for A little Bell rope 2s 6d
for five New Bell ropes 16s 0d

Apart from the frequent replacement of bellropes and the occasional new bellwheel, very little work was carried out on the bells and frame for the next one hundred years, and by 1859 the bells and fittings were in poor condition.

Two estimates dated 26 December 1859 and supplied by John Taylor & Co give some idea of the state of the bells at this time and the work required to put them in good order:

... for the recasting of the third Bell in the peal of five (which is cracked) belonging to the parish church of Lyddington the said Bell supposed to weigh 7cw^t — 2qrs little more or less - the new Bell to be the same dimensions and note as the old consequently as near the same weight as is possible to cast — and warranted to possess a fine, clear and harmonious tone and in tune with the rest — also for the fitting up the said new Bell with new Hanging — viz wheel, headstock, wrought iron bowed gudgeon, clapper to work upon brass bushes, bearing brasses cast of the best wearing metal fitted in cast iron carriages, roller, straps, caps, cramps and all requisite bolts the whole fitted to Bell, also for taking down the old Bell, hoisting and fixing the new complete including carriage to & from, travelling and all other expenses in consideration for the sum of <u>Thirty Pounds</u> and all the old materials rendered useless and in place of which the contractor supply new.

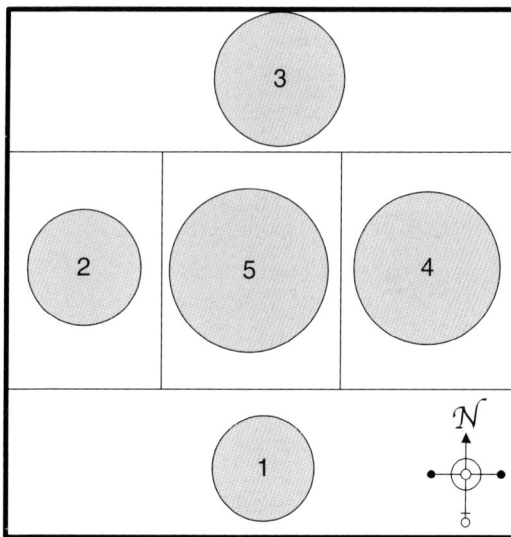

John Taylor surveyed the belfry at Lyddington in 1859 and this drawing is taken from his sketch of the layout of the old timber bellframe (closed archives of John Taylor)

Estimate by John Taylor & Co of Loughborough to new hang the four other bells (which are in a very bad and dilapidated state) ... also new clappers to strike on the opposite sides of the Bells being much worn where they now strike therefore being liable to crack — some of the Bells having very large and clumsy cannons or crown so as to make it an impossibility to hang them to ring and clapper properly — J T & Co engage to cut off those that are so — and drill through the crowns of the Bells which may then be securely fastened to the headstocks also to fix the whole of the above complete in the tower ... for the sum of <u>Forty Pounds</u> ....

The order for recasting the third bell was placed in November 1860, and inscriptions were ordered at an additional cost of four pence a letter. No charge was made for the date. The bell was returned by rail to Manton Station on 21 February 1861, the parish being responsible for transporting it to Lyddington. It appears that the work on the other four bells was never commissioned.

By 1931 the old oak frame of 1677 had become unsafe and the only way that ringing could be carried out was by using ropes attached to the clappers to pull them against the stationary bells. This situation prevailed until 1975 when the Rev Charles Wright organised a fund for retuning and rehanging the bells in a new frame designed for six. This work was carried out by John Taylor in the same year. Ray Richardson, a keen bell ringer and proprietor of Belton Garage, and Harold Sleath, also of Belton, removed the old frame on a voluntary basis. A sixth bell, a new treble, was added to the ring in 1977.

Access from vice

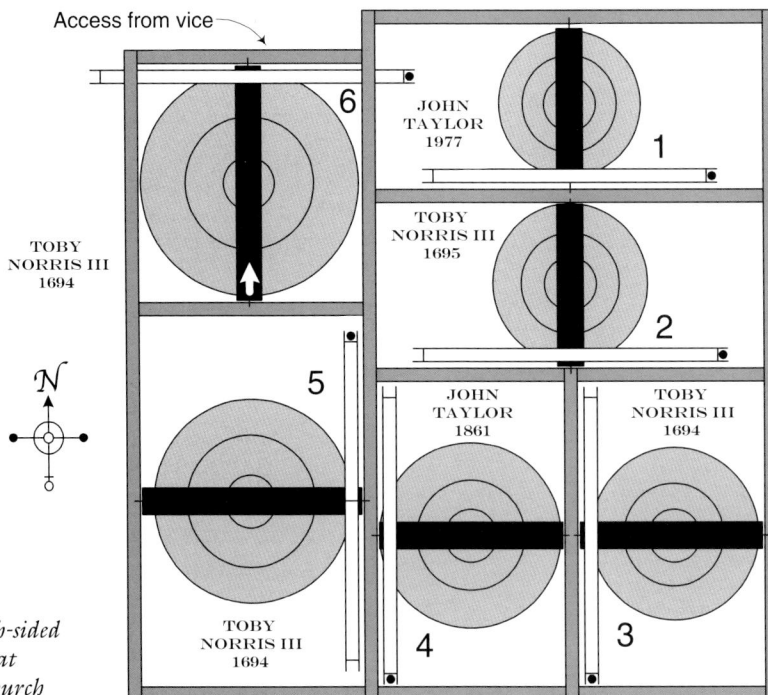

Plan of the high-sided steel bellframe at Lyddington Church

BELL DETAILS

1 1977. Treble. Diameter 749mm (29½in). Weight 245kg (4cwt 3qr 9lb). Cast without canons. Cast-iron headstock. Cast by John Taylor & Co of Loughborough.

I WILL SING UNTO THE LORD

On the waist:

REV'D CHARLES WRIGHT VICAR

On the waist opposite:

19 [47] 77

2 1695. Diameter 775mm (30½in). Weight 245kg (4cwt 3qr 8lb). Canons removed. Cast-iron headstock. Cast by Toby Norris III of Stamford.

[10] I [70] WARING [70] TOBY [70] NORRIS ~ [70] CASTE [70] ME [70] 1695 [70]
[10] [70] I [70] IRELAND [70] I [70] CRADIN ~ [70] C [70] W

3 1694. Diameter 826mm (32½in). Weight 298kg (5cwt 3qr 14lb). Canons removed. Cast-iron headstock. Cast by Toby Norris III of Stamford.

[10] [70] W [70] BROWNE [70]

4 1861. Diameter 876mm (34½in). Weight 371kg (7cwt 1qr 5lb). Canons removed. Cast-iron headstock. Recast by John Taylor & Co of Loughborough.

GREGORY BATEMAN M:A. CURATE.
~ TAYLOR & Cᵒ.. FOUNDERS. 1861.

On the waist:

T: J: BRYAN ⎤ CHURCHWARDENS
JOSEPH WRIGHT ⎦   ~ A:D 1861

Former bell

Cast in 1694 by Toby Norris III of Stamford. Diameter 883mm (34¾in). Weight 357kg (7cwt 0qr 4lb). Inscription unknown.

5 1694. Diameter 965mm (38in). Weight 459kg (9cwt 0qr 5lb). Canons removed. Cast-iron headstock. Cast by Toby Norris III of Stamford.

[10] [70] W [70] PRETTY [70]

6 1694. Tenor. Note F. Diameter 1086mm (42¾in). Weight 648kg (12cwt 3qr). Canons removed. Cast-iron headstock. Cast by Toby Norris III of Stamford.

[10] C PRETTY [70] TOBYAS [70] NORIS [70] ~ CAST [70] VS [70] ALL [70] ORE [70] IN 1694

Note that there is only one 'R' in 'NORIS' in this inscription. North (1880, 140) recorded two. The clock strikes the hours on this bell.

BELLRINGING CUSTOMS

From the beginning of the Churchwardens' Accounts in 1626 until the First World War it seems that the bells were well used. From 1824 and annually for the rest of the nineteenth century payments were made to bellringers at Christmas, and also for ringing on other occasions:

1727   paid for Ale on the King's Crownacon Day for the Ringers [Coronation of George II]   12s 0d

1761 to the Ringers the Crownation [Coronation of
George III] 6s
1804 Pd Midnight Peal to Ringers 6s 0d
1831 Sept 8 Pd ringers on the Coronation Day
[Coronation of William IV] £1 0s 0d

On Sundays a bell was rung at 8am and for Divine Service the bells were chimed. If Evening Prayer was to be said in the afternoon a bell was rung after Morning Service, but if it was later in the evening the bell was rung at 4pm. The Pancake Bell was still being rung at 11am on Shrove Tuesday in 1880, as was a Daily Bell at 1pm. Earlier this bell had been rung at 8am. The Gleaning Bell was rung at 8am and 5pm during the harvest for which women and older children paid a penny a week to the clerk. At the Death Knell there were thrice three tolls for a male and thrice two tolls for a female, both before and after the knell. At funerals the custom of chiming all the bells had died out by 1880, to be replaced by tolling the tenor (North 1880, 141).

Today Lyddington Church has an active band of bellringers. The treble is chimed for the Sunday Service held at 8.30am, 9.30am or 6pm. For the Family Service held at 11am the bell is either rung or chimed. The bells are also rung on New Year's Day, Easter Day and Armistice Day, and at weddings if requested.

RINGERS' RULES

The following notice of *circa* 1880 hangs in the ringing chamber:

The inverted scratch dial on a south aisle buttress at Lyddington Church

S. Andrew's Liddington
Rules for the Belfry.
1. The number of Ringers shall not exceed <u>eight</u>.
2. No one but a Ringer shall be allowed to be present in the belfry on any occasion on which the bells are being used. (N.B. This does not apply to the choir <u>passing through</u> to the belfry on Sundays on their way to the organ gallery).
3. No refreshments of any sort shall be allowed, nor any smoking in the belfry.
4. There shall not be more than two <u>practice</u> nights in each week and these shall not be either on a Saturday or Sunday.
5. A copy of these rules shall be kept hung up in the belfry and also in the safe in the Vestry.
6. Any Ringer breaking these rules will be at once dismissed.
7. On the occasion of any vacancy, a new ringer shall be appointed by the Vicar & Churchwardens.

Signed:- T N Gillham Vicar

Edwᵈ Sharman	Franˢ Stevenson
	Churchwardens
W Burnaston	Joseph Branston X
	[his mark]
Joseph Muggleton	Charles Frisby
Harry Sewell	James Brewster
George Wadland	
	Ringers

SCRATCH DIAL

There is a scratch dial on the east face of a south aisle buttress: the fourth buttress from the west. It is on a relocated stone and inverted. On the north wall of the chancel, near the vestry door is a résumé of a church history written in 1839 by Charles A Herbert which states that there was evidence of a 'mass clock' on the outside of the south porch. This porch was demolished

Location of the scratch dial at Lyddington Church

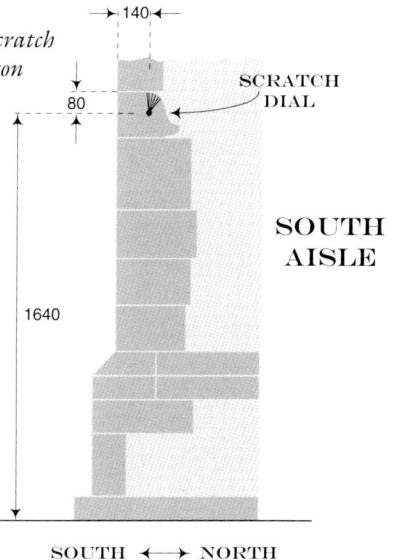

LYDDINGTON

Location	East face of fourth buttress from west of south aisle					
Condition	Average					
Gnomon Hole Diameter		25				
Gnomon Hole Depth		Filled in				
Height above ground level		1640				
Line Ref	a	b	c	d	e	f
Length	76	76	76	76	76	76
Angle (°)	173	180	192	204	214	235

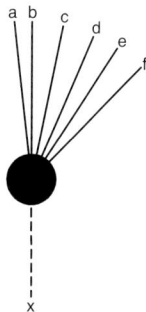

Note that this scratch dial is on a relocated stone and is inverted. Angles are therefore measured clockwise from datum line x.

Scratch dial details

about 1803 (Dickinson 1983, 71). It seems likely therefore that some of the stone was used in the restoration of this buttress, accounting for the present location of the scratch dial.

CLOCK HISTORY

There was a clock in Lyddington Church as early as 1626 as shown by the following extracts from the Churchwardens' Accounts:

1626	ffor keeping of ye Clocke	6s 8d
1627	payd for ye keepinge of ye Clocke	6s 8d
	Paid for oyle for ye Clocke	2d
1629	Paid to Bouth ye clock keper for kepinge ye Clock	6s 8d
1645	Payed to Roger Booth for lookinge to the Clock for the said 2 years by agreemt	13s 4d

This early clock probably had a verge and foliot escapement and a timber frame, and would have required winding every day. Although its timekeeping would have been poor compared to modern standards, it was quite adequate for the period.

By 1680 turret clocks with the new and far more accurate anchor escapement were being introduced. John Watts of Stamford is believed to have installed this new type of clock in many Rutland churches. A John Watts, clockmaker of Caldecott, is recorded in the Churchwardens' Accounts from 1661 to 1670 as working on the Lyddington clock. He may be the John Watts who later moved to Stamford:

1662	Spent with John Watts when he come to ye clok	6d
	payd to Steeven Manton for carrying ye clock and fetching of it hom from Colldicoat [Caldecott]	3s 0d
	payd to John Watts for mending ye clock	£2 3s 0d
1666 Nov	payd John Wattes for mending ye clocke	4s 0d

A William Fox was paid 6s 8d in 1703 for mending the clock. This William was probably the father of Robert Fox, an Uppingham clockmaker who worked on the clock forty years later. Hugh Sharpe was responsible for maintaining and repairing the clock from 1706 until 1738. He occasionally wound the clock, repaired locks and carried out work on the bells. He may therefore have been a village blacksmith:

1719	Expences when we bargained with Hugh Sharpe to look after the Clock	2s 0d
	For a new pendalum for the clock	5s 0d
	For mending the clock wheel	1s 0d
	To Hugh Sharpe for a new nutt and sockett for ye clock	2s 0d
	for looking after the clock from Michaelmas 1718 to Xmas 1718	3s 0d
	And for Cleaning the clock	1s 0d
	ffor a hook for the pendal [pendulum]	6d
1728 Jan 5	Expended when we bargained with Hugh Sharpe to mend the Clock	8d

| Feb 28 | Paid Hugh Sharpe for repairing the Clock | £2 7s 4d |

Thomas Warren, presumably the parish clerk, like his predecessor John Warren, was paid regularly from 1718 to 1744 'for looking after the church clock' and for 'clock winding'. His half yearly salary in 1726 for this duty was 6s.

In 1738 a new clock was supplied by Everard Billington of Market Harborough, Leicestershire. He probably acquired it from Thomas Eayre II of Kettering, the foremost turret clock maker in this area. Like its predecessor, the new clock did not have a dial:

1738	Spent when agreed for the Clocke	2s 6d
	pd E[illegible: Everard?] Billington for a new Clock	£10 10s 0d
1739	pd one shiling for ale when the Clock was done	1s 0d
1740	pd for Alle when the Clock was mended first time	1s 0d
1742 May 5	pd Robert ffox for mending the Clock	£1 11s 0d
	pd to Robert fox for one year looking after ye Clock from May ye 5 1742 to May ye 5 1743	2s 6d

From 1744, John Fox, son of Robert, was responsible for the clock, a duty which he and his son carried out until 1818. Another reference to work on the clock is when it was repaired by John Houghton of Uppingham:

| 1848 May | Houghton for repairing Church Clock | £3 3s 0d |

In 1890 a decision was made to replace the old clock as part of the general restoration of the church and an order was placed with William Potts & Sons of Leeds. The following is transcribed from Potts' Order Book (Michael Potts):

[Page]144   Lyddington Church
Or[der] 506
Small Gravity Striking Clock
One dial. 5ft skeleton fastened up to wall
Wall 5' 9" thick
Clock room 11' 6" square. 19ft high
Bell 3' 7" dia. 15cwts [the tenor bell]
Centre of dial 14ft above clock room floor
Flat weights to suit space available.
All gun metal wheels and steel pinions
Subsequently agreed to strike half hours omitting 12.30 & 1.30
All mason and joiner work to find
Try and fix dial without scaffold by hoisting from top of tower & fasten up from ladder
Going Cord 124ft
Striking Cord 114ft
See particulars sent and sketch
Belfry 14' 8" high. Total for weights 33ft
Rockingham Station for Goods, & passengers per G. N. Rly. via Grantham.
11 July 1890

The new clock was installed a month later. It was

The clock by W Potts & Sons at Lyddington Church, installed in 1890 (Brian Stokes)

donated by three sisters, the daughters of Major Henry Christopher Marriot (1826-1908) of Avonbank, Pershore. He lived at Lyddington House from about 1887 to 1890. This may have been his hunting base in Rutland. A brass plaque on the tower arch records this gift:

**TO THE GLORY OF GOD
THE CLOCK WAS PRESENTED TO
LYDDINGTON CHURCH ON ITS
~ RESTORATION
BY THREE SISTERS
JULIA, EDITH & ROSAMOND MARRIOTT
AUGUST 1890**

The reopening of the church was reported in the *Lincoln, Rutland & Stamford Mercury* on 5 September 1890:

LYDDINGTON — Church Re-Opening
The church of St. Andrew was re-opened on Friday, after being closed ten months for repairs and restoration ... During the work of restoration the services were conducted in the bede-house ... The old clock, which had no face and which made a mere pretence of keeping time, has been replaced by a new one made by Messrs. Potts, of Leeds. It is of the best gun metal, has a 5ft skeleton dial, and possesses all the latest improvements. The pendulum weighs 2cwt., and the clock, which requires winding once a week, strikes the hours and half-hours ... The clock was given by the Misses Marriott, and cost £110 ....

The parish was evidently very pleased with its new clock, and a testimonial from the Vicar was included in Potts' catalogue of 1894 (Michael Potts):

Caldecote Vicarage, Uppingham,
September 28th, 1893.

Gentlemen,
The churchwardens of Lyddington join with me in expressing approval of the Clock placed by you in our Church tower. It keeps accurate time, and its striking both hours and quarters [Note that the clock actually strikes hours and half-hours] is heard distinctly over our village (a very long one), as also over the surrounding district, proving a great convenience to the dwellers in scattered farm houses.

I am, Gentlemen,
Yours faithfully,
WILLIAM R. MANGAN
Vicar of Lyddington-cum-Caldecote

In 1990 the centenary of the clock was celebrated by an exhibition in the church, which included three clocks made by John Watts. The event was reported by the *Rutland Times* on Friday 31 August:

... visitors from far and wide had a lot of time for an exhibition at Lyddington parish church over the weekend. The event marked the centenary of the church clock and there was a theme of time. Numerous clocks and time-pieces were on display along with memorabilia of the late Victorian era when the church clock was installed.
In one corner of the beautiful church a Victorian drawing room was recreated, and in another there were costumes of the times.
One of the organisers, Janet Ingram, expressed gratitude to the people who had loaned items for the three day

exhibition ... and on Sunday there was Evensong during which the vicar, The Revd Bernard Taylor, gave a sermon based on 'Time'. Half-a-dozen clocks chimed while he was speaking.

The three hymns sung during the service were those which featured in a service in 1890 to celebrate the restoration of the church and the installation of the clock ....

CLOCK DETAILS

Details of the present clock:

Maker:	W Potts & Sons of Leeds
Signed:	Plate on the frame: W. POTTS AND SONS LEEDS. 1890
Installed:	11 July 1890
Cost:	£110
Donors:	Presented on the restoration of the church by Julia, Edith & Rosamund Marriot
Frame:	Large cast-iron flatbed
Trains:	Going and striking
Escapement:	Double three-legged gravity escapement
Pendulum:	Wooden rod and cylindrical cast-iron bob weighing 2cwt
Rate:	49 beats per minute
Striking:	Countwheel hour and half-hour striking on the tenor
Winding:	Weekly. Winding reduction gear on strike train
Weights:	Cast-iron
Dial:	Potts' skeleton dial with blue background and gilded roman numerals, minute ring and hands. 1.5m diameter

Location:	West face of the tower
Notes:	Replaces at least two previous clocks, one installed before 1626 and the other in 1738. The cost of the latter was £10 10s Clock set at a high level on the wall in the ringing chamber. A ladder is required to reach the platform for winding and maintenance purposes

LYDDINGTON MANOR HOUSE

A discarded early seventeenth-century sundial with the initials NH is built into the surface of the thrall in the cellar of Manor House, Lyddington. Part of the surface of the dial has been broken away along the line of the gnomon, and there is some evidence of numerals along the bottom edge. The deeds of the house show that the cellar was constructed in 1759.

A drawing of the discarded sundial in the cellar of Manor House, Lyddington

LYNDON

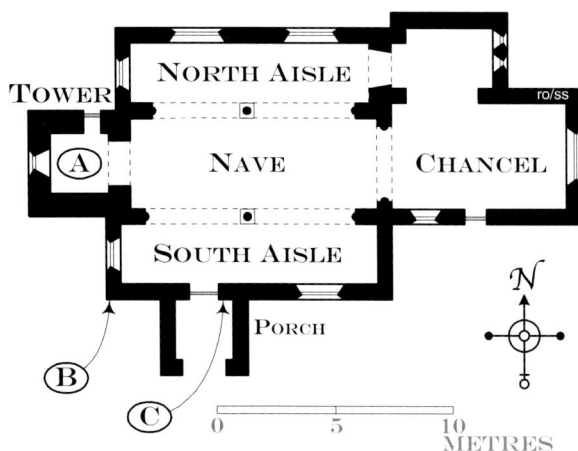

Plan of St Martin's Church
A: Four bells in the belfry. Ringing chamber at the base of the tower
B: Scratch dial 1
C: Scratch dials 2 and 3
Access to the belfry is by ladders

ST MARTIN   Grid Ref: SK 907044

PARISH RECORDS

There is a continuous run of Churchwardens' Accounts from 1726 to 1959 in four books (DE 1938/12-15). Also in the archive are loose financial and other documents from 1871 to 1937 (DE 1938/24/1-3), a Vestry Minute Book 1868-1921 (DE 1938/19) and two books containing PCC Minutes covering the period 1968-78 (DE 5163/5-6). A notebook/scrapbook compiled by the Rev T K B Nevinson contains amongst other things transcripts of registers, historical notes and newspaper cuttings. It was started 16 December 1888 and has additions through to 1979 (DE 5163/1).

BELL HISTORY

From 1716 and probably earlier, there has been a ring of four in the tower but unfortunately, due to lack of records, very little of their history can be chronicled. Irons' Notes record the following concerning the early bells at Lyndon (MS 80/1/3):

1613	**Oct 15**	They have a bell broken
1613	**Oct 29**	Ward [Churchwarden] to certify that the bells are repared

The entries probably refer to the repair of a broken bell fitting such as a headstock, bearing or wheel, rather than a broken bell. At this time there was one bell dated 1597 and one or more others for which details have been lost. New or recast bells were installed in 1624 and 1687 by the Norris foundry in Stamford and in 1716 by Henry Penn of Peterborough. The Churchwardens' Accounts include many references to the purchase of bellropes and the repair of bellwheels but only one item concerning the frame:

1729	The bell frame mending	1s 0d
1733	For a lather [ladder] and Bell whele mended	1s 4d
1740 March 27	paid for Beell Rope	5s 0d
	paid andrew hand for mending ye bell wheels	3s 6d
1768 March 24	pd Richard Tyler Bill for Iorn [iron] work for the Bells	3s 7½d
1788	pd Andw Hand for Repairing the Bell wheels & stopping out the pigeons	3s 0d

Andrew Hand was the village carpenter.

The parish clerk was responsible for ringing the bells and one clerk at Lyndon, who served for forty years, is

Thomas Cliffe, parish clerk at Lyndon during the nineteenth century (DE 5163/4/1)

remembered by a memorial, now almost indecipherable, on the outer east wall of the porch:

Sacred to the memory of John Barsby
who died Dec. 1, 1810, aged 87 years.
Also Eliz. his wife who died
May 24, 1788, aged 61 years.
He was forty years clerk of this place.
He sung his psalms, he's run his race.
He's closed his book, he's said Amen.
In Christ he hopes to rise again.
Harrison, Stamford

Other known clerks at Lyndon during the nineteenth century were William Hotchkin, Richard Barsby and Thomas Cliffe. By all accounts Thomas Cliffe was a strong-minded character and on 14 February 1893 he wrote to the rector concerning a disagreement and concluded, 'Therefore I cannot concientially pronounce the Amen anymore. I will light the fire as usual and chime the Bells and stop at the service but make no response to those horrid sentences in all other matters' (DE 5163/1).

The same source quotes under a list of 'Special peals':

1905 Nov 4 A muffled peal was rung, Thomas Cliffe ex-Parish Clerk having been buried this day.

Nevinson's notes (DE 5163/1) provide information on the state of the bells towards the end of the nineteenth century:

The Treble being much out of tune, the 2nd Bell badly cracked, and the framework of the entire Peal rapidly decaying, the Patron (Ed. N. Conant Esq.) undertook to put the whole into thorough repair. On Sept. 24th, 1889 the two smallest Bells were re-cast at the Bell Foundry of John Taylor & Sons, Loughborough, the Rector (T. K. B. Nevinson) being present. The old Inscriptions were reproduced, the words 'Recast September 1889' being added.

The sum paid by Mr Conant was £109 8s. This was irrespective of repairs to the masonry of the Tower, new flooring &c ... The cost of the new Bells, at £4 15s per cwt., amounted to £41 14s 7d. The price allowed for the old Bells, at £2 15s per cwt., amounted to £20 5s 7d.

All four bells were rehung with new fittings in the present composite oak and cast-iron frame.

The bells were dedicated on Sunday 24 November 1889 and the South Luffenham Ringers rang peals at intervals during the day. At the start of the Dedication Service, the bells were 'duly suspended' and a peal was rung after prayers. The ringers sounded another peal after singing Psalm 122 and the hymn 'In Sinai's dreary waste' was sung. This hymn appropriately contains the following:

Now, Lord, Thy people meet
Here where Thine honour dwells,
Invited by the tuneful peal
Of sweet melodious bells.

Oh! may those bells that call
To praise and worship here
In every breast an echo find,
To every heart be dear.

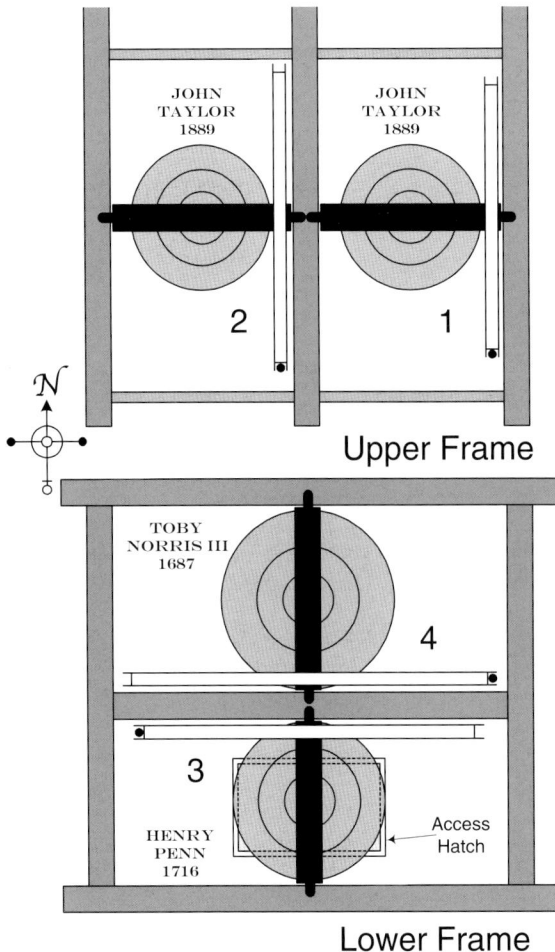

Plan of the bellframes at Lyndon Church

BELL DETAILS

1 1889. Treble. Diameter 686mm (27in). Weight 210kg (4cwt 0qr 15lb). Cast with canons. Timber headstock. Recast by John Taylor & Co of Loughborough.

NVNC MARTINE EGO CANA VOBIS ORE
~ IVCVNDO REMMEDG HVNTE 1597.

(Now O Martin, I sing to you with joyful voice.
Remigius Hunt)

On the waist:

[121]
RECAST SEPTEMBER 1889.

Former bell

Cast 1597. Ascribed to Matthew Norris of Leicester and possibly of Stamford (George Dawson). Diameter 660mm (26in). Weight 182kg (3cwt 2qr 9lb).

NUNC MARTИE EGO CAИA VOBIS ORE
~ IVCVNDO REMMEDG HVИTE 1597

Thomas North (1880, 141) omits the I in MARTINE. However, Nevinson (DE 5163/1) records that this I was present before the bell was recast. Remigius Hunt inherited the manor and advowson of Lyndon in 1586. He died in 1618.

2 1889. Diameter 711mm (28in). Weight 236kg (4cwt 2qr 17lb). Cast with canons. Timber headstock. Recast by John Taylor & Co of Loughborough.

[85]OMNIA FIANT AD GLORIAM DEI AD 1624.

(Let all things be done to the Glory of God)

On the waist:

[121]
RECAST SEPTEMBER 1889

Former bell

Cast in 1624 by Toby Norris I of Stamford. Diameter 711mm (28in). Weight 193kg (3cwt 3qr 5lb).

[26]OMNIA FIANT AD GLORIAM DEI AD 1624

3 1716. Diameter 759mm (29⅞in). Weight 223kg (4cwt 1qr 16lb). Canons retained. Timber headstock. Ascribed to Henry Penn of Peterborough (Pearson, 1989).

SAMVEL BARKER ESQR 1716

According to John Taylor, in a report dated 11 December 1889, this bell was 'smaller and consequently lighter than it should be, and is also rather thin in the crown' (DE 5263/1). Samuel Barker, Lord of the Manor, was the likely donor of this bell. He was buried at Lyndon in March 1759.

4 1687. Tenor. Note A. Diameter 864mm (34in). Weight 315kg (6cwt 0qr 23lb). Canons removed. Timber headstock. Cast by Toby Norris III of Stamford.

[10] SR [69] THOMAS [69] BARKER [69]
~ BARONET [69] OF [69] LINDEN [69]
~ MR [69] CLAYTON [69] RECTER
1687

The letters are very poorly formed and irregular in their positioning. North (1880, 141) did not record the initial cross and the date of this bell and he incorrectly recorded MP for MR. The date is incised deeply into the bell metal immediately below the initial cross.

The donor of this bell, Sir Thomas Barker, Lord of the Manor, represented the County in Parliament and inherited the manor and advowson of Lyndon from his father, Sir Abel Barker. Sir Thomas was buried at Lyndon 22 March 1707. The Rev William Clayton died whilst Rector of the Parish and was buried 20 October 1730. He was succeeded by the Rev W Whiston, the celebrated mathematician and philosopher.

SANCTUS BELL

In 1880 there was evidence of a former Sanctus Bell in the east window of the tower (North 1880, 141). No such evidence remains today.

Lyndon welcomes the newly married Rector and his wife in 1891 (DE 5163/4/1)

Feb. 5, 1891. The Rector of the Parish brought home his Bride. An evergreen arch had been erected at the entrance to the village, westward of the turn to Edith Weston, bearing a banner with the words,"Health & Happiness. From here the carriage was drawn by men to the Rectory: peals at the same time were rung on the bells of the Church.

BELLRINGING CUSTOMS

On Sundays the bells were chimed for Divine Service and the Sermon Bell was rung. At the Death Knell there were thrice three tolls for a male and thrice two for a female (North 1880, 143). This custom carried on into the next century and at 'a funeral the bell is tolled in single strokes without being raised' (DE 5163/1).

Nevinson records a list of special occasions when the bells at Lyndon Church were rung and it appears that the bells were well used, particularly at the end of the nineteenth century. Some of these occasions were for the celebration of local and national events. The Spur Bell was rung for the Rector (T K B Nevinson) when his first Banns of Marriage were published in 1890. After his marriage bells were rung when he returned to the village with his bride.

Bells were also rung to celebrate the marriage of Mr E W P Conant in 1898 and for the birth of a Conant son and heir the following year. A peal was rung after Evensong on 27 September 1896 'to commemorate the fact that during the preceding week Queen Victoria's reign had exceeded in length that of any other English sovereign'.

Muffled peals were rung on less happy occasions:

1892 Jan 20 From 3pm to 4pm the bells were rung, muffled from 3 to 3.40, half-muffled the rest of the time. This was the occasion of the funeral at Windsor of the Duke of Clarence and Avondale.

1896 Oct 16 A muffled peal was rung, this being the day of the funeral at Canterbury of Dr E W Benson, Archbishop of Canterbury.

1901 Jan 23 Muffled peals were rung at mid-day & in the evening, Q Victoria having died the previous evening.

Amongst other peals noted were those rung on the main Christian Festivals, on Lyndon Feast Day and New Year's Eve.

An interesting account in the *Grantham Journal* of 18 April 1903 records the welcome that was given to Mr E W P Conant and his family when they came to take up residence at Lyndon Hall. They were given enthusiastic receptions at Manton Station and on reaching the Hall. The church bells were rung on their arrival and during the evening.

Today a bell is chimed briefly before each service.

SCRATCH DIALS

Three scratch dials have been recorded. The first, which is covered in lichen and in poor condition, is on a soft sandstone quoin at the south-west angle of the early fourteenth-century south aisle. It may have been a full circle of twenty-four radial lines but most of the upper half has weathered away. The other two are on the thirteenth-century east jamb of the main door inside the porch.

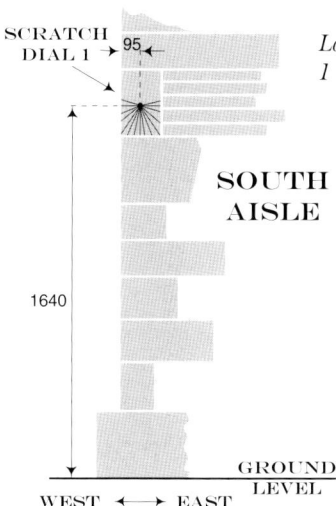

Location of scratch dial 1 at Lyndon Church

SCRATCH DIAL 1 → 95

SOUTH AISLE

1640

GROUND LEVEL

WEST ← → EAST

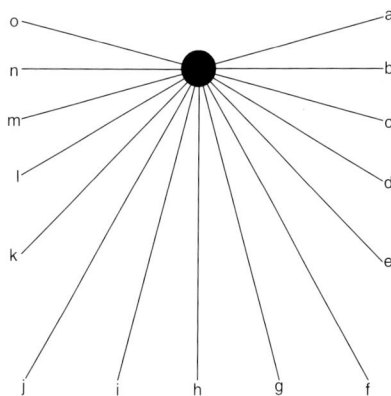

LYNDON 1								
Location	South west corner of south aisle							
Condition	Very poor							
Gnomon Hole Diameter			15					
Gnomon Hole Depth			35					
Height above ground level			1640					
Line Ref	a	b	c	d	e	f	g	h
Length	96	94	96	108	130	172	154	130
Angle (°)	75	90	105	120	135	150	165	180

Line Ref	i	j	k	l	m	n	o
Length	154	172	430	108	96	94	96
Angle (°)	195	210	225	240	255	270	285

Scratch dial 1 details

**SOUTH
DOOR**

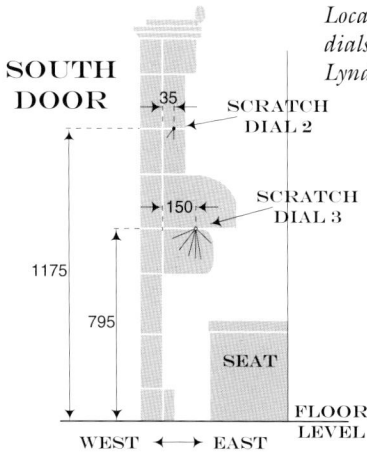

*Location of scratch
dials 2 and 3 at
Lyndon Church*

Scratch dial 3 details

35 ← SCRATCH
DIAL 2

→ 150 ← SCRATCH
DIAL 3

1175

795

SEAT

FLOOR
LEVEL

WEST ←——→ EAST

LYNDON 3						
Location	East jamb of south door inside south porch					
Condition	Average					
Gnomon Hole Diameter			Filled in			
Gnomon Hole Depth			Filled in			
Height above ground level			795			
Line Ref	a	b	c	d	e	f
Length	75	135	135	135	130	105
Angle (°)	140	170	180	197	215	238

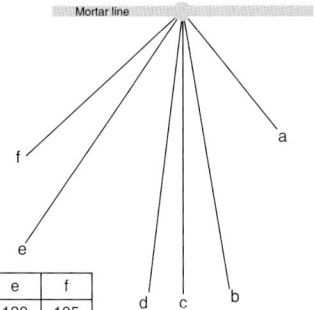

Mortar line

f

a

e

d   c   b

LYNDON 2		
Location	East jamb of south door inside south porch	
Condition	Average	
Gnomon Hole Diameter		8
Gnomon Hole Depth		15
Height above ground level		1175
Line Ref	a	b
Length	50	50
Angle (°)	180	220

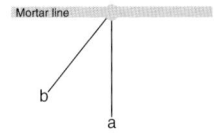

Mortar line

b

a

Scratch dial 2 details

LYNDON HALL

An eighteenth-century table sundial is located in the
grounds to the south-east of Lyndon Hall. There is also
an impressive clock turret with east and west facing dials
above the entrance to the Hall stables which were built

*The entrance to
Lyndon Hall
stables*

The clock movement at Lyndon Hall stables which was
supplied by Joseph Wilson of Stamford

in the 1860s. The clock movement is in a small room above the arch and a clock bell is suspended in a cupola above the dials.

CLOCK DETAILS
Detals of the present clock:

Maker:	John Smith & Sons, St John's Square, Clerkenwell, London
Signed:	On the setting dial: 'Wilson Stamford'
Installed by:	Joseph T Wilson, All Saint's Street, Stamford

Date:	*Circa* 1860
Frame:	Cast-iron plate and spacer
Trains:	Going and striking
Escapement:	Deadbeat
Rate:	60 beats per minute
Pendulum:	Wooden rod and cast-iron bob
Striking:	Rack striking
Clock Bell:	In cupola above clock turret
Weights:	Cast-iron
Winding:	Weekly
Dials:	Two skeleton dials with gilded roman numerals and hands on a blue background

MANTON

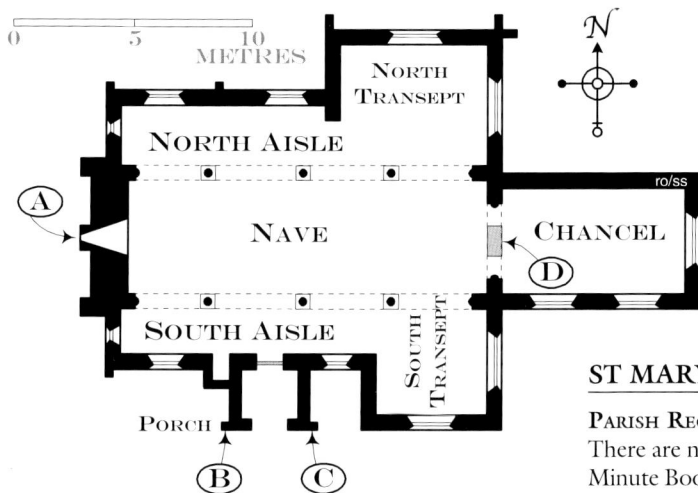

Plan of St Mary the Virgin's Church
A: Two bells in a double bellcote rung
from inside the church
B: Scratch dial 1
C: Scratch dial 2
D: Sanctus bellcote

From a drawing of Manton Church of circa 1793 (RCM F10/1984/35) showing the ridged roof over the bellcote. The structure behind the bellcote would have provided weather protection to the bellropes and access to the bells from inside the church. It appears to be of timber construction. A beam immediately below this position in the present roof is dated 1804, probably indicating the date when this structure was removed

ST MARY THE VIRGIN   Grid Ref: SK 881047

PARISH RECORDS
There are no available parish records prior to 1920. PCC Minute Books exist for 1920-30 and 1930-48 (DE 1939/9-10).

BELL HISTORY
The thirteenth-century double bellcote at the west end of the church is the largest in Rutland. This massive structure is supported by a wall some 1.5m (5ft) thick and is further strengthened by three buttresses, the centre one being carried up in stages nearly to the top of the bellcote. The bellcote terminated originally in a single gable but the upper part has been removed and replaced by a ridged roof. The date 1780 engraved on the top of the central buttress probably indicates when the alteration was carried out. There is also a Sanctus bellcote over the east end of the nave.

The two bells of *circa* 1550 and 1610 survived until 1920 when they were recast and rehung by John Taylor. This work was paid for by subscription and according to the Minutes of the Church Council Meeting on 2 March 1921 the cost was £143 14s 2d. This included repairs to the roof and spouting.

There is a story told in the village concerning the church bells. Apparently in about 1907, the Turkish Ambassador was on a secret visit to Lord Asquith at

Manton Grange. The village youths got to know about this and rigged ropes from the church bells to a nearby field. When the Ambassador arrived the bells were rung vigorously, much to the consternation of the police (Traylen nd, Pt 1).

BELL DETAILS

1 1920. Diameter 568mm (22⅜in). Weight 115kg (2cwt 1qr 2lb). Cast without canons. Metal head-stock. Recast by John Taylor & Co of Loughborough.

CUM VOCO AD ECCLESIAM VENITE 1610 T. S.

(When I call come to church)

On the waist: [56]
                         RECAST 1920

Former bell

Cast in 1610 by Toby Norris I of Stamford. Diameter 508mm (20in).

CUM VOCO AD ECCLESIAM VENITE 1610 T. S.

T S are probably the initials of Thomas Smythe who was a churchwarden at this time. North (1880, 143) incorrectly recorded this bell as the tenor and the date as 1619.

2 1920. Diameter 635mm (25in). Weight 156kg (3cwt 0qr 9lb). Cast without canons. Metal headstock. Recast by John Taylor & Co of Loughborough.

[2] ABCDEFGHI [11]

On the waist: [56]
                         RECAST 1920

The lettering is like [108].

Former bell

Circa 1550. Ascribed to Robert Newcombe I of Leicester. Diameter 572mm (22½in).

[2] ABCDEFGHI [11]

North (1880, 143) incorrectly recorded this bell as the treble.

The Sanctus bellcote at Manton Church

Plan of the bellcote at Manton Church

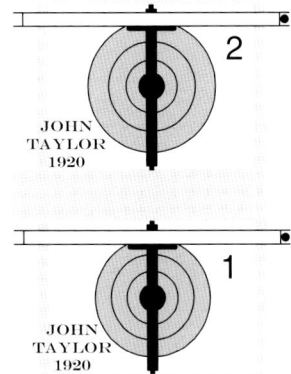

(below) The massive double bellcote at Manton. This is now the only Rutland bellcote with bellwheels, but full circle ringing is not possible as there are no stays. The ropes hang inside the church at the western end of the nave

SANCTUS BELL

The medieval Sanctus bellcote over the east end of the nave is one of only two surviving in Rutland, the other being at Market Overton. In neither case has the bell survived. The late eighteenth-century drawing of the church shows that the bell was missing then. There used to be fragments of a small bell with a diameter of about

273mm (10¾in) in the church chest, but it had no inscription or other marks. This may have been a former Sanctus Bell but Thomas North in 1880 believed that it was post-Reformation.

BELLRINGING CUSTOMS

On Sundays a bell was rung at 9am and for Divine Service both bells were chimed after which the tenor was rung. The Pancake Bell and the Gleaning Bell were rung

Location of the scratch dials at Manton Church

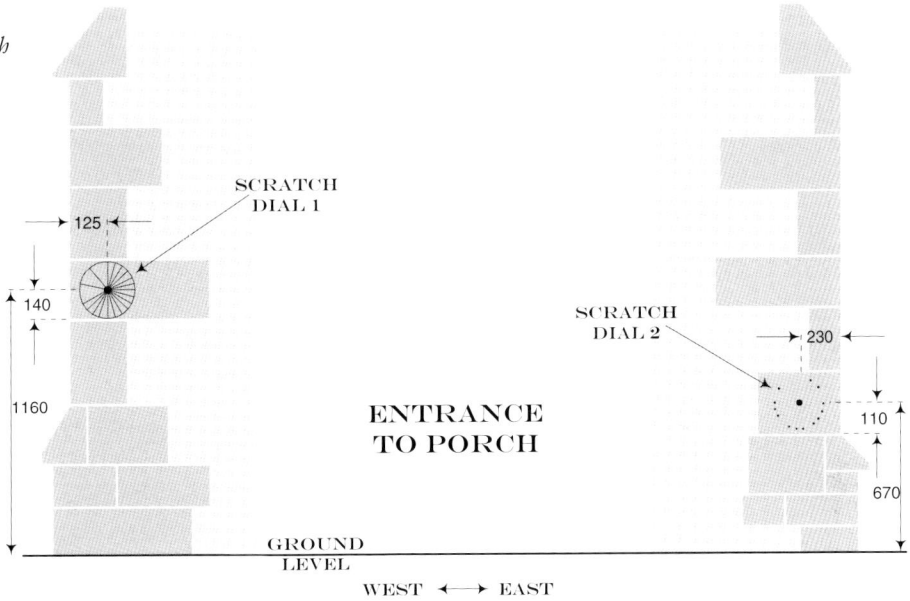

The scratch dial on the west front of the porch at Manton

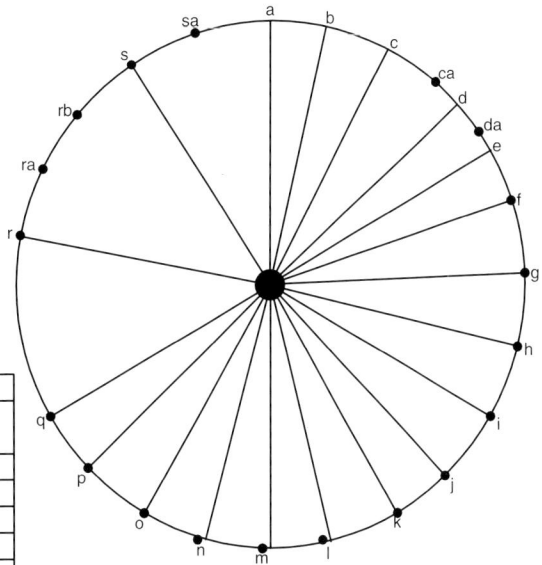

Scratch dial 1 details

MANTON 1											
Location	West front of the south porch										
Condition	Average										
Gnomon Hole Diameter		15									
Gnomon Hole Depth		30									
Height above ground level		1160									
Pock circle diameter		256									

Line/Pock	a	b	c	ca	d	da	e	f	g	h	i	j
Length	128	128	128	-	128	-	128	128	128	128	128	128
Angle (°)	0	13	28	41	48	56	60	72	88	103	120	135

Line/Pock	k	l	m	n	o	p	q	r	ra	rb	s	sa
Length	128	128	128	128	128	128	128	128	-	-	128	-
Angle (°)	150	166	180	195	210	226	240	281	297	311	327	343

Scratch dial 2 details. This dial is very weathered. It may also have had an outer circle

The scratch dial on the cottage at 20 St Mary's Road, Manton

MANTON 2														
Location	East front of the south porch													
Condition	Poor													
Gnomon Hole Diameter		12												
Gnomon Hole Depth		11												
Height above ground level		670												
Pock circle diameter		180												
Pock Ref	a	b	c	d	e	f	g	h	i	j	k	l	m	
Angle (°)	52	72	92	112	132	148	170	188	206	240	258	272	300	

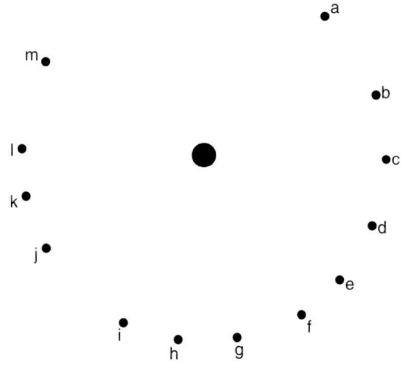

before 1880. At the Death Knell there were thrice three tolls for a male and thrice two tolls for a female, both before and after the knell. The tenor was tolled as a Call Bell shortly before a funeral and again when interment took place (North 1880, 143).

Today, prior to the start of Sunday services, both bells are chimed for five minutes and then one bell for the same time. They are rarely rung for weddings or funerals.

SCRATCH DIALS

There are two scratch dials on the front face of the fourteenth-century south porch, one either side of the entrance. It appears that both originally had twenty-four radial lines with pocks.

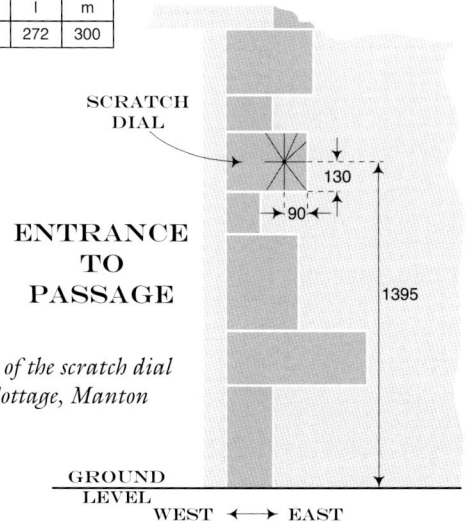

ST MARY'S ROAD, MANTON

Old Cottage at 20 St Mary's Road, Manton, has what is thought to be the only scratch dial on a secular building in Rutland. It is inverted and probably on a stone recovered from Manton Church following alterations, or from a demolished medieval church, the nearest being Martinsthorpe. The cottage is dated 1733.

Location of the scratch dial at Old Cottage, Manton

Scratch dial details

COTTAGE AT 20 ST MARY'S ROAD, MANTON										
Location	East jamb of passage door on south elevation									
Condition	Good									
Gnomon Hole Diameter		20								
Gnomon Hole Depth		20								
Height above ground level		1395								
Line Ref	a	b	c	d	e	f	g	h	i	
Length	130	130	80	130	120	130	110	90	150	
Angle (°)	0	20	90	140	180	220	245	280	320	

Note that this scratch dial is inverted. Angles are therefore measured from datum line x.

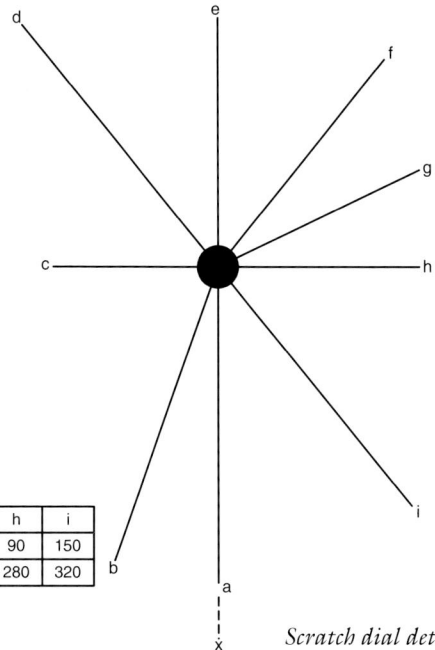

MARKET OVERTON

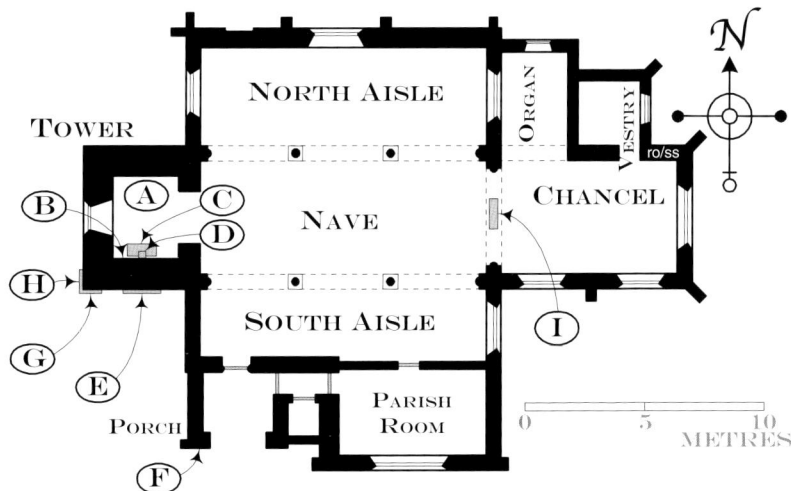

Plan of St Peter & St Paul's Church
A: Five bells in the belfry. Ringing
chamber at the base of the tower
B: Location of former Ellacombe
chiming frame at the base of the tower
C: Former clock movement on the first
floor of the tower
D: Synchronous electric clock
movement on the floor of the belfry
E: Clock dial
F: Possible scratch dial
G: South-facing sundial
H: West-facing sundial
I: Sanctus bellcote over the east gable
of the nave
Access to the clockroom and the belfry
is by ladders

ST PETER & ST PAUL   Grid Ref: SK 886165

PARISH RECORDS

The archive includes Churchwardens' Accounts 1810-89 (DE 4993/19) together with Vestry and PCC Minutes 1819-36 and 1861-1982 (DE 4993/20-22).

BELL HISTORY

A small bellcote over the east gable of the nave is evidence of a former Sanctus Bell but no records of it have survived. The bellcote is shown on a *circa* 1793 drawing of the church (RCM F10/1984/36). In 1880 there were three bells at Market Overton. The earliest surviving details relate to the present third bell that was probably cast in 1658 by Thomas Norris of Stamford. This bell was recast in 1885, but the inscription 'TD WP 1658' was lost. The second and fourth bells cast in 1737 by Thomas Eayre II are quite likely recasts of earlier Norris bells as the inscription used is one which was favoured by the Stamford foundry.

A new treble and tenor, both cast by John Taylor, were added in 1888. A composite oak and cast-iron high-sided frame for five bells was installed at the same time, together with an Ellacombe chiming frame and hammers. The bells were dedicated on 28 June 1889 when Archdeacon Lightfoot preached in the afternoon and Mr Lewis of Oakham in the evening. Dr Bridge, organist at Westminster Abbey, played the organ (DE 4993/21).

The bells were overhauled and rehung on ball bearings by John Taylor in 1935, and in November 1996 they were retuned and rehung with new headstocks, bearings, wheels and ropes, by the same company. This work also included the removal of the Ellacombe chiming frame and hammers. This last refurbishment was paid for by local fundraising, donations and grants. The re-dedication service was held on 26 January 1997 and the Rev Nick Denham led the celebratory evensong. Local choirs joined the congregation and Mike Davies captained a band of local ringers.

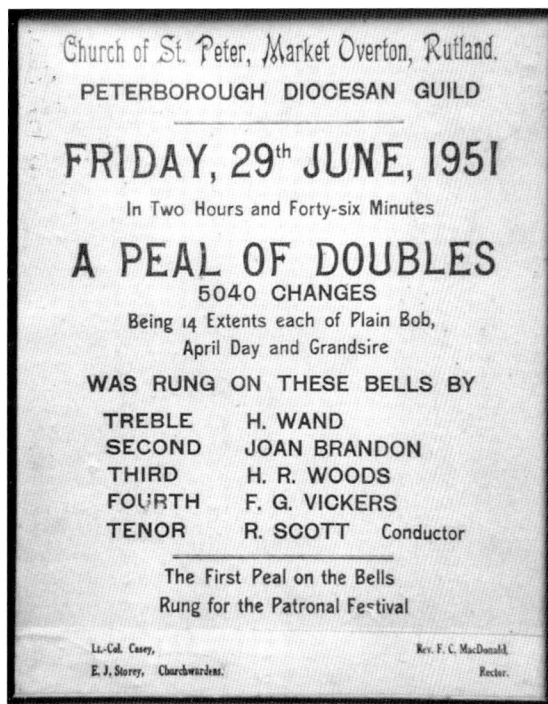

Church of St. Peter, Market Overton, Rutland.

PETERBOROUGH DIOCESAN GUILD

FRIDAY, 29th JUNE, 1951

In Two Hours and Forty-six Minutes

A PEAL OF DOUBLES
5040 CHANGES
Being 14 Extents each of Plain Bob,
April Day and Grandsire

WAS RUNG ON THESE BELLS BY

TREBLE	H. WAND
SECOND	JOAN BRANDON
THIRD	H. R. WOODS
FOURTH	F. G. VICKERS
TENOR	R. SCOTT Conductor

The First Peal on the Bells
Rung for the Patronal Festival

Lt.-Col. Casey,   Rev. F. C. MacDonald,
E. J. Storey, Churchwardens.   Rector.

This framed certificate at the base of the tower at Market Overton Church records a peal in 1951

*Plan of the bellframe
at Market Overton
Church*

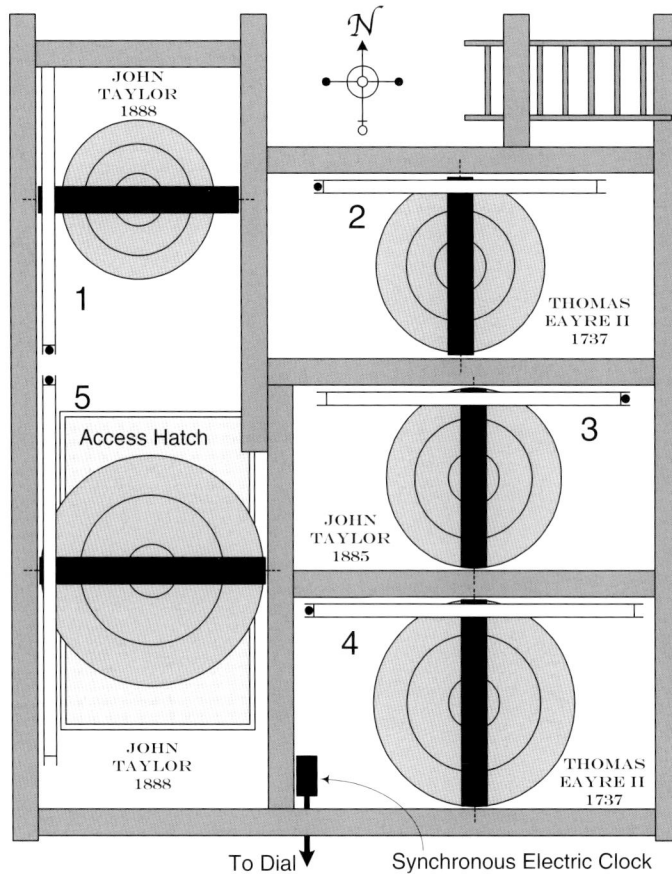

BELL DETAILS

1 1888. Treble. Diameter 762mm (30in). Weight 270kg (5cwt 1qr 8lb). Cast without canons. Cast-iron headstock. Cast by John Taylor & Co, Loughborough.

SIT LAUS DEO

(Let there be praise to God)

On the waist:

H. L. WINGFIELD. RECTOR
E. COSTALL. CHURCHWARDEN
A.D. 1888.

On the waist, opposite:

[120]

Harry Lancelot Wingfield was Rector of Market Overton from 1857 to 1892. Edward Costall was a farmer (White 1877, 684).

2 1737. Diameter 826mm (32½in). Weight 322kg (6cwt 1qr 10lb). Canons removed. Cast-iron headstock. Cast by Thomas Eayre II of Kettering.

OMNIA FIANT AD GLORIAM DEI ∴ [51] ∴
~ GLORIA PATRI FILIO ET SPIRITUI SANCTO
∴ T. E ∴ 1737:

(Let all things be done to the Glory of God. Glory be

to the Father, and to the Son, and to the Holy Ghost)

3 1885. Diameter 902mm (35½in). Weight 428kg (8cwt 1qr 20lb). Cast without canons. Cast-iron headstock. Recast by John Taylor & Co of Loughborough.

J. . TAYLOR AND C⁰. . FOUNDERS
~ LOUGHBOROUGH 1885.

Former bell

Probably cast in 1658 by Thomas Norris of Stamford. Diameter 864mm (34in).

TD WP 1658

This bell was cracked in 1885 (Powell 1938, 23). TD and WP were probably Thomas Draper and William Porter. Both are recorded as paying Hearth Tax in 1665 (Bourn 1991, 33).

4 1737. Diameter 978mm (38½in). Weight 542kg (10cwt 2qr 20lb). Canons removed. Cast-iron headstock. Cast by Thomas Eayre II of Kettering.

OMNIA FIANT AD GLORIAM DEI ∴ [51] ∴
~ GLORIA PATRI FILIO ET SPIRITUI SANCTO
~ ∴∴ THO ∴ EAYRE ∴ 1737 ∴

(Let all things be done to the Glory of God. Glory be to the Father, and to the Son, and to the Holy Ghost) A crack in this bell has been successfully repaired by 'stitching'. This repair was carried out *circa* 1895 (information from George Dawson) (see Bell Repair below).

5 1888. Tenor. Note F sharp. Diameter 1105mm (43½in). Weight 745kg (14cwt 2qr 18lb). Cast without canons. Cast-iron headstock. Cast by John Taylor & Co of Loughborough.

GLORIA PATRI FILIO ET SPIRITUI SANCTO
(Glory be to the Father, and to the Son, and to the Holy Ghost)
On the waist:

A. D. 1888.
H. L. W. [85] [85] E. C.
On the waist, opposite:
[120]

The initials are those of the Rector and churchwarden named on the treble. The clock struck the hours on this bell from 1912 until 1969 when a single train synchronous electric clock was installed.

SANCTUS BELL
A Sanctus Bell used to hang in the small bellcote over the east gable of the nave. There are no references to it in the surviving records. This Sanctus bellcote is one of only two surviving in Rutland, the other being at Manton.

BELL REPAIR
Modern welding techniques ensure that most cracked and broken bells are repaired rather than recast (see Brooke — Bell 3). Although this method was used in the

The Sanctus bellcote on the east gable of the nave at Market Overton Church

1920s it was then the exception rather than the rule. Prior to this mechanical solutions were adopted, the simplest being to drill a hole at the end of a crack to prevent it from spreading. The best known example of this is Big Ben. Another method was developed by M & O Ohlsson, bellfounders of Lübeck, Germany, and used on a small number of cracked bells in the United Kingdom in the 1890s and early 1900s. One of these was the fourth bell at Market Overton. In this technique, often referred to as 'stitching', the crack is clamped by butterfly-shaped plates let into the surface of the bell. The completion of the repair was 'effected by fusing & filling'. This work was carried out in the belfry, avoiding the need to remove the bell. Correspondence between Herr Ohlsson and A H Cocks, author of *Church Bells of Buckinghamshire*, published in 1897, is preserved in the Library of the Central Council of Church Bell Ringers. It refers to the repair of the Market Overton bell (Eisel 2001, 909).

The 'stitching' just above the soundbow inside the fourth bell at Market Overton. Two butterfly-shaped plates have been carefully let into the surface and the crack filled

M. & O. Ohlsson, Lübeck
Hof-Glockengiesser.

M & O Ohlsson's letterhead advertising that the company casts, hangs and repairs bells (Central Council of Church Bell Ringers)

Bellringing Customs

The Pancake Bell was rung at noon on Shrove Tuesday and the Gleaning Bell at 8am and 6pm during harvest. On Sundays a bell was rung at 8.30am and again after Morning Service when Evening Prayer was to be said. At the Death Knell there were thrice three tolls for an adult male and twice three tolls for an adult female, all before the knell. The variations on these for children were thrice two tolls for a boy and twice two tolls for a girl (North 1880, 144).

Churchwardens' Accounts only exist from 1810 to 1889 and they show that the bells were rung on 5 November in the years from 1811 to 1816 and 1838 to 1840.

Today Market Overton has a band of bellringers which practises regularly. They ring all five bells every Sunday from 9am to 9.30am before the service. The bells are also rung on New Year's Day, for the morning service on Easter Sunday, on Christmas Eve and in the morning of Christmas Day. On Armistice Day there is a full ring, half muffled and if requested the bells are rung half muffled for funerals. If bells are required at a wedding they are always rung for thirty minutes before and twenty minutes after the ceremony.

Scratch Dial

There are faint traces of a possible scratch dial on the west front of the thirteenth-century south porch.

Sundials

The two seventeenth-century sundials abutting each other on the south-west corner of the tower are said to be the gift of Sir Isaac Newton (*see* Chapter 3 — Clockmakers).

The double sundial at Market Overton Church today

The double sundial at the south-west corner of the tower at Market Overton. From a drawing of circa 1839 (Uppingham School Archives)

The lines and numerals of both dials were originally painted on the wall surface but these have now weathered away. The gnomon is missing on the west-facing dial.

Clock History

The Parish Records contain no references to an early clock here and the clock by James Gent & Son of London, installed in 1912, may be the first in the tower. At a Vestry Meeting in the schoolroom on 11 April 1912 it was recorded that Mr Wing should be thanked for presenting the village with a church clock. 'Mr Wing in his reply desired that if a record was to be made it should be stated that his sister & brother joined him in giving the clock as a memory to Mrs Wing.' William H Wing FSA was a man of antiquarian interests. He was involved in the investigation of the Roman and Saxon sites at Market Overton in the early 1900s. At the same meeting it was agreed that Mr Walker, sexton, should be 'paid a little' for winding the church clock.

At a Vestry Meeting in 1919 Mr Fred Freeman was appointed as sexton to replace a Mrs Walker, who had presumably taken over the duty from her husband for the duration of the Great War. She had been paid £1 for winding the clock.

In 1969, a new synchronous electric clock was installed by John Hartshorn of Badby, Northamptonshire. This clock does not strike the hours. The old clock by James Gent & Son is preserved in the clockroom.

Clock Details

Details of the former clock preserved in the clockroom:

Maker:	James Gent & Son, London
Signed:	On the setting dial
Installed:	1912
Cost:	Not recorded
Memorial:	Given to the memory of Mrs Wing by her family
Frame:	Cast-iron flatbed
Trains:	Going and striking

Escapement:	Deadbeat
Rate:	54 beats per minute
Striking:	Countwheel hour striking on the tenor
Weights:	Cast-iron
Winding:	Weekly
Dial:	Skeleton dial — black background with gilded roman numerals and hands. Installed in 1912
Location:	South face of the tower
Notes:	Overhauled by John Smith of Derby in 1936. Replaced by a synchronous electric clock in 1969

Details of the present clock:

Maker:	John Hartshorn of Badby, Northants
Signed:	Unsigned
Installed:	1969
Type:	Synchronous electric motor
Trains:	Going only
Location:	Belfry floor
Note:	Motionwork replaced in November 1998 by nylon gears with toothed belt drives to the hour tube and minute shaft

The setting dial of the clock by James Gent at Market Overton Church

MARKET OVERTON MANOR HOUSE

As a boy, Isaac Newton lived for some time with his grandmother, Harriet Ayscough, at the Manor House in Market Overton. Whilst there he painted a sundial on one of the ceilings, probably using a piece of glass to reflect the sun. The house was pulled down many years ago, but the piece of plaster with the dial on it was preserved and kept in the new house built on the site. It is understood that there is no trace of this sundial now.

This carved head on a building in the garden of the present Manor House in Market Overton is believed to be of Isaac Newton

Market Overton Church. The clock case provided essential protection for the mechanism from dust and dirt

MARTINSTHORPE

There is no surviving church at Martinsthorpe.

The first recorded rector of Martinsthorpe was a William de Aldesworth who was presented in 1258 (Phillips 1911-12, 230). We can be fairly certain therefore that there was a medieval church here and it would have had at least one bell. The village was depopulated by 1522 and it is assumed that the church then fell into decay, its valuable stone being reclaimed for use elsewhere. A stone marked with a scratch dial on a cottage in Manton may have originated from Martinsthorpe Church. No structural evidence or ground markings of the church have survived, and its exact location within the village is unknown.

The first Earl of Denbigh built Martinsthorpe House in the early 1620s on land at the centre of the deserted medieval village site (SK 868046). It saw little use after the Civil War and was eventually demolished in 1755.

However, the integral private chapel at the north-east corner of the house was saved and re-roofed to become St Martin's Chapel. This in turn became derelict and was eventually demolished in the early 1900s. There were no bells at this chapel.

PARISH RECORDS

There are no parish records apart from a few baptisms and marriages at the chapel between 1728 and 1746 which are recorded in the Uppingham and Wing Parish Registers.

MARTINSTHORPE HOUSE

An engraving of Martinsthorpe House in James Wright's *History and Antiquities of the County of Rutland* of 1684 shows that there was a sundial on the south elevation.

An enlargement of the sundial on the engraving of Martinsthorpe House

Martinsthorpe House from a plate engraved at the cost of William, the first Earl of Denbigh (Wright 1684, 90)

MORCOTT

ST MARY THE VIRGIN   Grid Ref: SK 925008

PARISH RECORDS

The following are the only relevant records in the parish archive:

Churchwardens' Accounts 1686-1773 (DE 2876/16/1-11, 17 & 18).
Town Book of the Churchwardens', Overseers' and Constables' Accounts 1638-1820 (DE 2876/20).

BELL HISTORY

Irons' Notes (MS 80/1/3) include the following references to the bells at Morcott:

1627 April 24	Gard (Churchwarden) ... admit that their Sts [Sanctus] bell is ryven [split]
May 10	Gard to certify the repaire of their Sts bell
1636 March 17	There is a bell riven at M [Morcott]
April 26	One of the bells in the steeple are broken
1637 Aug 2	Will [William] Angell warden not notifying the amending of the bells in the steeple

N

NORTH AISLE

ORGAN

ro/ss

(A) (B) NAVE CHANCEL

TOWER

(D)

SOUTH AISLE

0 5 10
METRES

PORCH

(E)

(C)

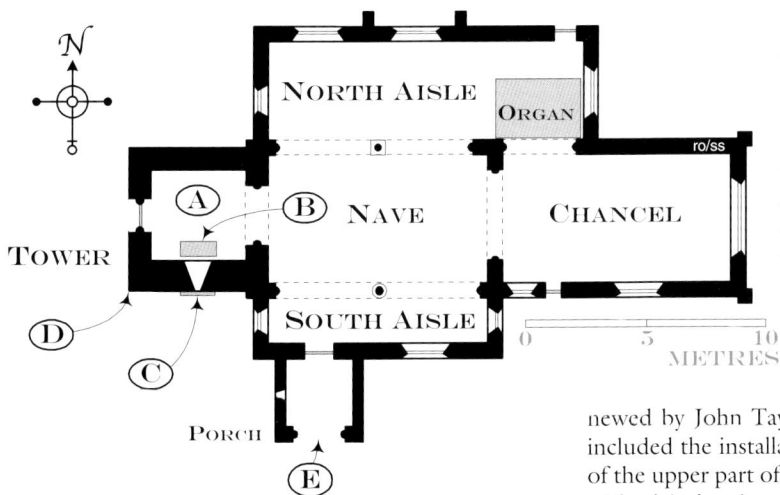

Plan of St Mary the Virgin's Church
A: Four bells in the belfry. Ringing
chamber at the base of the tower
B: Clock on the first floor of the tower
C: Clock dial
D: Former location of clock bell over
parapets
E: Sundial
Access to the clockroom and belfry is
by ladders

There is currently a ring of four bells in the tower at Morcott and two of these were cast in the sixteenth century. Thomas Norris of Stamford supplied a bell in 1637 and this is possibly a recast of the 'bell riven' as noted by Irons in 1636. The fate of the Sanctus Bell is unknown.

Entries in the Churchwardens' Accounts between 1688 and 1698 indicate that there were then four bells in the tower, as now, confirming that the tenor by Thomas Eayre II dated 1726 was recast from an earlier bell:

1688	Feb 24	Itm paid for fore bell ropes	10s 0d
1695	June 4	Itm pd for the ffastning the first bell gudgin	6d
		Itm pd for the second bell claper mending	6d
1696	May 25	Paid for making the third bell badrick [baldrick]	1s 0d
1698	July 30	Paid to Thomas Law of Cliff [Kings Cliffe] for 4 bellropes	8s 6d

When the new church clock was installed in 1921 it was arranged to strike the hours and quarters on the bells in the belfry. As a result the old clock bell, which was fixed externally on the south-west corner of the tower, became redundant. It is believed that from then until it was removed in 1940, it was used as a Priest's Bell. As it had no clapper it was probably rung by pulling on the wire connected to the old clock hammer. The inscription on the bell is translated as 'I admonish when I move' or 'While on the swing I warnings bring', implying that it was originally cast for swing ringing. Before this bell was utilised as a clock bell in the seventeenth century its original intent is not known. It is possible that it was not cast for Morcott Church but brought in from elsewhere when the clock was installed. It is now retained inside the church.

The bells are hung in an early seventeenth-century low-sided oak frame that was renovated and partly re-

newed by John Taylor & Co in 1930. The renovation included the installation of angle plates at each junction of the upper part of the frame. New cast-iron headstocks with plain bearings were installed at the same time. All the bellwheels have been replaced in recent years.

BELL DETAILS

1 1637. Treble. Diameter 692mm (27¼in). Weight 197kg (3cwt 3qr 14lb). Canons removed. Cast-iron headstock. Cast by Thomas Norris of Stamford.

**[26] [52] THOMAS [52] NORRIS [52]
~ MADE [52] ME [52] 1637 [52]**

2 16th Century. Diameter 708mm (27⅞in). Weight 192kg (3cwt 3qr 4lb). Canons retained. Cast-iron headstock. Cast by Watts [possibly Francis] of Leicester (Pearson 1989).

ABCDE FGHIK LMNO

The three blocks of letters are spaced evenly around the inscription band. All letters are like **[16]** and **[17]**. Scheduled for preservation (Council for the Care of Churches). The clock chimes the quarters on this bell and the third.

3 16th Century. Diameter 794mm (31¼in). Weight 292kg (5cwt 3qr). Canons retained. Cast-iron headstock. Ascribed to Newcombe, Leicester (George Dawson).

Part of the inscription on the second bell at Morcott

Plan of the low-sided timber bellframe at Morcott Church

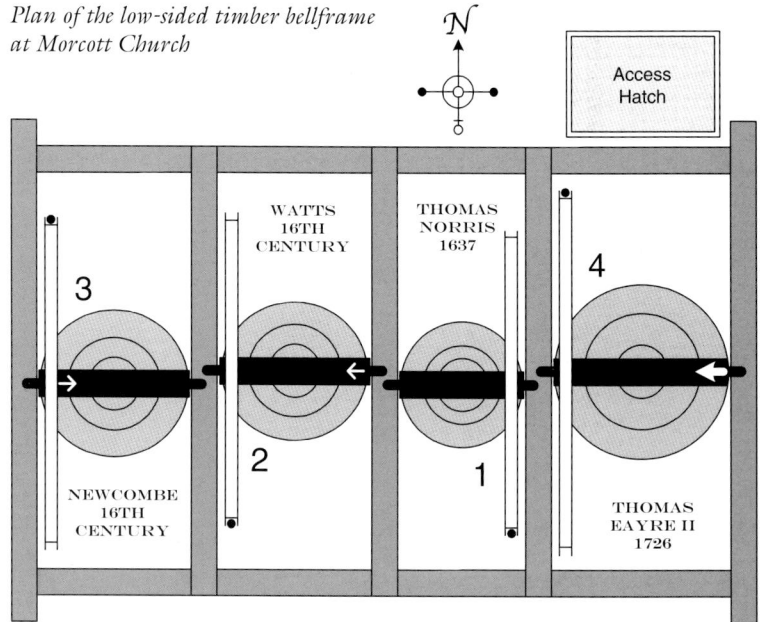

From a photograph of Morcott Church, taken in 1914 by George Henton, showing the clock bell above the south-west corner of the tower. Note the octagonal clock dial supported by two stone corbels. This single-handed dial was replaced by the present skeleton dial in 1921 (Henton 1071)

ʃ M A R I A **[13]**   **[9]**
(Saint Mary)

The inscription is spaced evenly around the bell. The lettering is like **[108]**. Thomas North records this bell and the tenor in the wrong order. Scheduled for preservation (Council for the Care of Churches). The clock chimes the quarters on this bell and the second.

4 1726. Tenor. Note A. Diameter 838mm (33in). Weight 305kg (6cwt). Canons retained. Cast-iron headstock. Cast by Thomas Eayre II of Kettering. **[49]**

IHS NAZARENE REX IUDÆORUM FILI DEI ~ MISERERE MEI [49] GLORIA PATRI FILIO ~ ET SPIRITUI SANCTO: 1726.

(Jesus of Nazareth King of the Jews, Son of God, have mercy on me. Glory be to the Father, and to the Son, and to the Holy Ghost)

The clock strikes the hours on this bell.

FORMER CLOCK BELL

Circa **1620**. Diameter 352mm (13⅞in). Weight approximately 34kg (2qr 20lb). Canons retained. Unknown founder.

AD MONEO **[123]** CVM **[123]** MOVEO **[123]**
(I admonish when I move
[or] While on the swing I warnings bring)

There are indentations on the soundbow where it has been struck by a clock hammer.

In 1880 Thomas North (1880, 144) described it as 'a small clock-bell now without clapper placed under a bell-cote at an angle of the battlements of the tower'. A bell can

The former clock bell at Morcott. Note the indentation as a result of being struck by the clock hammer

just be seen in this position in a photograph of Morcott Church taken by George Henton in 1914 (Henton 1071).

This bell is now preserved in the church. Scheduled for preservation (Council for the Care of Churches).

BELLRINGING CUSTOMS

On Sundays the treble was rung at 8am and the treble and second at 9.30am. For Divine Service the bells were chimed prior to the tenor being rung. The treble was rung after Morning Service.

The Gleaning Bell was rung at 8am and 6pm during harvest and the Pancake Bell rung on Shrove Tuesday. At the Death Knell thrice three tolls were given for a male and thrice two tolls for a female, both before and after the knell (North 1880, 145). Two isolated entries in the accounts are examples of when the bells were rung on other occasions. Thomas North recorded that a peal was still being rung on 5 November in 1880:

| 1696-97 April 15 | Paid to the ringars on the thanks-giving day | 1s 0d |
| 1772 Nov 5 | Gave the Ringers | 1s 6d |

Today the tenor is usually chimed for two minutes before the 11am and 6pm Sunday services. The bells are chimed on Armistice Day. When a peal is required for a wedding a band of ringers is provided by the Bellringers' Guild.

SUNDIAL

The following extract from the accounts confirms that the sundial in the gable of the south porch is late seventeenth century, or earlier, and that it was painted rather than incised. At this time a sundial would have been required to regulate the church clock and it may have been installed at the same time as the clock:

| 1695-96 Jan | Item Paid to Solomon Wing for the painting of the ... Porch dial ... [and other work] | 6s 6d |

The sundial is not shown on the *circa* 1793 (RCM F10/1984/40) or *circa* 1839 (Uppingham School Archives) drawings of the church. All that now remains is the dial stone on which the two fixing holes for the gnomon can be seen. The numerals and lines have entirely weathered away.

CLOCK HISTORY

Entries in the Churchwardens' Accounts confirm that there was a clock in the tower in the late seventeenth century. In the following examples John Lambard was probably the clock winder. William Fox was an Uppingham clockmaker:

1686-87 Nov 5	Item Paid for wiar for the clock	4d
Feb 6	Item paid for a Clock Rope	1s 11d
	Paid to John Lambard for looking to the clock	13s 4d
1689 March 20	Paid to William Fox for mending the clock	2s 6d
	Item Spent in ale when we paid for the clock mending	6d
March 29	Item Paid to John Lambard for looking after the clock	13s 4d

The following extract confirms that the clock then had a dial. At this time it would probably have been a wooden dial with a single hour hand:

| 1695-96 Jan | Item Paid to Solomon Wing for the painting of the hand dyal ... and new lodging [re-fixing] of it | 6s 6d |
| 1696-97 May 25 | Spent on Solomon Wing when he finished the dials [sundial and clockdial] | 4d |

There are no clues as to the age of the clock or its maker, but the reference to 'balance' in the following extract from the accounts may imply that it was the early type of verge and foliot movement. If this were the case, it would almost certainly have been converted or exchanged about this time for a movement with the more accurate pendulum control:

| 1696-97 June 9 | Paid for mending the balance of the clock | 1s 2d |

The only entries in the accounts for the eighteenth century are as follows:

| 1772 April 28 | The Church clock a year | 2s 6d |

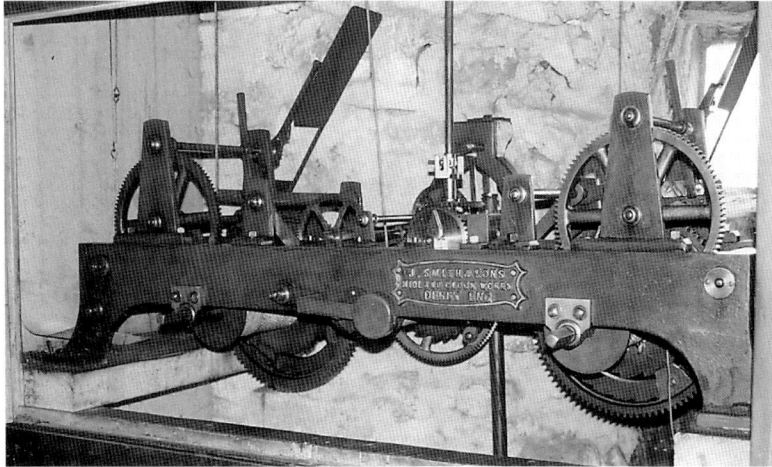

Morcott Church. The three-train flatbed clock installed in 1921

1773 May 13 To Mr Payne for Clock looking after
13s 4d

There are no further accounts available that might have provided further details of the clock. However, it seems that the seventeenth-century clock survived until the present one was installed, as the church leaflet of 1922 states that the former clock 'had only an hour hand, and the works were roughly made'. The mechanism of the old clock was stored in the clockroom until *circa* 1990 and one of the stone weights remains there. It is 330mm (13in) high with an iron loop held in by lead: a typical seventeenth-century clock weight.

The photograph taken by George Henton on 16 June 1914 (Henton 1071) shows that the dial was then octagonal and that it was in a lower position than the present skeleton dial. It was supported by two corbels in a similar manner to the dial at Glaston.

The present clock and skeleton dial were installed in 1921 as a memorial to the men of Morcott who died in the First World War. It is a three-train flatbed movement by John Smith & Sons of Derby and strikes the hours and chimes the ting-tang quarters on the church bells. A brass plaque attached to the weight duct in the ringing chamber records the installation of the clock:

**THE CLOCK WAS PLACED IN THE TOWER
OF THIS CHURCH BY PUBLIC SUBSCRIPTION
EASTER 1921
IN MEMORY OF THE MEN OF THIS PARISH
WHO FELL IN THE GREAT WAR 1914-1918**

CLOCK DETAILS
Details of the present clock:

Maker:	John Smith & Sons, Midland Clock Works, Derby
Signed:	On the setting dial and on the frame
Installed:	1921
Cost:	£190

Memorial to:	The men of Morcott who died in the First World War
Presented by:	The parishioners of the village
Frame:	Cast-iron flatbed
Trains:	Going, striking and chiming
Escapement:	Pinwheel
Pendulum:	Wooden rod and cast-iron cylindrical bob
Rate:	53 beats per minute
Striking:	Countwheel hour-striking
Chiming:	Ting-tang quarter chimes
Weights:	Cast-iron
Winding:	Hand wound weekly
Dial:	Skeleton dial with a blue background and gilded roman numerals and hands
Location:	South face of the tower
Note:	The clock is in a pine case with a glass door

SUNDIAL HOUSE

The south-east declining vertical sundial above a window gable of Sundial House (SK 924008) in Church Lane appears to be a converted datestone. It is dated 1627.

The converted date stone above a window gable at Sundial House in Church Lane, Morcott

NORMANTON

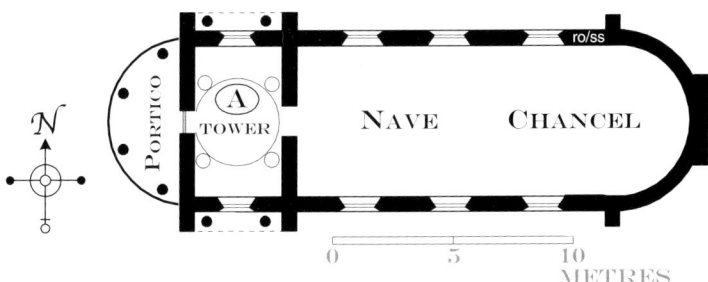

Plan of St Matthew's Church prior to deconsecration
A: One bell in the tower
The bell was rung from the western vestibule below the tower

ST MATTHEW
Grid Ref: SK 933063

Church closed and deconsecrated in 1970. Now a Water Museum.

PARISH RECORDS
The Parish Records include Churchwardens' Accounts 1769-1813 and 1813-36 (DE 1579/5-6) and a faculty of 1911 for rebuilding the nave (DE 1579/12).

BELL HISTORY
The only relevant expenditure in the Parish Records is the occasional purchase of a bellrope, the last entry being:

1825 Jan 11   New Bell Rope & putting up   9s 0d

The original church, possibly dating from the fourteenth century, was described in 1579 as being in a very

Normanton Church following the rebuilding of 1911
(Canon John R H Prophet)

ruinous condition. A new bell was cast or recast for the church by Thomas Hedderly of Nottingham in 1749. Fifteen years later Sir Gilbert Heathcote of Normanton Hall rebuilt the church as a plain building with a square chancel and aisleless nave, but the existing tower was left intact. Laird, in the early 1800s, speaks of 'its little Gothic turret peeping out from a shrubbery' (*VCH* **II**, 87).

In 1826 the present tower and portico replaced the former tower, when it is assumed the old bell was rehung in a new wooden frame. This new work was designed by Thomas Cundy of London, architect to the Grosvenor estates, and is said to be a copy of one of the towers at St John's Church, Westminster. In 1911 the nave and chancel were rebuilt in a style to match the work of 1826, as a memorial to the first Earl of Ancaster, by his widow (*VCH* **II**, 87).

In 1970 the Bishop of Peterborough deconsecrated Normanton Church as it stood below the eventual water line of the new Rutland Water. Two years later a group of volunteers, known as the Normanton Tower Trust, began to raise money for its preservation.

Normanton Church as it was after the rebuilding of 1764 (Canon John R H Prophet). *Based on a drawing of circa 1793* (RCM F10/1984/41)

The final scheme adopted was to raise the internal floor level and build a bank and causeway to give protection and access. In 1983 it became a Water Museum in the care of Anglian Water. Just after deconsecration all the graves, memorials, stained glass and fittings were removed. The bell was sent to Wakerley Church which was then being used as a repository for redundant church furniture. Both the bell and the glass were later installed in the new church of St Jude's in Peterborough. The old wooden frame, headstock and bellwheel, however, remain in the tower at Normanton.

Thomas North (1880, 145) recorded the bell as being cast by [Thomas] Hedderly of Nottingham in 1749. However, a survey carried out in 1975, when the bell was in storage in Wakerley Church, revealed that it was in fact cast by Joseph Eayre of St Neots in 1766. This later bell may have been a replacement when the church was rebuilt in 1911, having been hung elsewhere until this date.

BELL DETAILS
(prior to deconsecration of the church in 1970)

1 **1766**. Diameter 591mm (23¼in). Weight approximately 140kg (2cwt 3qr). Canons removed. Timber headstock. Cast by Joseph Eayre of St Neots.

CVM VOCO VENITE [45]
~ JOSEPH EAYRE S^T. NEOTS FECIT 1766 [45]
(Come when I call)

Removed in 1970 and subsequently transferred to St Jude's Church, Peterborough.

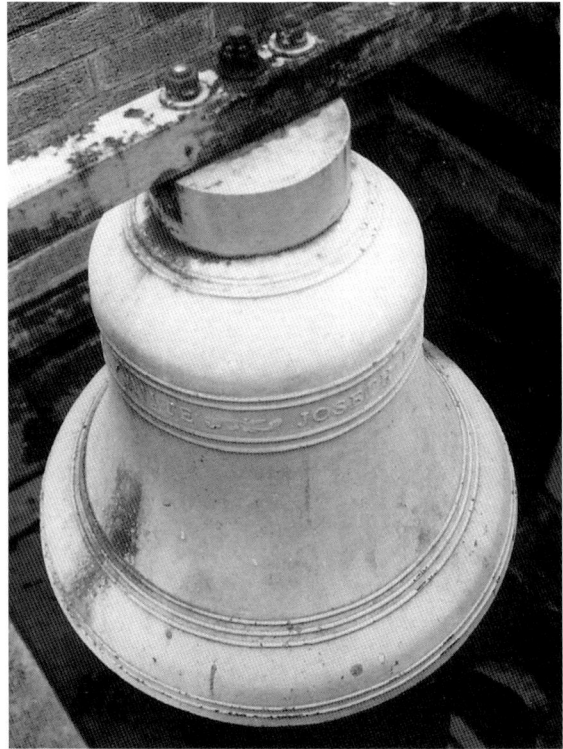

The former Normanton bell hanging at St Jude's Church, Peterborough. Note that the canons have been removed and that it is hung from a steel headstock with a lever for chime ringing. Following the installation of three other bells in a new tower in 1984, this bell is again redundant

Possible former bell
Cast in 1749 by Thomas Hedderly I of Nottingham.
GOD BE OVR 1749 SPEED. HEDDERLY

BELLRINGING CUSTOMS
No bellringing customs have been recorded.

CLOCK HISTORY
Irons' Notes (MS 80/5/22) record that in 1681 there was 'A dial for clock'. No other details have been found.

NORMANTON HALL

Before Normanton Hall was demolished in 1925 a late eighteenth-century table sundial stood in the centre of the circular lawn at the front of the house at SK 935063.

NORMANTON HALL STABLES

Normanton Hall Stables were left standing when the house was demolished in 1926. They now form part of

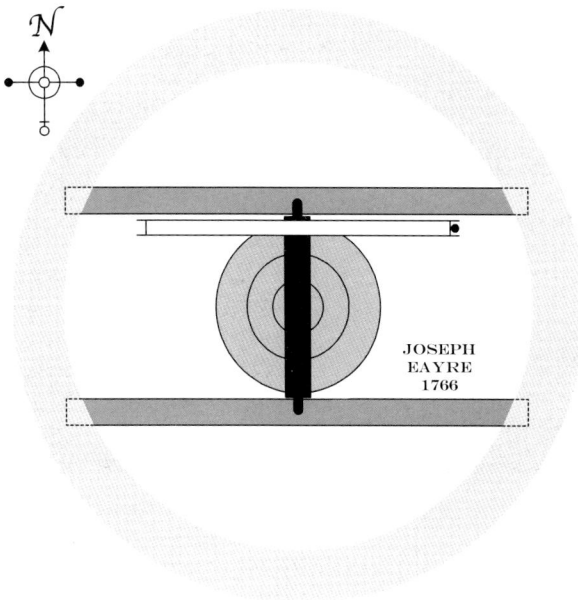

Plan of the bellframe at Normanton Church prior to deconsecration

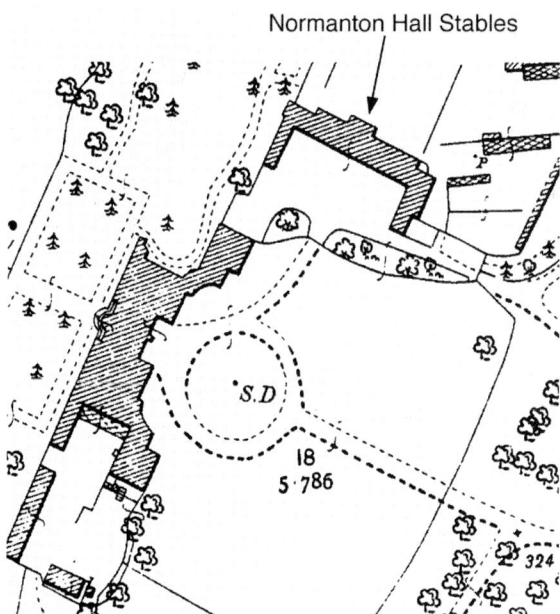

Normanton Hall Stables

'S.D'

18
5·786

324

The table sundial (SD) at Normanton Hall is marked on the early 1900s OS Second Edition 25 inch map

Normanton Park Hotel. The original clock turret with its clock and bell remain. The bell, which has no date or inscription, hangs in the cupola above the dial. Although the stables were built in 1856, the clock movement appears to be *circa* 1750.

Clock Details

Maker:	Possibly Thomas Eayre II of Kettering
Signed:	Unsigned and undated
Installed:	Unknown
Frame:	Birdcage
Trains:	Going and striking
Escapement:	Anchor
Pendulum:	Offset pendulum with wooden rod and 305 mm (12 in) diameter lenticular bob
Rate:	41 beats per minute
Striking:	Countwheel hour striking
Clock bell:	In cupola above turret
Weights:	Cast-iron
Winding:	Weekly
Dial:	Single dial with stone chapter ring and glazed centre. Black roman numerals and gilded hands
Notes:	The pendulum hangs in a separate lath and plaster case. The clock has an electric night silence unit

Will's Cigarettes.

Normanton Park.

Normanton Hall sundial as illustrated here was No 20 in a series of twenty-five cigarette cards of Old Sundials published by W D & H O Wills in 1928

The clock turret at Normanton Park Hotel

The setting dial and first wheel of the going train of the Normanton Park Hotel clock. Note the interesting wheel crossing

NORTH LUFFENHAM

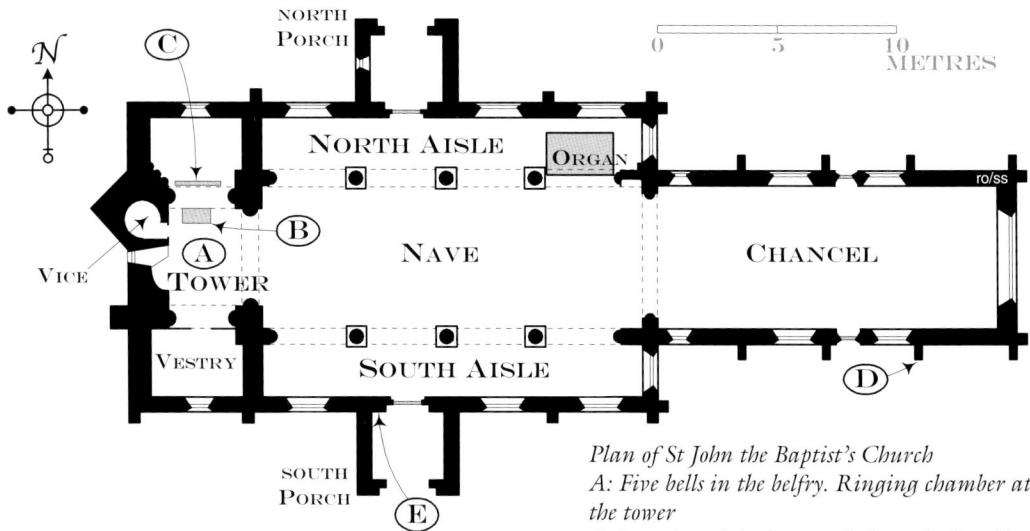

Plan of St John the Baptist's Church
A: Five bells in the belfry. Ringing chamber at the base of the tower
B: Location of the former clock on the first floor of the tower
C: Location of the former clock dial
D: Sundial
E: Scratch dial
The vice provides access to the belfry

ST JOHN THE BAPTIST   Grid Ref: SK 934033

PARISH RECORDS

The following documents in the parish archive were searched for historical details concerning the clock and bells:

Churchwardens' Accounts Book 1810-1903, including a church inventory dated 1877 (DE 1940/55).
Vestry Minute Book 1833-1938 (DE 1940/70).
Notes on the history and architecture of the church and church bells and lists of rectors *circa* 1880 to 1988 (DE 1940/103/1-18).
Notes on the history of North Luffenham from *circa* 1860 to 1931 (DE 1940/104/1-14).
Newspaper cuttings relating to the church and village 1861-1950 (DE 1940/105/1-11).
Memorandum book and notes from *circa* 1882 to 1955 including press cuttings, sketches and photographs compiled by the Rev P G Dennis (DE 1940/102).

BELL HISTORY

The tower was built in the late 1200s and this probably indicates the date of the earliest bells. Evidence of a very early bellframe includes a drawing of a truss from an old frame scratched onto the south reveal of the west window in the belfry. Crescent shaped floorboards discov-

ered when the old oak frame was removed in 1989 were from the trusses of a *circa* 1500 frame.

A new oak frame was made in 1701 by John Browne. This was installed diagonally across the tower and had pits for five bells on the lower level and two above. Two side members from this bellframe are retained in the church. Carved into one of them is:

These floorboards were revealed when the old oak bellframe at North Luffenham was removed in 1989. They are the curved side members from the trusses of a circa 1500 frame (Nicholas Meadwell)

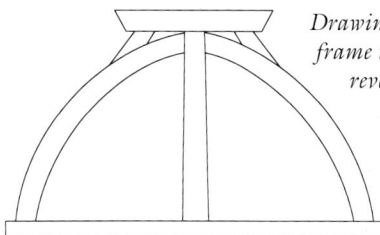

Drawing of an early bell-frame truss on a window reveal in the belfry at North Luffenham

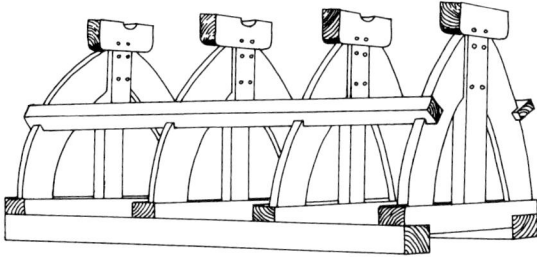

A typical circa 1500 three-bell bellframe with curved trusses

North Luffenham Church. When the 1701 frame was removed in 1984 it was temporarily reassembled in the churchyard (Nicholas Meadwell)

THO': MVNTON
IOHN PITTS
CHVRCHWARDENS
1701
IOHN BROWNE
FECIT

IOHN BROWNE FECIT 1677 was found on the old timber bellframe at Lyddington (see Lyddington, Bell History). A John Brown is also recorded as working on the frame at Donington, Lincolnshire, in 1695 (North 1882, 386), and as rehanging the bells at Stamford All Saints' in 1710. The entry in the Churchwardens' Books at All Saints', which had five bells and a 'Sancte bell' in 1727 (North 1882, 666), provides a good indication of the cost of rehanging bells at this time:

1710 Paid John Brown for hanging the Bells as p. Bill £10 12s 0d

The earliest bell of the present ring of six at North Luffenham is the third, which was cast at the Newcombe foundry in Leicester during the sixteenth century. This bell, together with the present second and fourth and the present sixth before it was recast, were four of the bells present in 1701. Two plain bells which were reported as 'formerly resting in the framework above the present five bells' were given, or sold, sometime before 1880 to Pilton Church (Phillips 1907-08, 38). However, this may be incorrect in part as John Taylor's survey of 26 June 1909 reported that there were two bells 'at the foot of the spire' (closed archives of John Taylor). It may be that these two bells were not in fact transferred to Pilton until after this date. Interestingly, although Thomas North (1880, 139) did not record the actual bells, he did confirm that there were 'two vacant places for bells, showing that there were probably once seven bells here'. The one bell unaccounted for was probably the present fifth before being recast by Thomas Eayre II of Kettering in 1742. According to the *Rutland Magazine* (1907-08, 38) there were then two stories current in the village regarding the church bells: first 'a piece of land belonging to the town was sold, in years gone by, in order to add one Bell to the existing peal'; secondly the following couplet commonly quoted in the village was thought to

relate to the present tenor bell:

Hinman he loved ringing well,
So he sold the land and bought a bell.

The Hinman family lived in North Luffenham in the early seventeenth century and were said to have inhabited a cottage to the south of the Wesleyan Meeting House.

A plain 457mm (18in) diameter clock bell was installed about 1780 (Phillips 1907-08, 38). It almost certainly hung away from the main frame, probably in the lower of the two north lights of the spire. From this location it would be heard by most of the village.

Churchwardens' Accounts between 1810 and 1901 contain very few references to the bells other than the usual replacement of ropes and purchase of oil, although £12 3s 9d was paid to John Stafford in 1811 for repairing the bells. Amongst those named for supplying bellropes from the 1820s to the end of the century were John Lawe, Darnell, Charles Wymond, Christian and Little-dyke.

Although much of the Victorian restoration work was completed by 1874, repairs to the tower did not proceed due to lack of funds. In 1880 Thomas North reported that the tenor was cracked and therefore unused.

John Taylor's survey of the belfry in 1909 noted that the timber frame was fixed diagonally across the tower and that it was 'no good' (closed archives of John Taylor). In 1938 the bells were recorded as being unringable and the clock bell was no longer in use (Powell 1938, 23). In 1943 the Rev Lawrence Field noted that the clock bell was stored in the vestry together with the clock.

A 1949 report on the bells by John Taylor concluded that the installation was unsafe, even for chime ringing. Chime ringing, however, continued on one bell for special occasions until the mid 1950s. For a period an amplified recording of a peal at St Mary the Virgin, Edith Weston, was used whenever the bells were required at North Luffenham.

The installation remained in this condition until 1989

when the ring was retuned and rehung in a new high-sided steel frame with new fittings. The cracked tenor was recast and a new treble added by John Taylor at the same time. The Rector and the Church Council of North Luffenham succeeded in raising a considerable sum of money towards the cost of this work. The Aerial Erector Flight of the Mechanical Engineering Squadron based at RAF North Luffenham donated their time to the removal of the old bells, making and installing the new frame, and rehanging the bells, all under the guidance of Nicholas Meadwell. This reduced the total cost by about fifty per cent. The work was completed in time for the Christmas services of 1989.

AND EVERY TONGUE

Part of the inscription from the treble at North Luffenham illustrating the lettering used on this bell and the tenor

BELL DETAILS

1 **1989**. Treble. Diameter 699mm (27½in). Weight 239kg (4cwt 2qr 24lb). Cast without canons. Cast-iron headstock. Cast by John Taylor & Co of Loughborough. Decoration [54] around the inscription band. On the waist:

AND EVERY TONGUE
CONFESS THAT
JESUS CHRIST IS LORD
TO THE GLORY OF
GOD THE FATHER
(PHIL. 2.11.)

On the waist opposite:

19 [56] 89

The lettering is like [137].

2 **1630**. Diameter 730mm (28¾in). Weight 218kg (4cwt 1qr 4lb). Canons retained. Cast-iron canon-retaining headstock. Cast by Thomas Norris of Stamford.

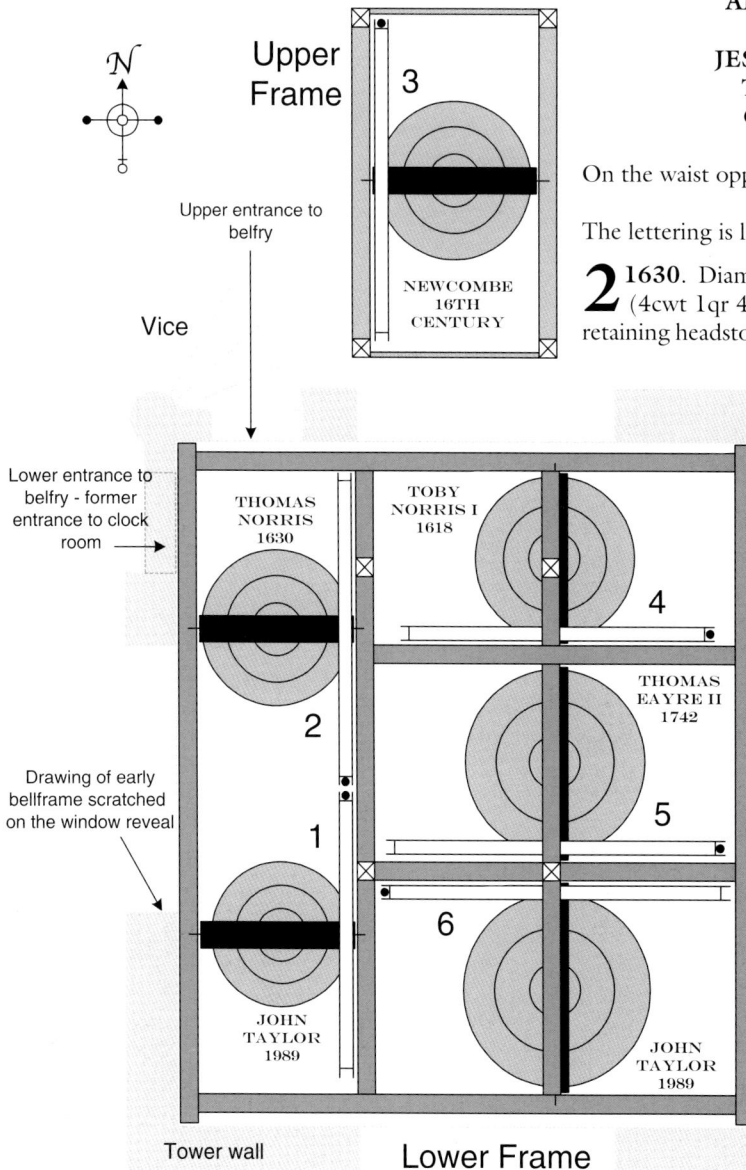

Plan of the bellframe at North Luffenham Church

**[14] IO [53] EXTO⋀ [53] ED [53] HV⋀T [53] RO
~ [53] MV⋀TO⋀ [53] HE [53] LAW [53] CHWA**

The date **1630** is immediately below **HE** of **HE LAW**. John Exton and Edmund Hunt may have been the donors of this bell. Edmund Hunt is also named on the fourth and the tenor. Robert Munton and Henry Lawe, churchwardens in 1630, were buried at North Luffenham on 16 November 1658 and 30 April 1653 respectively.

3 **16th Century**. Diameter 756mm (29¾in). Weight 244kg (4cwt 3qr 6lb). Canons retained. Cast-iron canon-retaining headstock. Cast by Newcombe, Leicester.

[106]

𝔐𝔢𝔩𝔬𝔡𝔦𝔠 𝔊𝔢𝔯𝔢𝔱 𝔑𝔬𝔪𝔢𝔫 𝔆𝔞𝔪𝔭𝔞𝔫𝔞

[103]

(This bell shall melodiously bear the name ...)

The inscription and founder's marks are almost illegible. This seems to be an incomplete inscription as early bells often included the name of the Saint to which the bell was dedicated. There is an intrusive mark after **Campana** in the shape of **I**. It may have been the first letter of IOHANNIS. Scheduled for preservation (Council for the Care of Churches).

4 **1618**. Diameter 800mm (31½in). Weight 268 (5cwt 1qr 2lb). Canons retained. Cast-iron canon-retaining headstock. Cast by Toby Norris I of Stamford.

**[26] OM⋀IA [5][5] FIA⋀T [5][5] AD [5][5]
~ GLORIAM [5][5] DEI [5][5] 1618
[26][5] E [5] HV⋀T [5] H [5] STAFFORDE
~ [5] GARDIA⋀ [5][5]**

(Let all things be done to the glory of God)

Henry Stafford is also named on the tenor.

The canon-retaining headstock and decorated canons on the fourth bell by Thomas Eayre II at North Luffenham

5 **1742**. Diameter 899mm (35⅜in). Weight 402kg (7cwt 3qr 18lb). Canons retained. Cast-iron canon-retaining headstock. Cast by Thomas Eayre II, Kettering.

**THO. EAYRE FECIT . 1742 ⦂ [51] ⦂ OMNIA
~ FIANT AD GLORIAM DEI •⦂•• [51]
~ • •⦂•• • GLORIA DEO SOLI •⦂•**

(Let all things be done to the glory of God.
Glory to God alone)

6 **1989**. Tenor. Diameter 978mm (38½in). Weight 578kg (11cwt 1qr 14lb). Cast without canons. Cast-iron headstock. Recast by John Taylor & Co of Loughborough.

**[34] 𝔦𝔄 𝔇𝔦𝔤𝔟𝔶 𝔦𝔬 𝔅𝔄𝔰𝔰𝔢𝔱 𝔦𝔷 𝔧𝔬𝔥𝔫𝔰𝔬𝔫
~ 𝔢𝔡 𝔥𝔳𝔫𝔱 𝔥𝔢 𝔰𝔱𝔄𝔣𝔣𝔬𝔯𝔡 𝔡𝔄 𝔤𝔦𝔟𝔰𝔬𝔫
~ [34] 1619 [7]**

On the waist:

**[42] RECAST 1989 [56] [42]
RESTORED BY THE
PEOPLE OF THE
VILLAGE WITH
R. A. F.
NORTH LUFFENHAM**

The retained lettering of the former inscription is like **[116]**. The rest of the lettering is like **[137]**.

Former bell

Cast in 1619 by Henry Oldfield II of Nottingham. Diameter 978mm (38½in). Weight approximately 521kg (10cwt 1qr).

**[34] 𝔦𝔄 𝔇𝔦𝔤𝔟𝔶 𝔦𝔬 𝔅𝔄𝔰𝔰𝔢𝔱 𝔦𝔷 𝔧𝔬𝔥𝔫𝔰𝔬𝔫
~ 𝔢𝔡 𝔥𝔳𝔫𝔱 𝔥𝔢 𝔰𝔱𝔄𝔣𝔣𝔬𝔯𝔡 𝔡𝔄 𝔤𝔦𝔟𝔰𝔬𝔫
~ [34] 1619 [7]**

Thomas North (1880, 137) reported that this bell was cracked and not used. He did not record this inscription as being in gothic. James Digby was the son of Roger Digby of North Luffenham. Isaac Johnson was born in 1601. He was a grandson of Archdeacon Robert

*Isaac Johnson of
North Luffenham*
(DE 1940/103/13)

Johnson, the founder of Oakham and Uppingham Schools. Isaac was brought up by his grandfather at North Luffenham Rectory and inherited most of his grandfather's estates in 1625, by which time he was living at Boston, Lincolnshire. In 1630 he sailed to America where he was one of the founders of the City of Boston, then a colony in Massachusetts Bay. He died on 1 October 1630 (Field, *circa* 1950).

Edmund Hunt was baptised 30 September 1593 and buried 18 October 1666. John, the son of Henry Stafford was baptised on 4 April 1619 (DE 1940/102).

This bell was often referred to as the 'Isaac Johnson Bell' and an appeal was made in 1989 to the Governors and Citizens of Boston, Massachusetts, to help with the cost of recasting it. The outcome of this appeal is not recorded.

CLOCK BELL

Circa **1780**. Diameter 457mm (18in). Weight approximately 76kg (1cwt 2qr). Canons retained. Unknown founder. No inscription or decoration. 'The Bell connected with the clock was fixed about the year 1780, but is without inscription' (Phillips 1907-08, 38). It was removed from the tower *circa* 1940 and stored together with the clock in the vestry. Sold together with the clock to a private collector in 1997. For further details see under Clock History.

BELLRINGING CUSTOMS

On Sundays and other days when a service was to be held

The cracked tenor at North Luffenham before being recast in 1989 (Nicholas Meadwell)

the second bell was rung at 8am. The bells were rung for half an hour, then the fourth bell was rung for five minutes before a service commenced. The bells were always rung or chimed immediately after a service at which Banns had been called for the first and third times. This was referred to as the Spur Peal. The fourth bell was the Passing Bell. At the Death Knell there were thrice three tolls for an adult male, thrice two tolls for an adult female, twice three tolls for a young man, twice two tolls for a young woman, three tolls for a boy and two tolls for a girl. The fourth bell was also tolled at funerals (DE 1940/102).

At the beginning of the nineteenth century the churchwardens paid for ringing the Gleaning Bell during harvest. The following is the last recorded payment for this custom:

1830 Sept 13   Ringing the Bell in Harvest   12s 0d

The accounts include payments for ringing a bell at the death of George III who died on 29 January 1820 and for the funeral of George IV who died on 26 June 1830:

1820   Passing Bell for the King   1s 0d
1830   Funeral Bell for his late Majesty 2s 6d

Today North Luffenham has its own band of bellringers which practises on Thursday evenings. The bells

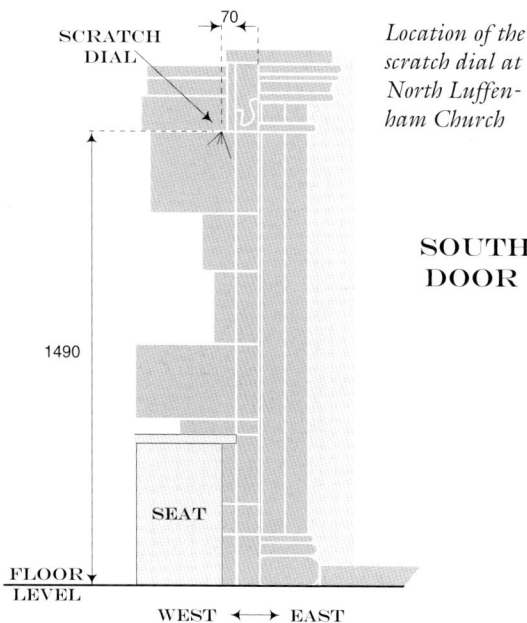

Location of the scratch dial at North Luffenham Church

NORTH LUFFENHAM				
Location	Inside south porch on wall west of main door			
Condition	Good			
Gnomon Hole Diameter			Filled in	
Gnomon Hole Depth			Filled in	
Height above ground level			1490	
Line Ref	a	b	c	d
Length	135	50	60	85
Angle (°)	169	188	214	236

Scratch dial details

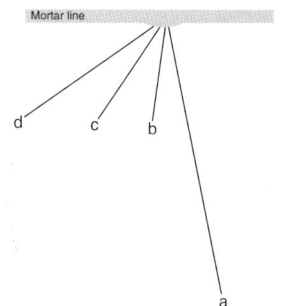

are rung before all of the Sunday services, at Christmas, on Armistice Day, and if requested at weddings. They are also occasionally rung by visiting bands during the year. At funerals the bells are chimed.

SCRATCH DIAL

There is a small scratch dial on the late thirteenth-century wall to the west of the south door, inside the south porch.

SUNDIAL

There is a mid eighteenth-century sundial on a south buttress of the chancel. It cannot be seen from the village and its main purpose was probably as a source of time for regulating the church clock. It originally had a painted dial. Most of the markings have weathered away but some roman numerals can be seen along the bottom edge when the sun is at an oblique angle to the dial.

CLOCK HISTORY

There is little evidence of an early clock at North Luffenham. However, on the west reveal of the north window of the former clockroom the initials 'HE' and the date '1699' are carefully incised into the stonework, together with some geometric decoration. These could be the initials of a clock winder, or of an otherwise unknown clockmaker who installed or maintained an early clock here. Also, 'IW 1717' is scratched on a stone near the top of the vice. The 'IW' is in the style of the initials of John Watts the Stamford clockmaker as seen on the wooden frames of several of his clocks. A stone weight and an early type of lead pendulum bob found in the church point to the existence of an early clock here.

An unsigned and undated two-train hour-striking

The incised initials and date found in the former clockroom at North Luffenham Church

Initials scratched on a stone near the top of the vice at North Luffenham Church. Possible evidence that John Watts, clockmaker, visited an early clock here

birdcage clock was installed in the tower *circa* 1750. It has a number of features which suggest that it was made by Thomas Eayre II of Kettering (*see* Chapter 3 — Clockmakers, Thomas Eayre II). A drawing of *circa* 1839 (Uppingham School Archives) shows that it then had a circular dial with two hands on the north face of the tower. The church was struck by lightning on 10 June 1822 and the top part of the spire was destroyed. 'The parish clock was stopped by the concussion, and the strong iron spindle of the weathercock was bent and precipitated amongst the bells' (Phillips 1907-08, 36).

The surviving Churchwardens' Accounts include many references to the clock. The annual maintenance and repairs were generally entrusted to Uppingham clockmakers. William Aris II is the first to be mentioned and his annual fee for maintenance work was 12s:

1810 Oct 19   Aris a Bill for Repairing the Clock
  £4 10s 0d

He attended the clock quite regularly and his last visit was paid for the year after he died:

1843 July 31   To Aris repairing the Clock   10s 0d

In 1810 and 1822 John Houghton was summoned to repair the clock. The next Uppingham clockmaker noted in the accounts is Mark Flint:

1860 April 10   Mr Flint, for cleaning the clock as
  per bill   15s 0d

Thomas Aris, son of William, carried out repairs to the clock and fitted new lines in 1861 at a cost of 10s. Mark Flint's last recorded visit to the clock was in 1864. Further work was carried out on the clock around the time of the church restoration:

1868   A Scotney for regilding Clock & Hand   5s 0d
  J Hicks for repairing Church clock   £1 0s 0d
  R Littledike new Clock Line   3s 0d

J Penney was paid £1 11s 6d in 1810 for winding the clock. John Bolland was named as the clock winder for twenty-five years until 1884 when Fred Price, the new Parish Clerk, took over. The accounts show that he carried out this duty until 1886 and may have done so until 1897, after which there is no further mention of the clock. The clock was removed from the tower sometime after this date. It was stored in the vestry together with the clock bell in 1943 and later in a barn at Manor Farm. Both clock and bell were sold to a private collector in 1997. The clockroom was lost when the bells were rehung at a lower level in 1989.

CLOCK DETAILS

Details of the former clock:

Maker:	Thomas Eayre II of Kettering
Signed:	No name or date on the clock. Setting dial missing
Installed:	Circa 1750
Cost:	Not recorded

North Luffenham church clock just prior to its sale in 1997 after being out of use for nearly a century

(below) The small two-train flatbed clock movement at The Pastures, North Luffenham, was supplied by William Potts of Leeds and is located immediately behind the dial. Despite the nameplate on the frame it was actually made by Haycock of Ashbourne. This is one of sixty or so clocks made by Haycock for William Potts. In 1902, of forty-one clocks installed by Potts, five were made by Haycock (information from Michael Potts)

Frame:	Wrought-iron birdcage frame, 914mm (36in) wide, 457mm (18in) deep and 610mm (24in) high
Trains:	Going and striking
Escapement:	Anchor
Pendulum:	Iron rod and lead bob
Rate:	40 beats per minute
Striking:	Countwheel hour striking. Missing countwheel replaced in 2001
Clock Bell:	Unknown founder, but probably by Thomas Eayre II of Kettering
Weights:	Lead
Winding:	Hand wound daily. Going train barrel turns once in two hours
Dial:	Originally a circular dial on north face of the tower. Dial and motionwork missing

CLOCK DETAILS

Details of the present clock:

Maker:	Haycock of Ashbourne
Signed:	'W Potts & Sons, Makers, Leeds, 1903' on the frame
Installed:	Supplied and installed by William Potts of Leeds in 1903
Frame:	Cast-iron flatbed
Trains:	Going and striking
Escapement:	Deadbeat
Pendulum:	One-second pendulum
Rate:	60 beats per minute
Striking:	Countwheel hour striking
Clock Bell:	Cast by Robert Newcombe I of Leicester *circa* 1550. Hangs above the clock dial (see below)
Weights:	Cast-iron
Winding:	Weekly
Dial:	Copper skeleton dial by C F A Voysey

THE PASTURES

Pevsner (1984, 491) describes The Pastures, in Glebe Road, North Luffenham, as follows: 'By C F A Voysey 1901, with alterations of 1909. A charming composition on three sides of a quadrangle ... Close to the junction of the W and N ranges a tower with clock and bell and saddleback roof.' C F A Voysey was a well-known and much respected architect at the beginning of the twentieth century. He supplied the copper skeleton dial which may have been to his own design.

The clock dial and bell at The Pastures in Glebe Road, North Luffenham

CLOCK BELL DETAILS
Circa **1550**. Diameter 660mm (26in). Weight approximately 195kg (3cwt 3qr 10lb). Canons retained. Hung dead from a beam projecting from the ridge of the gable. No clapper. Cast by Robert Newcombe I of Leicester.

[11] ꞄOBꞄꞀⰌꞄ [2] ꞀⰌⱲⱲOⰌꞄ

The lettering is like [108]. This is the former treble from the church at Little Bowden, near Market Harborough, Leicestershire. It was removed and sent to Taylor's Bell Foundry for scrap when the wooden tower was replaced by a bellcote in 1902. Fortunately it was saved from this fate at the last minute and subsequently acquired by Halliday of Stamford for £10. Halliday was the builder involved both at Little Bowden and The Pastures. Miss Conant, for whom the house was built, did not like the harsh tone of the bell, so a boxwood plug was inserted into the nose of the bell hammer. It now has a very mellow tone. Scheduled for preservation (Council for the Care of Churches).

The base of a table sundial remains in the gardens of The Pastures. This sundial, like the house, was designed by C F A Voysey.

LUFFENHAM HOUSE

An engraving of Luffenham House (Wright Additions 1788, 7) shows that there was a sundial on the south elevation. The house was demolished in 1806.

OTHER SUNDIALS IN NORTH LUFFENHAM

Two early sundials remain in the village: one at Sundial Cottage (SK 937032) in Digby Drive, the other on the front of Manor Farm (SK 932035). There used to be a sundial on the barn next to Dovecote House (SK 934034), and two table sundials are shown on the early 1900s OS Second Edition 25 inch map, one in the grounds of Luffenham Hall (SK 935032) and the other in a paddock to the north-east of Sundial Cottage (SK 937032).

This early 1900s OS Second Edition 25 inch map shows the location of Sundial Cottage and two table sundials (SD), one in the grounds of Luffenham Hall and the other near Sundial Cottage in Digby Drive, North Luffenham

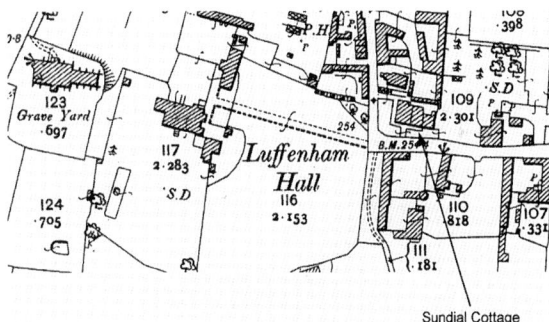

(left) Detail from an engraving in Wright (Additions 1788, 7) of Luffenham House showing the sundial

OAKHAM

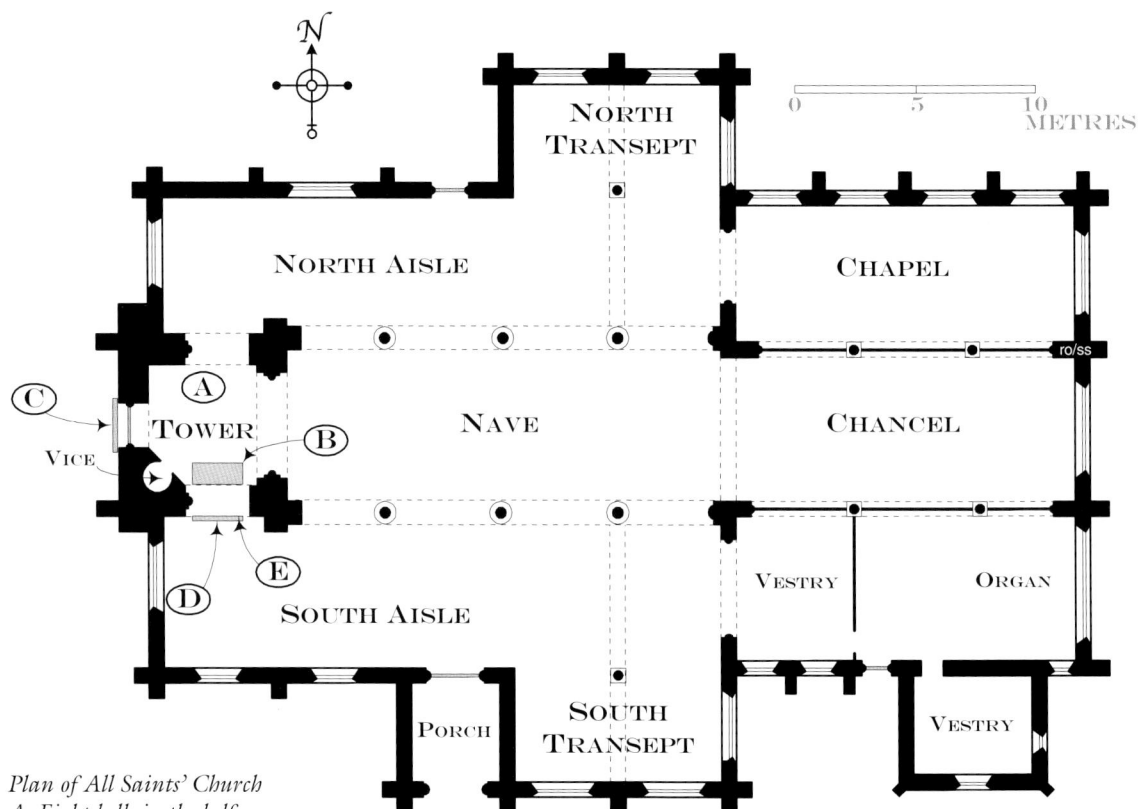

Plan of All Saints' Church
A: Eight bells in the belfry
Ringing chamber on the first floor of the tower
B: Clock in the ringing chamber
C: West-facing clock dial. D: South-facing clock dial
E: Location of former clock dial
The vice provides access to the ringing chamber and belfry

ALL SAINTS   Grid Ref: SK 861089

PARISH RECORDS

There is an extensive collection of parish records and the following have been searched for this project:

Churchwardens' Accounts 1735-1805 (Deanshold) (DE 2694/274) and 1909-20 (DE 2694/275). The gaps in the bound accounts are to some extent made up by bundles of loose bills, receipts and correspondence that cover the period 1726 to 1862 (DE 2694/183-7 and 702-07, and DE 3178/2).

Register book of parish apprentices 1807-31 (DE 2694/477).

Historical notes on the bells by George Phillips covering the period 1807-20 (DE 3178/40).

A list of subscribers to the church restoration *circa* 1858 (DE 2694/236).

Oakham Church Restoration 23 February 1860 (DE 2694/238) and correspondence (DE 2694/239).

The Order of Service for the dedication of the bells dated 24 February 1911 (DE 2694/138).

A photograph of the restored west-facing clock dial *circa* 1925 (DE 2694/241).

The Story of Oakham Church, School & Castle by A E Fraser, vicar, dated 1932 (DE 2694/660).

A report on the church tower 1935 (DE 2694/710).

A plan of the church tower *circa* 1935 (DE 2694/711).

A report on the condition of the tower dated 3 June 1943 (DE 2694/245).

Dimensions and weights of the bells *circa* 1950 (DE 2694/246).

A faculty of 1979 regarding work on the bells (DE 2694/224).

An Archdeacon's Certificate authorising work on the bells dated 25 November 1981 (DE 2694/248).

BELL HISTORY

The tower of Oakham Church was added in the fourteenth century and it is highly likely that at this time there would have been at least two bells hanging in the belfry.

The following three extracts from Irons' Notes (MS 80/1/3) are the earliest references relating to the bells.

The first is a bequest taken from the will of Hugh Butler. A Thomas Crowsforth is also recorded as leaving a bequest for 'repairs of the bells' in 1531. The seventeenth-century entries are taken from Visitations:

1501 March 7   to the ch [church] of Alholowas [All Hallows] in O ... [Oakham] to the bying of a ... [illegible] belle to the stepell [steeple]   20s

1615 May 11   Our churche leads & bell frames to be in decay

1616 July 12   Gard [Guardians] to certify regarding the repair of their bells

Bequests must have been a welcome source of revenue for the repair of church property and 'Mr. Robert Blackburn, who died early in the sixteenth century, left by his will 3s 4d to the bells of Oakham Church' (North 1880, 146). The 'repair of their bells' in 1616 refers to the frame and fittings and not the bells themselves. Another source gives details of the 1605 Visitation, including the fact that two bellwheels were broken and that they were being mended (Haddelsey 1972, 18).

Several entries in the Churchwardens' Accounts for 1793 imply that a new bell may have been supplied:

1793 Jan   paide the Clarke fo 6 Days worke at the Bell   6s 0d
           caarg [carriage] of the New hanings for the Bell   6s 0d

Jan 10   paid the men for helpin with the Bell   2s 0d
           paid Mr Arnolds Bill   £9 12s 3d

The Mr Arnold noted here was the Leicester bell-founder Edward Arnold, who had cast bells for six other churches in Rutland by the end of 1793. These included a Priest's Bell at Barrowden and a clock bell at Glaston. The amount paid to Edward Arnold indicates that it was a small bell, possibly the former Priest's Bell. The disbursements show amounts being paid for unspecified services to a Mr Toon in the 1790s and it is possible that these were for bellropes. A Mr Toon of Oakham regularly supplied bellropes and clocklines to other churches in the area at this time.

Various repairs were carried out to the bell fittings and frames. In July and November 1828 John Lewin submitted a bill to the churchwardens for providing amongst other items 'a new slider to the Bell', '36 feet of Board capping for the Bell frames', 'Board for the floor — 12 feet', '2 levers for the Bells' and 'a New Door'. He had worked eight days in November 'at the Bell frames' at a total cost of £1 4s 0d. More work was obviously required in the belfry during December of that same year.

Oakham Church was evidently in a poor state of repair by the mid 1850s and a Restoration Committee was formed. The architect Sir Gilbert Scott, in his report to the committee, 'recommended a tier of strong iron ties for the tower, the floor immediately over the church to be renewed and the other floor and the bell timbers to be substantially repaired'. The intention to 'repair the tower and rearrange the floors' was accepted at a meeting held on 30 April 1857 (Rooksby 1995, 10-11). Work on the general restoration commenced on 7 September 1857. However the bell timbers in the tower were unexpectedly found to be beyond repair and as a

This loose bill of 1808 (DE 3178/2/5) confirms that there was a 'ting tang bell' [Priest's Bell] at this date which was evidently recast in 1840

(above) John Lewin's bill as submitted to the Oakham churchwardens in December 1828 (DE 2694/184/22)

(left) George Toon of Oakham was regularly supplying ropes to Oakham Church at the beginning of the 1850s (DE 2694/706/7)

consequence required complete renewal. This setback, together with the rehanging and recasting of two of the bells, meant that the church could not be reopened until 10 November 1858. Funds to defray the cost of the restoration, which totalled £6086 0s 4d, were almost entirely raised by local subscribers. The committee recorded (DE 2694/238) what they paid out with regard to the bells:

Bells, recasting and rehanging	£105 11s 2d
Bell Timbers and Floors renewing	£73 14s 6d

Further details are recorded in the Restoration Accounts as follows (DE 2694/708/31):

The New Bell Frames
and the assisting of Man to hang the Bells.
Time and Materials
Mason 5 days. Laborers 25 days, taking down the Bells

The New Bell Timbers as agreed	£50 0s 0d
Carpenters Time and Materials	£29 8s 3d
The Ironmongers Bill at Oakham	£11 7s 6d

From information supplied by Thomas North (1880, 146) it is apparent that Oakham had at least three bells in 1677, one cast in 1618 by Hugh Watts II of Leicester and two cast in 1677 by Toby Norris III of Stamford. One of the Norris bells, the tenor in 1880, was recast by John Taylor in 1875. It is possible that the bell supplied by Henry Penn in 1723 and the two cast by Mears in 1858 were recasts of earlier bells.

The remaining two bells hanging in the tower in 1880 were newly cast by John Taylor in 1860. C A Stevens mentions these in his *Notes on Oakham Church Restoration* [the notes are dated 1 October 1904]. He says:

In May 1860, I had obtained an estimate from Taylor, of Loughborough, for the addition of two Treble Bells. It was kindly proposed to recognise my work in the Schools

and Church by putting my name upon them. But this, as well as any other testimonial, I declined. My wife and I had not worked for that. It was, however, known that we were especially desirous of having the parvise screens, north and south of the Chancel, completed, according to drawings I had obtained from Mr. Scott. It was therefore proposed to gratify us by a special subscription for the purpose. We could not but accept the compliment so put, and in May, 1861, Dr. Wood wrote informing me that both the Bells and Screens were fixed, by a joint subscription of £250 (Stevens 1905-6, 27).

The following report from the *Lincoln, Rutland & Stamford Mercury* comments on the recasting of the Norris bell by John Taylor in 1875:

November 1874 Oakham bazaar held to raise funds for the restoring by means of casting the great bell [the tenor bell] in the peal of bells at the parish church which has long been irremediably cracked and has been a melancholy sound day and night to the ears of the inhabitants, for it is the bell upon which the clock strikes (Traylen 1982a, 19).

In 1910 the

bell frames were in such a condition that the PCC thought steps should be taken to put them in order. The first proposal was to put in new steel frames. This broadened out into a scheme for recasting the whole peal and adding Cambridge chimes to the clock [this latter statement is misleading, as the clock as supplied in 1858 was already capable of playing Cambridge chimes]. The total cost has been within a few pence of £500 (Phillips 1911-12, 41).

Gillett & Johnston of Croydon undertook the work and much of the cost was covered by public subscription. The old bells and wooden frames were removed from the tower in October 1910 and the eight recast bells were hung in a new metal high-sided frame in February 1911. Because the bells were located in different positions within the new frame the Croydon firm had to adapt the clock so that it continued to chime the Westminster Quarters on the bells. The new bells were dedicated by the Lord Bishop of the Diocese on 24 February 1911.

The Priest's Bell hangs above the fourth bell on the north side of the frame.

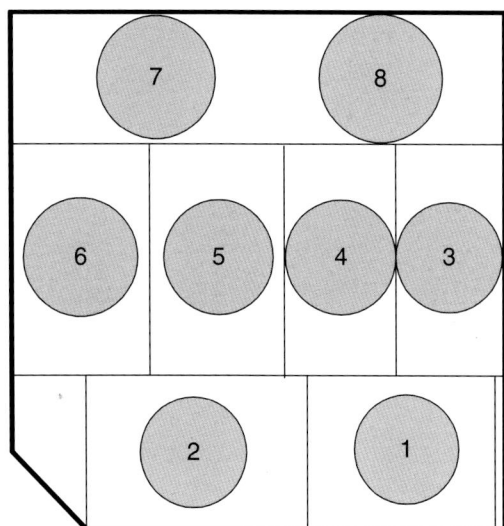

From a sketch of the layout of the bells and timber bellframe at Oakham Church, prepared when John Taylor surveyed the belfry on 27 January 1875. Presumably this is when he visited the church to prepare a quotation for recasting the tenor bell. Notes taken on this visit include: 'The 3rd & 4th ought to be where the 1st & 2nd are as the pits are too narrow to receive them ....'
(closed archives of John Taylor)

The first major peal on the new bells is recorded on a tablet in the ringing room: 'A complete peal of Grandsire Triples - 5050 changes. On Saturday May 6th 1911 in 3 hours and 10 minutes.' The tablet also includes the names of the ringers.

A remarkable tale regarding an accident to the tenor bell in 1924 is told by William Higgs, in his unpublished memoir entitled *My Connection with the Grand Old Church of All Saints', Oakham*:

A very alarming incident happened one Sunday morning. We were pulling the bells up to ring for morning service and I was ringing the Tenor. When about half way up the Tenor rope suddenly stopped. The other bells were raised as soon as possible and I went up to the bell chamber to see what had happened. A very unusual sight met my eyes.

The Tenor bell lay horizontally in the frame and I could see straight through the bell, which had broken off round the crown, leaving the top still attached to the headstock. The bell luckily jammed in the frame, or it might have cleared us all to the bottom ... This bell and the seventh were taken back to the Foundry and recast, more metal being added. From then on the bells went very well, except that we had two clappers break and fall out.

In the 1930s bellringing was suspended in the belief that it was causing the tower to deteriorate. The suspension continued for the first few years of the Second World War when all bellringing throughout the country had to cease. During this time the ringing of church bells was reserved for use as a warning of invasion. When this ban was lifted Charles Peas RIBA was asked to inspect the tower and recommend what remedial action was required. In his report dated 3 June 1943 he concluded that the tower was stable and that there was no reason why bellringing should not resume. No strengthening

The sorry state of the bell in 1924. The boy behind the bell is Robert Hoy who was later to become a keen bellringer (Mrs Hoy)

The damaged bells were taken to Oakham Station by Midland Railway Company horse and dray to begin their journey back to Croydon for recasting (Mrs Hoy)

work was required, but he did recommend that the installation should be serviced before the bells were again put into use (DE 2694/245).

The bells were rehung with new bearings in 1979 by Eayre and Smith of Kegworth. This work was carried out as a memorial to Kenneth Marshall, master baker of Daventry and Newnham, Northamptonshire, who died in Oakham on 4 September 1978. He was the father-in-law to the then vicar, Canon Alan Horsley. A small aluminium plaque recording this donation is attached to the frame under the tenor bell.

BELL DETAILS

1 **1910**. Treble. Note E flat. Diameter 724mm (28½in). Weight 268kg (5cwt 1qr 3lb). Cast without canons. Cast-iron headstock. Recast by Gillett & Johnston of Croydon.

RECAST BY GILLETT & JOHNSTON ~ CROYDON 1910

There is a band of decoration [58] below the inscription band.

FORMER BELL

Cast in 1860 by John Taylor & Co of Loughborough. Treble. Diameter 762mm (30in).

JOHN TAYLOR & CO FOUNDERS ~ LOUGHBOROUGH. ~ AGO GRATIAS HUMILLIME H. F. 1860

(I render thanks most humbly)

H. F. was the Rev Heneage Finch, vicar of Oakham at the time. This and the second bell were installed in recognition of the work of C A Stevens, late Curate of Oakham.

2 **1910**. Note D. Diameter 740mm (29⅛in). Weight 257kg (5cwt 7lb). Cast without canons. Cast-iron

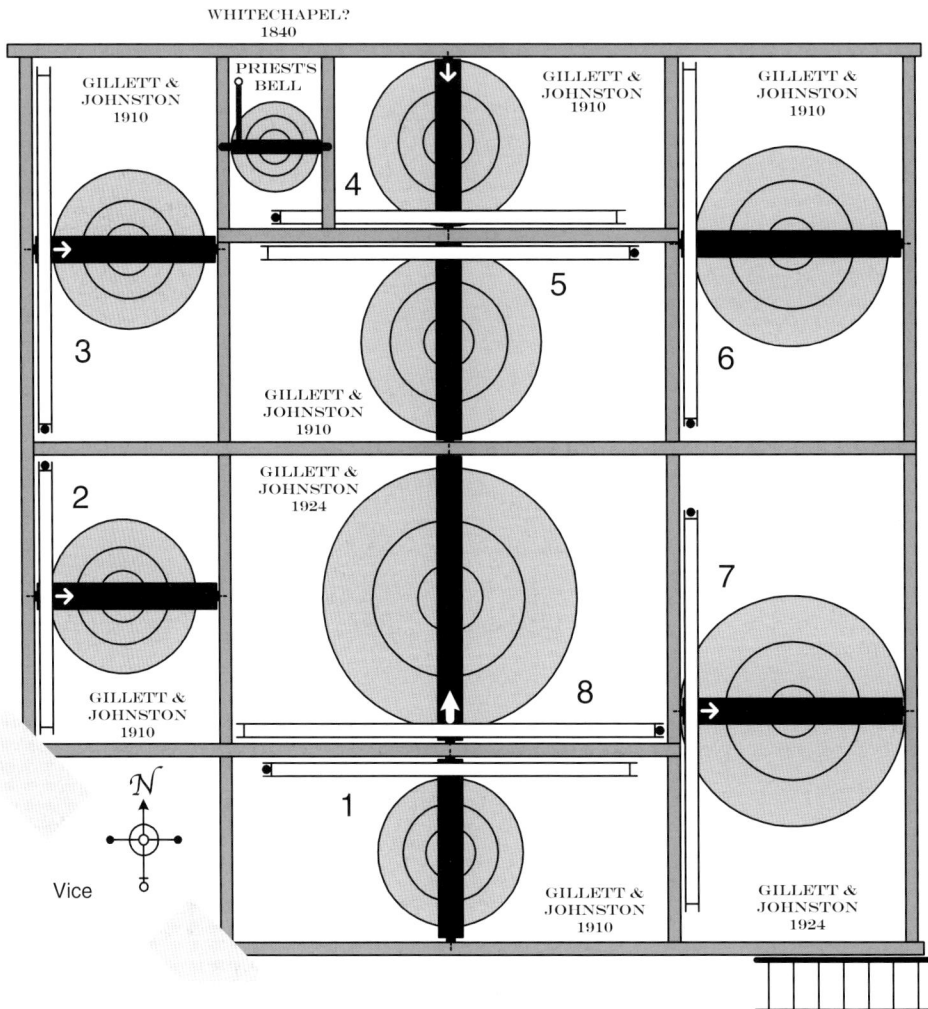

Plan of the high-sided metal bellframe at Oakham Church

headstock. Recast by Gillett & Johnston of Croydon.

RECAST BY GILLETT & JOHNSTON ~ CROYDON 1910

There is a band of decoration [58] below the inscription band. The clock strikes the quarters on this bell.

Former bell

Cast in 1860 by John Taylor & Co of Loughborough. Diameter 787mm (31in).

JOHN TAYLOR & CO FOUNDERS ~ LOUGHBOROUGH 1860

This and the treble were installed in recognition of the work of C A Stevens, late Curate of Oakham.

3 1910. Note C. Diameter 794mm (31¼in). Weight 307kg (6cwt 4lb). Cast without canons. Cast-iron headstock. Recast by Gillett & Johnston of Croydon.

RECAST BY GILLETT & JOHNSTON ~ CROYDON 1910

There is a band of decoration [58] below the inscription band. The clock strikes the quarters on this bell.

Former bell

Cast in 1677 by Toby Norris III of Stamford. Diameter 826mm (32½in).

[26] GOD SAVE THE KING T MEKINGS ~ TOBIE ИORRIƧ CAƧT ME 1677

4 1910. Note B flat. Diameter 851mm (33½in). Weight 366kg (7cwt 23lb). Cast without canons. Cast-iron headstock. Recast by Gillett & Johnston of Croydon.

RECAST BY GILLETT & JOHNSTON ~ CROYDON 1910

There is a band of decoration [58] below the inscription band. The clock strikes the quarters on this bell.

Former bell

Cast in 1858 by Mears, Whitechapel Bell Foundry, London. Diameter 864mm (34in).

G. MEARS FOUNDER LONDON 1858

5 1910. Note A flat. Diameter 937mm (36⅞in). Weight 483kg (9cwt 2qr). Cast without canons. Cast-iron headstock. Recast by Gillett & Johnston of Croydon.

RECAST BY GILLETT & JOHNSTON ~ CROYDON 1910

There is a band of decoration [58] below the inscription band.

Former bell

Cast in 1858 by Mears, Whitechapel Bell Foundry, London. Diameter 914mm (36in).

G. MEARS FOUNDER LONDON 1858

6 1910. Note G. Diameter 991mm (39in). Weight 527kg (10cwt 1qr 14lb). Cast without canons. Cast-

Details of the former fourth and fifth bells from the Sales Day Book of Whitechapel Bell Foundry (Whitechapel Archives)

iron headstock. Recast by Gillett & Johnston of Croydon.

RECAST BY GILLETT & JOHNSTON ~ CROYDON 1910

There is a band of decoration [58] below the inscription.

Former bell

Cast in 1618 by Hugh Watts II of Leicester. Diameter 978mm (38½in).

IHƧ : NAZARENVS REX : IVDEORVM FILI : ~ DEI MISERERE : MEI 1618 [3]

(Jesus of Nazareth King of the Jews, Son of God, have mercy on me)

The letering was like [142]

7 1924. Note F. Diameter 1149mm (45¼in). Weight 876kg (17cwt 1qr). Cast without canons. Cast-iron headstock. Recast by Gillett & Johnston of Croydon.

RECAST BY GILLETT & JOHNSTON. ~ CROYDON. 1924.

On the waist: [76]

GILLETT & JOHNSTON. CROYDON. 1910.

There is a band of decoration [58] below the inscription band. The clock strikes the quarters on this bell.

Former bells

Cast in 1723 by Henry Penn of Peterborough. Diameter 1067mm (42in).

FRANCIS CLEEVE : WILL. MAIDWELL : ~ CHURCHWARDENS. ~ HENRY PENN MADE ME 1723 O O O

O O O indicates the impressions of three coins.

Recast by Gillett & Johnston of Croydon in 1910. Diameter 1113mm (43⁷⁄₁₆in). Weight 796kg (15cwt 2qr 19lb).

RECAST BY GILLETT & JOHNSTON
~ CROYDON 1910

This bell was cracked when the tenor broke away from its headstock in 1924.

8 1924. Note E flat. Diameter 1302mm (51¼in). Weight 1249kg (24cwt 2qr 9lb). Cast without canons. Cast-iron headstock. Recast by Gillett & Johnston of Croydon.

RECAST BY GILLETT & JOHNSTON.
~ CROYDON. 1924.

On the waist:

TO THE GLORY OF GOD THIS PEAL WAS RECAST
BY PUBLIC SUBSCRIPTION DEC. 1910
J. H. CHARLES. M.A., R.D. VICAR.

W. M. KEAL ⎱
G. PHILLIPS ⎰ CHURCHWARDENS

There is a band of decoration [58] below the inscription.

George Phillips, who lived adjacent to the church in Church Passage, died on 31 May 1924 at the age of 67. On the evening of his funeral the bellringers of All Saints' Church rang a 'date' peal of 1924 changes

> as a tribute to the efforts of their late friend and colleague in connection with the re-casting and re-hanging of the ring of eight bells. The ringers were A Lee, A Ward, J Wheeler, W E Higgs, R Grinter, T H Wheeler, S Towell, and T Scott. For all who heard it, the beautiful, sombre, echoing effect of the half-muffled peal must have come as a heart-warming tribute to George Phillips's thirty-three years of unselfish service on behalf of the county he had adopted as his own (Coyne 2000, 443-4).

The clock strikes the hours on this bell.

Former bells

Cast in 1677 by Toby Norris III of Stamford. Weight 1016kg (20cwt) (Whitechapel Archives).

GOD SAVE THE KING.
~ TOBIE NORRIS CAST ME 1677

Recast in 1875 by John Taylor & Co of Loughborough. Diameter 1238mm (48¾in).

J TAYLOR & CO BELLFOUNDERS
~ LOUGHBOROUGH 1875.

C A Stevens (1905-06, 27) confirmed that this bell 'was perfectly sound in 1860'. Recast in 1910 by Gillett & Johnston of Croydon. Diameter 1245mm (49in). Weight 1159kg (22cwt 3qr 8lb).

TO THE GLORY OF GOD THIS PEAL WAS RECAST
BY PUBLIC SUBSCRIPTION. DEC. 1910.
J. H. CHARLES. MA. RD. VICAR.

W. M. KEAL ⎱
G. PHILLIPS ⎰ CHURCHWARDENS

The crown broke away from this bell in 1924.

PRIEST'S BELL

1840. Diameter 438mm (17¼in). Canons retained. Timber headstock. Hung for chime ringing. Founder un-

The new tenor at the Croydon Bell Foundry in 1910 (Phillips 1911-12, 41)

known, but probably recast by the Whitechapel Bell Foundry, London.

H STINSON ⎱
J RUDKIN ⎰ CHURCH WARDENS 1840

Former bell

Possibly cast by Edward Arnold of Leicester in 1793. A notice in the ringing room reads:

> The ninth bell, known as the Sanctus bell, sits on the frame above the 4th, is for chiming only - in the past the bell would have been chimed 33 times before a service, to indicate the number of years that Christ lived on earth. The treble bell is now used.

HANDBELLS

There is a set of nineteen handbells in the ringing room.

BELLRINGING CUSTOMS

There was scarcely a public or private event on which the bells were not rung, and the anniversary of what were considered red letter days in the history of the country was always marked by a rousing peal. Among the Churchwarden's Accounts for Oakham are to be found numerous items for payments for bell-ringing, all having historical associations (Phillips 1911-12, 41).

The available accounts enumerate regular payments made to the ringers during the eighteenth century, many of which continued through to the beginning of the 1900s. The following extracts highlight some of the bell-ringing customs at Oakham during the nineteenth century. Those listed from 1805-14 are from *Rutland Magazine* **V** (Phillips 1911-12, 41):

1805	Nov 7	Paid ringers for Nelson's victory [Battle of Trafalgar]	£1 1s 0d
1806	June 4	His Majesty's Birthday [George III]	6s 0d
	Sept 22	His Majesty's Coronation Day [George III]	6s 0d
	Oct 25	His Majesty's Accession Day [George III]	6s 0d
	Nov 5	Powder Plot	6s 0d
	Dec 25	Christmas Day	6s 0d
1807	May 29	To ringers on Restoration [of Charles II]	6s 0d
1809	June 6	Paid ringers for Good News	£1 1s 0d
1813	June 21	Paid for Lord Wellington's Victory [over the French at Vitoria]	£1 1s 0d
1814		Paid Ringers for News of Peace [First Peace of Paris]	£2 2s 0d
1828	June 5	[At a Vestry Meeting] 'that the Ringers be allowed for Ringing upon the Kings Birth Day [George IV] one Sovereign.'	
1830	June 26	Passing and funeral Bell for the King [George IV]	3s 0d
1831	April 6	Ringing the 8 O'clock Bell	£1 5s 0d
		Tolling the Bell for 3 Vestry Meetings	3s 0d
1861	Sept 30	Recd of the Churchwardens the sum of Ten Shillings for half years Chiming on Saints Days and during Lent	
	Oct 7	Recd of the Churchwardens the Sum of Four Pounds for half a years chiming	
	June 25	Received of Churchwardens of Oakham the sum of Twenty one shillings for Ringing at Confirmation	
1862	March 1	Received of Mr Burn the sum of £1 for Bryan ringing gleaning Bell two years	
1881	April 14	Received of the Churchwardens the Sum of one pound fifteen shillings for Ringing Curfew Bell & Chiming During Saints Days & Lent	

On Sundays the third bell was rung at 9am and again after Morning Service. For Divine Service the bells were chimed for twenty minutes, the Sermon Bell for seven minutes and after that the Priest's Bell for three minutes. The seventh bell was called the Meeting Bell as this was rung to call town meetings. The Pancake Bell was rung on Shrove Tuesday, the Gleaning Bell at 8am and 6pm during harvest, and the Curfew rung from Old Michaelmas Day to Old Lady Day. Peals were rung on Christmas Eve and New Year's Eve. At the Death Knell

thrice three tolls were given for a male and thrice two tolls for a female, both before and after the knell (North 1880, 147).

The Curfew, Gleaning and Pancake Bells were still being rung *circa* 1910. The Meeting Bell had been discontinued except that it was still rung for Vestry Meetings. The bell noted by Thomas North as being rung after the Morning Service was popularly called the Pudding Bell, but it had been discontinued by 1910. It is said that it was rung to give notice to the housewives so that they might have dinner ready by the time the congregation reached their homes (Phillips 1911-12, 42-3).

The Pudding Bell was certainly being rung in 1820. At this date there was a large clock in front of the single gallery at Oakham Church and when the vicar saw the time approaching one o'clock he abruptly concluded his sermons. One of the parishioners asked why he did this and his short answer was: 'Sir, I'll spoil no poor man's pudding' (Traylen 1982a, 19).

The earliest date recording the possible use of the Gleaning Bell in Oakham is in 1748 and found in an extract taken from the Court Rolls. Amongst the Ancient Pains, Orders and Bye-Laws established by the Manor and Castle or Lordship of Oakham is this order: 'that no person shall go into the Wheat Field to glean before seven o'clock in the morning or continue therein after six o'clock in the evening' (Phillips 1909-10, 31).

At the very end of the nineteenth century William Higgs joined a newly-formed bellringing band under the captaincy of Mr Needham, and this band took up change-ringing for the first time in Oakham. Mr Higgs records:

The first most important event I rang for was the relief of Mafeking, then the peace rejoicing at the end of the Boer War, the funeral of Queen Victoria, and the Coronation of King Edward VII. I have rung for most of the great Royal and National occasions since.

There is a memorial to him in the ringing chamber:

> **Memorial to**
> **W^m. Edgar Higgs**
> **by his fellow bellringers and friends**
> **to commemorate 60 years devoted service**
> **to the church, as chorister and bellringer.**
> **Died 17th JAN 1956**

At the time of writing the treble is chimed thirty-three times for the Sunday 8am and 6pm services. For the Main Eucharist at 10.30am the bells are rung for forty minutes, then the treble is chimed for five minutes before the service. The treble is chimed at the Elevation of the Host during the Eucharistic Prayer. Handbells are rung at 'Gloria' on Maundy Thursday and Easter Day.

Bells are rung on Christmas Day, and on New Year's Eve at midnight the tenor is tolled then all the bells are rung for twenty minutes. For the Armistice Day Service the bells are rung muffled, and they are also rung for the Parade Service at 3pm.

The bells are occasionally rung for funerals and then only by request, when the seventh bell is tolled. For weddings the bells are rung for approximately twenty minutes when the bride and groom leave the church. Other bellringing occasions include Feast Days, such as Ascension and All Saints, and the Civic and Memorial Services.

RINGERS' RULES

Bell Ringers' Rules dated 17 December 1897 and titled **ALL SAINTS' OAKHAM, BELFRY BYE-LAWS** list seven rules, two of which are:

5. Any Member absent from, or more than 10 minutes late for, practice without reasonable excuse, or not punctually attending Sunday Chiming, in his own person or by an approved substitute when summoned, shall be fined 3d.

6. The Fees received for Ringing on any Special Occasion shall be equally divided amongst the Ringers on that occasion, after 5 per cent has been deducted and paid to the Foreman.

BELLMAN

The following entry is taken from notes on the bells by G Phillips (DE 3178/40):

Circa 1815
Paid for making the Bellmans fresh coat   6s 6d

By 1910 the official position of Bellman had almost

disappeared in England but in Oakham it was noted that 'we have the Bellman still in evidence' (Phillips 1909-10, 28). Members of the Ellingworth family were town criers of Oakham for over three hundred years. Documents retained by the family include a list of locations where the crier used to cry. These included: Market Place against the Pump, the centre of Penn Street, at the back of the Crown, Mill Street at the South Street end, and Northgate Street at Church Street corner (*Rutland Times* 1 June 2001, 9).

CLOCK HISTORY

The clock dial shown in an engraving of *circa* 1684 of Oakham Church (Wright 1684, 99) is the only evidence of an early clock here. Entries in the Churchwardens' Accounts, which begin in 1776, provide some details but it is more than likely that the early clock had already been replaced by then. The diamond-shaped dial supports the possibility that this early clock may well have been of the type installed by John Watts of Stamford. The following extracts from the accounts confirm that there was a working clock here at the end of the eighteenth century and that the dial only had an hour hand:

1786	April 29	paide Francis Dunstone for Looking after the Clock and chimes and washing the linen and surplus	£4 12s 6d
1792	July 7	paid Clark for gildin the balls [?] and clock Hand	10s 6d
	Oct 29	paid Mr Burton for his Opinion on Dial plate	5s 0d

The earliest known illustration of a Rutland church with a clock dial is an engraving of Oakham Church in James Wright's The History and Antiquities of the County of Rutland *first published in 1684. This illustration is taken from the engraving and shows a diamond-shaped dial with one hand on the south face of the tower. Inset: an enlargement of the dial*

The timber diamond-shaped dial with its single hour hand may have survived until this date. A drawing of *circa* 1839 shows that by then there was a solid round dial with a bezel, and both hour and minute hands (Uppingham School Archives). The following extracts from the Churchwardens' Accounts show that there was a chime barrel at All Saints' prior to the installation of the present clock in 1858:

1783	June 24	Paid Sam Read for the Clock and Chimes for 10 Weeks from Lady Day to the 4 of June	11s 0d
1793	July 31	paid for loadin chimes	1s 0d
	Aug 8	paid Mr Adcock for carrige of chimes to Leicester	£1 11s 6d
	Aug 17	help for unloadin the chimes and getting them up into the loft	6s 0d
	Sept 9	paid Mr Arnold's Bill	£24 0s 0d

Specific details of the chime barrel have not survived, but normally tunes would be played on the church bells every three hours, the tune being changed automatically every day. The only other chime barrel recorded in Rutland was at Cottesmore which was removed in 1867. There is a chime barrel at Stapleford Park Church, Leicestershire, which is some eight miles due north of Oakham, made and installed in 1785 by Edward Arnold, a bellfounder and clockmaker of Leicester. It is entirely separate from the clock except that originally it was tripped by a wire from a lever on the clock. It can play six tunes and the names of these are recorded on a brass plaque on the frame.

The Stapleford chime barrel is almost exactly the same as that at Kings Norton in Leicestershire, built in 1765 by the clockmaker and bellfounder Joseph Eayre of St Neots and Kettering, except that it plays nine tunes. Edward Arnold took over the Eayre business in 1772 before moving to Leicester. The fact that the chime barrel was sent for repairs in 1793 to Edward Arnold of Leicester suggests that either he, or his predecessor, was the maker. It is also quite possible the chime barrel was supplied together with a new clock sometime before the beginning of the surviving accounts.

John Simpson was responsible for repairing and maintaining the church clock and chimes in 1793. Vouchers made out by his son, Stephen Simpson, a watch and clockmaker of Oakham, show that the chime barrel was being used until just before the present clock was installed in 1858. Stephen Simpson's shop was in Market Place and he is known to have been working from 1828 until 1867:

1853		to the Chimes repairing Chimes [and other work]	£2 0s 0d
		To Winding up the Clock & Chimes for twelve months	£4 0s 0d
1854	Feb 20	to mending the wires of the Chimes & oiling all over	2s 6d

| **March 25** | To Winding up the Clock & Chimes for twelve months | £4 0s 0d |

In 1858 the present clock, together with two new skeleton dials, was supplied and installed by Frederick Dent of London. It was part of the general restoration of the church then being carried out by Messrs Ruddle & Thompson (Traylen 1982a, 19) under the direction of the architect Sir Gilbert Scott. This three-train flatbed movement, with 'quarter chimes of Great St Mary's, Cambridge' [more often referred to as Westminster Chimes] was financed partly by rate and partly by subscription. The original bill for the clock has not survived but it is known that in June 1858 the churchwardens of Oakham were presented with £120 from the Vestry Meeting for the purchase of a new clock 'to chime the quarters and to have two faces, one on the south and one on the west of the tower' (Traylen 1982a, 19). Some details of the installation work, mainly in connection with the stone corbels which support the clock, are contained in the Church Restoration Accounts (DE 2694/708/31):

Time and Material Connected with the Clock
Making Scaffolds, restoring of string courses, fixing & working of stone corbels, cutting holes through Walls and attending & assisting Man to fix the Clock.
Mason 13 days 7 hours. Lab [Labourer] 11 days   £5 1s 6d

Stephen Rodely's bill for repairs to the chiming train of the new clock at Oakham Church (DE 2694/276/63)

Cube of Mansfield Stone Corbells	£1 4s 9d
2½ Bushells of Cement	10s 0d
Cube of Ketton Stone	1s 6½d
	£6 17s 9½d

Carpenter attending on Clock man Making Stages
for Cranks [bell hammers]
Time and Materials   **£11 2s 0d**

Edward Dent (1790-1853) and his nephew Frederick Rippon Dent (1808-60) had both been involved in making the Great Clock for the Palace of Westminster [often referred to as Big Ben]. This clock, although completed by 1858, was still in the Dent workshops when the Oakham clock was being assembled. Sir Edmund Beckett Denison, Lord Grimthorpe, was technical adviser to the Dents and many of his ideas were incorporated into the Westminster clock. He was also involved with the Oakham clock and this is confirmed in *Notes on Oakham Church Restoration* by C A Stevens, a former Curate of Oakham. Of the clock, he says:

Partly by Rate and partly by Subscription, a new Clock with skeleton faces by Dent was provided, with the chimes of St. Mary's, Cambridge. These chimes should not be called by the somewhat contemptuous term of 'Quarter jacks,' a name usually confined to 'ding dongs.' I should much like to know whether the clock, which was of special construction recommended by Mr. E. Beckett Denison (Lord Grimthorpe), has done justice to its inventor and maker during its 45 years use (Stevens 1905-06, 25).

By 1861 Stephen (Simpson) Rodely had taken over responsibility for maintaining and repairing the church clock. His premises were in Market Place (White 1863, 839). His bill of 1861 confirms that the new clock had quarter chimes.

By 1910, Charles Payne, another Oakham clock-maker, was looking after the clock and by 1913 his fee had more than doubled:

1910   March 30   C Payne Clock winding etc £2 10s 0d

The All Saints' clock by Frederick Rippon Dent. The weights, which are suspended from pulleys above the clock, cannot fall below the level of the clockroom floor. As a result of this limited fall the clock has to be wound every 48 hours. The single four-legged gravity escapement developed by Lord Grimthorpe was adopted for this clock. The particular advantage of this escapement is that it isolates the varying effects of wind on the hands from the clock and hence improves time keeping. A number of versions were tried on the Westminster clock whilst it was still in Dent's workshops, and the double three-legged gravity escapement subsequently became the standard for all good quality turret clocks

This photograph shows the newly regilded west-facing clock dial being hoisted back into position circa 1930 (DE 2694/241). As the handcart shows, the work was carried out by Billows & Son who had premises at 20 Northgate Street (Kelly 1925, 736). The dials were again regilded in 1993. This work was paid for by a grant from the Town Council as it is the only public clock in Oakham

CLOCK DETAILS

Details of the present clock:

Maker:	Frederick Dent, London
Signed:	Plate on the frame: Dent London 1858
Installed:	1858
Cost:	Not recorded, but in June 1858 the church-wardens of Oakham were presented with £120 from the Vestry Meeting for the purchase of a new clock
Frame:	Cast-iron flatbed
Trains:	Going, striking and quarter chiming
Escapement:	Single four-legged gravity
Pendulum:	Wooden rod and cast-iron cylindrical bob
Rate:	52 beats per minute
Striking:	Countwheel hour-striking
Chiming:	Cambridge quarter chimes
Weights:	Cast-iron
Winding:	Hand wound every 2 days owing to the limited fall of the weights. Winding reduction gear on all three trains
Dials:	Two skeleton dials by Dent with gilded roman numerals and hands
Location:	South and west faces of tower

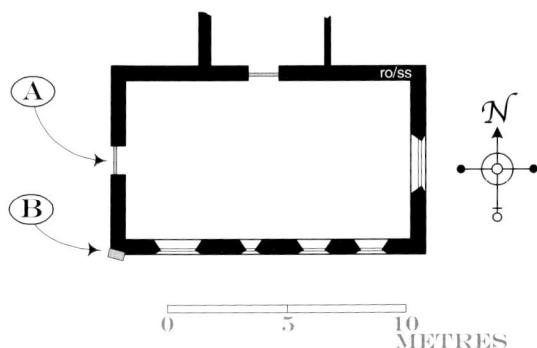

Plan of the Chapel of St John & St Anne
A: One bell in a bellcote rung from inside the chapel.
B: Sundial

CHAPEL OF ST JOHN & ST ANNE

The early fourteenth-century chapel of the Hospital of St John and St Anne was in existence when the Charity was founded in 1399 by William Dalby, a wealthy wool merchant. Most of the other buildings on the hospital site were demolished in 1845 when land was acquired for the construction of the Syston to Peterborough Railway. The chapel was renovated in 1983 when the Charity's Westgate flats were erected on adjacent land.

CHAPEL RECORDS

Details concerning the present and former bells are contained in the Book of Decrees 1663-1774 (DE 2694/813). Later accounts are held by the Charity.

BELL HISTORY

There is a single bell in a bellcote over the west wall of the chapel dated 1744. The Book of Decrees shows that there was a bell prior to this. Entries for the period 1723-7 include 'mending the bell 4s 0d' and the purchase of two bellropes. Bellropes were also purchased in 1739 and 1741. The accounts include the following, confirming that the old bell was recast:

1744	Carriage of the bell to Kettering	1s 8d
	Mr Eayres as per bill	£2 15s 0d

At about this time the buildings started to fall into disrepair, as there were insufficient funds for their proper maintenance. However, the accounts for Exposita Extraordinaria [additional expenses] show that bellropes were purchased up to 1785 despite the lack of funds, and the chapel continued to be used (Parkin 2000, 5 & 19):

1784 Jan	Smith for Bell Rope	1s 4d

The bell was rehung by Nicholas Meadwell in 1983 as part of a renovation programme.

BELL DETAILS

1 1744. Diameter 318mm (12½in). Weight approximately 25kg (1qr 27lb). Canons retained. Timber headstock. Cast by Thomas Eayre II of Kettering.

T •: EAYRE •: A •: D •:• 1744: [51] [51]

SUNDIAL

There is a seventeenth-century limestone sundial at the south-west corner of the Chapel of St John & St Anne, Oakham. It is about 2.5 metres above ground and is set at an angle so that it faces due south. The gnomon was missing in 1935 (*VCH* **II**, 23) but the sundial was restored when the chapel was renovated in 1983.

The sundial at St John & St Anne's Chapel

BELL RINGING CUSTOMS

Although the ringing of the Curfew Bell was abolished by Henry I in 1100 the custom still continued in Oakham up to 1881 as shown in the Chapel Accounts. Another such bell could be heard in the town until about 1910 as confirmed by Miss D Ellingworth, BEM, of Oakham. She recorded her memories of Oakham at that time including: 'The sounds of the day were the song of the birds, the rhythm of the anvil at the Blacksmith's shop next door (Wiggintons), the 'Angelus', the Crown Hotel fly passing to meet the London slip train at 7 o'clock, and the curfew from St. John's Church at 8 p.m.' (Traylen 1982a, 19).

ST JOSEPH'S ROMAN CATHOLIC CHURCH

A Roman Catholic church dedicated to St Joseph & St Edith was built in 1883 by the third Earl of Gainsborough in Mill Street, Oakham. In 1975 it was replaced by a larger church, dedicated to St Joseph, in Station Road. This new church has a small bell in a tubular steel tower. The bell appears to have no inscription and is chimed by an electric clapper.

OAKHAM CEMETERY

Oakham Cemetery, in Kilburn Road, was built in 1860. At its centre are two chapels linked by an arch over which is a slender tower. There is a bell in the upper part of the tower but its details have not been recorded.

Oakham Cemetery has a bell in the central tower

CATMOSE HOUSE & THE COTTAGE

A small limestone bellcote on the south face of a west wing at Catmose House (now Rutland County Council Offices) once supported and protected the house bell. The early 1900s OS Second Edition 25 inch map shows that there used to be table sundials in the grounds of Catmose House (SK 863086) and in the garden of the property then known as The Cottage (SK 862085).

The bellcote on Catmose House, Oakham

From a postcard dated 1907 showing a table sundial on the circular lawn at Catmose House, Oakham (RCM H15.1981)

The arch of Stephen Simpson's long-case clock (RCM 1969.413). It has a painting of the Oakham Union Workhouse, with a central bell turret. This building is now part of Oakham School

OAKHAM UNION WORKHOUSE

Stephen Simpson, a clockmaker whose premises were in the Market Place, was commissioned by the Guardians to make a longcase clock for the Oakham Union Workhouse (Clough 1981, 82-3). This clock, which is now in Rutland County Museum (1969.413), has a painting of the former workhouse in the arch. It shows that there used to be a central bell turret over the front facade and it is known that this was rung to call the poor of the town for their soup at meal times (information from David Bland). The building is now known as Schanschieffs, part of Oakham School.

OAKHAM SCHOOL

Over the western entrance to the former canal wharf buildings, now known as Old Stables, and overlooking the sports field, there is an illuminated clock dial with black roman numerals and the motto QUASI CURSORES [Like Runners]. This motto is also included in a carved plaque over an entrance to the School in Market Street. The Gent master clock sends a pulse every thirty seconds to each of two slave dials and this can be ob-

served in the movement of the minute hands. The date of installation is not known, but it is probably late 1950s. Also at Oakham School there is a louvred bell turret over School House in the Market Place.

HIGH STREET

The vertical sundial with the motto 'Tempus Fugit' [Time Flies] located on the former Matkin's building at 13 High Street, Oakham, is perhaps the best known sundial in Rutland.

MARKET PLACE

The late sixteenth-century Buttercross in Market Place has a cuboid sundial. It consists of a block of limestone with dials facing the four cardinal points of the compass.

One of two slave dials of a Gent electric master clock system located in the physics laboratories of Oakham School, the other dial being located in the physics quadrangle

The vertical sundial in High Street, Oakham

When this photograph of the Buttercross in
Oakham Market Place was taken circa
1920 the dials were probably in working
order (RCM 1984.51.35)

The Buttercross sundial in 2000 is in
poor condition. All the gnomons
are missing. The dials originally
had painted numerals and lines

PICKWORTH

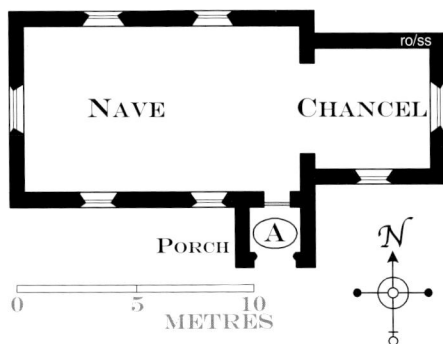

Plan of All Saints'
Church
A: One bell on the
porch roof rung from
inside the porch

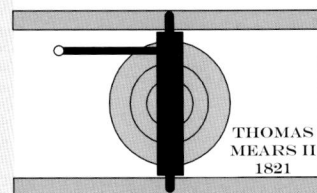

Plan of the
bellframe at
Pickworth
Church

ALL SAINTS   Grid Ref: SK 992138

PARISH RECORDS
There are no relevant records available.

BELL HISTORY
The present church was built in 1821 and was conse-
crated in 1824. In 1880 Thomas North recorded that it
had 'one small bell in a gable'.

It is said that the original fourteenth-century church
was destroyed during the Battle of Losecoat Field in
1470 and finally demolished in 1731 when the tower
and spire were taken down. Only an arch remains in a
nearby private garden. The tower of this early church
would have had at least one bell.

BELL DETAILS

1 **1821**. Note G. Diameter 559mm (22in). Weight
about 127kg (2cwt 2qr). Canons retained. Timber
headstock. Cast by Thomas Mears II, Whitechapel Bell
Foundry, London (Pearson 1989). No inscription.

BELLRINGING CUSTOMS
The bell used to be rung a few times at the end of the
Morning Service and after the first publication of the
Banns of Marriage (North 1880, 147). Today the bell is
chimed before Holy Communion.

SUNDIAL COTTAGE
The Bluebell Inn closed in the late 1950s. It is now known
as Sundial Cottage after the sundial over the front door.

PILTON

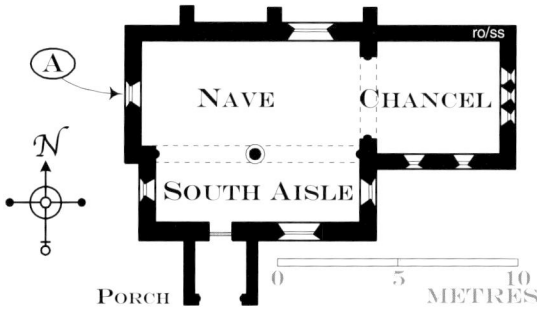

Plan of St Nicholas's Church
A: Two bells in a double bellcote rung from inside the church

Pilton bellcote. Inverted cones prevent water running down the wires into the church

ST NICHOLAS Grid Ref: SK 915029

PARISH RECORDS

The only relevant documents are the Churchwardens' Accounts 1877-1901 and 1902-42 (DE 2505/4-5).

BELL HISTORY

The bellcote is of thirteenth-century origin, indicating that there were medieval bells at Pilton. It was rebuilt during the general restoration of 1877-78 when some of its original character may have been lost (*VCH* II, 212).

An article in *Rutland Magazine* (Phillips 1907-08, 38) points to the origin of the bells at North Luffenham:

... the two Bells formerly resting in the framework above the present five bells were sold, or given, many years ago to Pilton Church. As a matter of fact, there are two small Bells in the Pilton turret. There are no inscriptions on them.

They may have been installed at the 1877-78 restoration when the bellcote appears to have been rebuilt (Dickinson

1983, 88). However, a survey of the North Luffenham belfry by John Taylor suggests that the installation date was after 1909 (*see* North Luffenham, Bell History).

The following are the only items of note in the accounts relating to the bells:

1877	Oct 13	R Littledyke for Bellropes	10s 6d
1911		Extra for Tollng Bell	1s 0d
1913		Mears for repairing bells	£4 10s 6d

William Mears was an ironmonger with a shop in the

From a drawing of circa 1793 showing that the upper part of the bellcote at Pilton was then missing. There appears to be only one bell and it had a bellwheel (RCM F10/1984/44). A drawing of circa 1839 (Uppingham School Archives) shows the bellcote to be in much the same condition but two bellwheels are evident

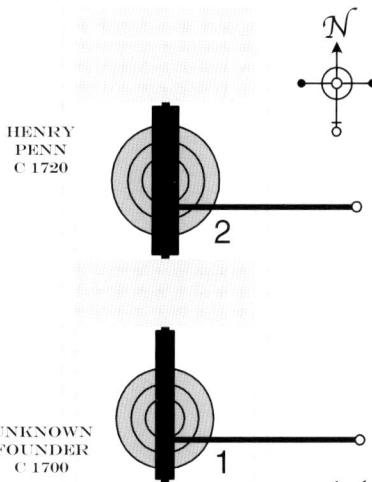

HENRY PENN C 1720

UNKNOWN FOUNDER C 1700

A plan of the bellcote at Pilton Church. The bells are hung for chime ringing

Market Place at Uppingham (Kelly 1912, 681).

By 1984 the bells were in a poor state, particularly the treble which was hanging out of true and its clapper missing. They were fully restored by Nicholas Meadwell in 1990, and this included fitting new headstocks and clappers. Both headstocks are inscribed **NJM 1990**.

BELL DETAILS

1 *Circa* **1700**. Diameter 483mm (19in). Weight approximately 89kg (1cwt 3qr). Canons retained. Timber headstock. Unknown founder. No inscription or decoration. Thomas North incorrectly reported that this bell was cracked (North 1880, 147).

2 *Circa* **1720**. Diameter 533mm (21in). Weight approximately 114kg (2cwt 1qr). Canons retained. Timber headstock. Ascribed to Henry Penn of Peterborough, based on the width of the inscription band (George Dawson). No inscription or decoration.

BELLRINGING CUSTOMS

On Sunday a bell was rung at 8am and for Divine Service both bells were rung followed by the Sermon Bell. At the Death Knell there were thrice three tolls for a male and thrice two tolls for a female, before the knell (North 1880, 148).

Today the bells are rung for Sunday services although there is no fixed order or time. They are also rung for weddings and funerals if requested.

The treble at Pilton in its bellcote just before a new headstock was fitted in 1990 (Nicholas Meadwell)

PRESTON

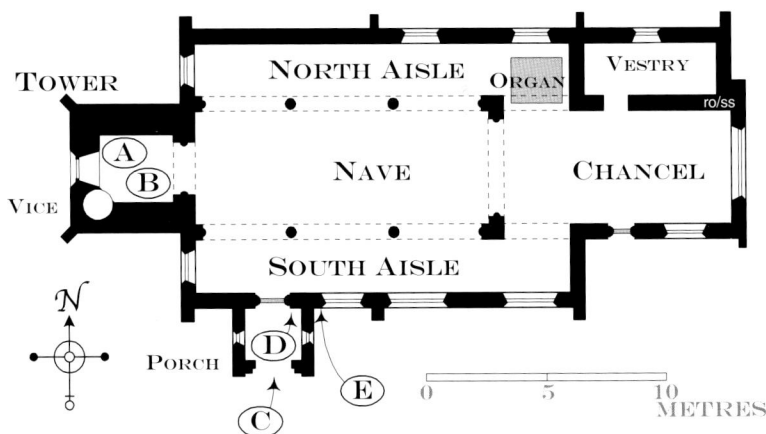

Plan of St Peter & St Paul's Church
A: *Six bells and a Sanctus Bell in the belfry. Ringing chamber at the base of the tower*
B: *Former clockroom on the first floor of the tower*
C: *Sundial*
D: *Scratch dial 1*
E: *Scratch dial 2*
The vice provides access to the former clockroom and belfry

ST PETER & ST PAUL   Grid Ref: SK 870024

PARISH RECORDS

The following documents in the parish archive have been searched:

Tithes and fees payable to the Rector of Preston (DE 2641/31).
Churchwardens' Disbursements from April to August 1738 (DE 2461/40).

Churchwardens' Accounts 1596-1792 and 1826-1921 (DE 2461/39 & 41).
Churchwardens' Vouchers 1812, 1854, 1861, 1866-8 (DE 2461/42).
A faculty for erecting two bells in the belfry dated 5 Jan 1909 (DE 2461/10).
Installation of a new treble, with covering letter 16 Dec 1963 (DE 2461/25/1-2).
Order of Service for the dedication of the new treble dated 31 May 1964 (DE 2461/29).

BELL HISTORY

An inventory attributed to the first half of the eighteenth century (Parkes 1984, 156) lists 'four bells and a clock' in the church, and Thomas North's survey (1880, 148) records that there were three large bells and a Sanctus Bell. The then second, known from its inscription as the 'Gabriel Bell', and the small Sanctus Bell are both thought to date from *circa* 1400. They are consequently two of the earliest bells in Rutland. Of the remaining bells, one was added at the end of the sixteenth century bearing a loyal inscription to Elizabeth I, and the other was cast in 1717.

A payment recorded in the Churchwardens' Accounts of 1726 to Thomas Slater 'for framing ye bells' confirms that a new bellframe was installed at this time. Timber for this frame was covered by a separate receipt:

1726 July 28
Receiv'd of Jon Bains Church Warden ye sum of four pounds ten shillings & eight pence for timber for ye bell frames by me Thomas Slater

Payments were also made to Thomas Slater, Benjamin Broome and a Mr Turner for 'Ironwork for the Bells'.

The surviving accounts show that during the second half of the seventeenth century various members of the Rawlings and Pulford families were paid for 'mending the bells' but no details are given of the actual work carried out.

Nineteenth-century disbursements record expenditure on bellropes supplied by Christian of Uppingham, S B Dennison of Oakham and Dexter of unknown location. Three villagers, Robinson, Charles Manton and William Sharpe, who were boot and shoe makers, were paid for mending the bellropes at various times.

John Taylor refurbished the three early bells in 1901, and although not specifically recorded, it is understood that they were rehung in a new metal high-sided frame at the same time. This now forms the lower level of the present frame. At a Vestry Meeting on 27 April 1908 the Parish accepted an offer by General Codrington to donate a new bell. The Meeting also agreed that the £51 already raised for this purpose should be used to acquire a second new bell. Although both bells were cast in 1908, the faculty for erecting them was not granted until 5 January 1909. A plan attached to the faculty shows that they were to be hung in a new low-sided metal frame for two above the existing bells.

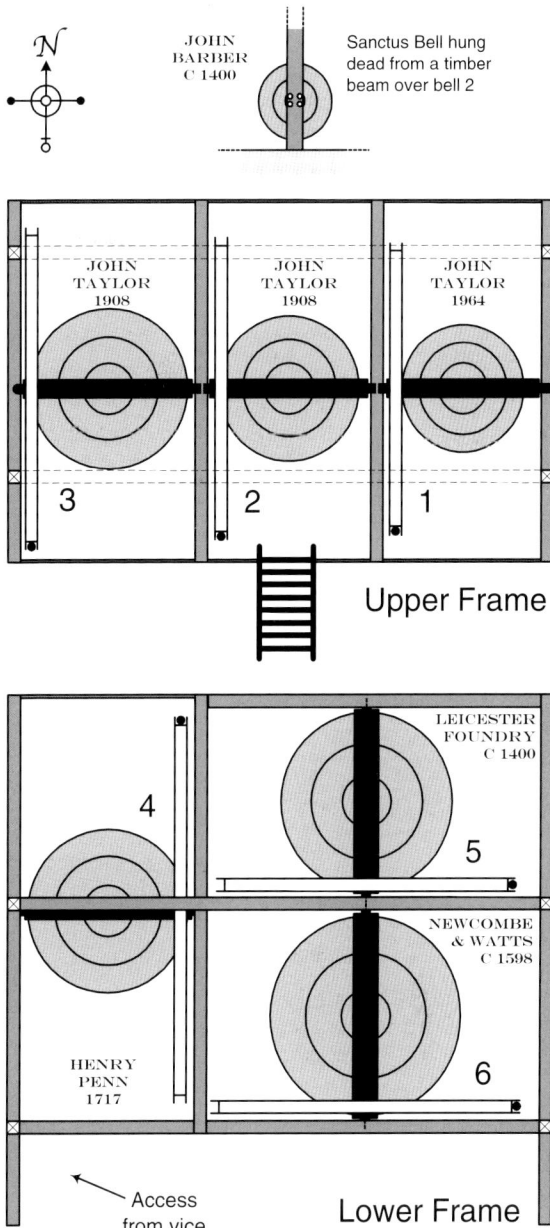

Plan of the bellframe at Preston Church

The following testimonial is included in John Taylor's brochure of *circa* 1920 (DE 2520/2):

Preston Hall, Uppingham
5th Feby., 09
Dear Sir,
The new bells were dedicated to-day, and everyone is delighted with them.
I am, yours faithfully,
(Major General) A.E. CODRINGTON

A new treble by John Taylor was added early in 1964 to make the current ring of six. This was donated as a memorial to William Melville Codrington by his family and hung in an extension to the upper frame. The Dean of Peterborough conducted the Dedication Service on Sunday 31 May 1964.

The medieval Sanctus Bell, dedicated to St Mary, is the only such bell to survive in Rutland. It was originally suspended from the keystone of the belfry east window (North 1880, 148). In 1964, in order to ensure its more secure preservation, it was hung dead from a beam in the belfry high above the other bells. Its clapper has been missing since before 1880. The beam is dated 1897 but this does not appear to have any significance with regard to this or the other bells.

BELL DETAILS

1 **1964**. Treble. Note F. Diameter 645mm (25⅜in). Weight 183kg (3cwt 2qr 11lb). Cast without canons. Cast-iron headstock. Cast by John Taylor & Co of Loughborough.

JOHN TAYLOR & CO [65] FOUNDERS [65]
~ LOUGHBOROUGH [65] 1964 [65]

On the waist:

THE FAMILY OF
~ WILLIAM MELVILLE CODRINGTON
(1892-1963) GAVE ME IN HIS MEMORY
MY VOICE SHALT THOU HEAR BETIMES
~ O LORD.

2 **1908**. Diameter 711mm (28in). Weight 229kg (4cwt 2qr 1lb). Cast without canons. Cast-iron headstock. Cast by John Taylor & Co of Loughborough.

Goodwill toward men [126]

On the waist:

Preston Parishioners gave me
[65] 1908 [65]

On the waist opposite:

[94]

There is a band of decoration [54] below the inscription band. The capital G on the inscription band is like [61] and the lower case letters like [113]. The capitals on the waist are like [130] and the lower case lettering like [82].

3 **1908**. Diameter 762mm (30in). Weight 270kg (5cwt 1qr 7lb). Cast without canons. Cast-iron headstock. Cast by John Taylor & Co of Loughborough.

Glory to GOD in the highest [126]

Preston

One of the few examples of a village being named on a Rutland bell

Henry Penn's date on the fourth bell at Preston Church

On the waist:

Alfred E. Codrington gave me
[65] 1908 [65]

On the waist opposite:

[94]

The capitals and lower case lettering on the inscription band are like [61] and [113] respectively. The capitals and lower case lettering on the waist are like [130] and [82] respectively. There is a band of decoration [54] below the inscription band.

There is a memorial plaque on the nave wall, north of the tower arch, to Lieutenant General Sir Alfred Edward Codrington GCVO KCB of Preston Hall, who died 12 September 1945.

4 **1717**. Diameter 819mm (32¼in). Weight 293kg (5cwt 3qr 3lb). Canons removed. Cast-iron headstock. Cast by Henry Penn of Peterborough.

1717

North (1880, 148) recorded the date on this bell as 1771. He also recorded this bell and the present fifth, the then treble and second, in reverse order .

5 *Circa* **1400**. Diameter 899mm (35⅜in). Weight 398kg (7cwt 3qr 10lb). Canons removed. Cast-iron headstock. Ascribed to an unknown Leicester founder.

[33] GABRIEL

Letters are like [89]. The initial cross and letters are evenly spaced around the inscription band. Scheduled for preservation (Council for the Care of Churches).

6 *Circa* **1598**. Tenor. Note G sharp. Diameter 991mm (39in). Weight 518kg (10cwt 0qr 23lb). Canons removed. Cast-iron headstock. Ascribed to Newcombe and Watts of Leicester.

[23] GOD SAVE OUR QUEENE ELIZABETH [3]

All letters are like [16] and [17] with the exception of ʒ which is like [111].

This bell has indentations on the sound bow as a result of being struck by a clock hammer. Scheduled for preservation (Council for the Care of Churches).

SANCTUS BELL

Circa **1400**. Diameter 400mm (15¾in). Weight approximately 51kg (1cwt). Canons retained. No headstock. Clapper missing. Ascribed to John Barber of Salisbury (Pearson, 1989).

[15] S [132] MARIO

(St Mary)

The inscription on the medieval Sanctus Bell

The letters are crowned as in **[115]**. **O** appears to be an impression of an early lead token (information from Yolanda Courtney, Leicester City Museums). It may however be an Edward III [1327-77] groat as recorded on a bell at Swyncombe, Oxfordshire, which also had **[15]** and **[132]** (Ellacombe 1872, 137). On this basis the bell may be slightly earlier than indicated.

The only surviving medieval Sanctus Bell in Rutland. Scheduled for preservation (Council for the Care of Churches).

BELLRINGING CUSTOMS

An undated document, attributed to the first half of the eighteenth century, lists the tithes and fees payable to the rector of Preston. It also includes some of the clerk's bellringing duties:

> for ringing the Bell for a Funeral four Pence, for making a Grave a Shilling if they be Landholders or have a Coffin, but if they be Cottagers or have no Coffin they pay him four Pence for ringing the Bell and four Pence for a Grave, but if a Cottager have a Child die they pay him four Pence for ringing the Bell, and three Pence for making the Grave (Parkes 1984, 156).

At the Death Knell there were three tolls for a male and two tolls for a female, but this custom had ceased before 1880. On mornings when there was to be Divine Service and a sermon preached, the tenor bell was rung at 8am. If no sermon was to be preached it was rung at 8.30am. At the conclusion of the morning service the same bell was rung if Evening Prayer was to be said (North 1880, 148).

Another former custom was for a bell to be rung at 5am to call the parishioners to work (Traylen 1989, 34).

At the time of writing the bells are rung for ten minutes before the 11am and 6.30pm Sunday services. They are also rung for the Christmas services and for weddings and funerals if requested.

SCRATCH DIALS

Two scratch dials have been found at Preston Church. One is at a low level to the east of the main door and must pre-date the fourteenth-century porch that encloses it. The other is on the west side of the south aisle window immediately east of the porch. This scratch dial is on a fifteenth-century window frame.

SUNDIAL

There is a vertical sundial over the apex of the south porch. The dial is crudely engraved with arabic numerals on a 711mm (28in) square, 203mm (8in) thick limestone block. When it was restored by Nicholas Meadwell in the mid 1990s a new gnomon was fitted and the numerals repainted. It is probably mid seventeenth century and may have been installed at the same time as the church clock. It can be seen in both *circa* 1793 (RCM F10/1984/45) and *circa* 1839 (Uppingham School Archives) drawings of the church.

CLOCK HISTORY

Although there is no clock now, nor physical evidence of a former clock, a document attributed to the first half of the eighteenth century lists tithes and fees payable to the Rector of Preston and an inventory of the church furniture, including: 'four bells and a clock'. The Churchwardens' Accounts confirm that there was definitely a clock here in the middle of the seventeenth century:

1656-57   ffor a Clocke and setting it up   £3 5s 0d

In the same year an agreement to maintain this new clock was made with Henry Nicholls of Glaston. A copy of this interesting document is included in the accounts:

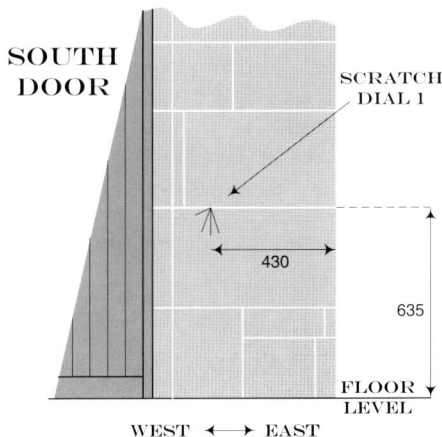

Location of scratch dial 1 at Preston Church

PRESTON 1				
Location	Inside porch on wall east of main door			
Condition	Good			
Gnomon Hole Diameter			Filled in	
Gnomon Hole Depth			Filled in	
Height above ground level			635	
Line Ref	a	b	c	d
Length	110	106	96	94
Angle (°)	164	180	200	232

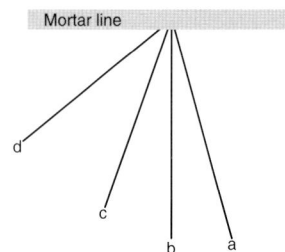

Scratch dial 1 details

1657 March 30

Agreed between the inhabitants of Preston, and Henry Nicholls of Glaston, that the said Henry Nicholls shall keepe the clock in good and sufficient repaire, and find all materials belonging to it and for the same he is to be payd five shillings yearly upon the Tuesday in Easter week.

Henry   Nicholls

HN

his   Marke

It seems quite likely that Henry Nicholls was also the supplier of the clock, but this is not confirmed in the records. The accounts confirm that Henry Nicholls was in fact paid five shillings each year for 'keeping the clock' until 1669. He was followed by William, Robert and then George Fox, all Uppingham clockmakers who were paid for 'keeping ye clock in repair' on various occasions between 1677 and 1738.

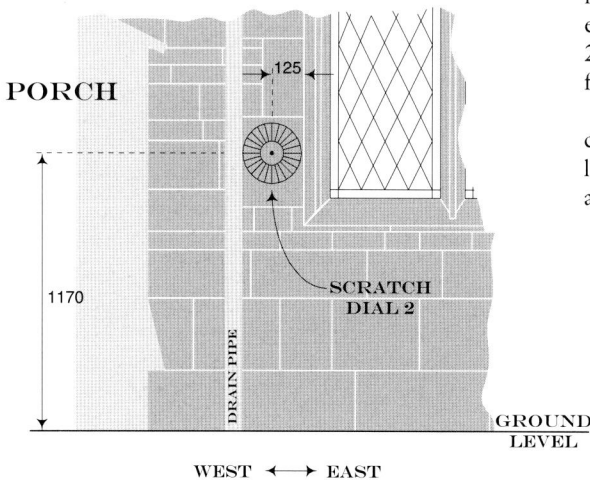

PORCH

Location of scratch dial 2 at Preston Church

It was generally the duty of the clerk to wind and lubricate the clock and amongst those recorded as having carried out this duty in the late seventeenth century were Thomas and Ralph Dale who were paid on separate occasions 'for keeping the clock in repair'.

Thomas Loseby, as clerk, looked after the clock from 1704 until 1719. From 1720 right through until 1793, this duty was carried out by Samuel and William Quenborough:

1690 April 24 To Thomas Dale for his clerks wages 10s
and for keeping ye clock 10s
£1 0s 0d

1710 April 14 Paid to Thos Loseby Clarke ... and for keeping ye clock oyl and wire for ye same 13s 4d

1769 March 28 Paid Wm Quenborough for Clarks wages washing ye surplice & looking after ye clock £1 7s 6d

The clock was repaired by William Aris II, of Uppingham in 1831 and 1832 but by 1835 the clock was evidently giving cause for concern. A Vestry Meeting on 28 April of that year records 'That the Church Clock be forth with put into good & sufficient repair'.

William, followed by his son Thomas, continued to carry out repairs until 1848. The following extract is the last reference to the clock and it is assumed that it was abandoned shortly after:

1848 Nov 15 Paid Mr Aris for repairing Clock 10s 0d

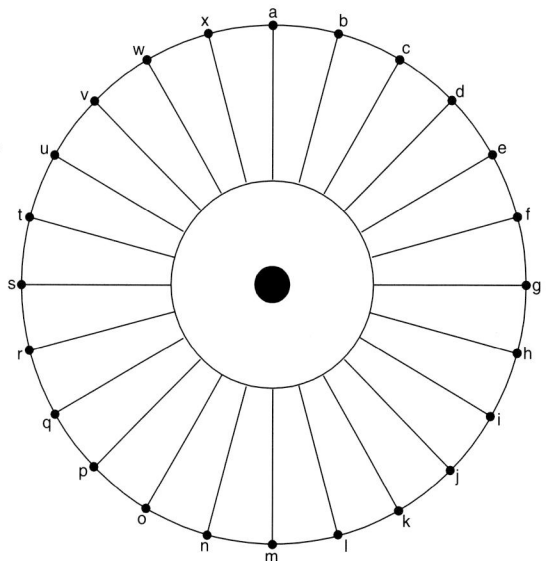

Scratch dial 2 details

PRESTON 2											
Location	West side of south aisle window east of the porch										
Condition	Average										
Gnomon Hole Diameter	18										
Gnomon Hole Depth	40										
Height above ground level	1140										
Inner circle diameter	100										
Outer circle diameter	250										
Pock circle diameter	250										

Line Ref	a	b	c	d	e	f	g	h	i	j	k	l
Length	75	75	75	75	75	75	75	75	75	75	75	75
Angle (°)	0	15	30	45	60	75	90	105	120	135	150	165

Line Ref	m	n	o	p	q	r	s	t	u	v	w	x
Length	75	75	75	75	75	75	75	75	75	75	75	75
Angle (°)	180	195	210	225	240	255	270	285	300	315	330	345

From a drawing of circa 1839 showing the porch sundial at Preston Church (Uppingham School Archives)

No details of this clock have survived. It was installed in 1656-57, just before the invention of the pendulum and anchor escapement. It would therefore have had the verge and foliot form of control and almost certainly a timber frame. It is difficult to believe that this clock would have survived in this form until the mid nineteenth century, as its inability to keep accurate time would by then have been unacceptable. However, there is nothing in the accounts to suggest that it was converted to the more accurate anchor escapement, and perhaps poor timekeeping and unreliability were the reasons for its eventual demise. It is unlikely that it had a dial, as there is no mention of repairing, repainting or regilding one in the accounts. Neither is a dial shown on any of the early drawings of the church.

There is an empty room on the second floor of the tower and it is thought that this was the former clockroom. When John Bains, churchwarden, replaced the bellframe

The restored sundial at Preston Church in 2000

in 1726 it was probably this room that was referred to when he was also paid for 'mending ye clock loft'.

SUNDIALS IN PRESTON

There is a seventeenth-century sundial on the gable end of a cottage at 10 Main Street, a former inn. Two other early sundials have been noted in the village, one on the south elevation of the Jacobean Manor House in Cross Lane, and the other on the front of Wings House at the north end of the village. When this latter dial was partially restored the missing gnomon was replaced by a replica of that on the church sundial.

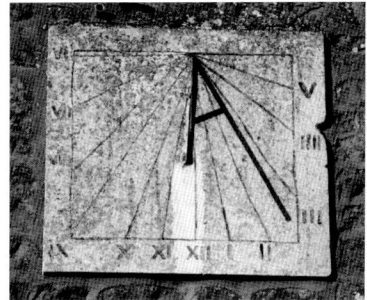

The seventeenth-century sundial at 10 Main Street, Preston

RIDLINGTON

ST MARY MAGDALENE & ST ANDREW
Grid Ref: SK 848027

PARISH RECORDS
The only relevant records are all dated 1903. They are: Accounts for rebuilding the church tower (DE 3468/ 31-3).

> Accounts and receipts from John Taylor for taking down and rehanging the bells, and for supplying a set of fifteen handbells (DE 3468/34-7).
> The Order of Service for the dedication of the church tower in July 1903 (DE 3468/30).

BELL HISTORY
The tower at Ridlington is mainly fifteenth century but

the lower part looks to be earlier. A middle buttress indicates that a bellcote was either built or intended (Pevsner 1984, 503). A former bellringers' doorway in the west wall of the ringing chamber is now blocked (*VCH* II, 94). In 1903 the upper stage of the tower was rebuilt, and there is a tablet on the west wall recording this.

In the same year, John Taylor & Co recast the three bells, but the original inscriptions were retained. Until this date the treble, cast *circa* 1510 by Robert Mellour of Nottingham, was one of the oldest bells in Rutland. The tenor was another early bell, cast *circa* 1595 by Watts of Leicester. The second bell by Thomas Norris of Stamford, was dated 1671. The recast bells were hung in a new two-level frame designed for four, and the ring of four

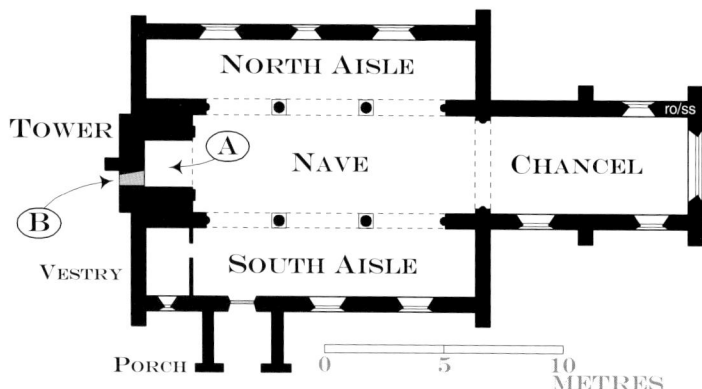

*Plan of St Mary Magdalene &
St Andrew's Church
A: Four bells in the belfry. Ringing
chamber at the base of the tower
B: Blocked bellringers' door
Access to the belfry is by two ladders*

*John Taylor's invoice for the bells and new frame at
Ridlington* (DE 3468/34)

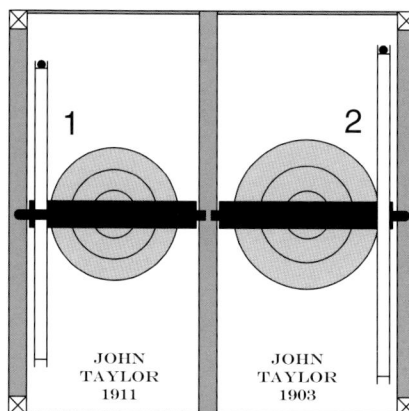

*Plan of the
bellframes at
Ridlington
Church*

Upper
low-sided
frame

Lower
high-sided
frame

was completed by the addition of a new treble by John
Taylor in 1911.

A card near the tower door records that it was not
until 1963 that the first quarter peal was rung on these
bells. It was a peal of 1296 Plain Bob Minimus completed in forty-three minutes for the Carol Service on
Sunday 22 December.

BELL DETAILS

1 1911. Treble. Diameter 660mm (26in). Weight
186kg (3cwt 2qr 18lb). Cast without canons. Cast-
iron headstock. Cast by John Taylor & Co of Loughborough.

> ELIZA. POWELL. TO GOD'S GREATER
> ~ GLORY FOR HIS MARVELLOUS
> ~ PRESERVATION.
> CHRIST THY GRACE UPON US POUR
> ~ [54] 1911 [54]

On the waist: [122]
The lettering is like [109].

2 1903. Diameter 699mm (27½in). Weight 213kg
(4cwt 0qr 21lb). Cast without canons. Cast-iron

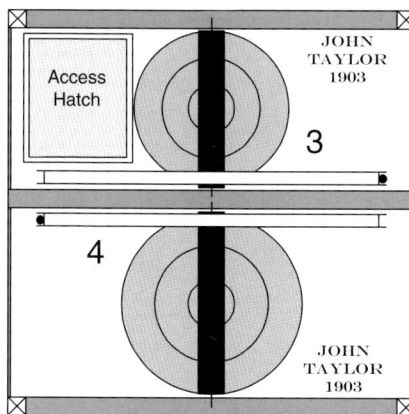

headstock. Recast by John Taylor & Co, Loughborough.

> [21]
> [6] ☙ [6] ☙ [6] ☙
> [9]
> MAY. WE. ALL. OUR. GOD. ADORE

On the waist:

> [105]
> 1903

The S in the inscription band is like [124]. The rest of

A drawing of the inscription and founder's marks on the second bell at Ridlington. The repeated S probably indicates that it was a pre-Reformation Sanctus Bell (Phillips 1905-06, 131)

the lettering is like [130].

Former bell
Circa 1510. Ascribed to Robert Mellour of Nottingham (George Dawson). Diameter 692mm (27¼in).

[21]

[6] ⌂ [6] ⌂ [6] ⌂

[9]

3 1903. Diameter 765mm (30⅛in). Weight 265kg (5cwt 0qr 25lb). Cast without canons. Cast-iron headstock. Recast by John Taylor & Co of Loughborough.

[52] 16 [10] 71 [52] I [52] WOODES
AND. FOR. MAD'S. GOOD. LABOVR. MORE
On the waist:

[105]

1903

The lettering is like [130]. A John Woods Sen and Jnr are listed as paying Hearth Tax in Ridlington in the year 1665 (Bourn 1991, 42). The lettering for Woodes is very irregular.

Former bell
Cast in 1671 by Thomas Norris of Stamford. Diameter 762mm (30in).

J. WOODES 16 [10] 71
Note that North (1880, 149) records **J** for **I**.

4 1903. Tenor. Note B flat. Diameter 864mm (34in). Weight 379kg (7cwt 1qr 24lb). Cast without canons. Cast-iron headstock. Recast by John Taylor & Co of Loughborough.

MASTER [3] TOMAS [3] HAXELRIG [3]
TILL. TO. THEE. OVR. SPIRITS. SOAR
On the waist:

[105]

1903

The lettering on the first inscription band is like [16] and [17]. The rest of the lettering is like [130].

Former bell
Cast by Watts [possibly Francis] of Leicester *circa* 1595 (Pearson 1989). Diameter 787mm (31in).

MASTER [3] TOMAS [3] HAXELRIG [3]
Thomas Hazelrigg was probably the donor of this bell. Thomas North in 1880 suggests that Thomas was the son of Miles Hazelrigg of Nosely in Leicestershire.

BELLRINGING CUSTOMS
On Sundays a bell was rung at 8am. For Divine Service the bells were chimed and the Sermon Bell, the tenor, was rung. At the Death Knell thrice three tolls were given for a male and thrice two for a female, both before and after the knell (North 1880, 149). The Gleaning Bell was rung morning and evening during harvest until about 1850. By this time more efficient mechanical reapers were being introduced and these left little in the fields for the gleaners to gather (Traylen nd, Pt 2). A Spur Bell which announced wedding banns was rung in the village up to the 1930s. The custom was known as Ringing the Spur (Traylen 1989, 35).

Today the bells are chimed for five minutes before every Sunday service and for the Harvest Festivals. Ridlington does not have its own band of bellringers but a local visiting band rang in the New Millennium. Outside ringers are also arranged for Carol Services, weddings and for the opening of the Street Fayre in June.

HANDBELLS
On Christmas Eve before the First World War, a group of handbell ringers would ring at strategic points around the village and in the surrounding locality. They usually walked to Coles' Lodge and then onto all the lodges along the Chater Valley through to Manton (information from Michael Gray).

SUNDIAL
A table sundial, dated 1614, in the grounds of a house in Ridlington (see below) is probably the original church sundial. The practice engraving on the reverse of the dial includes words to the effect that 'This dial belongs to Ridlington Church'.

John Taylor's invoice for a set of fifteen handbells for Ridlington Church (DE 3468/36)

According to Irons' Notes on the Visitation of 1681, the churchwardens were ordered by the Archdeacon to 'amend the dial in the churchyard'. This is probably a reference to the sundial (*VCH* **II**, 95).

CLOCK HISTORY

The earliest church clocks in Rutland had a verge and foliot escapement, no dial and probably a wooden frame. One of these clocks, and the earliest recorded in the county, was at Ridlington. Two items in Irons' Notes (MS 80/1/3) refer to it:

1618	The clock & chimes [chime barrel?] are out of repair and do not go
1620 Feb 22	Heny Gord Thos Hansworth to certifie the repaire of the clock

There is no longer a clock at Ridlington and there is no record of when it was removed.

RIDLINGTON SUNDIAL

A table sundial in the grounds of a house in Ridlington is thought to have been originally in the churchyard. The pedestal is part of a limestone window frame. The dial is 159mm (6¼in) square with a triangular gnomon 76mm (3in) tall, 70mm (2¾in) long and 3mm (⅛in) thick. The dial is inscribed 'Isaack 1614 Symmes' and 'The gift of Sir Willyam Bulstrode'. The engraver's initials in the border are 'TD' or 'DT'. The geometric design is carefully executed. The sun is personified by a simple face in the centre and the hours are divided into quarters.

The Symmes family is found locally in the sixteenth and seventeenth centuries, but no mention has been found of Isaack. Robert Symmys leased the lordship of the manor of Ridlington in 1517 and the wills of several other members of the family up to 1629 have survived. Sir William Bulstrode, described as 'of Ridlington Parva and Exton', was Sheriff of Rutland in 1604 and he was returned to Parliament as one of the two knights of the shire for Rutland, five times between 1620 and 1628.

The rear of the Ridlington dial has some very interesting practice engraving. Apart from decoration, letters and words there are two caricatures that appear to be contemporary with the other engraving. 'One is of a bald-headed man, whose neckwear may represent clerical or legal attire.' The other is of a man smoking a pipe. The dial is dated only thirty years or so after the general introduction to England of the custom of 'drinking' tobacco (Clough 1978, 69-72).

A drawing of the dial plate of the Ridlington sundial (Rutland County Museum)

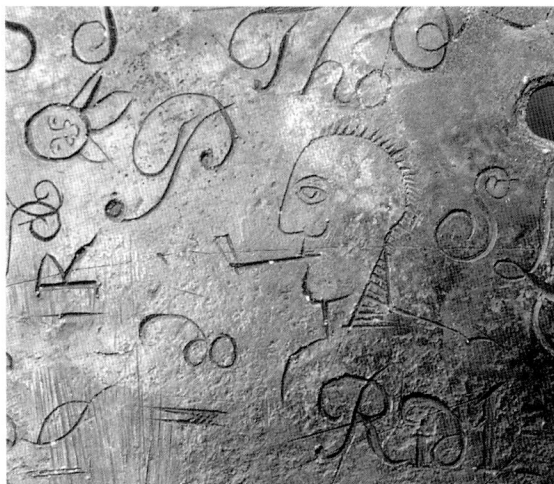

Some of the practice engraving on the rear of the Ridlington dial showing a bald-headed man in clerical or legal attire and an early reference to 'drinking' tobacco

RYHALL

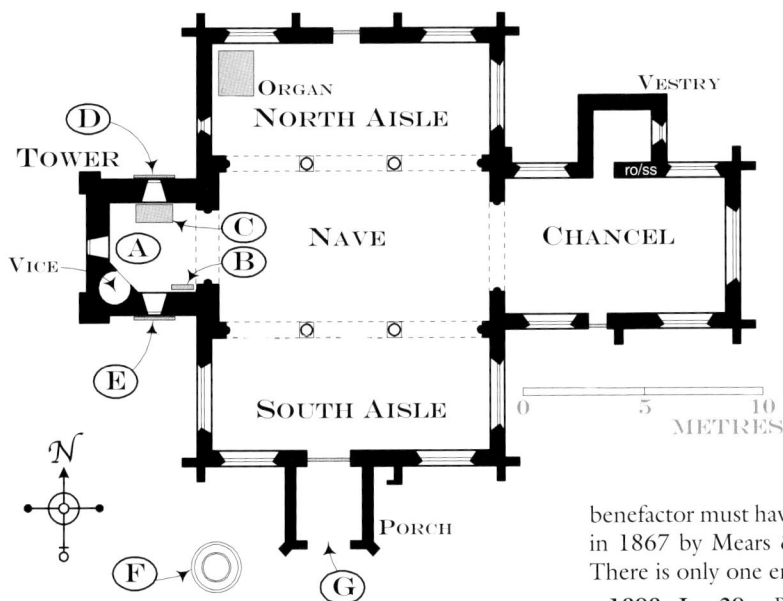

Plan of St John the Evangelist's Church
A: Five bells in the belfry. Ringing chamber on the first floor of the tower
B: Ellacombe chiming frame at the base of the tower
C: Clock in the ringing chamber
D: Nineteenth-century clock dial
E: Former location of eighteenth-century stone clock dial
F: Stone dial laid out in the churchyard
G: Location of former sundial
The vice provides access to the ringing chamber and the belfry

ST JOHN THE EVANGELIST   Grid Ref: TF 036108

PARISH RECORDS

The Churchwardens' Accounts survive from 1792 to 1915 (DE 2425/11-12) and contain many references to the bells and clock.

BELL HISTORY

The tower has a ring of five bells in a seventeenth-century low-sided oak frame with contemporary fittings. Three of these bells, the present third, second and tenor, were cast by Thomas Norris of Stamford in 1626, 1627 and 1633. It is known that the present treble, recast in 1790 by Edward Arnold was also originally a Norris bell of 1633 (Wright *Additions* 1788, **II**, 3). The present fourth may also have been cast at the Norris foundry as the seventeenth-century frame was made with pits to accommodate five bells.

A nearby public house, now a private dwelling, was known as the Five Bells. It closed in 1914.

Irons' Notes provide the following detail concerning the early bells at Ryhall (MS 80/1/3) and this may confirm that a new frame was installed at the time of the earliest Norris bell:

1625 Nov 25
The bells & bell frames there & the churche are oute of repaire & are not presented.

There are many entries in the accounts concerning payments to bellringers and for new bellropes but there is little detail concerning the bells and frame. A local

benefactor must have paid for the fourth bell to be recast in 1867 by Mears & Stainbank as it is not mentioned. There is only one entry concerning the frame:

1898 Jan 29   Repairing bell frame   £1 0s 0d

The following extract suggests that there may have been a separate clock bell, but North did not record it. Today the clock strikes the hours on the tenor and it is probably this bell that is referred to here:

1910 Jan 10   Repairing Clock Bell   3s 6d

In 2000 the bells had been unringable for many years due to the poor condition of the frame and fittings. However the bells can be chimed by means of an Ellacombe chiming frame at the base of the tower. The ropes are so arranged that the bells can also be chimed from the ringing chamber. This frame was recorded by Thomas North in 1880. Regular payments were made in the accounts to the ringers from 1865 to 1879, but from then until 1900 entries refer more to chiming. Very few specific details are given but the following serve as examples:

1880 May 24   G Towel for chiming on each
                   Sunday in the year   £1 4s 8d
1881 April 26 Carter for chiming Bells   £1 10s 0d
1895   Bell Chiming and ringers   £3 1s 0d

At the time of writing it seemed certain that a generous bequest by a local benefactor would ensure the bells were restored and rehung in a new frame. The proposal was to install the frame at a lower level to reduce the stress on the tower. An added benefit was that the old timber frame could be preserved in situ. The proposal also included the addition of another bell to make a ring of six.

BELL DETAILS

1 **1790**. Note C. Treble. Diameter 775mm (30½in). Weight approximately 305kg (6 cwt). Canons re-

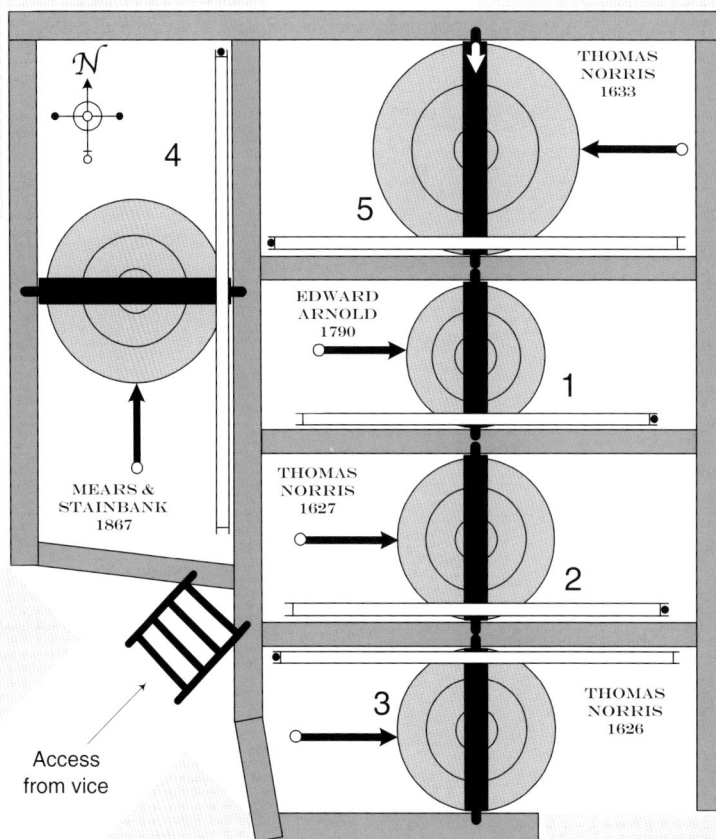

*Plan of the old bellframe at
Ryhall Church in 2000*

*Plan of the proposed layout
of the new bellframe at
Ryhall Church
(Hayward Mills Associates)*

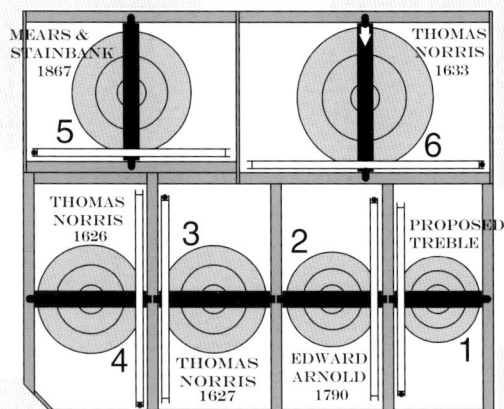

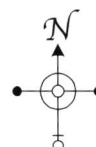

tained. Timber headstock. Recast by Edward Arnold of
Leicester.
EDW^D ARNOLD LEICESTER FECIT 1790
Decoration [**46**] is below the inscription band.

Former bell
Cast in 1633 by Thomas Norris of Stamford.
      THOS. NORRIS MADE ME 1633
(Wright *Additions* **II**, 1788).

2 1627. Note B flat. Diameter 838mm (33in). Weight
approximately 330kg (6cwt 2qr). Canons retained.
Timber headstock. Cast by Thomas Norris of Stamford.
    [26] OMИIA [53] FIAИT [53]
~ AD [53] GLORIAM [53] DEI [53] 1627 [53]
    (Let all things be done to the Glory of God)

3 1626. Note A. Diameter 889mm (35in). Weight
approximately 381kg (7cwt 2qr). Canons retained.
Timber headstock. Cast by Thomas Norris of Stamford.
[26] ИOИ [90] CLAMOR [90] SED [90] AMOR
~ [90] CAИTAT [90] IИ [90] AVRE [90] DEI [90]
(It is not noise but love that sings in the ear of God)

Below the inscription band:

THOMAƧ [53] ИORRIƧ [53] CAƧT [53] ME ~ [53] 1626 [53] C [53] B [53] I [53] W [53] CH WA [53]

This bell is cracked.

4 **1867**. Note G. Diameter 965mm (38in). Weight 501kg (9cwt 3qr 13lb). Canons retained. Timber headstock. Recast by Mears & Stainbank, Whitechapel Bell Foundry, London.

MEARS & STAINBANK. FOUNDERS LONDON.

The date of this bell is given in *VCH* (**II**, 74).

Former bells:

Cast in 1720 by Thomas Eayre II of Kettering.

I.H.S. NAZARENE REX JUDAEORUM FILI ~ DEI MISERERE MEI.
OMNIA FIANT AD GLORIAM DEI 1720

(Jesus of Nazareth King of the Jews, Son of God, have mercy on me.

Let all things be done in the Glory of God)

Note that W Harrod records the inscription on this bell as **GLORIA DEO SOLI 1729** (Wright *Additions*, **II**, 1788). Possibly a recast of a Norris bell of *circa* 1627.

The tenor at Ryhall. Note the massive timber headstock and the unusual design of the bellwheel

5 **1633**. Tenor. Note F. Diameter 1105mm (43½in). Weight 711kg (14cwt). Canons retained. Timber headstock. Cast by Thomas Norris of Stamford.

THOMAS [69] NORRIS [69] MADE [69] ME ~ [69] 1633 [69]

The clock strikes the hours on this bell.

BELLRINGING CUSTOMS

A Pancake Bell was rung on Shrove Tuesday. At the Death Knell there were thrice three tolls for a male and thrice two tolls for a female (North 1880, 150).

Today the bells are chimed before all Sunday, Easter and Christmas Services, as well as at midnight on New Year's Eve. Bells are also chimed for weddings if requested.

RINGERS' RULES

A warning to bellringers is painted on a board framing an arched and blocked doorway on the east wall of the clockroom. Two ledges at the base of the board are shelves for candles. It is likely that this wooden arch was placed in the ringing chamber as part of the 1857 restoration and it may be a replacement of an earlier board. Various members of the Holmes family were buried in the churchyard at the beginning of the eighteenth century.

The following words are laid out in two lines around the arch. There is a similar warning on a wall in the ringing chamber of Tinwell Church:

Whoever . Comes . Into . This . Place . His . Pleasure . ~ For . To . Take . And . Rings . A . Bell . To . Him ~ . We . Tell . This . Law . With . Him . Make

That . Every . Time . He . Turns . A . Bell . In . The . ~ Light . Or . Dark . He . Then . Shall . Pay . ~ Without . Delay . Two Pence . Unto . The . Clark

Augst. 31ʳˢᵗ 1857   **Cris. Holmes 1715.**

One of the candleholders on the ringers' warning board at Ryhall Church

SUNDIAL

Early drawings including one of *circa* 1793 (RCM F10/ 1984/48) show a sundial over the south porch. It was removed and lost during building work in the 1980s.

CLOCK HISTORY

Ryhall church clock is signed by William Bird of Seagrave, Leicestershire. It is unlikely that he was the maker as this is the only turret clock known by him. A headstone in Seagrave churchyard has the following inscription:

To the Memory of
Will[m]. Bird Sen[r].
He departed this life
Sep. the 9th 1767
Aged 52 years.

This date is four years before the date on Ryhall Church clock. However his son, another William, succeeded him, and although he was better known as a watch-maker, it was probably William junior who supplied this clock.

It was normally one of the clerk's duties to wind the church clock, and in the early Churchwardens' Accounts Matthew Bingham was clockwinder from 1794 up until 1808.

The Fenn family have been responsible for this clock for several generations, and amongst signatures found in

A drawing of William Bird's nameplate on Ryhall church clock

the clockroom are many belonging to this family, the earliest found being 'R Fenn 1881'. Before it was converted to automatic winding in 1979, Cyril Fenn who was parish clerk wound the clock daily at 8am for many years. At the time of writing his grandson Simon was still responsible for the clock:

Easter 1889 Mr W Fenn's account for new ladder & repairing Clock case, new frame and co £1 15s 0d
Bell Chiming 10/- Clock winding 7/- and Stove management 10/- 4 months £1 7s 0d

Included in the Parish Clerk's annual salary of £6 18s 9d were the following two items:

Easter 1891 11 months Bell Chiming at 30/- a year
11 months Clock winding at 21/- a year

John Brumhead of Stamford is the first clockmaker recorded as working on the clock, in 1838. He and William Brumhead, possibly his son, continued to be responsible for repairing and maintaining the clock until Francis Pinney, also of Stamford, took over this duty in the 1850s. During the next three decades three other Stamford clockmakers, H Warren, J Hicks and John Britton, also attended to the clock.

The conversion to two hands may have been at the same time as the installation of a new dial on the north face

Taken from a drawing of 1811 showing the original location of the clock dial on the south face of Ryhall church tower (Blore **I**, Pt 2, 56)

Francis Pinney & Sons' voucher for work on Ryhall church clock (DE 2425/11)

Ryhall church clock dial fell to the ground during a storm in 1940 and is now laid out in the grass near the south porch. It appears to have been carved from one piece of limestone and is approximately 125mm (5in) thick. The ring inside the numerals has divisions every fifteen minutes, a characteristic of single-handed dials. Dials with a minute hand have a minute ring outside the numerals

The fifteen-minute divisions of the original stone dial

Ryhall church clock

The present clock dial at Ryhall Church is on the north face of the tower. The dial surround was cast onto the face of the stonework

of the tower. This date is not recorded, but it was probably in the middle to late 1800s as indicated by the cast concrete dial surround. The work was probably paid for by a local benefactor as the conversion is not recorded in the accounts. It may have been part of the 1857 restoration.

Although the clock movement is now free-standing in the ringing room, it used to have a case:

1888-89 Mr W Fenn's account for new ladder & reparing Clock Case, new frame etc £1 15s 0d

The dial was refurbished in 1915:

July 6 Gilding Weather Cock & Clockface — Hail £3 15s 6d

CLOCK DETAILS
Details of the present clock:

Maker:	William Bird of Seagrave, Leicestershire
Signed:	'William Bird, Seagrave. 1771' engraved on brass plate on an upper bar of the frame
Installed:	1771
Cost:	Not recorded
Frame:	Wrought-iron birdcage
Trains:	Going and striking

Escapement:	Anchor escapement
Pendulum:	Offset pendulum, 1.84m (72½ in) long, suspended from a corner post of the frame. Wrought-iron rod and lenticular lead bob
Rate:	44 beats per minute
Striking:	Countwheel hour striking
Weights:	Cast-iron weights removed from the clock. One remains in the clockroom
Winding:	Originally wound daily. Both trains converted to automatic winding by Smiths of Derby in 1979

(left) Ryhall Church. The anchor escapement of William Bird's clock

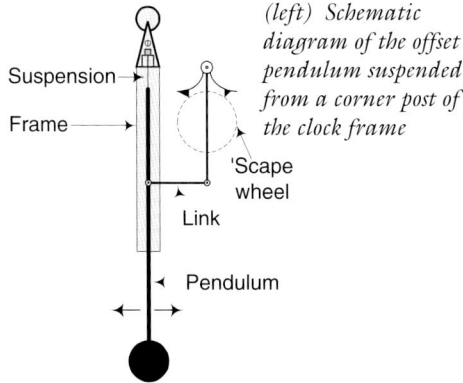

Suspension

Frame

'Scape wheel

Link

Pendulum

(left) Schematic diagram of the offset pendulum suspended from a corner post of the clock frame

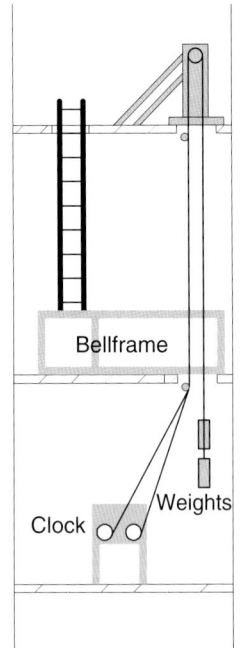

Bellframe

Weights

Clock

Dial:	Victorian convex copper dial for two hands. Black background with gilded roman numerals, minute ring and hands. Cast concrete bezel
Location:	North face of the tower
Notes:	Original octagonal limestone dial is marked out for a single hour hand. Probably contemporary with the clock. Preserved in the churchyard. Originally on the south face of the tower

(right) Ryhall church clock. Showing how the clock weights were suspended prior to the clock being converted to automatic winding in 1979

RYHALL VICARAGE

A sundial is shown in the garden of the Vicarage on the early 1900s OS Second Edition 25 inch map. It was a table sundial with three intertwined snakes forming the pedestal. It was stolen in the late 1970s when the Vicarage was unoccupied.

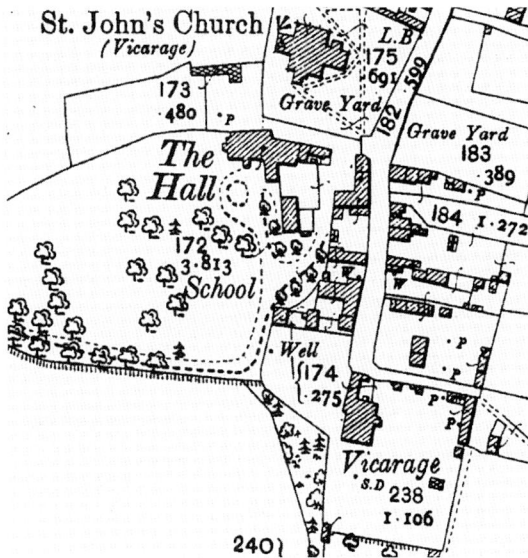

Early weight pulleys above the belfry at Ryhall. Because of the construction of the tower, the clock weights could not fall below the level of the clockroom floor. The weights were therefore wound through the belfry via a pair of wooden pulleys mounted in a frame on the floor above the bells. These early pulleys still exist in their original position in the north-east corner of the tower. This suggests that the clock movement has always been in its present location against the north window of the clockroom. Since the original south dial was at belfry floor level the leading-off work from the clock to the dial would have crossed the clockroom at a high level

Ryhall Vicarage. Location of the sundial (SD) on the early 1900s OS Second Edition 25 inch map

SEATON

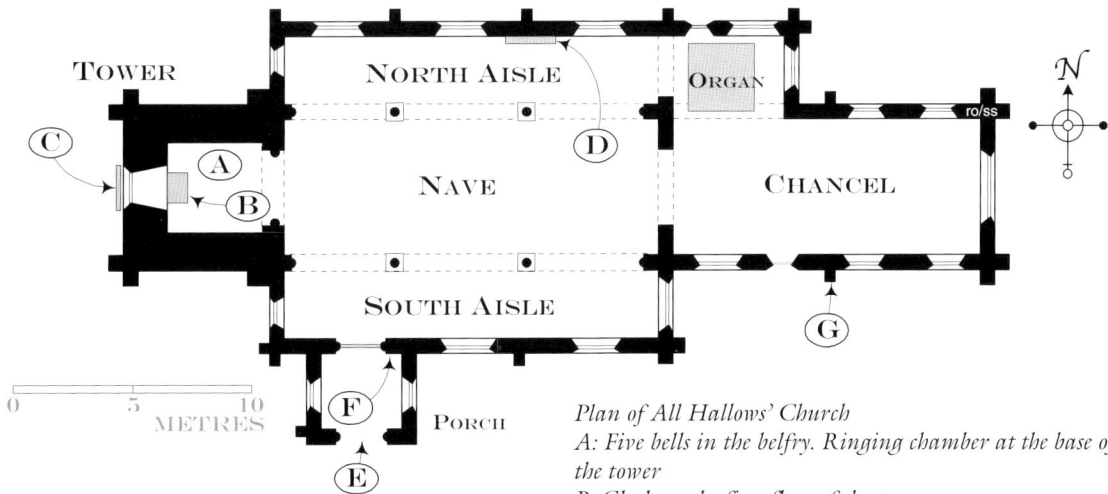

Plan of All Hallows' Church
A: Five bells in the belfry. Ringing chamber at the base of
the tower
B: Clock on the first floor of the tower
C: Clock dial. D: First World War memorial
E: Sundial
F: Scratch dial 1
G: Scratch dial 2
Access to the clockroom is by ladder from the base of the
tower. Access to the belfry is by ladder from the clockroom

ALL HALLOWS   Grid Ref: SP 904983

PARISH RECORDS

The following documents in the parish archive were searched:

Constables' Accounts of *circa* 1665-91 (DE 1883/53).
Glebe Terriers 1682-1777 (MF 495).
The Home Guard Fieldname Survey of Rutland 1943
(DE 1381/535).
The Enclosure Map of 1856 (DE 1381/523).
A letter concerning the title and possible sale of 'Bellrope
Field' dated 26/11/1941 (DE 1883/35/1/1-2).
Specification and correspondence concerning the church
clock 1919 to 1920 (DE 1883/33/1-4).
Notes on the history of Seaton during the nineteenth
and twentieth centuries (DE 1883/57).
Apprenticeship indenture of William Goodman of Seaton
to John Fox of Uppingham in the trade of a whitesmith
dated 5 July 1784 (DE 2417/22).

BELL HISTORY

Little is known of the early bells at Seaton but there were at least three in the tower by the end of the sixteenth century. They were cast by Richard Bentley, Matthew Norris and Richard Holdfield. All three remain in the tower today and they are scheduled for preservation. The inscription on the bell by Matthew Norris was possibly retained from an earlier bell. According to Irons' Notes (MS 80/1/3) the 'peals are all out of repaire' in 1612.

In 1669/70 the belfry was restored and a new tenor cast by Toby Norris III installed. Fourteen years later the treble by Henry Bagley completed the present ring of five.

During restoration work at the church in 1875 the belfry floor and access ladder were replaced. In 1914 John Taylor & Co rehung the bells in a new low-sided

metal frame and as a result the following testimonial was used by the company in their brochure *circa* 1920 (DE 2520/2):

SEATON, UPPINGHAM,
June 24th, 1914.

Dear Sirs,

I herewith enclose cheque for our Bells Restoration.
I also take this opportunity of expressing to you the perfect satisfaction the work has given to the Rector, Churchwardens and all officers concerned. Assuring you of future recommendations from our particular district.
I am, Sirs, yours faithfully,

ROBERT CROWDEN,
Churchwarden.

In 1993 Nicholas Meadwell refurbished this bellframe and its fittings.

There used to be an area of land in the parish called 'Bellrope Field' which was originally let out to raise funds for the purchase of new bellropes, or to pay for work in connection with the bells and fittings. A local benefactor possibly bequeathed the land to the church. It is referred to in correspondence of 1941 between the Rev C Escritt of Seaton Rectory and the Deputy Registrar of the Diocese of Peterborough. Mr Escritt was seeking evidence of title to 'Bellrope Field' with a view to its possible sale. There are no documents recording the outcome of this

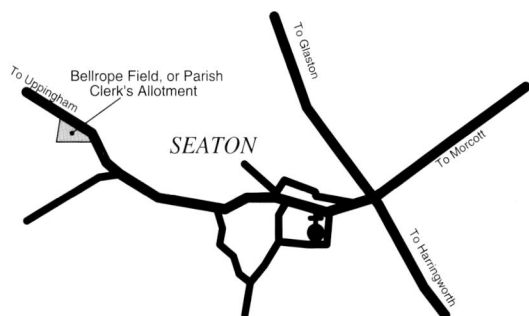

The location of the 'Bellrope Field' at Seaton

N

Plan of the low-sided metal
bell frame at Seaton Church

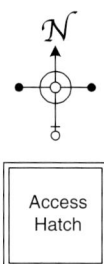

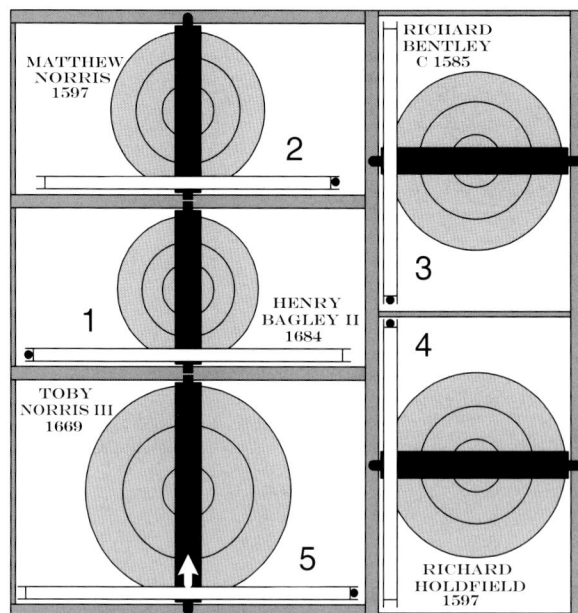

Access
Hatch

enquiry and it is presumed that the land was in fact sold. An area of land referred to as 'Bellrope Field', also known as the 'Parish Clerk's Allotment', is still remembered in the village (information from Hilary Crowden).

BELL DETAILS

1 **1684**. Treble. Diameter 702mm (27⅝in). Weight 210kg (4cwt 0qr 16lb). Canons removed. Cast-iron headstock. Cast by Henry Bagley II of Chacombe, Northamptonshire.

HENRY [100] BAGLEY [100]
~ MADE [100] MEE [100] 1684 [100]

A band of decoration [100] is below the inscription. The oak headstock was replaced in 1993 due to rusting of the retaining bolts.

2 **1597**. Diameter 759mm (29⅞in). Weight 222kg (4cwt 1qr 13lb). Canons removed. Cast-iron headstock. Ascribed to Matthew Norris of Leicester and possibly of Stamford. He may have cast this bell in association with Richard Holdfield (George Dawson) (*see* Chapter 3 — Bellfounders, Stamford).

ƧUM ROƧA PVLSATA MVИDIA
~ MARIA VOCATA 1597

(I being rung am called Mary, the Rose of the World) The inscription words are placed evenly around the bell. Probably a retained inscription from a former bell. Scheduled for preservation (Council for the Care of Churches).

3 *Circa* **1585**. Diameter 841mm (33⅛in). Weight 357kg (7cwt 0qr 3lb). Canons removed. Cast-iron headstock. Cast by Richard Bentley, ? of Buckingham.

ƦEDDИVOF LLEƁ [88] EVLTEИEƁ
~ EDƦAhCEVƦ [25] [88]

(The inscription reads backwards: Ryecharde Benetlye Bell Fovndder)
The letters are like [87]. Scheduled for preservation (Council for the Care of Churches).

4 **1597**. Diameter 879mm (34⅝in). Weight 361kg (7cwt 0qr 11lb). Canons removed. Cast-iron headstock. Ascribed to Richard Holdfield, an itinerant bellfounder, later of Cambridge. He may have cast this bell in association with Matthew Norris (George Dawson) (*see* Chapter 3 — Bellfounders, Stamford).

[8] CELOƦVM ChƦISTI PLATIAT
~ TIƁE ƦEX SOИVS ISTI 1597

(O Christ King of Heaven may this sound be pleasing to thee)

ƦIChAƦDE BƦOVGhTOИ
~ ƦOBAƦTVS ShEFELDE AƦMIGIƦE

(Richard Broughton Robert Sheffield Esquire)
Letters are like [125]. Richard Broughton and Robartus Sheffield were the probable donors of this bell. Richard Broughton was the son of Markes Broughton of Seaton who was descended from a Lancashire family. He was living in 1613. Robert Sheffield was the son and heir of

Part of the inscription on the fourth bell at Seaton

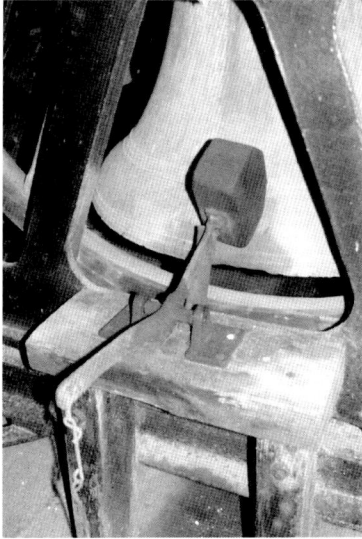

The tenor bell and clock hammer at Seaton

service. Each bell is rung in rotation week by week. For special services a band is assembled for full circle ringing. On Armistice Day the tenor is tolled half-muffled. Bells are rung for weddings on request.

BELLRINGERS' CANDLEHOLDERS

There are two ornate limestone candleholders in the ringing chamber. The oldest is in the centre of the north wall and there is a replica opposite.

SCRATCH DIALS

There are two scratch dials on the church. One is located inside the south porch on the east side of the twelfth-century door. This doorway was moved to its present position when the south aisle was built a century later. The second is on the south face of the buttress to the east of the chancel door. This scratch dial is very weathered and difficult to see. The chancel is late thirteenth century.

George Sheffield also of Seaton (North 1880, 151). Several members of the Sheffield family are recorded as Rector or Patrons of the church in the late sixteenth and early seventeenth centuries. Scheduled for preservation (Council for the Care of Churches).

5 1669. Tenor. Note G. Diameter 997mm (39¼in). Weight 517kg (10cwt 0qr 21lb). Canons removed. Cast-iron headstock. Cast by Toby Norris III of Stamford.
[26] [52] GOD [52] SAVE [52] THE [52] KING
~ [52] 1669 [52]
The clock strikes the hours on this bell.

BELLRINGING CUSTOMS

In 1880 a bell was rung on Sundays at 9am but prior to this date it was additionally rung at 7am. For Divine Service a bell was tolled for fifteen minutes then all the bells chimed for ten minutes. Finally, the treble was tolled for five minutes. After Morning Service three bells were chimed to give notice that Evening Prayer would be said. Before the Death Knell three tolls for a male and two tolls for a female were given. The Pancake Bell and Gleaning Bell had ceased to be rung by 1880 (North 1880, 151).

At the time of writing one bell is rung before each

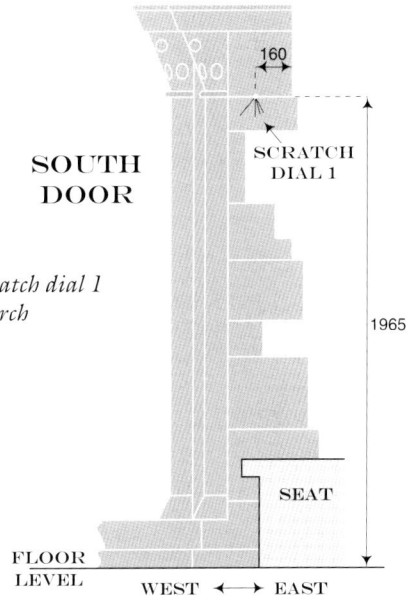

Location of scratch dial 1 at Seaton Church

The scratch dial to the east of the south doorway at Seaton Church

SEATON 1				
Location	Inside porch - on south aisle wall east of main door			
Condition	Excellent			
Gnomon Hole Diameter		10		
Gnomon Hole Depth		Filled in		
Height above ground level		1965		
Line Ref	a	b	c	d
Length	75	85	90	100
Angle (°)	161	168	185	214

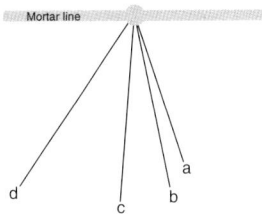

Scratch dial 1 details

SEATON 2			
Location	Central buttress of south wall of chancel		
Condition	Poor		
Gnomon Hole Diameter			12
Gnomon Hole Depth			Filled in
Height above ground level			1970
Line Ref	a	b	b
Length	135	135	135
Angle (°)	178	189	203

Scratch dial 2 details

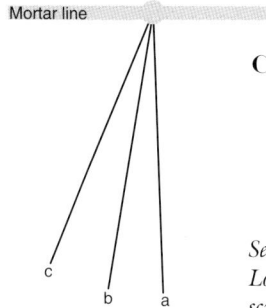

Seaton Church. Location of scratch dial 2

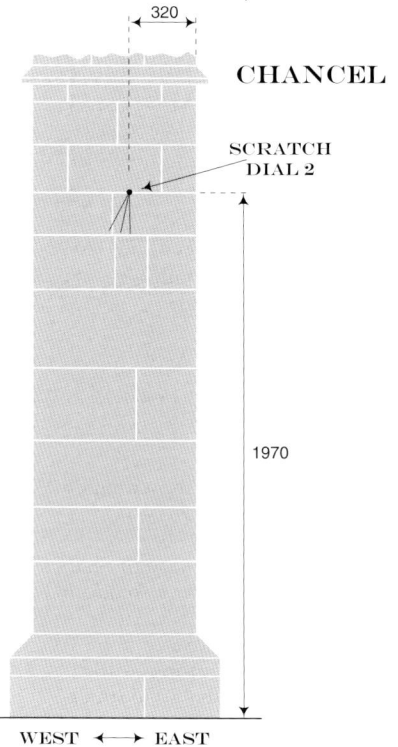

SUNDIAL

There are the remains of a vertical limestone sundial high in the gable of the south porch. Two marks indicate the location of the iron gnomon. There are no visible lines or numerals and the dial is generally in a poor condition. The porch is fourteenth century but the sundial is of a much later date. It is shown on a late eighteenth-century drawing of the church (RCM F10/1984/49) and on another of *circa* 1839 in Uppingham School Archives. This drawing clearly shows the sundial complete with its gnomon.

CLOCK HISTORY

It might be expected that there was an early clock at Seaton but the lack of any surviving Churchwardens' Accounts precludes any confirmation. Neither of the *circa* 1793 (RCM F10/1984/49) or *circa* 1839 (Uppingham School Archives) drawings of the church show a clock dial, but this is not conclusive as a dial may have been on an unseen elevation or there may have been a clock without a dial.

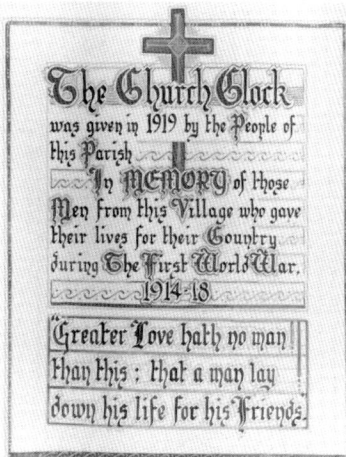

The gift of the clock is recorded by this framed notice on the west wall of the nave at Seaton Church

John Smith & Sons' invoice for Seaton church clock (DE 1883/33/4)

The present church clock was paid for by general subscription and installed by John Smith & Sons of Derby in December 1919. It is a memorial to the seven men of Seaton and Thorpe by Water who died in the First World War and was dedicated on 18 May 1920. A memorial tablet on the north wall of the nave was unveiled at the same service.

William Potts and Sons of Leeds also tendered for the supply and installation of this clock. A transcript of their unsuccessful tender and hand-written specification (DE 1883/33/1) is included as Appendix 5.

CLOCK DETAILS

Details of the present clock:

Maker:	John Smith & Sons, Midland Clock Works, Derby
Signed:	On the setting dial and on the frame
Installed:	1919
Cost:	£98 10s
Memorial to:	The men of Seaton and Thorpe by Water who died in the First World War
Presented by:	The parishioners of Seaton and Thorpe by Water
Frame:	Cast-iron flatbed
Trains:	Going and striking

Escapement:	Pinwheel
Pendulum:	Wooden rod and cast-iron cylindrical bob
Rate:	52 beats per minute
Striking:	Countwheel hour striking
Weights:	Cast-iron
Winding:	Hand wound weekly
Dial:	Skeleton dial with a blue background and gilded roman numerals and hands
Location:	West face of the tower

RAILWAY MISSION CHAPEL

Seaton Railway Mission Chapel was one of three chapels erected along the Rutland section of the Manton to Kettering railway line when it was being constructed in the mid 1870s. The chapel, built in 1876, was located amongst a settlement of navvies' huts, known as 'Cyprus', in a field overlooking the Welland valley at the northern end of Seaton viaduct. The chapel could be distinguished from the other buildings by its wooden steeple which housed a single bell. The other chapels were at Glaston and Wing (see under Glaston and Wing). At a meeting to 'forward the objects' of the Navvy Mission Society, the Bishop of Peterborough made the following comments:

Seaton church clock

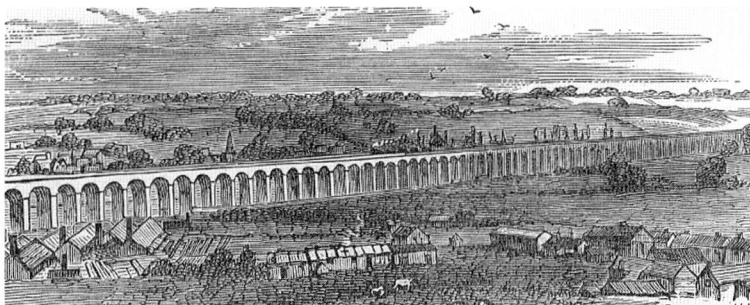

The navvy hut settlement known as 'Cyprus' at the northern end of Seaton viaduct in 1876 (Barrett 1880, after xv)

The bell turret on the former school at Seaton

I am deeply interested in the success of this Society ... A railway was to be made in my diocese, and suddenly four thousand new inhabitants were hutted in temporary villages in rural parishes. The clergy of these parishes - already fully occupied, were unable to do anything ... I myself directed the Mission. The contractors built three wooden churches at different points, and there I have preached to the most earnest and attentive congregations (Barrett 1880, 149).

All three chapels were closed on completion of the line in 1878.

SEATON VILLAGE SCHOOL

Seaton School was built in 1859 and closed a century later. The elegant bell turret has been retained but the bell is missing.

SOUTH LUFFENHAM

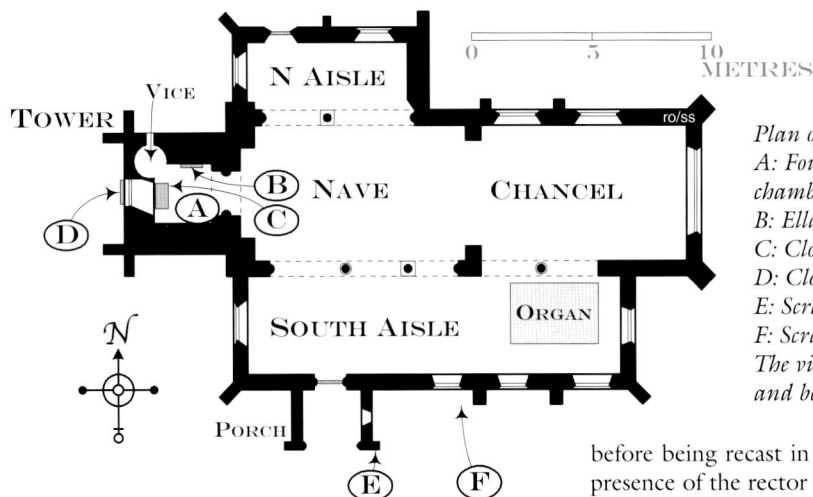

Plan of St Mary the Virgin's Church
A: Four bells in the belfry. Ringing
chamber at the base of the tower
B: Ellacombe chiming frame
C: Clock on the first floor of the tower
D: Clock dial
E: Scratch dials 1 and 2
F: Scratch dials 3 and 4
The vice provides access to the clockroom
and belfry

ST MARY THE VIRGIN   Grid Ref: SK 941019

PARISH RECORDS

Records searched include the Churchwardens' Accounts 1742-89 and 1877-1918 (DE 2540/10-11) and the Vestry Minute Book 1853-94 (DE 2540/12).

BELL HISTORY

Of the four bells currently in the tower two are scheduled for preservation. The earliest of these, the tenor, is *circa* 1510 and is ascribed to Robert Mellour of Nottingham. The other, the treble, was cast by Hugh Watts I of Leicester in 1593.

The earliest reference to the bells is in Irons' Notes (MS 80/1/3):

> 1577 **Nov 4** Churchwardens accused that they sell their belles and refuse to come unto aine [any] account for them. Charge to be considered when the Rector and Church wardens have given their version of the matter.

Another note from this source, dated December 1577, states that the accused defended themselves by saying that they had the full number of bells required (this appears to have been three), and that they had made a saving by using the metal from the tenor to cast a new treble (this new bell may have been the present second

before being recast in 1886). Seven parishioners in the presence of the rector testified to this story. The money saved, which appears to have been misappropriated, was ordered to be paid 'to no other purpose than the use and advantage of the churche'. The rector later informed the authorities that the money had been used to purchase 'one close and one house'.

A Bell Ringers' Field Charity is referred to in a terrier dated 1749 and the endowment consisted 'of a piece of land containing 1 acre and 6 poles let out at an annual rent of £4'. The net income was paid to the bellringer. This charity, 'regulated by a scheme of the Charity Commissioners dated 18 October 1921', has two stories relating to it:

First, it has been suggested that the events of 1577 may be the origin of a Bellringer's Close in South Luffenham. Secondly, tradition states that an acre of land on the Morcott Road, opposite Halfway House, was donated to the church by a lady who had lost her way on South Luffenham Heath. She was guided home by the sound of the church bells and she donated this land in gratitude, on the condition that a bell was rung at 5am and 8pm every day between 19 October and 25 March (*VCH* II, 207). North (1880, 139) gives these dates as 19 Setember and 25 March.

It seems that these two stories are linked and the following may provide an explanation:

The field [that donated] was originally at Foster's Bridge,

but the endowment was transferred to the Bellringers Field, also known as the Feast Field and Bell Field [possibly Bellringer's Close of 1577 as noted above]. It is sited opposite the Halfway House, and now has a bungalow built on it. [The ringing of the 5am and 8pm bells] ... continued for many years until the outbreak of the Great War (Traylen nd, Pt2).

The church bells were considered to be the responsibility of the churchwardens, and anyone interfering with them was severely reprimanded. Another extract from Irons' Notes records an incident which resulted in one Josia Tookye appearing before the Church Court:

1604 Dec 12 Josia Tookye for entering into our parishe church at xi of a clocke in the night upon Dec 1 being Satterday & there in most unruly manner rang the bells for the space of two hours ... after went to the alehouse & spent all the night in drinkinge & most terrible drunkenes.

The punishment, if any, is not recorded.

Further examples from Irons' Notes relate to the present third bell:

1618 April 29 one of their bels is lately cracked ...
June 4 Thos Tookey & Zach Wallbanks to certifie the castinge of one of their bells by St Joh Bapt

This bell was recast by Toby Norris I of Stamford in 1618. Details of it were recorded by Thomas North (1880, 139) before it was recast yet again by John Taylor in 1886.

Although the above notes indicate that both the second and third have been recast at least twice, the present complement of four bells has not changed since before the beginning of the seventeenth century.

It seems certain that a new timber bellframe was installed in 1681 as Thomas North (1880, 139) recorded that **AL . AW . CW . 1681** was carved on one of the beams.

The following are typical examples from the surviving eighteenth-century Churchwardens' Accounts. Throughout this period there are regular payments for bellropes indicating that the bells were well used, but very little seems to have been spent on their repair:

1758	Paid Thos Islip for the Bell Claper ...	1s 8d
	Paid John Payne for 2 Bellropes	6s 6d
1759	Paid Thos Islip Bill for the Bell well [wheel]	£1 4s 6d
	Paid John Willford for 2 Brasiss Runing [recasting] and other work for the Bell	5s 8½d
1769	pd for Ale when the Bell Claper was don	1s 0d

The bells remained in the timber bellframe of 1681 until 1886 when John Taylor recast the second and third

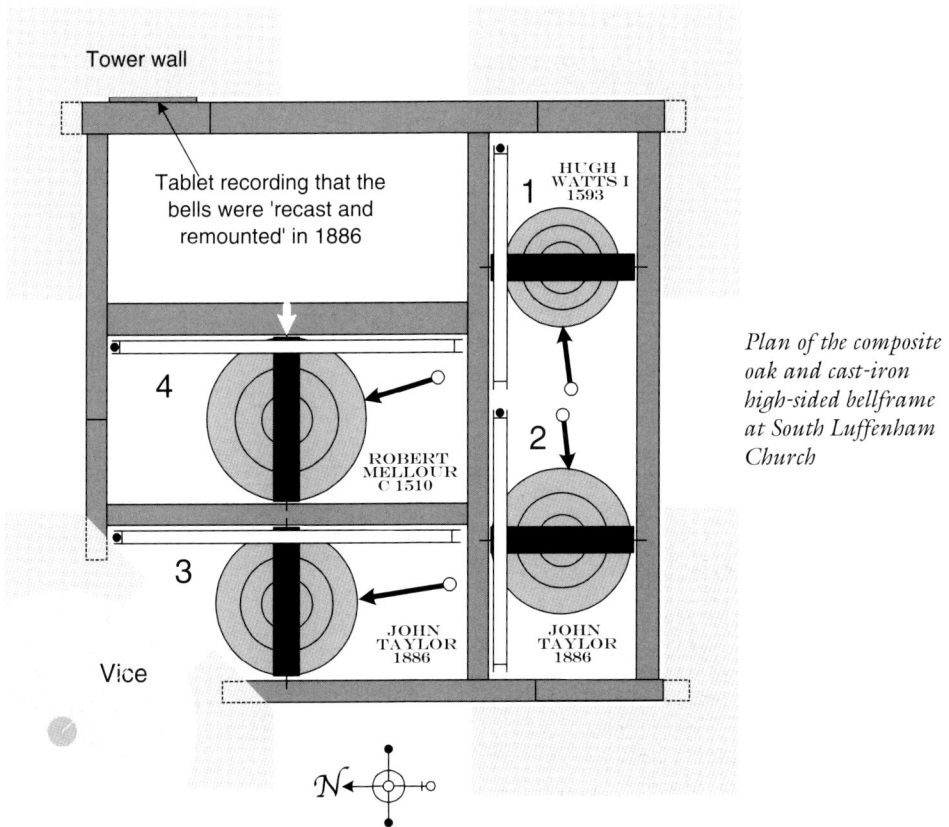

Plan of the composite oak and cast-iron high-sided bellframe at South Luffenham Church

bells and rehung all four in a new composite cast-iron and oak high-sided frame. A stone plaque let into the east wall of the belfry records this work:

**TO THE HONOUR OF GOD
THESE BELLS WERE RECAST
AND REMOUNTED AD 1886
HENRY HOLDEN D.D. RECTOR**

W WOOD }
J C TAILBY } CHURCHWARDENS

E. STAPLETON

This frame has five pits, but a fifth bell has never been added. The Ellacombe chiming frame on the north wall at the base of the tower, also designed for five bells, was probably installed at the same time.

BELL DETAILS

1 1593. Treble. Diameter 565mm (22¼in). Weight approximately 127kg (2cwt 2qr). Canons retained. Timber headstock. Cast by Hugh Watts I of Leicester.

ƎϩGI ƎϺ [13] ƐϤꓭϺ [13] Sꓧꓶꓭꟿ [13] ꟿꓭh
[3]

(Hew Watts made me 1593) [in reverse]
Thomas North mistakenly recorded the founder's mark [13] on this bell as [37]. There are unused bolt holes in the crown of this bell. Scheduled for preservation (Council for the Care of Churches).

The inscription on the treble at South Luffenham

2 1886. Diameter 692mm (27¼in). Weight approximately 229kg (4cwt 2qr). Cast with canons. Timber headstock. Recast by John Taylor & Co, Loughborough.

**J : TAYLOR & Cᴼ.. FOUNDERS
~ LOUGHBOROUGH 1886.**

Former bell

Founder and date not known. Diameter 616mm (24¼in). No inscription. This bell was cracked and had no clapper in 1880 (North 1880, 139).

3 1886. Diameter 711mm (28in). Weight approximately 241kg (4cwt 3qr). Cast with canons. Timber headstock. Recast by John Taylor & Co of Loughborough in 1886.

**J TAYLOR & Cᴼ FOUNDERS
~ LOUGHBOROUGH 1886.**

Former bell

Cast in 1618 by Toby Norris I of Stamford. Diameter 705mm (27¾in).
[14] OMꓤIA :: FIAꓤT :: AD :: GLORIAM :: DEI
~ ::::: 1618
(Let all things be done to the Glory of God)

A third of the bell rim was missing in 1880 (North 1880, 139).

4 *Circa* 1510. Tenor. Diameter 794mm (31¼in). Weight approximately 305kg (6cwt). Canons removed. Bolted through the crown to a timber headstock. Ascribed to Robert Mellour of Nottingham (George Dawson).

[21]
[6]   S   [6]   S
[9]

Letters and founder's marks are equally spaced round the inscription band. The S in the inscription band is like [124]. Scheduled for preservation (Council for the Care of Churches). The clock strikes the hours on this bell.

BELLRINGING CUSTOMS

The disbursements made by the churchwardens during the second half of the eighteenth century record that the bells were customarily rung to celebrate the failure of the attempt to blow up King and Parliament in 1605. Payments of 1s were made regularly from 1752 until 1797:

1752 Spent On Gunpowder Treson   1s 0d
1761 pad for Renging on the 5 of November   1s 0d

There is only one other entry, in 1880, for ringing on 5 November and the payment remained the same.

A number of parishes rang bells to celebrate Coronations. There is only one example in the South Luffenham accounts:

1761 Spent on the Kings [George III] Crownation 1s 6d

The duties of the parish clerk included ringing the bells for services. His annual wages also included ringing daily bells on special occasions although these are not specifically noted. The Curfew Bell, the treble, used to be rung at 8pm and the Morning Bell at 5am from 19 September until 25 March, except during a fortnight at Christmas (North 1880, 139). Both were rung until about 1912 (Traylen 1989, 34).

On Sundays a bell was rung at 8am and again after the Morning Service if Evening Prayer was to be said. Prior to the Death Knell thrice three tolls would be given for a male and thrice two tolls for a female. An hour before a funeral the tenor would announce the age of the deceased by tolling the number of years. The bell would also be tolled as the cortège processed through the village to the church (North 1880, 139).

In 1999 a band of ringers was formed to ring in the Millennium and the bells are now rung for all Sunday and Christmas Services. At the monthly Family Service all the bells are rung, followed by the chiming of the treble for five minutes before the service commences.

SCRATCH DIALS

There are two scratch dials on the south face of the porch

Details of scratch dial 1

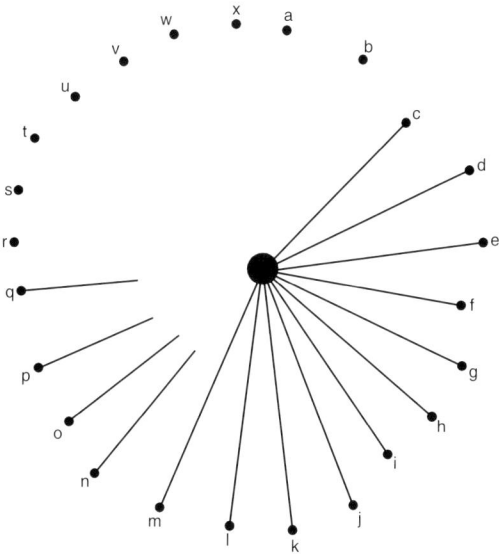

Location of scratch dials 1 and 2 at South Luffenham

SCRATCH DIAL 1

SCRATCH DIAL 2

ENTRANCE
TO
SOUTH
PORCH

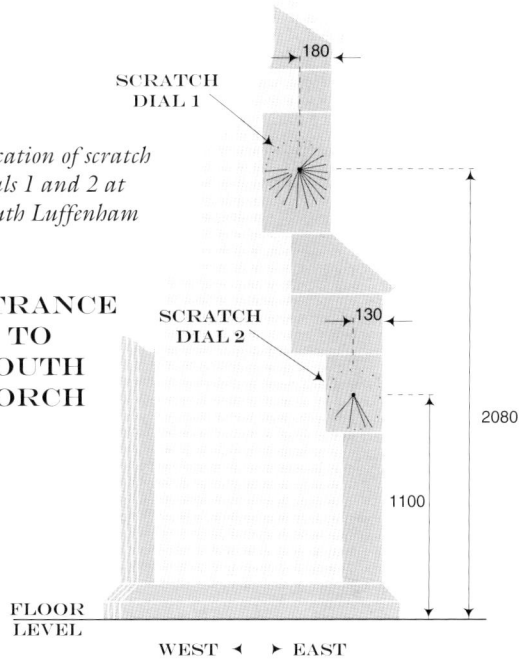

FLOOR
LEVEL

WEST ◄ ► EAST

SOUTH LUFFENHAM 1

Location	East front of south porch	
Condition	Poor	
Gnomon Hole Diameter		15
Gnomon Hole Depth		15
Height above ground level		2080

Line/Pock	a	b	c	d	e	f	g	h	i	j	k	l
Length/radius	58	112	100	112	112	100	110	110	110	122	125	125
Angle (°)	6	26	45	65	83	100	115	130	144	158	173	187

Line/Pock	m	n	o	p	q	r	s	t	u	v	w	x
Length/radius	125	80	70	65	60	124	126	130	125	122	122	120
Angle (°)	204	220	233	247	265	276	287	299	312	325	338	353

SOUTH LUFFENHAM 2

Location	East front of south porch	
Condition	Poor	
Gnomon Hole Diameter		10
Gnomon Hole Depth		10
Height above ground level		1100

Details of scratch dial 2

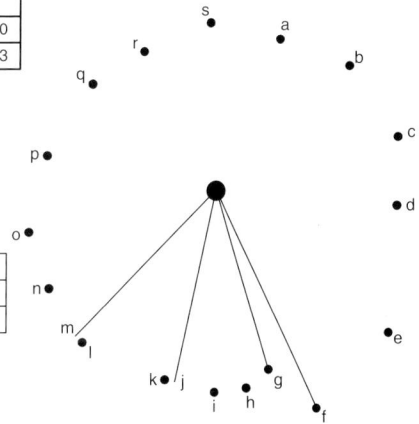

Line/Pock	a	b	c	d	e	f	g	h	i	j
Length/radius	80	90	95	90	100	116	90	95	95	95
Angle (°)	24	48	74	94	128	154	163	170	180	192

Line/Pock	k	l	m	n	o	p	q	r	s
Length/radius	95	100	100	95	90	85	80	70	80
Angle (°)	195	222	225	240	258	286	310	332	358

to the east of the entrance and the remains of two others on the face of a south aisle buttress. Both the porch and south aisle are fourteenth century.

CLOCK HISTORY

There was a clock at South Luffenham before 1742 as in that year, the earliest for which churchwardens' disbursements are available, there is a reference to the purchase of a clockline. After this date the clock was evidently well maintained for oil was regularly acquired and clocklines

replaced. Clockmakers Robert Fox, Joseph Furniss and William Aris I of Uppingham, and Richard Hackett of Harringworth, Northamptonshire, were all employed to clean, maintain and repair it. There are also many references to the parish clerk winding and oiling the clock.

In **1745** Robert Fox carried out repairs on the clock which involved it being taken to his workshop:

Spent when they took down the clock   6d
for fettching the Clock hom from Upinghm & beere
when they put It up   1s 6d

SOUTH
AISLE

SOUTH
AISLE

285

100

SCRATCH
DIAL 3

SCRATCH
DIAL 4

100

110

1450

1460

GROUND
LEVEL

WEST ←→ EAST

*Location of
scratch dials
3 and 4 at
South
Luffenham*

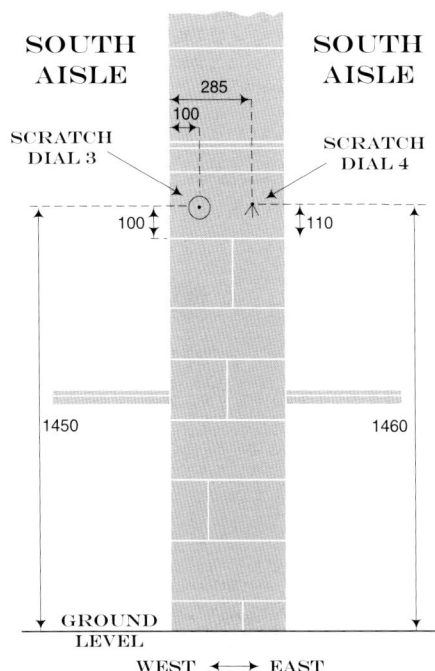

SOUTH LUFFENHAM 3		
Location	Central buttress of south aisle	
Condition	Poor	
Gnomon Hole Diameter	12	
Gnomon Hole Depth	12	
Height above ground level	1450	
Circle Diameter	150	
Pock Ref	a	b
Radius	75	75
Angle (°)	110	140

*Details of
scratch dial 3*

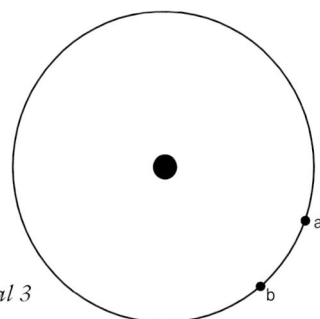

a

b

SOUTH LUFFENHAM 4		
Location	Central buttress of south aisle	
Condition	Poor	
Gnomon Hole Diameter	10	
Gnomon Hole Depth	12	
Height above ground level	1460	
Line Ref	a	b	c
Length	58	66	42
Angle (°)	145	180	230

Details of scratch dial 4

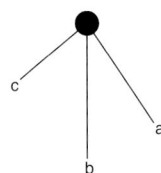

c

a

b

pad to Mr Fox for putin the clocke to rights   10s 0d

There is no definite evidence as to the maker of this clock. The necessity to remove it for repairs suggests that it may have been in the tower well before 1745. Close inspection of the window sill behind and below the present dial reveals that there is a groove in the centre which may have been made for or by the hand spindle of an early one-handed clock. This is centred on the faint shadow of a former diamond-shaped dial. Another groove can be seen in the drip moulding where the bottom corner of this dial stood. These observations are confirmed by Michael Lee (2000, 53).

Robert Fox was paid 2s 6d 'for the Cloke clingin' [cleaning] in 1750, and then in 1753 and 1755 there are two references to a 'Mr Ayres'. He may well have been William Aris I who later worked with Richard Hackett at Harringworth:

| 1753 | pd Mr Ayres Bill | £1 3s 6d |
| 1755 | paid Mr Ayres for cleaning the Clock | 5s 0d |

The following are some later examples of the work carried out on the clock:

1767	pd for 3 pullies for the Clock	3s 0d
	Oyl for the Clock	6d
1769 & 70	pd for wood when the Clock was Cleaned	6d
	pd for a Clock Line for the striking part	2s 6d
	spent in Ale when the line was put up	8d

Two years later Richard Hackett of Harringworth was engaged to repair or modify the clock, which again involved its removal from the tower:

1772	pd George Cave for a pulley wheel mendg and a shaft for the Clock	1s 0d
	spent on Mr Hackits account	4s 2d
	for Carying the Clock to Harringworth	3s 0d
1773	paid Mr Hackett Bill	£3 17s 6d

In 1782 Joseph Furniss was paid 10s 6d for cleaning the clock, a task which he had repeated on several occasions by 1790, and it was probably Joseph who carried out what appears to be another extensive repair two years later:

| 1792 | pd for Repairing the Clocke | £5 5s 0d |
| | for cariage of the Clocke to Uppingham and bringing back | 5s 0d |

William Aris I, who had moved to Uppingham following the death of Richard Hackett in 1782, was the next clockmaker to be engaged, and he maintained and repaired the clock from 1794 until 1798, the year in which he died.

Nothing of the clock's history is recorded for the next eighty years owing to a gap in the church accounts. From 1877 until 1882 James Pridmore, the parish clerk, was paid one guinea annually for winding the clock, followed by W Pridmore from 1889 until 1895. Payments for sundry repairs to the clock were made to John Hicks, a Stamford clockmaker, in 1881, 1882, 1884 and 1888.

In 1907 a new clock and skeleton dial were supplied and installed by William Potts & Sons of Leeds. There are no references to this installation in the accounts and it is assumed that a local benefactor was responsible. A Mr Spring is recorded as the winder of the new clock from 1909 until 1920.

South Luffenham church clock installed in 1907 by William Potts of Leeds

William Potts' dial of 1907 on the west wall of the tower at South Luffenham

South Luffenham Church
Rutland

Order 1491

Hour striking clock with 12" brass main wheels, gravity escapement & compensated pendulum. To show time on one external skeleton dial 4ft dia.

The dial has 4 lugs at 10 - 2 - (7 & 8) and (4 & 5) also a plain star centre with 6 small brass pieces ⌒ thus fastened to centre at 1 - 3 - 5 - 7 - 9 & 11.

The dial has also 4 cast iron ornaments thus ⚜ fastened on.

Hour hand ordinary spade & minute hand:

Dial hands & ornaments gilded but not centre.

Started June 25 / 07.

Cords [weight lines] each 100ft ⅝ circ.

South Luffenham church clock as described in William Potts & Sons' Order Book (Michael Potts)

CLOCK DETAILS

Details of the present clock:

Maker:	William Potts & Sons, Leeds
Signed:	On the front of the frame: W. Potts & Sons, Leeds, 1907
Installed:	1907
Frame:	Cast-iron flatbed
Trains:	Going and striking
Escapement:	Double three-legged gravity

Pendulum:	Wooden rod and cast-iron cylindrical bob
Rate:	49 beats per minute
Striking:	Countwheel hour striking
Weights:	Original cast-iron weights
Winding:	Weekly. Winding reduction gear on strike train
Dial:	Cast-iron skeleton dial by Potts. External

fleur de lys decoration at the quarters. Gilded roman numerals, hands, minute divisions and decorations

Location: West face of tower

SOUTH LUFFENHAM NATIONAL SCHOOL

When South Luffenham National School was rebuilt in 1872 a bellcote was included in the new structure. The school was closed in 1969 and subsequently converted into a private house but the bellcote and bell were retained. The bell has no inscription but the date 1872 is on the waist.

SOUTH LUFFENHAM HALL

A table sundial is shown in the garden of South Luffenham Hall on the early 1900s OS Second Edition 25 inch map.

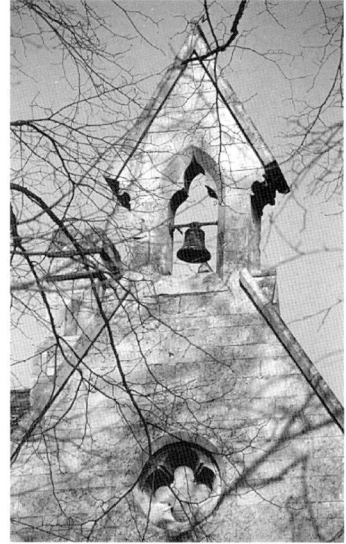

The bellcote and bell of the former National School at South Luffenham

STOKE DRY

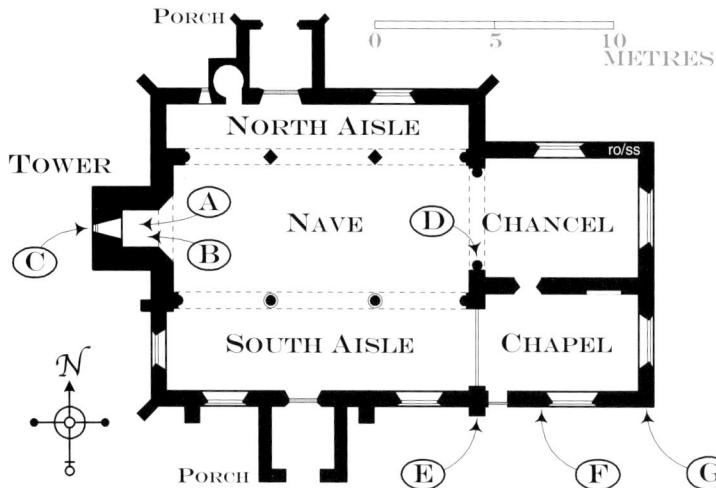

Plan of St Andrew's Church
A: One bell in the belfry. Ringing chamber at the base of the tower
B: Location of former clock on the first floor of the tower
C: Clock dial
D: Bellringer carved on Norman column
E: Scratch dial 1
F: Scratch dial 2
G: Scratch dial 3
Access to the former clock room and the belfry is by ladders

ST ANDREW   Grid Ref: SP 855968

PARISH RECORDS

There are no references to the bell or former clock in the surviving parish records.

BELL HISTORY

The earliest reference to bellringing in Rutland, and one of the earliest in the country, is in Stoke Dry Church. It is a carving on the Norman south column which supports the chancel arch. It depicts a man ringing a bell which is hanging in a bellcote. Below is a cowering animal which

has been interpreted as Satan trying to shut out the sound of the Sanctus Bell (church leaflet).

The original church, built in the twelfth century, consisted of a small nave and chancel. The north and south aisles, the chapel and the tower were added between 1200 and 1330. It is generally accepted that there was a bellcote over the west wall of the nave prior to 1300.

According to tradition the church formerly had two bells within a steeple of wood (North 1880, 152). The upper part of the present tower was built in the eighteenth century with only sufficient space for the present

From a rubbing of the Norman bellringer carved on a column to the chancel arch at Stoke Dry

Tower wall

Earlier access via nave roof

Access Hatch

Access now via ladder from former clock room below

THOMAS EAYRE III 1761

Plan of the bellframe at Stoke Dry Church

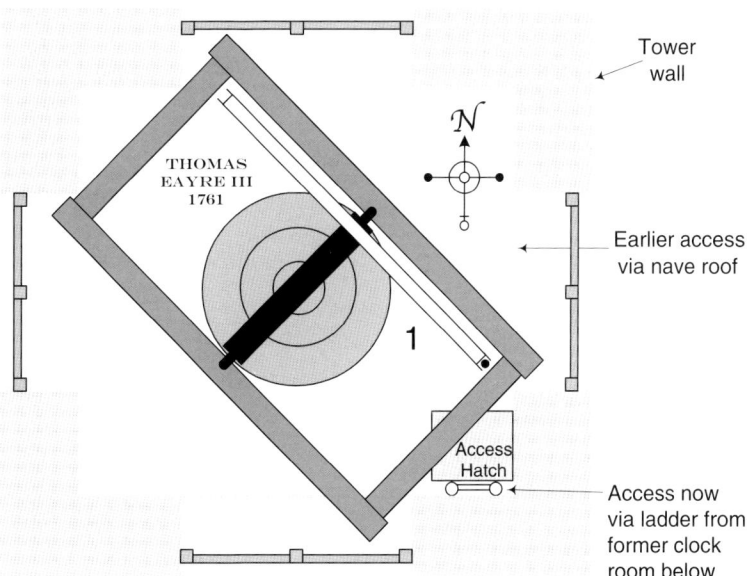

single bell. This was cast by Thomas Eayre III of Kettering in 1761, presumably for the new belfry. It hangs in a low-sided timber frame from a timber headstock which is inscribed **J. R. CHAPMAN 1912**.

Before the present internal ladders were installed, access to the belfry was via the nave roof and through a removable louvre in the belfry window. There is a story in the village (Traylen nd, Pt 2) that in 1890, at the marriage of the parson's daughter, there was a desire to create the effect of five bells being rung. To achieve this one man used the clapper to ring the bell and four others struck the outside of it with hammers. The five ringers were seen to be hauling up several barrels of ale after them and it is said that the subsequent 'ringing' was somewhat enthusiastic!

Bell Details

1 1761. Diameter 953mm (37½in). Weight approximately 483kg (9cwt 2qr). Canons retained. Timber headstock. Cast by Thomas Eayre III of Kettering.
THOS : EAYRE, DE KETTERING. FECIT. 1761:.
~ OMNIA FIANT AD GLORIAM DEI: [101]
~ LAUDATE ILLUM CYMBALIS SONORIS. •⋰•••
(Let all things be done to the Glory of God. Praise him upon the high-sounding cymbals)
The Latin inscription is based on Psalm 150 verse 5.

Bellringing Customs

On Sundays a bell was rung at 9am if Morning Prayer was to be said or at noon if Divine Service was to be in the afternoon. For this latter service the bell was tolled for fifteen minutes and then rung for a further fifteen minutes if a sermon was to be preached. At the Death

Knell there were thrice three tolls for a male, thrice two for a female, both before and after the knell (North 1880, 152).

At the time of writing the single bell is chimed for ten minutes before the 11am service held on the first and third Sunday of each month. It is also rung before the morning service on Easter Sunday and Christmas Day and to herald in the New Year. The bell is tolled when a parishioner dies and at a funeral if requested. The bell is also rung for weddings by request.

Scratch Dials

There are three scratch dials, two on the south wall of the chapel and one on the east buttress of the south aisle. The south aisle was added in the thirteenth century and the chapel a century later. The scratch dial on the west jamb of the chapel window is considered to be in its original location. The heights of the other two suggest that they are on relocated stones.

Clock History

There is now no clock movement in the tower but evidence of a former clock remains in the form of a diamond-shaped, Swithland slate dial on the west face of the tower. The single hour hand is mounted on a shaft which terminates in a room on the first level of the tower. This is still referred to as 'the clock room'. The dial is dated 1849 and is possibly a replacement of an earlier wooden dial.

Access to the clockroom was via a ladder behind the organ in the nave. This led to a door above the tower arch. The doorway has been blocked up but the old doorframe remains. The present access ladders in the

SCRATCH
DIAL 1

350

80

CHAPEL

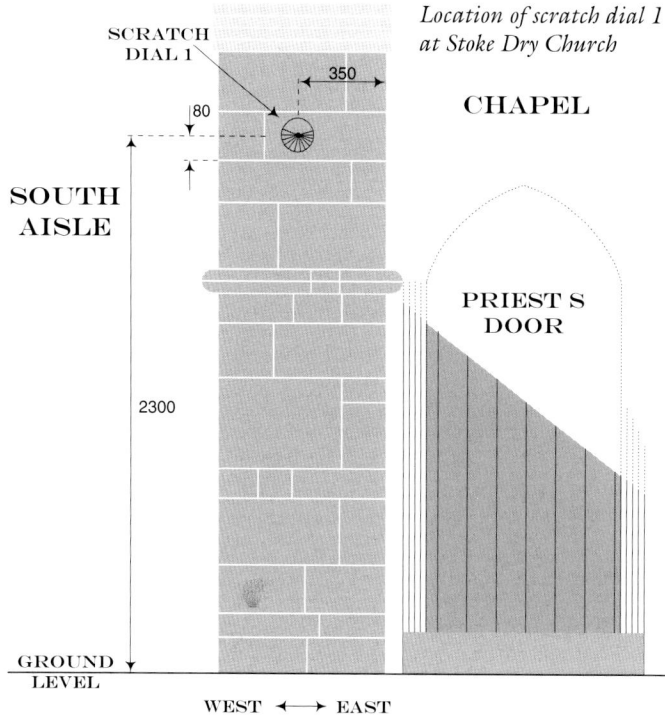

*Location of scratch dial 1
at Stoke Dry Church*

SOUTH
AISLE

PRIEST'S
DOOR

2300

GROUND
LEVEL

WEST ←——→ EAST

*Stoke Dry Church. Scratch dial 1 on the
east buttress of the south aisle*

Scratch dial 1 details

STOKE DRY 1													
Location	East buttress of the south aisle												
Condition	Very good												
Gnomon Hole Diameter			17										
Gnomon Hole Depth			20										
Height above ground level			2300										
Circle diameter			160										
Line Ref	a	b	c	d	e	f	g	h	i	j	k	l	m
Length	80	80	80	80	80	80	80	80	80	80	80	80	80
Angle (°)	72	95	115	127	145	158	175	190	205	223	242	264	283

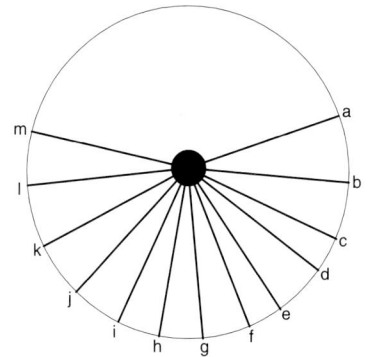

*Location of scratch dial 2
at Stoke Dry Church*

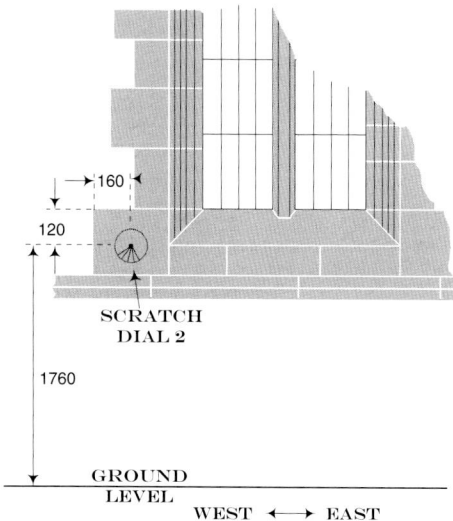

160

120

SCRATCH
DIAL 2

1760

GROUND
LEVEL

WEST ←——→ EAST

Scratch dial 2 details

STOKE DRY 2									
Location	West jamb on the south window of the chapel								
Condition	Poor								
Gnomon Hole Diameter			15						
Gnomon Hole Depth			10						
Height above ground level		1760							
Line/Pock	a	b	c	d	e	f	g	h	i
Length/radius	78	80	80	83	86	84	75	80	83
Angle (°)	62	80	89	115	140	165	167	184	188

Line/Pock	j	k	l	m	n	o	p	q
Length/radius	85	84	85	75	83	75	82	80
Angle (°)	202	215	224	224	238	242	266	292

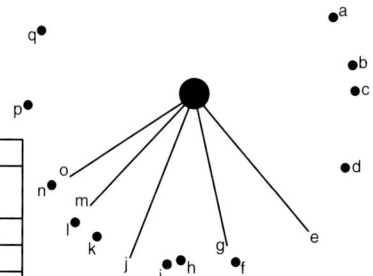

CHAPEL

SCRATCH
DIAL 3

120

280

2860

RETAINING
WALL

GROUND
LEVEL

WEST ←→ EAST

*Location of scratch
dial 3 at Stoke Dry
Church*

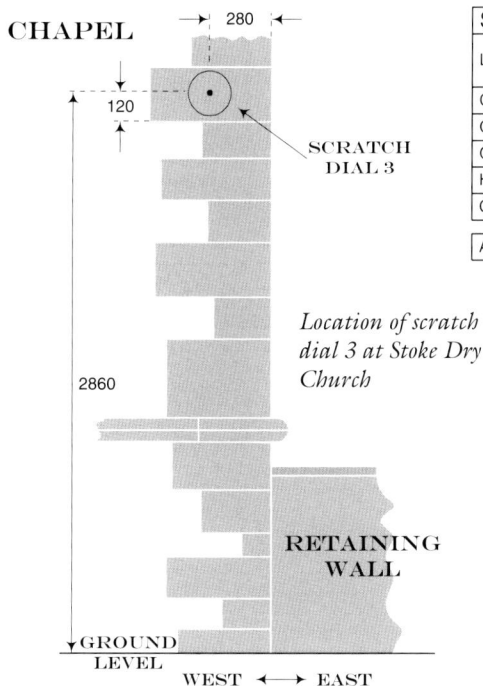

STOKE DRY 3		
Location	South-east angle of the chapel	
Condition	Very poor	
Gnomon Hole Diameter		8
Gnomon Hole Depth		10
Height above ground level		2860
Circle diameter		170

All lines have been weathered away

Details of scratch dial 3

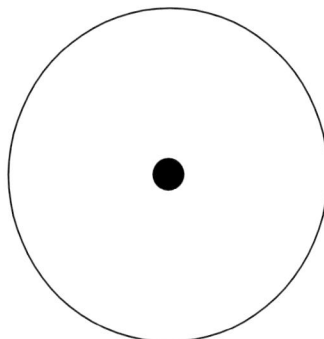

clock had been placed in the font (information from Ron Pace), and more recently, parts of the clock were found when an area of the churchyard was cleared. The weather vane was also found and this was replaced on the tower (information from Bob Salmon).

The slate clock dial on the west face of the tower at Stoke Dry. Note that it is marked out for a single hour hand, having four divisions between the hours

tower which lead to the clockroom and belfry are recent additions. A clock could well have been installed when the new tower was built in the eighteenth century. Thomas Eayre III of Kettering who cast the bell in 1761 may also have supplied the clock. There are, however, no surviving records to substantiate this theory. The clock was removed sometime in the late 1800s by the then vicar who was annoyed by its striking. In the 1960s it is remembered that wheels, arbors and levers from the old

STRETTON

ST NICHOLAS   Grid Ref: SK 949158

PARISH RECORDS
The parish records searched include the Vestry Minutes and Accounts for 1858-94 (DE 5189/5), 1895-1904 (DE 5189/6), and 1906-41 (DE 5189/7), and a faculty to restore the church dated 1881 (DE 5189/9).

BELL HISTORY
The double bellcote is early thirteenth century, confirming that there have been bells here from that time. The earliest of the present two bells is by Thomas Norris dated 1663, but references to earlier bells are made in Irons' Notes (MS 80/5/24). In 1607 it is reported 'the bells out of order & some of them broken'. In 1618 it is revealed that 'they have 3 bells & but one rope'. The third bell was a Sanctus Bell, and the bellcote for this is

shown over the east gable of the nave on a *circa* 1793 drawing of the church. This drawing also shows a structure with a ridged roof behind the double bellcote which presumably provided some weather protection to the bells and bellropes. Both this structure and the Sanctus bellcote had been removed by *circa* 1839. They are not shown on a drawing of this date in Uppingham School Archives.

There is little historical detail concerning the bells in the Vestry Minutes and Accounts. Most of the relevant entries refer to replacing the bellropes:

1855 June 22 Paid for Bell Ropes and putting up 15s 0d
Easter 1918 to Easter 1919
                    Mr Robert Littledyke (bellrope)   16s 0d

Early in the nineteenth century it was found that the west wall of the nave was leaning into the church. In an

Plan of St Nicholas's Church
A: Two bells in a bellcote. Bells rung from the west end of
the nave
B: Location of former Sanctus bellcote
C: Sundial
D: Scratch dials 1 and 2

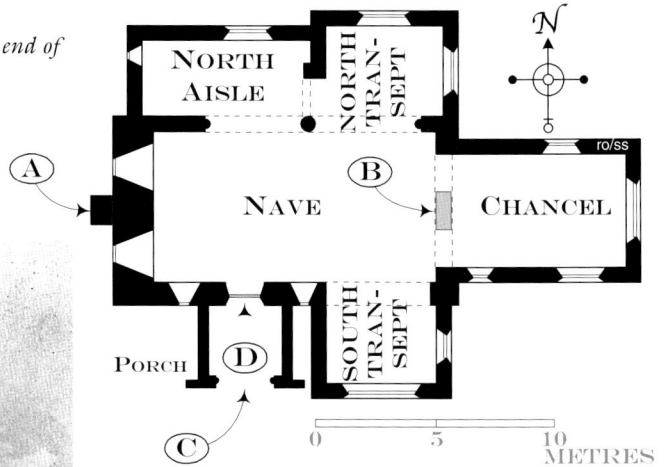

A late eighteenth-century view of Stretton Church (RCM F10/1984/51)

attempt to solve this problem an inner wall was built three feet from the old wall, the intention being to support the bellcote above. By 1873 the original wall was again causing concern. A faculty was granted in 1881 to restore the church and the work included the removal of the inner wall, considered to be a great eyesore, and the rebuilding of the west wall and the double bellcote. The stones were numbered, carefully preserved and replaced in their original position, new stone being used only where necessary. The church re-opened in 1882 but in 1885 the bellcote was severely damaged by a storm and it was again taken down and rebuilt.

An entry in the Vestry Minutes and Accounts shows that the parish was ready to contribute towards the war effort, for at a meeting held on Monday 5 April 1915 it was decided 'for the sake of War Economy' not to engage a Sexton at that time. The organ blower would take on the Sexton's duty of ringing the bells.

BELL DETAILS

1 **1710**. Diameter 559mm (22in). Weight approximately 127kg (2cwt 2qr). Canons retained. Timber headstock. Cast by Henry Penn of Peterborough.

[96] HENRY PENN FVSORE [96] 1710
(Henry Penn founder)

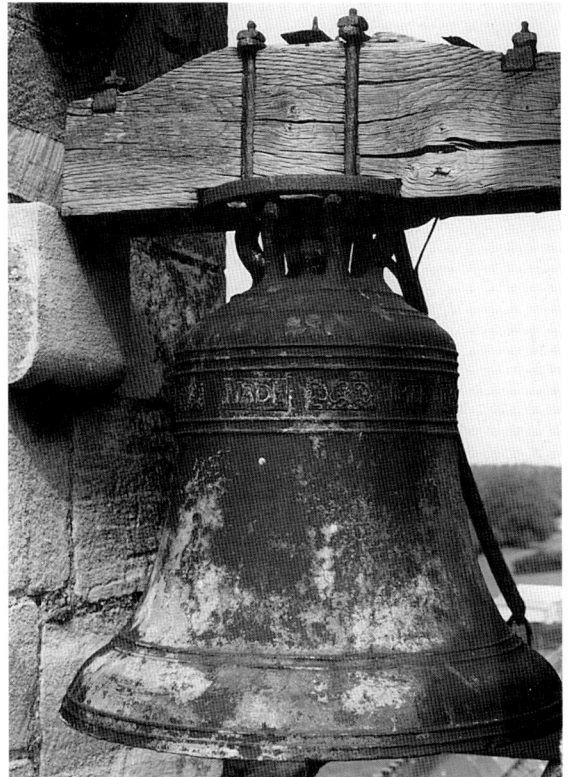

The Norris bell at Stretton Church showing the old headstock. In 1988, Nicholas Meadwell rehung both bells with new headstocks and bearings. The new headstocks are engraved **NJM 1988** (Nicholas Meadwell)

Decoration [96] may also be between the words.

2 **1663**. Diameter 584mm (23in). Weight approximately 140kg (2cwt 3qr). Canons retained. Timber

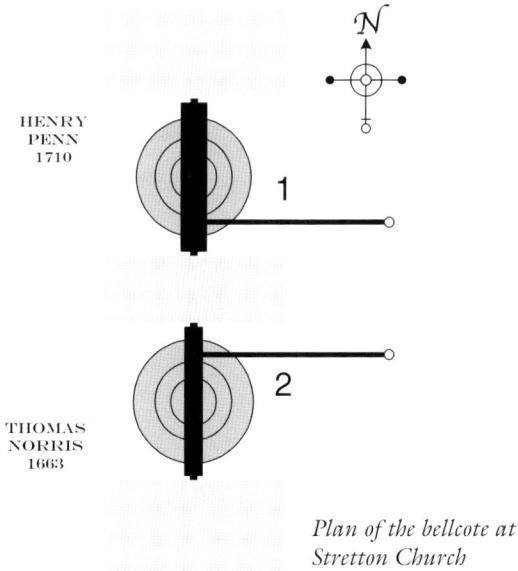

HENRY
PENN
1710

1

THOMAS
NORRIS
1663

2

*Plan of the bellcote at
Stretton Church*

*Stretton Church.
The bellcote and bells*

was added. Masons' marks, in the form of a group of circles in a geometric pattern, are evident on the wall near the capital of the west column, as shown in the scratch dial location drawing.

SUNDIAL

A crude sundial is marked on a stone in the gable of the thirteenth-century south porch. The angled gnomon is

headstock. Cast by Thomas Norris of Stamford.

**[10] THOMAS [52] NORRIS [52]
~ MADE [52] MEE 1663**

BELLRINGING CUSTOMS

No early customs have been recorded. Today one of the two bells is rung for five minutes prior to a service, which is usually held once every three weeks. Although the bells are not generally used on any other occasion they were rung on 1 April 1997 when Rutland regained its county status and at midday on the first day of the new Millennium.

SCRATCH DIALS

Two scratch dials have been recorded. Both are on the tympanum above the twelfth-century south doorway inside the porch, an unusual location for scratch dials. A raised ridge on the rear of the tympanum suggests that it may originally have been a coffin lid.

The upper scratch dial would have been in shadow for much of the day due to the semicircular moulding which envelopes it. However it is possible that this large stone was in a different and lower location before the porch

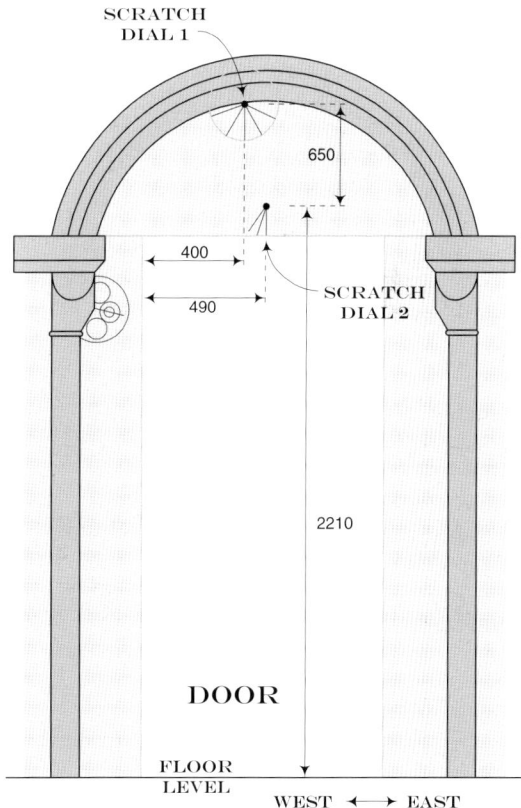

Location of scratch dials 1 and 2 at Stretton Church

STRETTON 2			
Location	On tympanum inside the south porch		
Condition	Poor		
Gnomon Hole Diameter		25	
Gnomon Hole Depth		15	
Height above ground level		2210	
Line Ref	a	b	c
Length	110	120	130
Angle (°)	173	198	222

*Scratch dial 2
details*

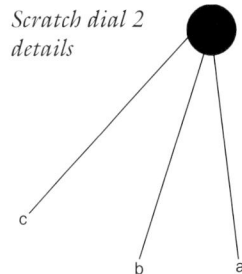

Scratch dial 1 details

STRETTON 1					
Location	On tympanum inside the south porch				
Condition	Average				
Gnomon Hole Diameter		25			
Gnomon Hole Depth		Filled in			
Height above ground level		2860			
Circle diameter		155			
Line Ref		a	b	c	d	e
Length		160	160	160	160	160
Angle (°)		118	148	180	215	251

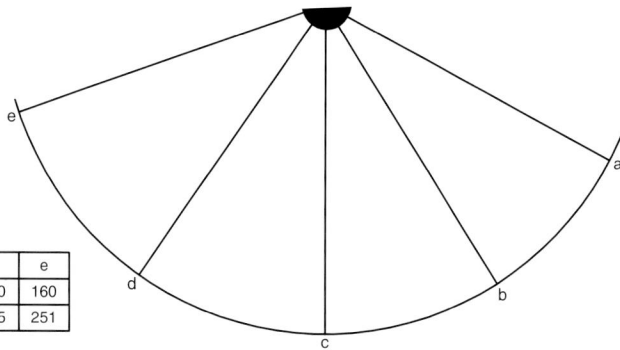

*Stretton church sundial. Note the two holes for the gnomon and the use of **VIIII** for nine and a cross for noon* (Walter Wells)

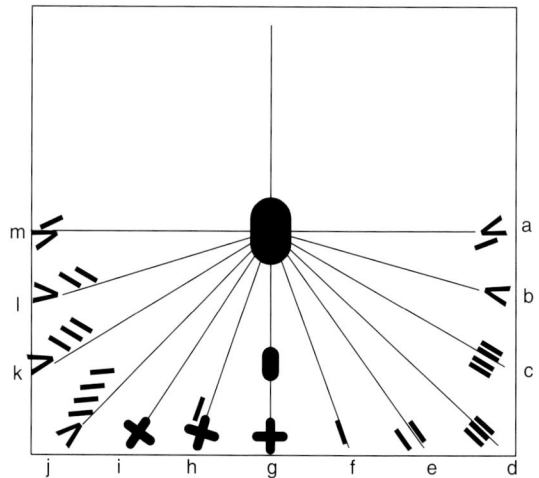

Details of the sundial at Stretton Church

Line Ref	a	b	c	d	e	f	g	h	i	j	k	l	m
Length	112	112	170	180	150	140	120	140	150	150	160	120	120
Angle (°)	90	105	119	132	144	159	180	199	214	225	239	253	270

missing but the roman numerals can still be seen. It has been suggested that this was originally a scratch dial with the gnomon and numerals added at a later date (information from Walter Wells) (*see* Chapter 1.4 — Scratch Dials). It is not shown on early drawings of the church.

STOCKEN HALL

Major Charles Hesketh Fleetwood-Hesketh purchased the estate of Stretton and Stocken from Lord Ancaster in 1907 (VCH **II**, 148). In 1914 he installed a clock in the Coach House at Stocken Hall. The large high-quality flatbed movement was supplied by John Walker two years after he had moved into his South Molton Street premises in London (AHS **23**, 338) (*see* Chapter 3 — Clockmakers). It was restored *circa* 1990 when the Coach House was converted into a dwelling.

The clock dial and bell over the converted Coach House at Stocken Hall

CLOCK DETAILS

Details of the present clock:

Maker: Possibly John Walker, London, although this company also installed clocks by other makers

Signed: On the frame 'JOHN WALKER 1, SOUTH MOLTON ST. LONDON. TO H.M. THE KING', and on the setting dial JNO WALKER, LONDON. 9341.

Installed: 1914

Frame: Large cast-iron flatbed

Trains: Going and striking

Escapement: Double three-legged gravity

Pendulum: Wooden rod and cast-iron cylindrical bob

Rate: 47 beats per minute

Striking: 24 hour rack striking. A double snail has adjustable cover plates to provide night silence. Plates can be removed to allow twenty-four hour striking, or replaced by a disk to prevent striking altogether

Clock Bell: Hung from a beam in the turret over the Coach House. Cast by John Warner in 1914 (see below)

Weights: Cast-iron

Winding: Weekly

Dial: One convex copper dial with gilded roman numerals and hands on a black background

Note: Restored *circa* 1990

The Stocken Hall Coach House clock movement, signed by John Walker of South Molton Street, London

The double three-legged gravity escapement of the Stocken Hall Coach House clock

Stocken Hall Coach House clock. The cover plates on the double snail, which turns once in 24 hours, prevents the clock from striking through the night. This night-silence period is adjustable

The rack on the Stocken Hall Coach House clock. It works in conjunction with the snail to ensure that the clock strikes the correct number at each hour.

CLOCK BELL DETAILS

1914. Diameter 965mm (38in). Weight approximately 508kg (10cwt). Cast without canons. Hung dead from a beam in the turret over the Coach House. No clapper. Cast by John Warner & Sons of London.

[134] [140] C. h. F.h. [140] [134] [140]
~ A. D. F.h. [140]
[134] WARNER. LONDON.

On the waist:

A. D.
1914.
STOCKEN HALL.

On the waist opposite:

[139]

[139] is the Fleetwood-Hesketh coat of arms. Apart from the lettering, [134] fills both inscription bands. The initials are those of Major Charles Fleetwood Hesketh and his wife Anne (information from Sue Howlett).

Major Hesketh regularly attended Stretton Church. His pew was immediately in front of the pulpit and if the sermon lasted more than ten minutes he would pull out his gold watch and place it prominently on the book rest in front of him (Howlett 1998, 12).

STRETTON SCHOOL

Stretton School, although now a private house, still retains the old school bell in its bellcote. The Vestry Minutes of 1873 record that the school and a house for the schoolmistress were provided by the second Lord Aveland of Normanton Hall who then owned the estates of Stretton and Stocken. The school finally closed in 1960. The bell is dated **1879** on the waist, above which is a decorative **A** [135], possibly for Aveland. The name

The Fleetwood-Hesketh family crest on the Stocken Hall Coach House clock bell. The motto is interpreted as 'What is [done] to you [do] this to another'

of the founder, **JOHN WARNER & SONS**, is on the soundbow.

The following note regarding the school clock is of interest:

The Rev. Barry visited to say that the school clock was 10 minutes slow, and that this was serious as in the absence of a church clock, the whole village went by that at the school. Stamford Post Office were responsible for checking the clock every week, and word was sent urgently that the Post Office verifying clerk should call immediately (Traylen 1999, 1950).

The bellcote and bell at The Old School, Stretton

TEIGH

HOLY TRINITY

Grid Ref: SK 865160

PARISH RECORDS

Apart from 'The Town Bill' for the year 1754 (DE 3855/9) no Churchwardens' Accounts have survived. Fortunately, the Churchwardens' and Constables' Accounts 1703-1809 were available in 1926 when the Rev E B Redlich, late Rector of the parish, published *The History of Teigh*. This book includes many examples from these accounts and other documents which were still available at the time. Other examples were transcribed in *Rutland Magazine* (Phillips 1907-08, 30-1).

BELL HISTORY

Two of the three bells in the belfry were supplied by Thomas Eayre II of Kettering in 1746. The tenor is the only bell in Rutland cast by Henry Oldfield I of Nottingham.

In *The History of Teigh*, the Rev E B Redlich comments on the accounts of 1703/4: 'The great event this year was the repair of the Church Bell. The details are amusing, and of course the great event had to be celebrated on six occasions, even when the parish bargained, at the expense of the parish. In the 1703 accounts we have read that 1s 4d was spent when the bell was taken down' (Redlich 1926, 104). The following extracts are from the accounts of churchwarden John Edgson:

1704

pade to widdow Hinman for caring the Bell to Stamford	5s 0d
pade to Edward Billins for bringin the Bell hom	2s 6d
pade for my hors when ye Bell came home	1s 0d
Spent ye same time	2s 0d
pade to the Bell founder for runing [casting] ye Bell and 27 pound of mettell	£6 16s 0d
spent when ye Bell was houng up ...	1s 10d
spent whe ye Bell founder was pade his muney	1s 0d
spent when ye Bell founder came ouer the first time	6d
spent when we bargind with him to run the Bell	1s 0d
spent at Stanford [Stamford] when ye Bell went	6d

The bellfounder referred to here was Alexander Rigby who had by this time taken over the Stamford bell foundry from Toby Norris III. Widow Hinman and Edward Billings, who transported the bell to Stamford and back, are both recorded as being farmers in the parish of Teigh (Redlich 1926, 126).

'ye Bell Player' mentioned in the following extract from the same year suggests that the church then had a clavier:

| **1704** | pade to peter Blackband and James Ketell for mending ye Bell Player | 7s 0d |
| | pade to William Wyare for minding ye grate Bell Player and other worck | 10d |

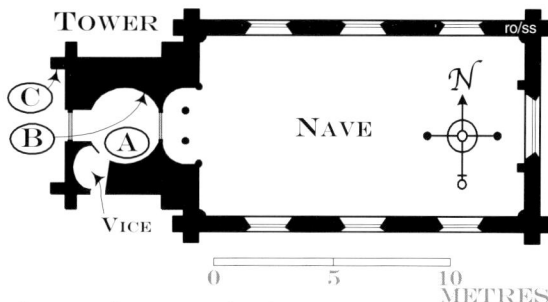

Plan of Holy Trinity Church
A: Three bells in the belfry. Ringing chamber on the first floor of the tower
B. Chiming frame in the ringing chamber
C. Scratch dial
The vice gives access to the ringing chamber

Robert Sherard, the fourth Earl of Harborough, rebuilt his medieval estate church at Teigh in 1782, and those at Saxby and Stapleford in 1783. In 1792-93 Teigh spire was removed and the upper stages of the tower rebuilt. A new oak bellframe with pits for three bells was installed two years later. It is thought that prior to this Teigh had five bells, as there is 'a tradition that two bells were removed from this Church to that at Stapleford' (North 1880, 153). No other information regarding this transfer has come to light although there is an unused bell in the vestry at Stapleford Church. This bell, which has lost its clapper, is dated 1733 but there is no other inscription. It has been confirmed that it was cast by Thomas Eayre II of Kettering (information from George Dawson).

The other transferred bell may have been that recast by Alexander Rigby which is referred to in the accounts of 1704 (see above). However, Thomas North does not list a Rigby bell at Stapleford or Saxby in his 1876 survey of Leicestershire. It may alternatively have been recast into one of the two bells supplied to Teigh by Thomas Eayre II in 1746.

The low-sided oak bellframe installed in 1794 remains in the belfry today. **I x H x 1794 x C W** is carved on a diagonal timber facing the vice and **W x S x 1794 Fecit** is carved on a central strut. **I H** was J [Iohannes or John] Hinman, a churchwarden in 1794-95, and **W S** was William Sims, a local carpenter. The following items from the accounts provide some details of the cost of this undertaking:

1794 April	Paid the Men Bord at the Church	4s 6d
	Paid W. Sims on account to wardes the Bell frames	£10 0s 0d
	Sims bill	£7 3s 0d

1795 April 11 Sims Bill (Credit. W. Sims old wood. 7s.)
£15 19s 2½d
April 4 Sims Bill   £6 9s 8d

There are very few items in the accounts which refer to the repair and maintenance of the bells. Examples include 'keeing' the bells [a key was a brass wedge used to locate and adjust the headstock bearings], mending the bellwheels and the repair and purchase of new bellropes. In 1814 'the three bell ropes weighed 15lbs. and cost 1s 2d a yard'. Very little other information is available.

In 1958 all three bells were removed from their headstocks and hung dead from new timber beams laid across the frame. This work was carried out by John Taylor & Co who also installed a chime hammer to each bell and an Ellacombe-type chiming frame on the north wall of the ringing chamber.

BELL DETAILS

1 1746. Treble. Diameter 819mm (32¼in). Weight about 305kg (6cwt). Canons retained. Hung dead from a timber beam. Cast by Thomas Eayre II, Kettering.

OMNIA FIANT AD GLORIAM DƎI ⁘ [117] ⁝
~ GLORIA DEO SOLI ⁚• [99] •⁚ T •⁚• EAYRE
~ •⁝• A : D •⁚• 1746 •⁚•• [117] ••⁚••

(Let all things be done to the Glory of God.
Glory to God alone)

2 1746. Diameter 841mm (33⅛in). Weight about 317kg (6cwt 1qr). Canons retained. Hung dead from a timber beam. Cast by Thomas Eayre II, Kettering.

OMNIA FIANT AD GLORIAM DEI •⁚• [99] ••⁚
~ GLORIA PATRI FILIO ET SPIRITUI
~ SANCTO •⁚• 1746 •⁚•

(Let all things be done to the Glory of God. Glory be to the Father, to the Son and to the Holy Ghost)
The canons are decorated with cross-hatching and vertical beading.

3 *Circa* 1540. Tenor. Diameter 876mm (34½in). Weight about 356kg (7cwt). Canons retained. Hung dead from a timber beam. Ascribed to Henry Oldfield I of Nottingham.

IN [*] NOIE IhS MARIA [19] [18] [118]

(In the Name of Jesus. Mary)

There are abbreviation marks above the **IƐ** and **hS** (see illustration). Spaced around the soundbow:

O O O O

W. H . 1702 is scratched on the bell in position [*]. The gothic lettering is like **[136]**. **O O O O** are four impressions of a 1526-40 silver groat, two of the obverse and two of the reverse. The canons have a banded decoration with claws at the base. Scheduled for preservation (Council for the Care of Churches).

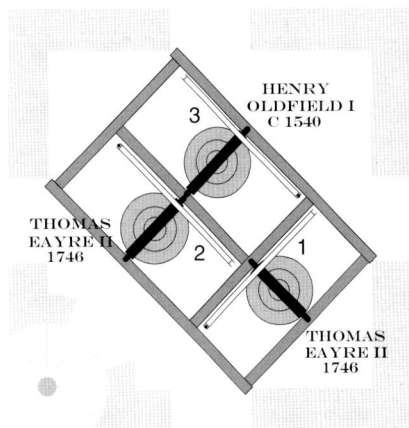

Plan of the low-sided timber bellframe at Teigh Church

The arrangement of the bells at Teigh prior to 1958 when they were hung for full-circle ringing

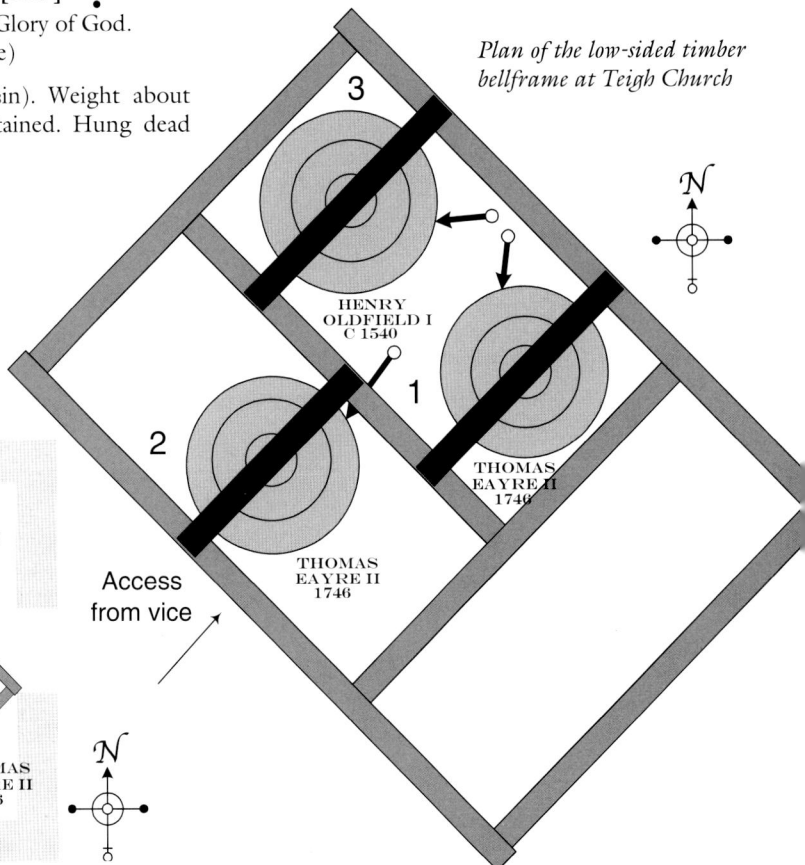

The inscription on the tenor by Henry Oldfield I at Teigh

BELLRINGING CUSTOMS

Ringing and looking after the church bells were important duties for the parish clerk and they were tasks he took seriously. Details from the seventeenth century Registers of the Diocese of Peterborough reveal that Teigh had its share of problems regarding the morality and scandalous activities of its inhabitants. The following account relates an incident concerning the bells at Teigh (Redlich 1926, 88):

> **1614** Thomas Smith goeth oute of the churche in prayer time to ffeast twice and did threaten the clarck in the church and brawled with him in the churchyard because he would not let him ringe.

The Churchwardens' Accounts include numerous allusions to public events and it seems that few were allowed to pass without a peal from the church bells. For example, in common with many other parishes, the bells at Teigh were rung in celebration of the overthrow of the Gunpowder Plot. Other examples of ringing for such events are included in an article in *Rutland Magazine* (Phillips 1907-08, 30):

Location of the scratch dial at Teigh Church

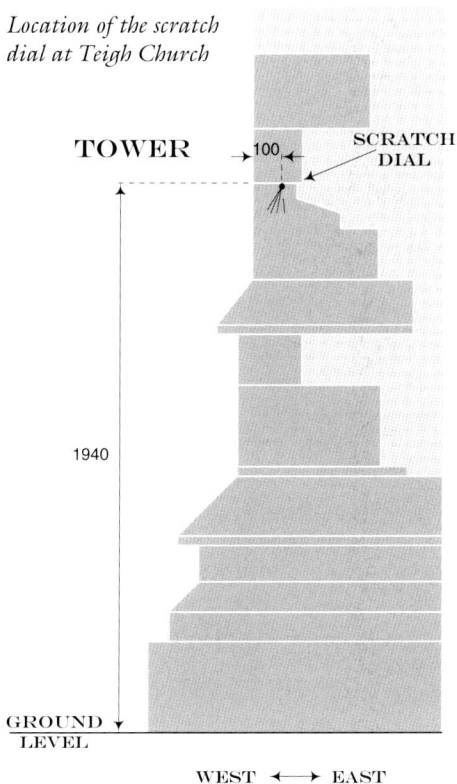

1706	spent that day ye news came ye french was beaten [victory at Ramillies]	1s 6d
1708	given to the Ringers St Georges day	1s 6d
	given to the Ringers the Thanksgiving day	1s 6d
	given to the Ringers ye 5 of November	1s 6d
1710	Given to ye Ringers on Quen Ann Berth Day	1s 6d
	Giving to ye ringers on ye 8 of March ye prockelamashon of Quen Ann	1s 0d
	Givens to ye Ringers on ye Quens Crounashon Day	1s 6d

1762 July 13

> To ye Ringers on ... ye birth of a young Prince 5s 0d

The following entry records when 'Old John Squire [the parish clerk] died and the Parish bore the expenses of his funeral and got 11s. 7d by the sale of his little possessions':

| 1707 | for the passing bell and grave | 1s 0d |

In 1926 the parishioners could remember other former customs. These included the 8am bell rung on Sundays to remind people of the 11am service, the Pancake Bell which was rung at 11am on Shrove Tuesday to summon all apprentices to a meal of pancakes, the ringing of the bells on New Year's Eve, the Gleaning Bell, the Curfew Bell, and the Passing Bell. The bells were also chimed after a funeral. Customs which were still current at that time included the ringing of the Death Knell when three tolls were given for a male and two for a female, followed by a number of tolls to indicate the age of the deceased. The Warning Bell was rung an hour before a funeral to prepare the bearers, followed by the Invitation Bell rung half an hour later to warn the bearers and others of the forthcoming service. For Divine Service all the bells were chimed and the Sermon Bell was rung in single strokes immediately before the service (Redlich 1926, 54).

Today the bells are chimed for fifteen minutes before each service.

SCRATCH DIAL

A scratch dial on the south face of the fourteenth-century north-west buttress to the tower is almost certainly on a relocated stone as it is in shadow until after midday. The lines are very shallow and can only be seen in any detail when the sun is in the west.

TEIGH				
Location	South face of north-west tower buttress			
Condition	Poor			
Gnomon Hole Diameter			10	
Gnomon Hole Depth			Filled in	
Height above ground level			1940	
Line Ref	a	b	c	d
Length	62	90	100	85
Angle (°)	170	188	206	222

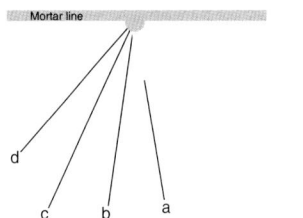

Scratch dial details

CLOCK HISTORY

Teigh does not have a clock now, but extracts from the accounts recorded by Mr Redlich provide definite evidence of a former clock. The earliest reference is in 1703 when John Squire was paid 14s 8d 'for looking after the clock and the surplis washing ...'. John was the parish clerk whose duties would have included the daily winding of the clock and its general maintenance. Payments were made 'for ye clock rope' in 1706 and 'oyle for ye Clock' the following year. With the death of John Squire, John Stanger became the clerk and in 1710 he reported that a board was required for repairs 'for ye clock hous windo'. At this time the payment for winding the clock was 10s a year.

The following extract gives the names of two men, possibly local blacksmiths, who carried out repairs on the clock (Phillips 1907-08, 30):

1710 p'd to Peter Blackbond for mending ye clock   6d
p'd to Phillpot for mending ye Clock & spout 3s 6d

No other facts concerning the church clock have come to light. There are no references to a dial and it probably did not have one. How long the clock continued to function is unknown. One possibility is that it was removed when the church underwent considerable rebuilding and restoration in the late eighteenth century. Evidence of the former clockroom can be seen in the ringing chamber. Here the level of a former floor can be identified, and the window and door to the clockroom still exist.

THISTLETON

Plan of St Nicholas's Church
A: One bell in the belfry. Ringing chamber at the base of the tower
B: Clock on the first floor of the tower
C: Clock dial
The vice provides access to the clockroom and there is a ladder from the clockroom to the belfry

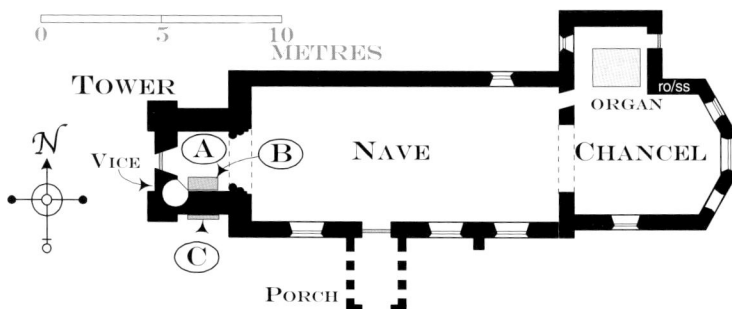

ST NICHOLAS   Grid Ref: SK 913180

PARISH RECORDS

The only relevant records are the correspondence, meeting reports and estimates concerning the new church clock which was installed in 1887 to celebrate Queen Victoria's Golden Jubilee (DE 5188/12/1-13). There are no surviving Churchwardens' Accounts.

BELL HISTORY

Little is known of the history of the bells at Thistleton. There were formerly two bells here but as both were cracked they were cast into the present bell in 1793 (North 1880, 153). Wright's *Directory* of 1880 records that the church was 'recently restored at the expense of the patron ... amongst other things the bell was rehung'. This bell was originally hung in the present timber frame for full circle ringing but the lower half of the wheel has been removed to clear the clock hammer. Consequently, only chime ringing is now possible.

BELL DETAILS

1 1793. Diameter 772mm (30⅜in). Weight approximately 292kg (5cwt 3qr). Canons retained. Timber

headstock. Recast by George Hedderly of Nottingham.
W: FREAR CH. WARDEN G. H. NOTT[N] 1793.
W: FREAR was incised on this bell after it had been cast. The remaining space on the inscription band shows signs that an inscription and decoration have been removed. It is possible that this bell was cast for another

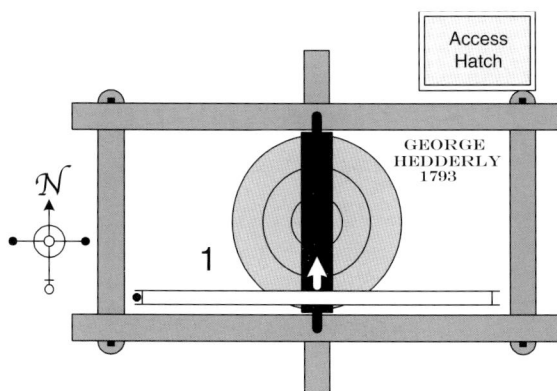

Plan of the bellframe at Thistleton Church

Unusually, the slider is pivoted outside the bellframe at Thistleton Church

church and adapted for use at Thistleton. The clock strikes the hours on this bell.

FORMER BELLS

No details of the two earlier cracked bells have survived but it is known that a Mrs Fludyer generously paid for these to be recast (North 1880, 153). It is thought that this Mrs Fludyer was Lady Mary Fludyer (1773-1855), the wife of George Fludyer of Thistleton (information from Betty Finch). They were married on 16 January 1792 (*Burke's Peerage* 1967, 2633). He built Ayston Hall in 1810 and their third son, John Henry, born at Thistleton in 1803, was later to become Rector of both Ayston and Thistleton (Waites 1987, 274).

BELLRINGING CUSTOMS

A bell used to be rung daily at 8am and 1pm and the Pancake Bell was rung at 11am on Shrove Tuesday. At the Death Knell there were thrice three tolls for a male and thrice two tolls for a female (North 1880, 153). An entry in Thistleton School Manager's Book states:

1888 October and the annual absence as all go out to the fields, gleaning (Traylen 1999, 200).

At the time of writing the bell is chimed before the Sunday Service which is held fortnightly.

CLOCK HISTORY

Wright's *Directory* (1880, 526) records: 'There is a clock, by which the hours are struck on the bell, but there is no dial'. This is the only evidence of a clock found prior to the present installation, which was commissioned to celebrate the Golden Jubilee of Victoria in 1887. The following notes are from the minutes of the 'Jubilee Meeting' held on 30 March 1887:

The question whether or not any steps should be taken to celebrate the Jubilee was decided in the affirmative. Two schemes found favour. One that a new Parish well should be sunk, found considerable support, but a fatal objection was raised with regard to it, the matter of expense, though perhaps rather exaggerated reactions prevailed on this point. But another scheme proposed by Mr. J Goodacre and seconded by Mr. W Towell, that a clock should be erected in the church Tower, received practically unanimous support, and it was at length decided that the Jubilee should be celebrated by the erection of a Church Clock. And a committee was formed to carry out this scheme. Composed of the following:

Revd Mr Thomson	Mrs Thomson
Mr W Hardy	Mrs Hardy
Mr Towell	Mr Goodacre
Mr E Edwards	

There was also a general feeling that there should be a day of rejoicing and that that day should be when the church clock was finished.

William Towell was the village schoolmaster, William Hardy was a farmer and John Goodacre was a grazier (White 1877, 699). At this time the lord of the manor was Sir John Henry Fludyer, Bart, MA (1803-96). By inheritance from his brother, General William Fludyer of Ayston Hall, John Henry owned most of the land in Thistleton. As already noted, he was Rector of both Thistleton and Ayston and in 1879-80 he rebuilt Thistleton Church, with the exception of the tower, in memory of his three elder children who had all died of scarlet fever.

On the death of William in 1863, Sir John moved from Thistleton Rectory to Ayston Hall (Waites 1987, 274) but he still maintained an active interest in village matters as indicated by the following letter (DE 5188/12/13):

Ayston Hall
Uppingham
April 15th [1887]

Dear Mr Thomson

I had heard that you Thistleton "good people" intended doing something to celebrate the Jubilee, & proposed getting a Church Clock. I have been so completely laid up with a nasty cold & cough that I had been unable to do or think about anything. I expected to hear from some of you & then to have told you I was quite ready to join in your scheme, & will subscribe £10. I almost fear you will not get a clock for the £52 you mention — but perhaps Willie

Thistleton church clock is maintained in excellent condition

The pinwheel escapement of Thistleton church clock

Hardy knows more about prices than I do. I shall hope to get to Thistleton when the weather is warmer, & shall hear from you how you get on. You must write to me again.

I am

Very Truly Yours

J. Henry Fludyer

At a meeting on 16 April the chairman, presumably Mr Thomson, stated that '... promises and payments have now reached a total of more than £40'. This, together with Henry Fludyer's donation, meant that the clock could be ordered. It was further commented '... that both the men and women of this village, have in this undertaking manifested a self denying enthusiasm beyond all expectation'.

Quotations had been sought from J B Joyce & Co of Whitchurch and Gillett & Co of Croydon, as well as John Smith & Sons of Derby who were to get the order.

On 19 April Henry Fludyer again wrote to Mr Thomson (DE 5188/12/12):

Towell tells me how very enthusiastic all Thistleton is about the clock. I think you have done well in selecting the Derby people, & I feel sure that they will make you a good Clock. It looked like "business" sending one of the Firm to look at the building. I have no doubt you will get

money enough but I am quite ready to add a bit when you come to the finishing up if necessary. A good clock will finish off the little Church nicely & you will be able to let your Neighbours know the "time".

CLOCK DETAILS
Details of the present clock:

Maker:	John Smith & Sons of Derby
Signed:	On the setting dial and on the frame
Installed:	1887 to celebrate Queen Victoria's Golden Jubilee
Cost:	£52
Frame:	Cast-iron flatbed
Trains:	Going and striking
Escapement:	Pinwheel
Pendulum:	Wooden rod and cast-iron cylindrical bob
Rate:	54 beats per minute
Striking:	Countwheel hour striking
Weights:	Cast-iron
Winding:	Hand wound weekly
Dial:	Blue skeleton dial with gilded roman numerals, minute ring and hands
Location:	South face of the tower
Note:	The clock was overhauled and repaired by Smiths in 1936

THORPE BY WATER

This parish does not have a church.

SUNDIAL
The weathered remains of a declining vertical sundial can be seen on the south elevation of 1 Main Street (SP 892965). Two holes indicate the angle of the missing gnomon.

The remains of the sundial on the south elevation of 1 Main Street, Thorpe by Water

TICKENCOTE

ST PETER Grid Ref: SK 991095

PARISH RECORDS

The only relevant records are a bundle of documents, including a faculty, dated 1933-34, concerning work carried out on the church bells (DE 2345/25/1-20).

BELL HISTORY

The lack of archival material prevents a detailed account of the history of the bells. A sketch dated 1731 and reproduced in the current Church Guide shows a double bellcote over the west end of the chancel. There is an additional arched opening set into the gable of this bell-cote and it probably housed a Sanctus Bell. A sketch by Stukeley of *circa* 1780 shows a bell in all three openings although one executed by Carter in 1785 shows the upper bell missing. This bellcote, like others in the county, was probably of the thirteenth century. Apart from the Norman chancel arch the whole church was rebuilt in 1792. The bellcote was removed and a low tower incorporating a belfry was built over the south porch.

In 1880 Thomas North recorded the two bells, but in the wrong order. No reference is made to a Sanctus Bell. At this date only the bell cast in 1630 by Thomas Norris of Stamford was being used as the undated bell was 'cracked and useless' (North 1880, 154).

In 1934 the Norris bell was sent to Taylors to be half-turned, tuned and cleaned, and a new bell ordered to replace the cracked tenor. Both bells were hung in cast-iron headstocks. The estimated cost for the 'rehanging of the smaller bell with entirely new fittings throughout' was £33. For 'supplying an entirely new bell to replace the cracked larger bell' the cost was £51 10s. An additional charge made for the inscription on the new bell was £3.

1785 engraving of 'Tickincourt Chapel, Rutlandshire' by Carter and published by S Hooper 20 July 1787

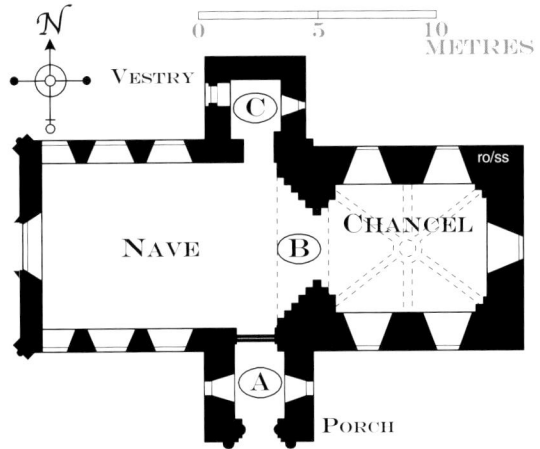

Plan of St Peter's Church
A: Two bells in the belfry. Ringing chamber at the base of the tower
B: Location of the former bellcote
C: Access to the belfry is by ladder through a trap door in the vestry and along a passage above the chancel arch

The new tenor was dedicated by the Lord Bishop of Peterborough on Sunday 25 March 1934. The cracked 'pre-Reformation Bell' has been preserved and is currently displayed in the church below the west window. Until 1990 it hung in an oak frame outside the church near the south porch door.

BELL DETAILS

1 1630. Diameter 594mm (23⅜in). Weight 123kg (2cwt 1qr 19lb). Canons retained. Cast-iron headstock. Cast by Thomas Norris of Stamford.

THOMAꙄ [53] ИORRIꙄ [53] CAꙄT [53] ME ~ [53] 1630 [53]

2 1934. Diameter 686mm (27in). Weight 204kg (4cwt 0qr 2lb). Cast without canons. Cast-iron headstock. Cast by John Taylor & Co of Loughborough.

O WORSHIP THE LORD ~ IN THE BEAUTY OF HOLINESS [71]

On the waist:

W. ST. GEORGE COLDWELL [65] RECTOR CHARLES BERTRAND MARRIOT K. C.] CHURCH ALFRED HOWARD } WARDENS

1934

On the waist opposite:

[47]

Charles Bertrand Marriot lived at Tickencote Hall. There is a memorial to him on the south wall of the nave.

Access

The pre-Reformation bell hung in an oak frame outside Tickencote Church until 1990 (Nicholas Meadwell).

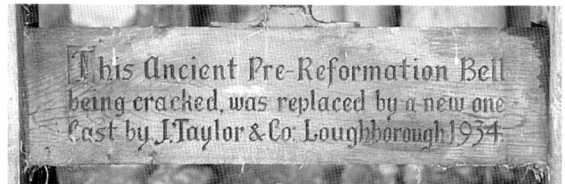

2   1

JOHN
TAYLOR
1934

THOMAS
NORRIS
1630

N

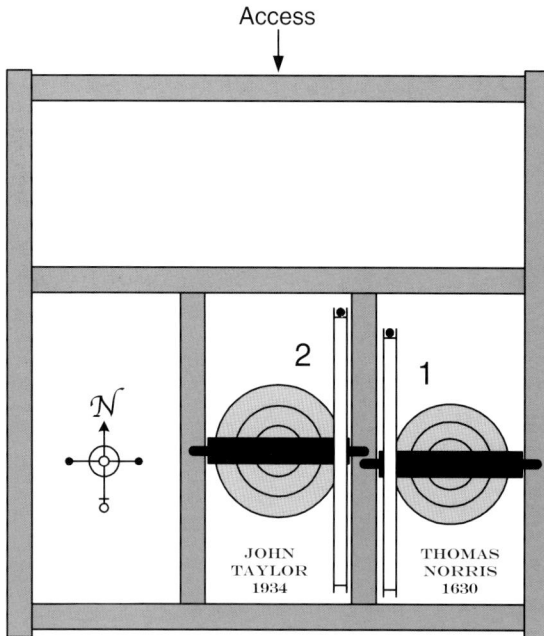

Plan of the low-level timber bellframe at Tickencote Church which was installed in 1792. It has pits for three additional bells. In 1934 the framework was strengthened with steel brackets

Former bell

Cast *circa* 1500 probably by a London founder (Council for the Care of Churches). Diameter 673mm (26½in). Weight approximately 203kg (4cwt). Canons retained. This bell was replaced by the present tenor and is now preserved in the church. It has no date or inscription. Scheduled for preservation (Council for the Care of Churches).

BELLRINGING CUSTOMS

On Sundays a bell was rung at 8am and again at 9am. For Divine Service the tenor was tolled for fifteen minutes and tolled again after the Morning Service to announce a second service. At the Death Knell thrice three tolls were given for a male and thrice two tolls for a female. It was the custom to toll the tenor for an hour while preparations were being made to convey the corpse to the church. The Gleaning Bell used to be rung prior to 1880 (North 1880, 154).

Today one of the bells is rung for two minutes before the 9.15am Sunday Service. Both bells are rung for weddings.

TINWELL

ALL SAINTS   Grid Ref: TF 006064

PARISH RECORDS

Documents searched include Churchwardens' Accounts 1781-1884 (DE 2271/29), a Vestry Minute Book covering the period 1845-85 (DE 2271/30) and an undated history of the church and parish (DE 2271/87).

BELL HISTORY

In the seventeenth century Tinwell had three, possibly four bells, all cast at the Norris foundry in Stamford. The earliest bell known is the former tenor cast by Toby Norris I which was dated 1620.

The Churchwardens' Accounts contain many interesting details of work carried out on the bells and the following are a few examples. The Mr Taylor noted in the 1811 entry was undoubtedly Robert Taylor of St Neots, whose descendants later established the well-known bell foundry at Loughborough. It is possible that the old oak frame, part of which still exists today, was installed at this time:

Plan of All Saints' Church
A: Four bells in the belfry. Ringing chamber at the base of the tower
B: Ellacombe chiming frame at the base of the tower
C: Clock on the first floor of the tower
D: Clock dial
The vice provides access to both the clockroom and the belfry

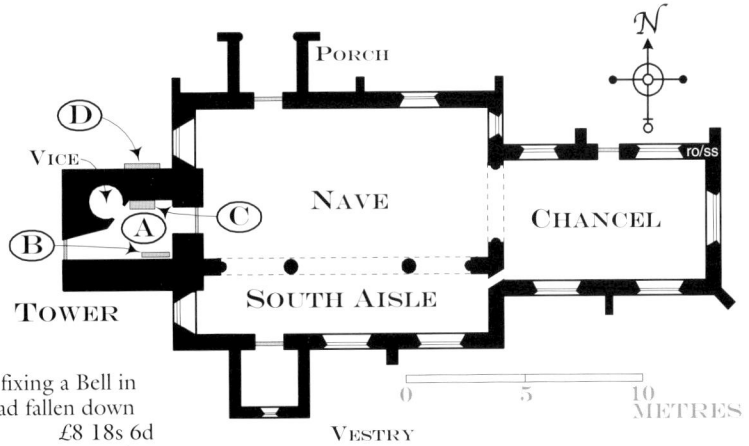

Easter 1794-95	Paid Smiths Bill for fixing a Bell in the Steeple which had fallen down £8 18s 6d
	Paid Joles Bill for Work done at the Bells and Church £9 14s 10d
Easter 1795-96	Pd Bonners Bill for Victuals and Drink whilst hanging a Bell £2 12s 2d
Easter 1799-1800	Pd Redmile [the village blacksmith] a Bill for repairing the Bells £2 8s 0d
Easter 1811-12	Pd Mr Taylor for Hanging the Bells and Carridge £28 5s 0d
To Lady Day 1880	Repair of bell, frames etc £4 13s 0d

There are frequent entries for the replacement of bellropes and amongst the suppliers named are Mr Looe [Lowe] 1781, Mr Glenn 1788, Robert Pepper 1823, Mr Palmer 1824 and Littledyke in 1834. The parish clerk had a number of duties in connection with the clock and bells, and the following examples are taken from an itemised breakdown of his salary:

1803	Pd Gouldins Bill for remander of his clarks fee	£1 13s 6d
	Do [ditto] for looking after the clock	12s 0d
	Do for putting up the bellrops	3s 6d
	Pd for Ringing	2s 0d

An inspection report of 11 December 1882 carried out by John Taylor & Co recommended that the bells should be rehung with new fittings and that the poor second and cracked tenor should be recast (closed archives of John Taylor). This work was carried out in 1883.

The bells were hung in a high-sided composite frame consisting of cast-iron 'A' frames with oak headers and base timbers. The canons were removed from both the treble and third bell. The second and tenor were cast without canons. All four bells have timber headstocks and are hung by four bolts through the crown. An Ellacombe chiming frame was also installed at this time.

Taylor's inspection report of 1882 shows that the old timber frame was in two parts. The cracked tenor was hung in a frame consisting of two beams at roof level in the tower, just below the saddleback. These beams still exist. This frame and the former frame at a lower level for the treble, second and third bells were probably installed by Robert Taylor in 1811-12.

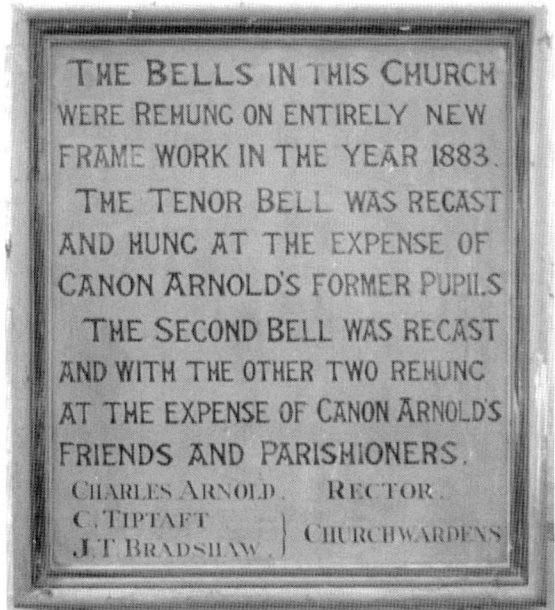

BELL DETAILS

1 1654. Treble. Diameter 762mm (30in). Weight approximately 279kg (5cwt 2qr). Canons removed. Timber headstock. Cast by Thomas Norris of Stamford.
[26] [52] THOMAS [52] NORRIS [52] ~ MADE [52] ME [52] 1654 [52]

A notice in the ringing chamber recording the work carried out in the belfry at Tinwell Church in 1883. Charles Arnold MA, Rural Dean and Honorary Canon of Peterborough, was Rector from 1827 to 1884. He died on 2 October 1884 aged 82 years. Headstones in the churchyard record that Charles Tiptaft, a farmer and auctioneer, died in 1891. John Thomas Bradshaw, farmer and grazier, was a churchwarden for forty-five years and he died in 1926, in his ninetieth year

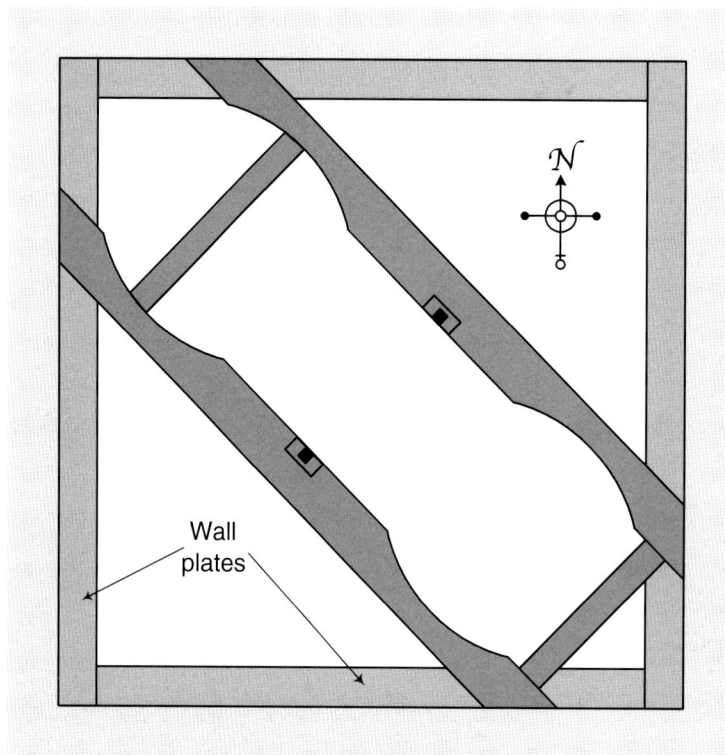

A plan of the old tenor bellframe at Tinwell which remains in the tower. The beams are 914mm (36in) apart. As the bell was 978mm (38½in) in diameter it was necessary to cut away part of the beams for full circle ringing

Wall plates

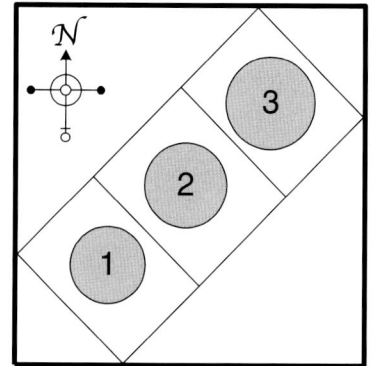

Taylor's inspection report on the bells at Tinwell also included a diagram showing the layout of the lower frame (closed archives of John Taylor)

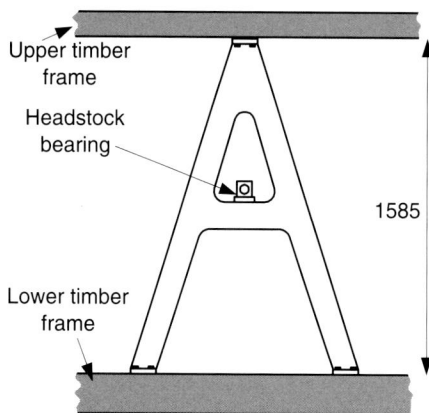

Upper timber frame

Headstock bearing

1585

Lower timber frame

A drawing of one of the cast-iron 'A' frames in the belfry at All Saint's Church, Tinwell

Thomas North reported that the canons on this bell were broken and that it was attached to the headstock by bolts. He also recorded NORRIS with one R (North 1880, 154).

2 1883. Diameter 800mm (31½in). Weight 307kg (6cwt 0qr 4lb). Cast without canons. Timber headstock. Recast by John Taylor & Co of Loughborough.

**RECAST BY JOHN TAYLOR AND C⁰.
~ LOUGHBOROUGH 1883.**

On the waist:

**RECAST AT THE EXPENSE OF
CANON ARNOLD'S FRIENDS
~ AND PARISHIONERS.
CHARLES TIPTAFT | CHURCH
JOHN THOMAS BRADSHAW |~ WARDENS.**

Former bell

Cast in 1708 by Henry Penn of Peterborough. Diameter 832mm (32¾in). Weight approximately 330kg (6cwt 2qr).

**THOMAS JOHNSON   GEORGE ALLIN
~ JOHN SISSEN HENRY GOODLAD
~ CHURCH WARDENS
{ HENRY 1708 }
{PENN MADE ME}**

According to Taylor's inspection report of 11 December 1882 this bell was recast because of its poor tonal qualities (closed archives of John Taylor).

3 1639. Diameter 889mm (35in). Weight approximately 381kg (7cwt 2qr). Canons removed. Timber headstock. Cast by Thomas Norris of Stamford.

**[10] [52] THOMAS [52] NORRIS [52]
~ MADE [52] ME [52] 1639 [52]**

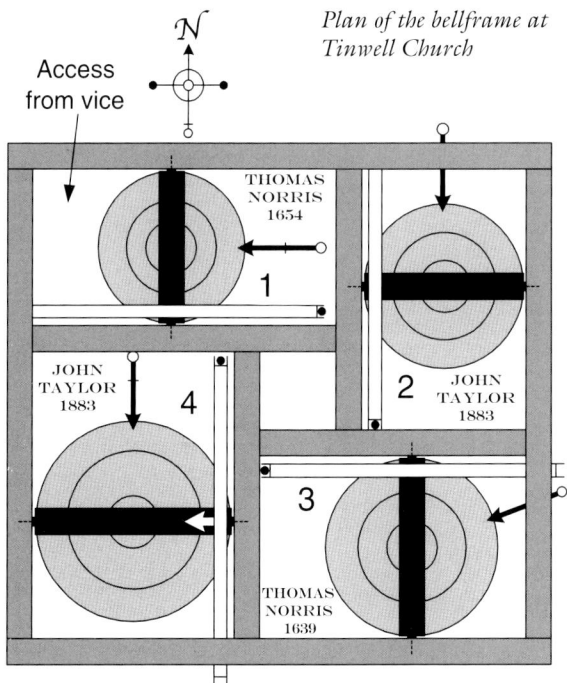

Plan of the bellframe at Tinwell Church

Access from vice

THOMAS NORRIS 1654 — 1

JOHN TAYLOR 1883 — 4

JOHN TAYLOR 1883 — 2

3

THOMAS NORRIS 1639

4 1883. Tenor. Note G. Diameter 978mm (38½in). Weight 503kg (9cwt 3qr 17lb). Cast without canons. Timber headstock. Recast by John Taylor & Co of Loughborough.

RECAST BY JOHN TAYLOR AND C°
~ BELLFOUNDERS LOUGHBOROUGH 1883
On the waist:
NON SONO ANNIMABUS MORTUORUM
~ SED AURIBUS VIVENTIUM 1620
(I sound not for the souls of the dead, but for the ears of the living)
RECAST AT THE EXPENSE OF
~ CANON ARNOLD'S PUPILS.
The clock strikes the hours on this bell.

Former bell
Cast in 1620 by Toby Norris I of Stamford. Diameter 978mm (38½in). Weight approximately 533kg (10cwt 2qr).
[10] ꝹON SONO ꜲꝹꝹIMABꝒS
~ MORTꝒORꝒM SEꝹ ꜲꝒRIBꝒS VIVEꝹTIꝒM
~ 1620

BELLRINGING CUSTOMS
The Gleaning Bell was rung during harvest and the Churchwardens' Accounts show that a payment of 5s a year was made for ringing this bell from 1855 to 1876. Throughout the period of the accounts the ringers were paid for ringing the bells at Christmas but there are no records for ringing on other specific occasions.

On Sundays the bells were chimed for Divine Service and a Sermon Bell was rung. At the Death Knell there were thrice three tolls for a male and thrice two tolls for a female. For a child there were thrice two tolls on the treble (North 1880, 154).

Today the bells are sometimes chimed for ten minutes before the 9.30am Sunday Service. If requested ringers from Stamford ring for weddings.

RINGERS' RULES
A warning to illicit ringers is painted in black on the wall at the base of the tower at Tinwell. There is a board with the same warning in the ringing chamber at Ryhall Church.

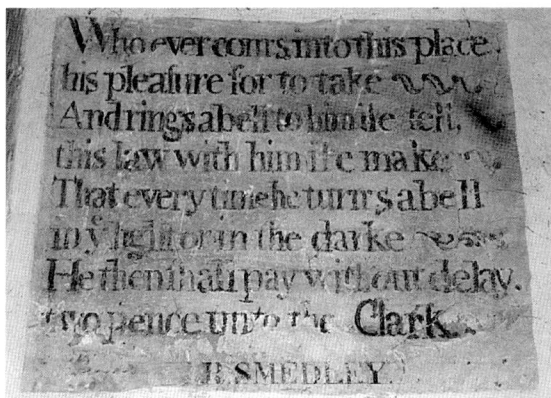

Ringers' rules painted on the wall in the ringing chamber at Tinwell Church. Robert Smedley was parish clerk from 1848 until 1875

CLOCK HISTORY
Confirmation that there was no early clock at Tinwell is provided in Irons' Notes (MS 80/5/22):

1619 They want [lack] a clock

There is however a theory, based mainly on the style and positioning of the present dial, which suggests that a John Watts' clock was installed here sometime between the years 1680 and 1710. The close proximity of Tinwell to John Watts' workshop in Stamford makes this a distinct possibility. Unfortunately, there are no surviving parish records which might contain evidence to support this hypothesis. The earliest Churchwardens' Accounts show that there was a clock here in 1781 when a Mr Looe [Lowe] was paid 5s for new clock lines. A *circa* 1793 drawing by Fielding (RCM F10/1984/59) clearly shows a diamond-shaped dial on the north face of the tower. A payment of £2 14s 4d was made in 1805, 'for painting the Dial & Wanscoting and other jobs at the Church'.

The church accounts cover a period of just over a hundred years and indicate that the clock was regularly maintained and repaired by Stamford clockmakers for the whole of this time. Some typical entries are given below. The first clockmaker mentioned is James Wilson in 1781:

Year ending 1795 Paid Wilson's Bill for Cleaning and
Repairing the Clock   7s 6d
Easter 1801-02 Paid Mr Wilson's Bill for looking after
& repairing The Clock   £1 11s 0d

James Wilson retired in 1803 and Thomas Haynes, his nephew, took over this duty. James had taken Thomas into partnership in 1799 (Tebbutt 1975, 68). Thomas, and later his son Robert Broughton Haynes, maintained and repaired the clock until 1859:

To Lady Day 1832 Paid Haynes and Son for reguilding
a cup [chalice?] ... inside and out
£5 5s 0d
Attending the clock   10s 6d

The basic annual payment of 10s 6d for the clock maintenance varied very little over the hundred years of the accounts.

The last clockmaker mentioned is John Hicks, who was for many years an assistant to Robert Haynes. He worked on the clock from 1862 until at least 1884.

As in most other Rutland villages, the parish clerk was responsible for winding the church clock. Thomas Goulding was the clerk at Tinwell from 1783 until about 1805 and he was paid 12s a year for this duty.

In 1929 a decision was made to replace the old clock. The successful installation of a new clock by G & F Cope & Co of Nottingham at Edith Weston Church in 1928 was to lead to a movement by the same maker being installed at Tinwell (DE 5032/39). The dial was restored in 1964 and the clock movement in 1990, both by residents of the village.

CLOCK DETAILS
Details of the present clock:

Maker:	G & F Cope & Co of Nottingham
Signed:	On the setting dial and on the frame
Installed:	1929
Cost:	Not recorded
Trains:	Going and striking
Escapement:	Pinwheel escapement
Pendulum:	Wooden pendulum rod and cast-iron cylindrical bob
Rate:	51 beats per minute
Striking:	Countwheel hour striking
Weights:	Cast-iron
Winding:	Hand wound weekly
Dial:	Diamond-shaped timber dial with blue background and gilded numerals and hands
Location:	North face of the tower
Note:	The movement is mounted on wall brackets and totally enclosed in a box

A drawing of circa 1793 of Tinwell Church by Nathan Fielding showing the clock dial on the north face of the tower (RCM F10/1984/59)

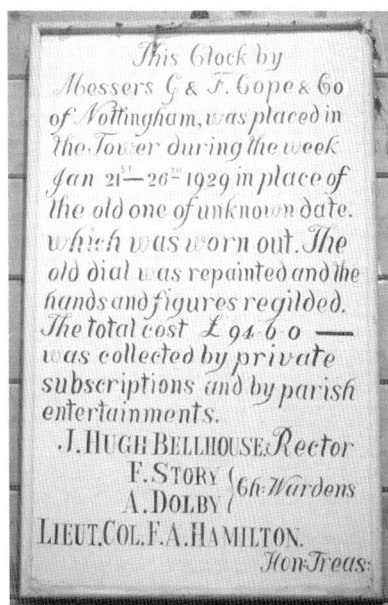

The installation of the clock at Tinwell is recorded on this board at the base of the tower. James Hugh Bellhouse MA was Rector from 1920 to 1932. Frank Story, a farmer, lived at The Gables and Agnes Dolby [Dalby] was a shopkeeper (Kelly 1925, 742). Lt Col Francis Alexander Chetwood Hamilton MC lived at Tinwell House (Kelly 1928, 798)

The timber dial on the north wall of the tower at Tinwell Church

TIXOVER

ST LUKE   Grid Ref: SP 971998

PARISH RECORDS

The Churchwardens' Accounts are available for 1749-1899 and 1900-64 (DE 2346/9-10) but they contain very little information concerning the bells.

BELL HISTORY

There is no indication that Tixover Church ever had more than a single bell. The bell listed by Thomas North in 1880 remains in the tower. It was originally hung in a timber frame with a wheel for full-circle ringing. The existing records make just one reference to the wheel:

1768 April 18 for 2 new stays for the Bellweel   2s 0d

The only other payments made in the accounts for bell maintenance are for the occasional replacement of a bellrope, mending the clapper in 1792, and for repairs made by Mr Tipping, a blacksmith from Barrowden, in 1825 and 1829. The belfry was cleaned in 1812.

At an unknown date, but probably within the last fifty years, the frame was removed. The bell is hung dead by its canons from a steel beam 4.1m (13ft 6in) above the belfry floor.

BELL DETAILS

1 *Circa* **1430**. 845mm (33¼in) diameter. Weight about 330kg (6cwt 2qr). Canons retained. Cast by Richard Hille of London (Pearson 1989).

[32] Sancta Fides Ora Pro Nobis [12] [29]
(Saint or Holy Faith pray for us)

Scheduled for preservation (The Council for the Care of Churches).

The medieval bell at Tixover is hung dead from a steel beam

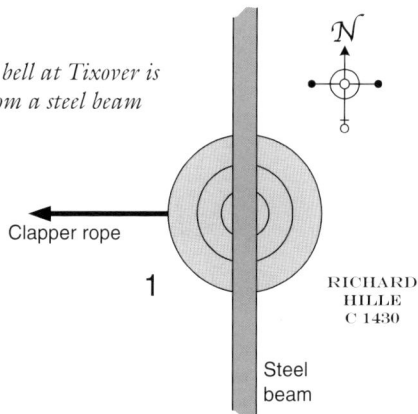

Clapper rope

1

RICHARD HILLE C 1430

Steel beam

The inscription on the Tixover bell

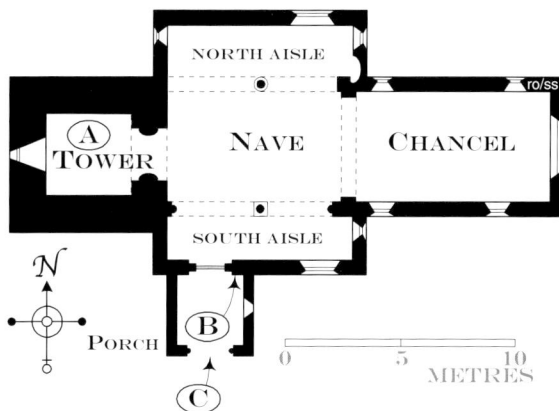

Plan of St Luke's Church
A: One bell in the belfry. Ringing chamber at the base of the tower
B: Scratch dial
C: Location of former sundial
Access to the first floor belfry is by a ladder

BELLRINGING CUSTOMS

At the Death Knell there were thrice three tolls for a male and thrice two tolls for a female (North 1880, 155). Today the bell is chimed prior to each service.

The Tixover bell is chimed from the base of the tower by a rope attached to the flight of the clapper. This is known as 'clocking'

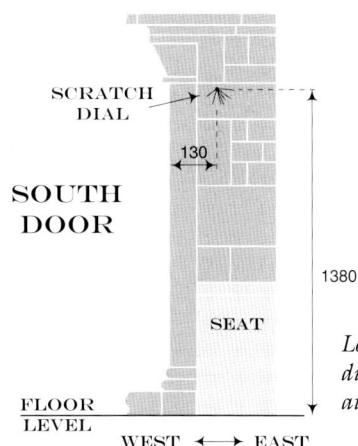

SCRATCH DIAL

130

SOUTH DOOR

1380

SEAT

FLOOR LEVEL

WEST ←→ EAST

Location of the scratch dial inside the porch at Tixover Church

TIXOVER							
Location	Inside porch on the wall east of the main door						
Condition	Excellent						
Gnomon Hole Diameter	10						
Gnomon Hole Depth	10						
Height above ground level	1380						
Line Ref	a	b	c	d	e	f	g
Length	85	75	60	70	65	70	60
Angle (°)	94	132	144	158	166	192	223

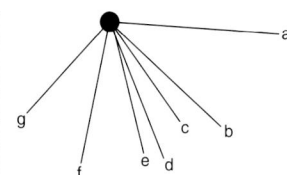

Scratch dial details

SCRATCH DIAL

There is a scratch dial inside the south porch to the east of the main door. This wall is early thirteenth century and the dial is on a stone adjacent to the top of the column.

SUNDIAL

A drawing of the church *circa* 1793 by Nathan Fielding (RCM F10/1984/61/2), shows a sundial set into the porch gable. It is not shown on a *circa* 1839 drawing (Uppingham School Archives), and there is no evidence of it today.

TIXOVER GRANGE

There is an eighteenth-century direct south limestone sundial with roman numerals and an iron gnomon across the south corner of Magnolia Cottage at Tixover Grange.

From a drawing of circa 1793 by Nathan Fielding showing a sundial in the porch gable at Tixover Church (RCM F10/1984/61/2)

UPPINGHAM

ST PETER & ST PAUL   Grid Ref: SP 867996

PARISH RECORDS

The archive includes an extensive collection of relevant documents and the following were searched for details concerning the former and existing bells and clocks:

Churchwardens' Accounts 1633-1871 (DE 1784/17-22) and 1871-1939, including Vestry Minutes (DE 4905/9).
Loose bills, vouchers and notes for 1806 and 1812 (DE 1784/21/2-23), 1832-59 (DE 1912/1/1-31 to DE 1912/19/1-44), and 1860-62 (DE 1912/22/1-5).
Restoration Committee Account Books 1860-66 and 1862-1900 (DE 1784/23-4).
Vestry Minute Books 1818-96 (DE 1784/60-3) and 1925-66 DE 4998/1-4).
Church Inventories dated 25 March 1825 and 4 April 1831 (DE 4796/59-60).
Letter from Smiths of Derby regarding the overhaul and

repair of the church clock dated 22 January 1930 (DE 4905/10).
Letter, report, estimate and faculty for converting the church clock to synchronous motor dated 1963 (DE 4796/96-8).
Correspondence regarding the bells including an estimate for a new bell ringers' floor 1965-68 (DE 4796/114-6).

BELL HISTORY

The will of Clement Brittayne of Uppingham dated 10 April 1589 provides an early reference to the church bells. It states 'I give and bequeath unto the said Church [Uppingham Parish Church] and Bells there ten shillings' (Prerogative Court of Canterbury 84, Sainberbe, RN **iv** 18-21).

According to Irons' Notes (MS 80/1/3) a parcel of church land known as Bellrope Land was granted to

Plan of St Peter & St Paul's Church
A: Eight bells and an Angelus Bell. Ringing chamber on the first floor of the tower. The Angelus Bell can be rung from the Choir Vestry at the base of the tower
B: Location of the clock on the second floor of the tower
C: Clock dial on the north face of the tower
The vice provides access to the clockroom and belfry

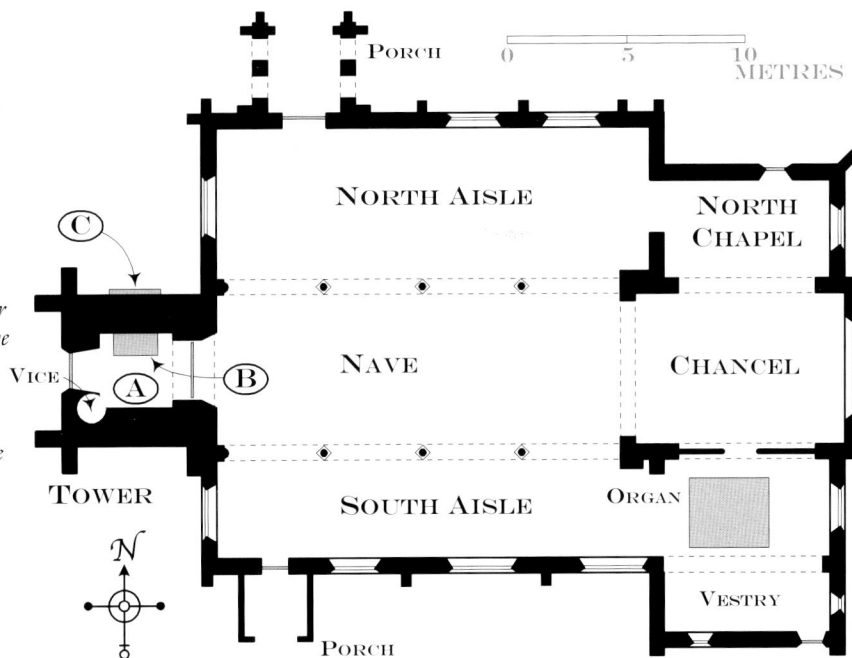

PORCH

0   5   10
METRES

NORTH AISLE

NORTH CHAPEL

VICE

NAVE

CHANCEL

TOWER

SOUTH AISLE

ORGAN

VESTRY

N

PORCH

Edward Wymarke and others in 1588. The resulting rental was presumably used for the purchase of bellropes or the maintenance of the bells. It was still owned by the church fifty years later but according to the following extract the revenue was misappropriated:

> **1638 May 10** Robert Sewell to acct for the rent & profit of the bell rope land by him held in the year 1637 being 40s which ought to be employed in the use of the church & hath been otherwise by him employed.

Robert Sewell may have been one of several disgruntled tenants who resented the levies raised to pay the high cost of the church restoration between 1632 and 1638. A petition was made to the Archbishop of Canterbury against these heavy taxes, but the outcome is not known (*VCH* II, 102).

The last mention of Bellrope Land is in the Churchwardens' Accounts:

> **1661** John Wells for ye towne land called
> Bellrope land   £3 0s 0d

The very complete and detailed Churchwardens' Accounts provide a great deal of information about the bells. A good example of this is the detailed accounts relating to repairs to or replacement of the bellframe during the restoration of the 1630s.

A payment of £36 5s 0d was made on 5 July 1634 'for hanging the bells and other payments about the steeple' and the bell hanger was paid a total of £18 1s 8d. His actual labour expenses 'ffor fifty 7 [57] days work for himself' were £4 15s 0d [20d per day] and for the same period his own man was paid 16d and two additional

men 15d per day. At this time there were four bells and a Sanctus Bell in the tower. As new floors were put in the tower it seems likely that a new bellframe was installed. Some examples of repairs to the bells at this time have been selected from the accounts:

> **1634**
> It. ffor 4 wheels making and the timber & nails £4 13s 4d
> It. ffor timber for the Saincts bells frame & the yoke 3s 0d
> It. ffor the 4 yokes [headstocks]   £1 0s 0d
> It. ffor ... leather for the Baldracks [baldricks]   4d
> It. ffor iron work for the Saints Bell [and other items] 4d
> It. ffor three new clappers weighing five score
>   & 3 pound at 8d the pound   £3 8s 8d
> It. ffor mending the second bell clapper being too light
>   2s 8d
> It. ffor carrying the same to the smith   2s 0d

Even after this restoration there were regular payments for the repair and maintenance of the bells and bellframe. The following examples give the names of local suppliers and craftsmen:

> **1636** Item to Anto. ffeilding & Wm king for worke
>   about the fore-bell   1s 1d
>   Item payd to John Sitton for a Balderike for the
>   same   1s 4d
> **1637** Item for a piece of wood to Mr Orme for the
>   great Bell yoke   2s 6d
>   It. to Antho. ffalkner [joiner and churchwarden]
>   for keeping the bells wch was left unpaid in the
>   yeare of our lord 1636 [no amount]
>   pd to Robt Sherwood for mending the bells 2s 2d
>   pd to willm yates for 7 days for the bells   10d
> **1639** Itmd to Dan Makereth for 3 new ropes & 2
>   shootings   12s 0d

Itmd to Jo Gray for Ironworke about the great
Bell [tenor]   8s 0d
1652   payd to Will Browne for keping ye bells
    & Roopes one yeare   £1 0s 0d

Although later extracts from the accounts are confusing it is evident that the Norris foundry at Stamford was employed to cast one or more bells for the Church. 'Bells' are mentioned as going to Stamford in 1662/3, but in 1667 'ye bell' only. It is possible that at least two of the four bells were recast in 1662/3 and that Thomas Norris recast another bell and supervised the hanging of it in 1667. There is no payment for recasting the bells in 1662/3 but this may have been defrayed by a donor. An entry for 1669 reveals that the churchwarden 'Spent at stanford with ye bell founder 6d'.

1662   Paid Mr Norris in earnest [to confirm a contract]
    ffor casting the bells   10s 0d
    pd ffor Mr Norris charges   8s 10d
1663   To Robert Pakeman for carrying ye bells to
    Stamford   4s 0d
1666   The bell charges cometh to   £9 7s 0d
    Spent at Stamford   3s 6d
    for caridg   4s 0d
    for ale at upingam on lodin the bell   1s 4d
1667   paid to a man that went to Mr Norrice   8d
    paid for charges when mr norrice came to hange
    ye bell   2s 0d
    pd to men that helped wv [with] ye bell   1s 0d
    pd more to ye Bell Hanger   £3 10s 0d

Thomas North (1880, 160) suggests that the following entry for 1702 records the fate of the Sanctus Bell. Payments were made for a 'litell bel rop' and 'for a Baldrick ffor ye little Bell' prior to this date, but there were none after:

1702   Recd for Bell Mettle   £1 2s 0d

The following extracts from the accounts imply that the Church had five bells by 1712, a new bell probably being acquired as a gift. There is no record of the founder. However the accounts two years later indicate that one of the older bells had to be recast, as it was taken down 'By ye order of the neighbours'. The 1714 accounts point to a written contract being drawn up for this work, the founder undoubtedly being Thomas Clay of Leicester:

1712   Paid John Green for 5 Bell Ropes   13s 0d
    for Ale when ye new bell was hung   1s 0d
1714   pd Mr How for making ye Articles
    [an agreement] wth ye Bell Hanger   2s 0d
    pd Job Swayen for taking down ye Bell By ye
    order of the neighbours   12s 0d
    pd for Ale for earnest for ye Bell   2s 6d
    pd expence at ye Running [casting] ye Bell at ye
    Bargaine making   13s 0d
    pd Mr Palmers Bill for drawing Bell-founders
    security   14s 4d
    pd Wm Burton for carrying ye bell to & from
    Leis' [Leicester]   £2 2s 6d

The agreement made between Alwyn Bradley and the Uppingham churchwardens in 1735 (DE 1784/18)

pd Wm Burton extraordinary on same occasion
    16s 0d
Expences at Leicester for Horses & charges £1 13s 6d
Half ye charges for Running ye Bell   £12 4s 6d
1 part of 3 for Hanging ye Bells   £10 0s 0d

The accounts from 1728 to 1764 show that little work was carried out on the bells other than in 1762 when the second, third and tenor were taken down. This was possibly for work to be carried out on the bellframe rather than on the bells themselves. However replacement of bellropes was a regular expense. In April 1735 a three-year agreement was made with Alwyn Bradley stating that he was to 'keep all the woodwork belonging to the Bells in the Steeple in good repaire'.

The following extract is taken from the minutes of a Vestry Meeting held on 8 November 1772:

The badness of the Bells being taken in Consideration it is a Greed by the Majority for to have the said five Bells to be taken Down and recast into Six, and signify our Assent to the Above.

There were seventeen signatories to this agreement and the work was entrusted to Pack and Chapman of the Whitechapel Bell Foundry, London.

In 1773 a Vestry passed the motion by a majority of four 'that two new Bells be added to the present Six in Order to Compleat a sett of Eight Bells for the Use of the Parish' and Walter Robart the churchwarden was to write to Pack and Chapman 'Ordering them to cast make and Send down to Uppingham two Bells called Trebles'. A Town Cottage was sold for eighty pounds to Robert Hotchkin and two hundred pounds paid by Mr Brown for the renewal of a lease to cover the costs involved. Other contributions included 'Old wood frames Sold for £3 0s 0d', evidence that a new frame was installed. Pack and Chapman's bill for recasting the bells was £210 6s

6d and this was paid in four annual instalments, the last being in 1776.

The following accounts provide some additional facts:

1773

Paid to Pack & Chapman on Acct of the Bells   £100 0s 0d
Pd Turner bell hanger his Bill   £58 10s 0d
To Wright carrier for carriage of the bells fm London
                                                                         £19 3s 6d
Gave to the Nottingham & Peterboro Ringers for
the Opening of the Bells   £5 5s 0d

The *Lincoln, Rutland & Stamford Mercury* reported the installation of the new bells in April 1773:

A complete new peal of bells cast by the eminent Messrs. Packe and Chapman of Whitechapel in London was opened at Uppingham church. 5040 grandsire triples was rung by a company of change ringers from Nottingham; it took 3 hours, 17 minutes in performing (Traylen 1982c, 13).

Few references to maintenance of the bells are made in later accounts. A Vestry of 5 August 1804 agreed that the church steeple was to be repointed and repaired and that the fourth bell was to be recast. Robert Taylor of St Neots recast this Pack and Chapman bell. The names of the two churchwardens at the time, Matthew Catlin and Richard Wade Junr, were placed on the inscription band. There is no record of payment for the recasting.

William Langley, carpenter, William Sharman, whitesmith, gunsmith and bellhanger, and Samuel Geeson, whitesmith, all of Uppingham, carried out repairs on the bellframe between 1835 and 1855. Samuel Geeson is also noted as a ropemaker in White's *Directory* of 1846 and Geeson & Kernick were bellhangers in High Street according to Barker's *Directory* of 1875.

In 1861 the estimates for the Church Restoration included for a new bellframe and for rehanging the bells. No evidence of a new frame has been found in the church accounts and the following extract from White's *Directory* (1877, 702) suggests that the old frame was retained: 'The tower and spire have also been restored, the fine west door re-opened, the bells re-hung, and two porches erected.'

John Taylor of Loughborough recast the fifth bell in 1895 when the ring of eight was rehung in a new two-level iron and steel frame with four bells at each level. The cost of this work was £313 14s 6d. These bells still

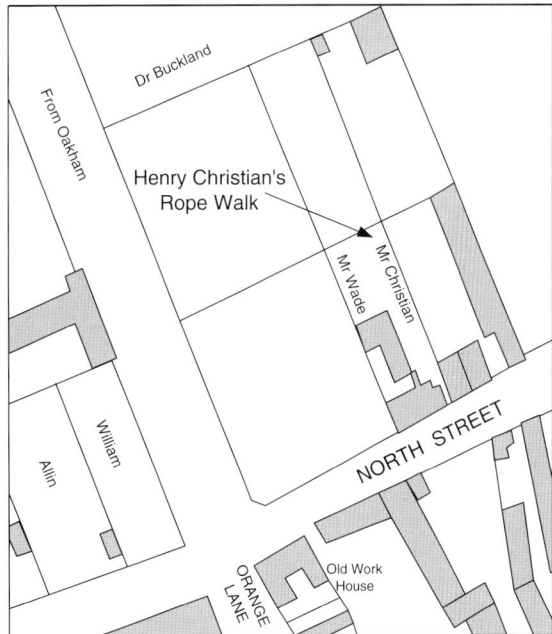

The location of Henry Christian's rope factory based on John Wood's plan of Uppingham dated 1839. The passageway, to the west of Tod's Terrace, is still remembered as 'The Rope Walk'

William Sharman's bill head (DE 1912/6/30)

remain in the tower today along with an Angelus Bell cast by the Loughborough bellfoundry in 1993. The Angelus Bell is in a small frame above the second level in the south-east corner of the belfry and it can be rung from the ground floor Choir Vestry. In 1992 the ringing chamber was moved to the first floor of the tower.

It is interesting to note some of the suppliers of bellropes to Uppingham Church. The earliest named are Solomon Day and Daniel Mackerith in the 1630s and Mr Green of Holt who supplied new ropes in 1666. A John Green provided bellropes in 1712 and a clockline in 1738, and a Robert Langton, his widow and John, presumably their son, supplied bellropes from 1728 to 1734. Other suppliers during the rest of the eighteenth century were Jacob Fowler, John Johnson, Edward and William Clarke. With the exception of Mr Green the locations of the suppliers have not been established. However Henry Christian supplied bellropes and clocklines exclusively to the church from the 1820s to the 1850s. He lived in Uppingham and in White's *Directory* of 1846 he was described as a 'Rope & sheep net maker, North St'. An extract from the *Rutland Magazine* of *circa* 1911 provides further information:

Many can still remember the Rope Walk as a rope walk, owned by a man named Christian. He made ropes for all Uppingham and the neighbourhood, and his "Walk" ran for a long distance parallel with Ayston Road (Bell 1911-12, 49).

Two former inns in Uppingham were named the Bell Inn and the Eight Bells. Vestry Meetings are known to have been held in the former in 1806.

BELL DETAILS

1 1773. Treble. Diameter 711mm (28in). Weight 251kg (4cwt 3qr 21lb). Canons removed. Cast-iron headstock. Cast by Pack and Chapman, Whitechapel Bell Foundry, London.
PACK & CHAPMAN OF LONDON FECIT 1773 [97]

2 1773. Diameter 740mm (29⅛in). Weight 276kg (5cwt 1qr 20lb). Canons removed. Cast-iron headstock. Cast by Pack and Chapman, Whitechapel Bell Foundry, London.
PACK & CHAPMAN OF LONDON FECIT 1773 [97]

3 1772. Diameter 787mm (31in). Weight 301kg (5cwt 3qr 20lb). Canons removed. Cast-iron headstock. Cast by Pack and Chapman, Whitechapel Bell Foundry, London.
PACK & CHAPMAN OF LONDON FECIT 1772 [97]
Quarter chime hammer on this bell.

4 1804. Diameter 864mm (34in). Weight 349kg (6cwt 3qr 14lb). Canons removed. Cast-iron headstock.

Recast by Robert Taylor of St Neots.
MATTHEW CATLIN AND RICHARD WADE
~ JUN^R CHURCHWARDENS. [67]
~ R. TAYLOR S^T. NEOTS FOUNDER 1804. [67]
Quarter chime hammer on this bell.

Former bell
Cast in 1772 by Pack and Chapman, Whitechapel Bell Foundry, London. Diameter 864mm (34in).
PACK & CHAPMAN OF LONDON FECIT 1772

5 1895. Diameter 899mm (35⅜in). Weight 397kg (7cwt 3qr 8lb). Cast without canons. Cast-iron headstock. Recast by John Taylor & Co of Loughborough.
PACK AND CHAPMAN OF LONDON FECIT
~ 1772 [65] [65] [65] [65] [65]
On the waist:
NUNC DENIQUE DISSONAM CONSONAM
~ ME REFICIT TAYLOR 1895 [65]
(When at last I became untuneful
Taylor now made me tuneful again)
Decoration [86] is below the inscription band. Quarter chime hammer on this bell.

Former bell
Cast in 1772 by Pack and Chapman, Whitechapel Bell

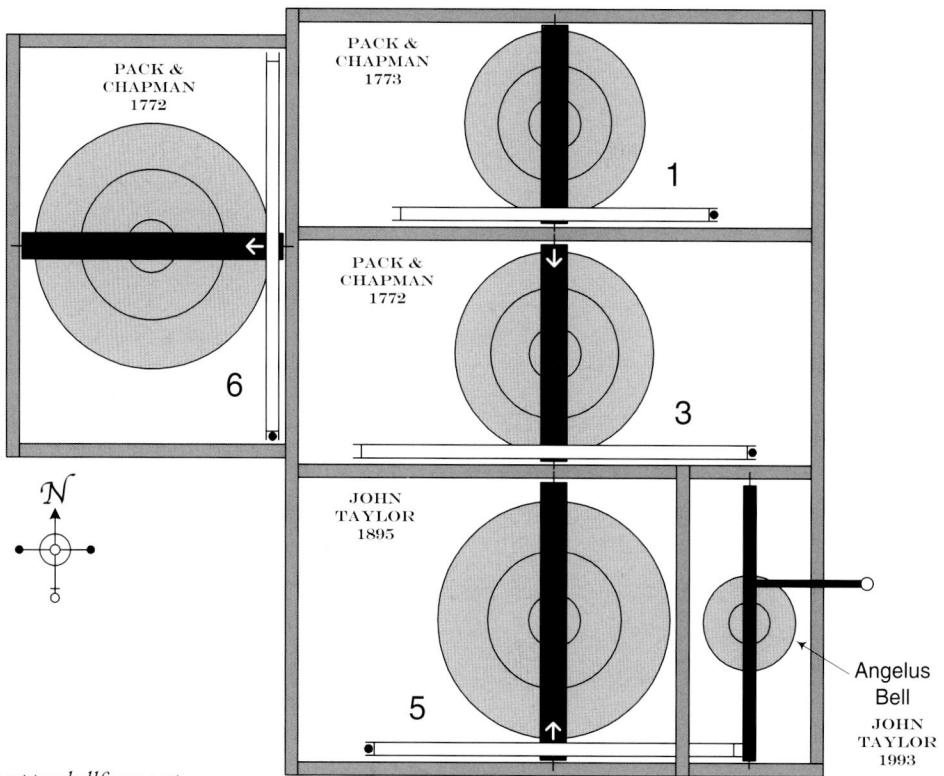

Plan of the upper bellframe at Uppingham Church

Upper Frame

Foundry, London. Diameter 883mm (34¾in).
PACK & CHAPMAN OF LONDON FECIT 1772

6 1772. Diameter 940mm (37in). Weight 428kg (8cwt 1qr 19lb). Canons removed. Cast-iron headstock. Cast by Pack and Chapman, Whitechapel Bell Foundry, London.
PACK & CHAPMAN OF LONDON FECIT 1772 [97]
Quarter chime hammer on this bell.

7 1772. Diameter 1038mm (40⅞in). Weight 549kg (10cwt 3qr 6lb). Canons removed. Cast-iron headstock. Cast by Pack and Chapman, Whitechapel Bell Foundry, London.
PACK & CHAPMAN OF LONDON FECIT 1772 [97]
Quarter chime hammer on this bell.

8 1772. Tenor. Note E. Diameter 1130mm (44½in). Weight 737kg (14cwt 2qr 2lb). Canons removed. Cast-iron headstock. Cast by Pack and Chapman, Whitechapel Bell Foundry, London.
PACK & CHAPMAN OF LONDON FECIT
Also on the inscription band, but incised:
WALTER ROBARTS CH WARDEN 1772
Below the inscription band:

YE RINGERS ALL THAT WHO PRIZE.
~ YOUR HEALTH AND HAPPINESS,
~ BE SOBER MERRY WISE
~ AND YOU'LL THE SAME POSSESS
North (1880, 155) omits THAT. Hour bell.

ᴀɴɢᴇʟᴜs Bᴇʟʟ
1993. Diameter 368mm (14½in). Weight approximately 38kg (3qr). Cast without canons. Fabricated steel headstock. Cast by John Taylor & Co of Loughborough.
On the waist:
ANGELUS AD VIRGINEM
(Angelus to the Virgin)
On the waist opposite:
19 [56] 93
Decoration **[98]** is around the inscription band. The Angelus Bell can be rung from the Choir Vestry at the base of the tower.

Bᴇʟʟʀɪɴɢɪɴɢ Cᴜsᴛᴏᴍs
Perhaps one of the earliest customs in Rutland is recorded at Uppingham in Irons' Notes (MS 80/1/3). It is dated 3 Dec 1593 and in essence it refers to a legacy left by Thomas Drake of Uppingham which specified the

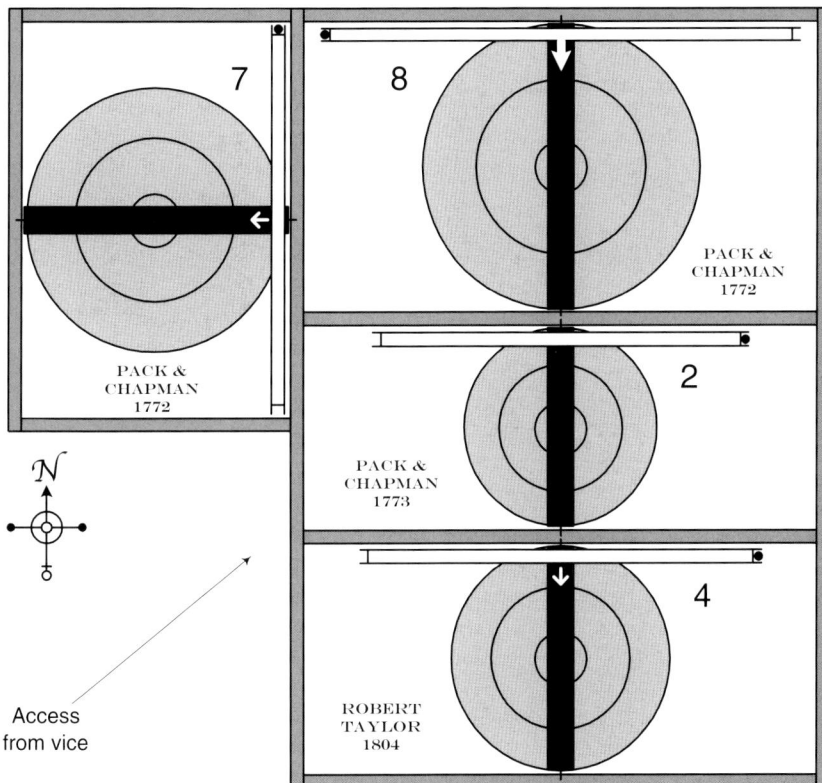

Plan of the lower bellframe at Uppingham Church

Lower Frame

ringing of an Obit Bell. This legacy decreed that 12d was to be given annually to the poor in the church of Uppingham on the Monday of Rogation Week. There was to be an anniversary service for himself and his two wives and this was to include Mass and a dirge with the ringing of bells. Everard Britten, an heir of Thomas Drake, was at this date accused of not fulfilling the terms of the will.

Irons' Notes reveal that a custom was not being adhered to at the beginning of the seventeenth century:

1628 Feb 5 Thos Rowlet Rector for not procuring the bell to be runge at 8 of the clocke at night & at 4 of the clocke in the morninge according to the custom of the parishe.

The bells were to be rung 'from the monday after Harborough fayre till Shrove Tuesday'. It was also noted that these daily bells had been rung for many years as a result of being set down in existing charities. Faulkner, the man responsible for ringing the church bells, refused to take on this duty:

1628 ... he answereth that he is not bound by lawe to doe it it beinge a mere secular service & hath no relation to is ministeriall function but contrary unto it.

There is an isolated reference in the Churchwardens' Accounts to a similar bell being rung in 1790:

1790 pd Jon Rudkin fro Ringing the 8 o clock &
5 o clock Bell to Lady Day   11s 0d

Mention is occasionally made to a 'Crier' being employed in the latter half of the eighteenth century and the beginning of the nineteenth:

1766 Paid for Crying the Bells   6d
paid the Cryer for Crying the town closes   4d
1769 pd Ed Barrattt for crying the Beasts to paster
[pasture]   4d

There are numerous references in the accounts, particularly those dating from 1663 to 1727, to the church bells being rung on occasions other than for calling parishioners to church services. There were peals for coronations and their anniversaries, for royal birthdays, for frequent thanksgivings and for celebrating events of national importance. Two popular celebrations were those of 'Guy Fawkes Night' on 5 November and 'Oak Apple Day' on 29 May. Both festivals no doubt delighted the band of ringers performing their services in the ringing chamber of Uppingham Church. It is evident that the town regularly followed these customs until the 1870s. The following is a selection of particular occasions when the bells were rung:

1638 It. given ye ringers for ye kings happy returne
from Spane   1s 0d
It. to ye ringers upon ye kings day   1s 0d
1664 pd ffor board ffor ye Ringers att ye bishops
visitation   1s 0d
1702 To ye Ringers 29th May & 5th Nov   4s 0d
To the Ringers on the day of Thanksgiving for
destroying 40 Ships Men of Warr of the ffrench &

Spanish Galleons at Vigo   10s 0d
1707 paid the Ringers for Ringing it being a day of
thanksgiving for ye Vnion   4s 0d
1727 Paid the Ringers it being the Queens birthday   2s 6d
1747 Ringers att the taking the French man of war   5s 0d
1817 Ringers at the funeral of the Princess Charlotte
£1 0s 0d
1826 Paid the Ringers, Duke of York's funeral   5s 0d

Two reports in the *Lincoln, Rutland & Stamford Mercury* (Traylen 1982c, 20 & 25) make interesting reading:

1855 Oct The news of the recent success of the allied armies before Sebastopol reached Uppingham and caused the greatest excitement. Bells rang throughout the day and at night a large bonfire was made in the Market Place.

1886 Dec The custom of ringing muffled peals on the church bells was again resorted to. After the midnight hour had struck, peals of a more joyous nature announced the advent of 1887.

Thomas North (1880, 161) recorded the following customs which were current in 1880. Before each service the fifth or the sixth bell was rung for ten minutes and then all of the bells chimed for five minutes before the commencement of Divine Service. The fifth or sixth was rung again, except when there was to be a celebration of the Holy Communion, when the seventh bell was tolled instead. On Sunday the treble was rung at 8am. At the Death Knell three tolls were given for a male and two for a female.

It was generally the duty of the Parish Clerk to supervise the ringing of the bells for services but there are very few references given in the Uppingham accounts that link them with this post. Two early references are noted:

1661   payd Phill ffleming in full for ringin ye
bell for ye yeare 1660   £2 0s 0d
1775 June 4 Pd Robt. Hills Salary for ringing the Bells
10s 0d

It is obvious from the accounts that Uppingham had an active band of ringers from the early eighteenth century and from 1747 such annual records as 'Pd for 5 days Rnging' became the norm. Only on rare occasions are these fixed days specified. Payments were made to 'Ringers' until 1930 but again few details are recorded.

At a Vestry dated 27 April 1878 it was resolved:

That the 16/- a quarter now paid to the Ringers be in future placed under the control of the Rector and Churchwardens, and that they be requested to make such regulation as to the appointments and duties of the Ringers as they may approve of.

Certificates and photographs held by the church record occasions when special peals were rung during the twentieth century. These include peals of Grandsire Triples of 5040 changes in 1952 and 1956 and commemorative peals to celebrate the Silver Jubilee of King George V in 1935 and the Coronation of Queen Elizabeth II in 1953. St Peter & St Paul's still has an active band of bellringers. At least four of the eight bells are rung for the 10.30am service on Sundays. The bells are rung for the 3pm Christmas Eve Crib Service, Midnight Mass at 10.30pm and on Christmas Day at 10.30am. At New Year the bells are rung half muffled before midnight and then open for twenty minutes afterwards. For the Millennium they were rung for a 12.15pm service. On Easter Day they are rung for the 10.30am service.

Bells are also rung for approximately thirty minutes after weddings but very rarely for funerals unless this is for a ringer or a person of local note. Quarter peals are rung on the occasion of Royal weddings, birthdays and funerals. On Armistice Day the bells are rung half muffled before the 10.30am service and on Remembrance Sunday.

The Uppingham Bellringers have an annual Bring and Buy Stall and have raised over six thousand pounds since 1986 for the upkeep of the bells and belfry. One recent major project was to move the ringing chamber up to the tower balcony to create a Choir Vestry below. This was completed in 1992.

CLOCK HISTORY

The following reference from Irons' Notes may point to the fact that Uppingham Church had a clock in the early seventeenth century:

1618 April 11 Wilbron (Everard, Guard) [Everard Wilbron, churchwarden] to certify that the dyall goeth well

A clock is mentioned in the Churchwardens' Accounts of 1633, the first year for which they are available. It would have had a verge and foliot escapement and possibly a wooden frame:

1633

It to Willm King for mending the clock	18s 0d
It to Hughe James for an handle [a winding handle?] for ye sayde Clock	2s 6d
It to Daniell Mackrith for 2 new Clock ropes	4s 0d

William King was probably a blacksmith/clockmaker. He is not recorded as working on any other clock in Rutland.

A new timber dial was erected in the following year and this may have been the first dial for the clock. The 'brass diall' was probably a setting dial [the setting dial on a clock movement indicates the position of the hour hand on the external dial]. Repairs were also carried out to the clock case and 'Wily' King repaired the clock again:

1634-35

It for the diall board	8s 0d
It for carrying board to ye Paintre	1s 0d
It for painting the diall board	£1 7s 8d
It for Iron for the Clock diall [probably support brackets]	8d
It for fetching the pulliws [pulleys for lifting the dial up the tower] from Liddington	6d
It for the use of the same	2s 6d
It for the brass diall	2s 4d
It for timber nayles and workmanship for the Clock Case	15s 0d

It for the two beams under the floor over the Clock	£1 12 0d
To Wily King for mending the clock	6s 8d

Repairs to the dial were carried out in 1638 at a cost of 2s. The following item may refer to a new sundial which would be required to regulate the clock:

1640 It pd Georg Bennitt for making the Diall and other work about the Church 5s 10d

From 1666 until 1698 William Fox was the local clockmaker who was responsible for repairing and maintaining the church clock and he probably installed a new clock during this time:

1686 paid to Wm ffox in part of the Clock £5 10s 0d

Clock winders noted during this period were Richard Hudson, William Hope, William Forman and John Taylor. John Boyer supplied a new clock rope for 2s 6d in 1686.

Robert Fox, another Uppingham clockmaker and possibly son of William, then took over responsibility for the clock. He continued with this duty until at least 1745, the year in which he supplied a new clock to Gretton Church, Northamptonshire, and also the year in which he became a churchwarden at Uppingham. For some of this time he was also the clock winder:

1707 Feb 6	gave the workmen when Robt Fox set up the clock ½ dozen of ale	6d
1720	To Robt Fox for work Due & Clock Keeping	£1 1s 1d
1725	pd Robt Fox for winding up & mending ye clock	£1 7s 6d

Clock rope suppliers noted during the first half of the eighteenth century include Robert and John Langton, John Johnson and John Green. The dial was painted in 1720:

1720 To John Billington for painting the Dial 15s 0d

John Fox, son of Robert, was responsible for the clock from about 1750 for the next twenty-five years or so, although it seems that he continued as a clockmaker in Uppingham until his death in 1802. In 1750 another new dial was provided for the clock and it may have been regilded in 1758:

1750	Hippisley for a Dial Board	£1 5s 0d
	Mr Billington's 1750 Church Clock Dial Plate	£2 18s ½d
	Toby Hipsly for Painting ye Dial Plate	£5 0s 0d
1758	Mr Cooke for Leaf Gold	£1 2s 0d
	Pulford for Gilding	13s 0d

In 1762 Thomas Rayment, a Stamford clockmaker, was asked to investigate an unspecified problem with the clock.

On 28 May 1776, 'A Meeting of the Vestry ... [was held] to consider of Putting up a New Town Clock ...'. It was resolved that:

The Churchwardens do forthwith Cause a New Clock to be made and Erected in the Place of the Old One - And that the old One being unfit for Use be sold for the purpose of defraying in part the Expence of Erecting a New One.

At another Vestry, held at the Unicorn Inn on 29 October 1776, it was resolved:

That The Churchwardens Do take up and borrow any sum of money not exceeding the Sum of One Hundred and Thirty Pounds for the purpose of defraying the Expences of Putting up a New Clock therein. And that the Parish shall be bound thereby to Repay the same with legale Interest.

It was also resolved that

Mr Joseph Furniss Clockmaker be appointed to the Care and Management of the Parish Clock in winding up and excepting always Repairs of Lines, Springs, Hammer work or unforseen Accidents, keeping the same in good Repair. And that the said Joseph Furniss be allowed the Sum of Two Pounds Twelve Shilings and Six pence yearly for the same, the said Allowance to commence at Michaelmas 1776.

The minutes were signed by Robert French, Richard Nevison [churchwardens], John Mosendew, James Hill [Overseers], Walter Robarts, Henry Bains, Johnathan Bullock, John Marriot and Richard Mills.

The new clock was purchased from John Whitehurst I of Derby. It is described in the Church Inventory of 1825-26 as a '30-hour Clock, with Quarter striking'. A new dial was provided at the same time:

1776
Paid John Wing as Rect. [receipt] for the Dial of
the Clock   £31 0s 0d
Do. Joseph Furness figureing the Dial with Gold Letters
                                                                                 £6 4s 6d
Do. James Holmes Joiner for making a Room for
the Clock   £21 10s 5d
Do. John Whitehurst for making the New Clock £65 0s 0d
Do. Panc Medmore for Scaffolding   £7 4s 6d

As the following extracts from the Vestry Minutes show, Joseph Furniss was contracted to maintain and repair the new clock. A Vestry held on 15 April 1777, 'adjourned from Easter Tuesday last', and at which there were ten people present, including Joseph Furniss, ad-

dressed the issue: 'That the sum of Seventy pounds was remaining due to sundry persons for Work done in the Compleating the New Clock and for the Repairs of the Church.' It was resolved that the churchwardens should borrow 'the Sum of Seventy Pounds upon Bond upon the Creditt of the Parish' and that this should be repaid 'with legal Interest'. It was also agreed that 'Mr Joseph Furness shall be Allowed the Sum of Three Guineas Per Annum for his Managing of the Town Clock instead of Two Guineas & a half Ordered at a former Vestry'. This payment was to commence as from the previous Michaelmas and for which he agreed 'to Wind up and keep the Town Clock in all repairs'.

At a Vestry of 26 Dec 1779, adjourned to 3 Jan 1780:

We the Churchwardens ... Do hereby Rate and Tax all and every Inhabitant of the said Parish of Uppingham with the following Sume of Money for and towards the Expences of the Town Clock and other Moneys paid laid out and expended in the necessary Reparations of the said parish Church of Uppingham ....

The rate was 6d in the £1 and applied to all houses and land within the Parish.

At a Vestry meeting held on 28 May 1807 it was decided to restore the clock dial:

It is agreed that Mr Francis Tyler is to Repair the Clock Dial in a Masterly manner Gilded &c for the sum of Four Pounds and to be compleated by the 2nd Sunday in July next ensuing.

By 1820 William Aris II was the Uppingham clockmaker responsible for the church clock. He was also the parish clerk. In 1830 his son Thomas Aris took over both roles and in the following year his annual salary as parish clerk was increased to £20. He retired from this post in 1842, but continued to look after the clock until he died in 1876. Throughout this period John Christian is recorded as the supplier of new lines for the clock.

A Vestry meeting of 12 April 1897, held in the 'Infant School Classroom', decided that the clock should be replaced. The following note, dated Easter 1898, is from the Churchwardens' Accounts:

The Parish Church Clock
The Committee appointed at the Easter Vestry 1897 to take steps to obtain a new Clock and Chimes at once took action. The work was instructed in the course of the following summer to Messrs John Smith and Sons, Midland Clock Works, Derby, and the clock, with "Tennyson" Chimes was erected in May, 1898, and was set going on Whitsun - Even, May 28.

The installation was reported by the *Lincoln, Rutland & Stamford Mercury*: 'The Diamond Jubilee clock in the tower of the parish church has at last been fixed' (Traylen 1982c, 25).

From 1877, Pinney and Sons were the clockmakers recorded for maintaining the clock until 1913 followed by George Cliff until the end of the available accounts in 1930.

From Uppingham Churchwardens' Accounts for 1829-30. Details of the payments made to William Aris II, parish clerk and clockmaker (DE 1784/22)

Although this clock remains in the tower, it underwent a major conversion in 1963. Details of this work are described in John Smith's quotation dated 18 April of that year:

> The existing clock, which was made and installed by us in 1898, has been well cared for through the whole of its life and in spite of its age it is generally in good mechanical order. It is however very dirty and several important repairs and improvements will have to be undertaken if the clock is to continue giving accurate and reliable service. The important repairs are the renewal of all the quarter chime barrel lifting cams (80 in number) and the renewal of the tube and spindle which carry the hands, and we advise also that the latest type of double safety clicks and coil springs should be fitted to replace the early type, the failure of which can cause a costly accident.
>
> This type of clock can be readily converted to the system of electric drive ... which we have adopted with such success. Such a conversion, while retaining the clock's original lay-out and all of its better features, completely eliminates the heavy and dangerous driving weights, the complicated rope and pulley system, and also all manual winding. The mechanism is also greatly simplified, making local care much easier, and the removal of the constant strain of supporting the weights will quite certainly prolong the working life of the clock very considerably.
>
> The modifications we advise provide for the quarter chimes and the hour striking to be powered directly by geared electric motors and the time part of the clock to be replaced by our latest type of synchronous-electric time unit.

A framed Illuminated Address in the Uppingham Church Records states that the clock 'was restored and electrified at a cost of £500' in January 1964. The sum 'was raised partly by public subscription, a gift of £50 from Uppingham Parish Council, collections in Uppingham School, collection boxes in shops, voluntary efforts and was completed by a family gift of £220 to the memory of John Adkins who was a boy in the School from 1894-1900, a Master in the Lower School from 1910-1937 and later a resident of the town'.

Before this conversion, Uppingham church clock was identical to that at Morcott church (see Morcott — Clock Details).

CLOCK DETAILS

Details of the John Whitehurst Clock:

Maker:	John Whitehurst of Derby
Installed:	1776
Cost:	£65
Frame:	Possibly chairframe
Trains:	Going, striking and chiming
Escapement:	Anchor
Pendulum:	Not recorded
Striking:	Countwheel hour striking
Chiming:	Countwheel quarter chiming
Weights:	Cast-iron
Winding:	Hand wound every day (30 hour clock)
Dial:	Timber dial with gilded numerals and hands

Notes:	Replaced a clock installed in 1686
	This in turn replaced an early seventeenth-century verge and foliot clock

Details of the Diamond Jubilee Clock as originally installed:

Maker:	John Smith & Sons, Derby
Signed:	On the frame and on the setting dial
Installed:	1898
Cost:	Not recorded
Frame:	Flatbed
Trains:	Going, striking and chiming
Escapement:	Pinwheel
Pendulum:	Wooden rod and cast-iron cylindrical bob
Striking:	Countwheel hour striking
Chiming:	Countwheel quarter chiming
Weights:	Cast-iron
Winding:	Hand wound weekly
Dial:	Cast-iron skeleton dial on the north face of the tower
Notes:	Clock converted to electric motor drive on all three trains in January 1964
	All train wheels including escapement removed and now lost

Details of the clock following conversion in 1964:

Escapement:	Removed and replaced by a synchronous electric motor
Pendulum:	Removed — not required
Striking:	Countwheel hour striking. Train wheels removed and barrel driven by electric motor and gearbox
Chiming:	Countwheel quarter chiming. Train wheels removed and barrel driven by electric motor and gearbox
Weights:	Removed
Winding:	Motor driven

UPPINGHAM SCHOOL CHAPEL

RECORDS

Correspondence concerning the original bells is held in the School Archives.

BELL HISTORY

The School Chapel was built in 1865, although the tower was not added until 1871. The tenor was cast in the same year but it was not installed until 1872 when two 'quarter bells' were added. All three bells were cast by John Taylor whose detailed account for the work, made out to Edward Thring, is preserved in the School Archives. The cost of the tenor was £163 14s 2d and that of the two quarter bells together was £147 8s 9d. The total cost including fittings was £352 16s.

A memorandum from John Taylor dated 25 October 1872 suggests that a fourth bell was being contemplated:

> We do not approve of adding the upper F as a 4th bell. [We] would rather hear the three as they are. Most decid-

*The clavier at
Uppingham
School Chapel*

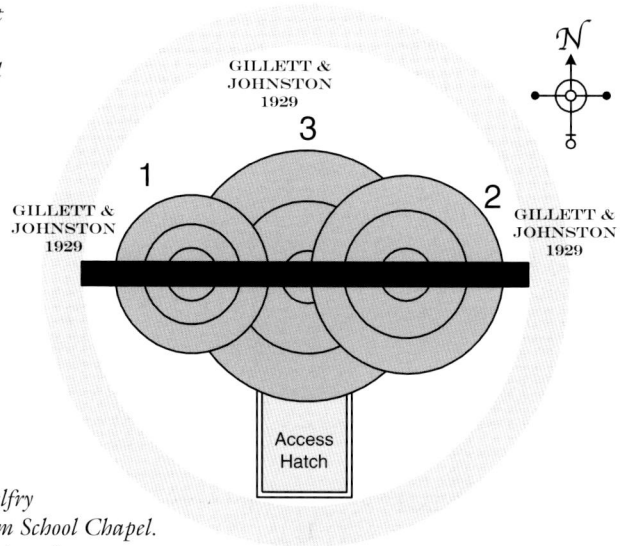

*Plan of the belfry
at Uppingham School Chapel.
1 and 2 are above 3*

edly the next bell to be added should be C - between B flat & D — we think a C bell would pass through the door ....

There is no record of this fourth bell being added and in 1929 all three Taylor bells were recast by Gillett & Johnston, but the reason for this is not recorded. As shown on the tenor, the recasting was paid for by Ralph Eastman, the father of William Marsden Eastman, a pupil at the school from 1925 to 1929 (information from Peter Lane). Owing to the confined space in the belfry they are hung dead from timber beams on two levels and are chimed by means of a clavier. There are no clappers and each lever of the clavier is connected by a wire to a striker inside one of the bells. Access to the belfry is by spiral stone staircase and ladder.

BELL DETAILS

1 1929. Note A. Diameter 737mm (29in). Weight approximately 254kg (5cwt). Cast without canons. Hung dead from a timber beam. Cast by Gillett & Johnston of Croydon. Decoration [58] is below the inscription band.

Former bell

Cast in 1872 by John Taylor & Co of Loughborough. Note B flat. Diameter 835mm (32⅞in). Weight 395kg (7cwt 3qr 4lb).

J TAYLOR & CO FOUNDERS
~ LOUGHBOROUGH 1872

Taylors described this and the second as quarter bells, indicating that they were intended for a clock with ting-tang quarters. However there are no records of a turret clock in the chapel, and the construction of the slender tower makes this an unlikely consideration.

*The tower of
Uppingham
School Chapel*

2 1929. Note F. Diameter 940mm (37in). Weight approximately 457kg (9cwt). Cast without canons. Hung dead from a timber beam. Cast by Gillett & Johnston of Croydon. Decoration [58] is below the inscription band.

Former bell

Cast in 1872 by John Taylor & Co of Loughborough. Note G. Diameter 949mm (37⅜in). Weight 522kg (10cwt 1qr 2lb).

J TAYLOR & CO FOUNDERS
~ LOUGHBOROUGH 1872

Taylors described this as a quarter bell (see note under Bell 1 — Former bell).

3 1929. Note C. Diameter 1187mm (46¾in). Weight approximately 965kg (19cwt). Cast without canons. Hung dead from a timber beam. Cast by Gillett & Johnston of Croydon.

RECAST BY GILLETT & JOHNSTON, ~ CROYDON. 1929.

On the waist:

THESE THREE BELLS WERE RECAST AT THE EXPENSE OF RALPH EASTMAN MARCH 1929

Decoration [58] is below the inscription band.

Former bell

Cast in 1871 by John Taylor & Co of Loughborough. Note D. Diameter 1241mm (48⅞in). Weight 1130kg (22cwt 1qr).

J TAYLOR & CO FOUNDERS ~ LOUGHBOROUGH LEICESTERSHIRE 1871

BELLRINGING CUSTOMS

Thomas North, in 1880, recorded that for Divine Service on Sundays, Saints' Days, and on Wednesdays during Lent, the bells were chimed, after which the Sermon Bell [the tenor] was sounded. On days when it was raining only two bells were chimed, to warn the boys to go direct into chapel without waiting for 'call-over'. At the end of term the occasion was 'celebrated by as merry a noise as these really fine bells can produce in their present position' (North 1880, 162).

UPPINGHAM SCHOOL TERCENTENARY BLOCK

The turret over the Tercentenary Block has a north-facing clock dial and two bells — a clock bell and the school bell — in the cupola above. The clock movement is in the roof-space below the turret.

SCHOOL CLOCK

The present two train flatbed clock movement was installed by John Smith of Derby in January 1930 (DE 4905/10). An unusual feature of this clock is that both trains have Huygens endless chain automatic winders which were fitted at the factory before the clock was installed. At the time of writing both are still in perfect working order after seventy years service.

It is assumed that the present clock bell was also installed in 1930.

CLOCK DETAILS

Details of the present clock:

Maker:	John Smith & Sons of Derby
Signed:	'John Smith & Sons Midland Clock Works Derby 1930' on the setting dial

The clock turret above the Tercentenary Block at Uppingham School

Installed:	1930
Frame:	Cast-iron flatbed
Trains:	Going and striking
Escapement:	Pinwheel
Pendulum:	Wooden rod and cast-iron cylindrical bob
Rate:	53 beats per minute
Striking:	Countwheel hour striking
Winding:	Huygens endless chain automatic winder on each train
Dial:	Skeleton dial with gilded roman numerals, minute ring and hands
Location:	North face of the turret over the Tercentenary Block

There was possibly an earlier clock here, as the Tercentenary Block complete with its clock turret was built in 1889 (*Uppingham School Magazine* **245**, 1893), some

Both trains of Uppingham School clock are wound automatically by electric motors via worm and wheel reduction gears on the front of the frame

forty years before the installation of the present clock. The early twentieth-century catalogue of J B Joyce & Co Ltd of Whitchurch records that this company supplied a clock to Uppingham (Pickford 1995, 136). This may have been installed in the Tercentenary Block. If this was the case its life was short, but it may have been replaced in favour of the Smith's clock because of the latter's automatic winding feature.

CLOCK BELL

Circa 1930 Diameter 533mm (21in). Bell inaccessible, but inscription includes:

TAYLOR FOUNDER LOUGHBOROUGH
May have replaced an earlier clock bell.

SCHOOL BELL

1929. Diameter 648mm (25½in). Cast by Gillett & Johnston of Croydon. On the waist:

**DONATED TO UPPINGHAM SCHOOL
RALPH EASTMAN JANUARY 1929**
Decoration [58] is below the inscription band. This bell can be tolled from the ground and first floors of the Tercentenary Block.

According to Bryan Matthews' history of the school (1984, 53), in 1927 the then headmaster, R H Owen, wanted a 'more imperiously compelling' school bell. As a result a 'more strident bell' was installed early in 1929 (see below). The donor of this bell was Ralph Eastman who also paid for the chapel bells to be recast in the same year.

Former bell

The former bell was the original bell from Archdeacon Johnson's Schoolroom and may have dated from 1584. It was transferred to the old chapel in 1771 and then to the new Tercentenary Block (see below). When the new

Archdeacon Johnson's Schoolroom at Uppingham prior to removal of the sundial. This photograph was taken by George Henton in 1916 (Henton 1145)

school bell was installed in 1929 the old bell was placed in the School Museum, but it disappeared from there at the beginning of the Second World War.

ARCHDEACON JOHNSON'S SCHOOLROOM

Archdeacon Johnson's Schoolroom, the original building of Uppingham School, was built in 1584. It still has a bellcote over the west gable but the bell has been missing beyond living memory. However, recent research in Uppingham School Archives by Peter Lane has unearthed the following which explains the fate of the bell:

Decrees of the Governors of the Free Grammar School

1771 Michaelmas That Mr Fancourt be allowed to remove the Bell from the School to the Chapel.

This note was in issue No 245 of *Uppingham School Magazine* dated August 1893, and an added note by the author states 'This is probably the same bell which now does duty in the Clock-tower over the Tercentenary Class Rooms'. If this was the case it was again removed when a new school bell by Gillett & Johnston was hung in the clock tower in 1929.

There was also a sundial over the south wall of Archdeacon Johnson's Schoolroom. It was taken down in 1990 as its weight was causing the wall to bow. Close inspection at that time revealed that there was one hole for the missing gnomon. There were no other marks on the dial, suggesting that it was originally painted rather than engraved. At the time of writing the dial stone and the ornamental pediments were in the small garden at the rear of the building. All were in a poor state and the dial stone was broken into two pieces and being used, face down, as a paving slab. It was, until recently, thought to be contemporary with the original school and as a result considered to be the oldest known sundial in Rutland. However, another note in *Uppingham School Magazine* (No 247, dated November 1893), again researched by Peter Lane, throws some doubt on this:

Decree of the Governors of the Free Grammar School

1741 Oct 2 That John Billington, be allowed seven guineas for the dial, to set it up and to preserve it twenty years for 6d a year to the Governors' satisfaction.

This may be a new or replacement sundial and the '6d a year' would probably be for annual repainting.

'UPPER' CRICKET PAVILION

The thatched cricket pavilion on the 'Upper' [playing field] was built in 1923, the gift of William Seeds Patterson, the School's cricket historian. It replaced the original pavilion which was built in 1864 on a site further to the south along Seaton Road. The clock by John Smith & Sons of Derby was originally located in the old pavilion which also had a bell turret. A tablet, which is dated

Uppingham School. The thatched cricket pavilion on the 'Upper' (Warwick Metcalfe)

The single train clock by John Smith and Sons of Derby at Uppingham School 'Upper' cricket pavilion

October 1897, records that it was 'erected in memory of work done for the school games by C W Cobb Esq. between 1873 & 1897'. C W Cobb MA, a Housemaster from 1873 until 1909, is credited with the reintroduction of hockey into the school. The clock is a small single train flatbed movement with a deadbeat escapement. The seconds pendulum has a wooden rod and cylindrical cast-iron bob, and the cast-iron skeleton dial on a hexagonal backing board overlooks the cricket pitch. There is a small plain undated bell in the thatched cupola above the pavilion roof. It is rung by a rope which hangs inside the pavilion.

CLOCK DETAILS

Detail of the present clock:

Maker:	John Smith & Sons, Derby
Signed:	On the frame
Installed:	1897 'in memory of work done for the school games by C W Cobb'
Frame:	Small cast-iron spacer frame
Trains:	Going train only
Escapement:	Deadbeat
Pendulum:	Wooden rod and cast-iron cylindrical bob
Rate:	60 beats per minute
Weight:	Cast-iron
Winding:	Hand wound weekly
Dial:	Skeleton dial facing cricket pitch
Note:	Transferred from old pavilion in 1923

CONSTABLES

Constables, prior to conversion to a school house, was originally Uppingham Union Workhouse, built in 1837. A postcard of the building of *circa* 1917 (RCM H36.1992) shows that, like the Oakham Union Workhouse, it had a bellcote and bell over the front façade. At this time it was known as Uppingham Auxiliary Hospital. One use of this bell when the building was a workhouse was to call the poor of the town for their soup at meal times.

THE HALL, BROOKLANDS & THE THRING CENTRE

The early 1900s OS Second Series 25 inch maps of Uppingham show that there were table sundials in the grounds of Uppingham Hall in High Street East and Brooklands in London Road. The Uppingham Hall sundial still exists, although it is now located in the south garden. The Hall, built in 1612, was acquired by the School on a ninety-nine year lease in 1891. It was subsequently converted into a school house to accommodate the ever-growing school roll.

There is also an early sundial on a south facing wall in the garden behind the Manor House, now the Thring Centre of Uppingham School.

From a circa 1917 postcard of Uppingham Auxiliary
Hospital (RCM H36.1992). The front section of the
building including the bellcote was removed in the later
conversion to a school house

THE MASONS' LAWN & THE LIBRARY

A new table sundial was presented to the school by the
praepostors [prefects] in the Millennium year. Its motto
is: 'Make time, save time. While time lasts. All time is no
time. When time is past.' This sundial was erected on the
Masons' Lawn in front of the School Library (*see* Chapter 1.5 — Sundials).

The School Library was originally erected by Archdeacon Johnson in 1592 as a hospital chapel. By 1771 it
was the School Chapel and in that year the school bell
was transferred here from Archdeacon Johnson's Schoolroom. By 1825 it was also used as School House dining
hall. In 1949 it was converted into an extension to the
library as a Second World War memorial. The cupola still
remains but the bell was removed to the new Tercentenary Block in 1889.

ODDFELLOWS HALL &
THE FORMER NATIONAL SCHOOL

Oddfellows Hall in North Street, now divided into apartments, has a bellcote over the east gable. The fate of the
bell is unknown. There was also a bellcote and bell at the
former National School in London Road.

MARKET PLACE

Inspection of the leaded lights above the display windows of the ironmongers on the corner of High Street
and the Market Place reveals that one of the many
services offered by William Mear was that of bellhanger,
but in 1925 (Kelly, 746) he is only listed as an ironmonger. A former proprietor of this business was Samuel

*(right) The bell-
cote on the former
Oddfellows Hall,
Uppingham*

Foster: 'Foster, Samuel — ironmonger, oil and colour
dealer, brazier, iron and tin plate worker, bellhanger,
gasfitter, and agricultural implement dealer, Market place'
(White 1877, 705). Charles White, who traded from
another ironmonger's shop in High Street East, also
advertised that he would hang bells (Harrod, 1870).

UNICORN INN

The former Unicorn Inn in High Street East has an
interesting sundial. The dial remains complete with its
gnomon, as does the relief image of a unicorn below, but
the dial furniture has been painted out for many years.
However, the hour lines and numerals can be seen when
the sun is at an oblique angle, and the dial details were
recorded by Mrs Gatty in 1890 (Gatty 1890, 152). It
was dated 1765 and its motto was 'Improve the Time'.

THE CROWN HOTEL

There is a magnificent well-preserved sundial high on the
south elevation of the Crown Hotel in High Street East.
Gatty's *Book of Sundials* dated 1890 states that it was
then in black and gilt. Its history is not known but it is
probably early nineteenth century. The motto is 'Non
Rego Nisi Regar' which is interpreted as 'I do not guide
unless I am guided ', apparently an acknowledgement of
submission to the sun (Gatty 1890, 207).

SUNDIAL HOUSE & SUNDIAL COTTAGE

Sundial House in High Street East has an eighteenth-
century sundial in the gable of a dormer window. A
plaque high on the south gable of 1 Stockerston Road
has the words 'Sundial Cottage', but there is no sundial
here now.

The sundial on the former Unicorn Inn in High Street East, Uppingham

A conjectural drawing showing the layout of the sundial on the former Unicorn Inn

PLEASANT TERRACE

Although the row of houses at Pleasant Terrace, Adderley Street, is late Victorian, a very early sundial is preserved on the south-facing wall of No 5. It is dated 1661 and has the initials R^{P}A in a circle about the gnomon root. R^{P}A may stand for Robert Pakeman and his wife. Their initials would have been RP and AP respectively. Alternatively, the RP could have been Richard Pepper. Both men are listed in the Uppingham Hearth Tax Assessment of 1665 (Bourn 1991, 40). The sundial was probably saved from an earlier cottage on this site which is mentioned in surviving deeds.

The sundial on The Crown Hotel in High Street East, Uppingham

The sundial in a dormer gable at Sundial House, High Street East, Uppingham

The sundial dated 1661 on the south elevation of 5 Pleasant Terrace, Uppingham

WARDLEY

ST BOTOLPH   Grid Ref: SK 832002

PARISH RECORDS
The only available records are a Vestry Minute Book 1869-80 (DE 1816/8) and some eighteenth-century Glebe Terriers (MF 495).

BELL HISTORY
A Glebe Terrier dated 10 July 1710 states that Wardley had 'A church consisting of one body of a church without Isles. A chancel, a low spire steeple wherein are 3 bells, & a sts bell standing in a churchyard ...'. This information is repeated in other terriers from 1702 to 1767. The 'sts bell' refers to a Saint's or Sanctus Bell but Thomas North made no reference to this in 1880.

Of the other three bells North reported that the old tenor was removed when it was found to be cracked. It was left at the bottom of the tower and eventually sold. The early seventeenth-century oak frame was described as

being 'much decayed and the ladders worm-eaten' (North 1880, 162). The wooden bellframe was replaced *circa* 1988 by a low-sided frame constructed from square section steel tubing. This is set diagonally across the tower as was the previous frame. The bells are hung in fabricated steel headstocks with levers for chime ringing only.

BELL DETAILS
1 1677. Diameter 610mm (24in). Weight about 152kg (3cwt). Canons retained. Fabricated steel headstock. Cast by Toby Norris III of Stamford.
GOD [52] SAVE [52] THE [52] KING [52] 1677

2 *Circa* 1529. Tenor. Note C sharp. Diameter 711mm (28in). Weight about 241kg (4cwt 3qr). Canons retained. Fabricated steel headstock. Cast by Thomas Bett of Leicester.

[2] ThOMA [77]
(Thomas)

Letters and marks are spaced evenly around the inscription band. All letters are like **[108]**. Scheduled for preservation (Council for the Care of Churches).

BELLRINGING CUSTOMS
On Sundays the two available bells were either chimed at 9am for a morning service or at noon for an afternoon service. The clerk commenced chiming for Divine Service when the vicar, who lived in Belton, had arrived at the

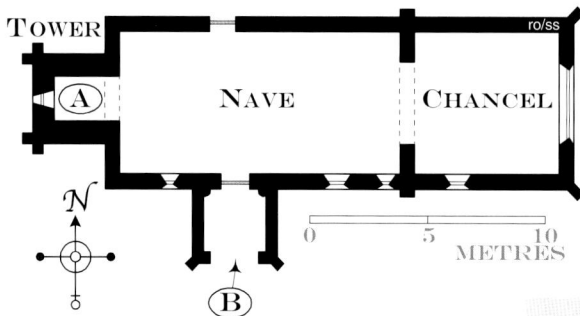

Plan of St Botolph's Church
A: Two bells in the belfry. Ringing chamber at the base of the tower
B: Sundial
Access to the belfry is by temporary ladders

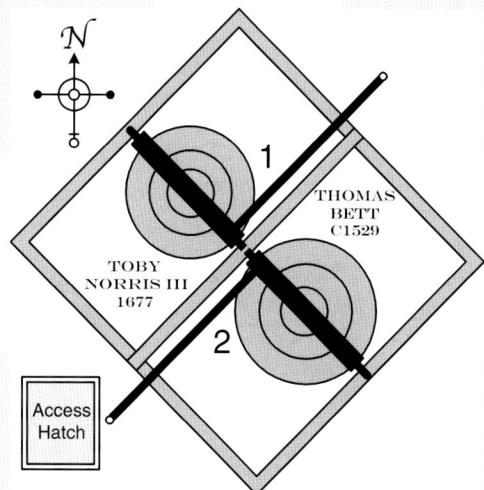

Plan of the bellframe at Wardley Church

From a drawing of circa 1793 of Wardley Church. The sundial is clearly seen over the south porch (RCM F10/1984/64)

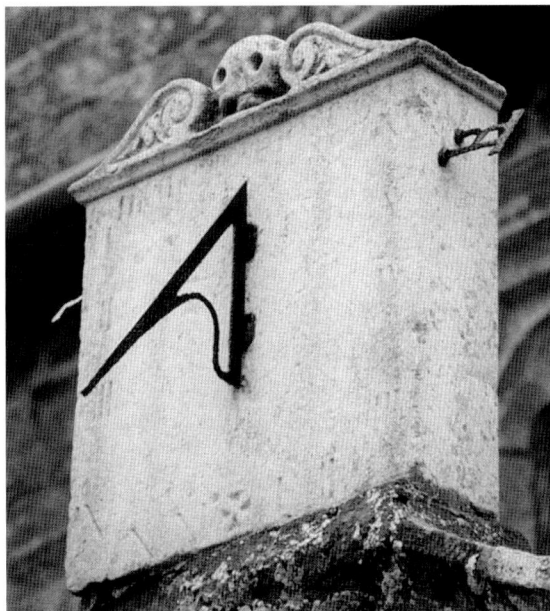

The sundial at Wardley Church

16 memento mori 94

The date and motto on the porch sundial at Wardley Church

church. The Pancake Bell used to be rung at 11am on Shrove Tuesday. At the Death Knell there were three tolls for a male and two tolls for a female. Prior to 1880, the bells were chimed at funerals when the procession entered the churchyard. By 1880, the second bell was tolled instead (North 1880, 162).

Today the tenor is chimed for five minutes before the start of Sunday Services.

CANDLE HOLDER

There are the remains of a possible bellringers' candle-holder on the north wall in the ringing chamber. Part of the date carved on it is missing.

SUNDIAL

The sundial placed above the gable of the south porch has three gnomons. It is a limestone block above which

is a pediment consisting of a skull between ornamental scroll decoration. It has a gnomon on each of its south, west and east faces. The roman numerals on the south face are worn but are still visible. The date and motto '**16 memento mori 94**' [remember death] are carved on the upper border but both are quite worn. The hour lines on the east and west faces have weathered away. The block is turned slightly to the east so that the south dial faces due south.

WHISSENDINE

ST ANDREW   Grid Ref: SK 833143

PARISH RECORDS

Relevant archival material includes the Churchwardens' Accounts & Levies 1695-1778 and 1781-1813 (DE 1831/15-16) and a Vestry Minute Book 1956-74 (DE 1831/17).

BELL HISTORY

Irons' Notes (MS 80/5/24) provide the earliest references to the bells at Whissendine:

1587	June 13	One of their bells is oute of repair
1696	Oct 3	A bell splitt

The earliest note probably refers to broken bell fittings. From details in the Churchwardens' Accounts and North's survey of 1880 it can be deduced that there were at least three bells in the belfry at the end of the seventeenth century. These were the two bells cast by Henry Oldfield II in 1609, which still remain in the tower, and the bell 'splitt' in 1696 as noted above. This latter bell could have been the present fifth which was recast in 1709,

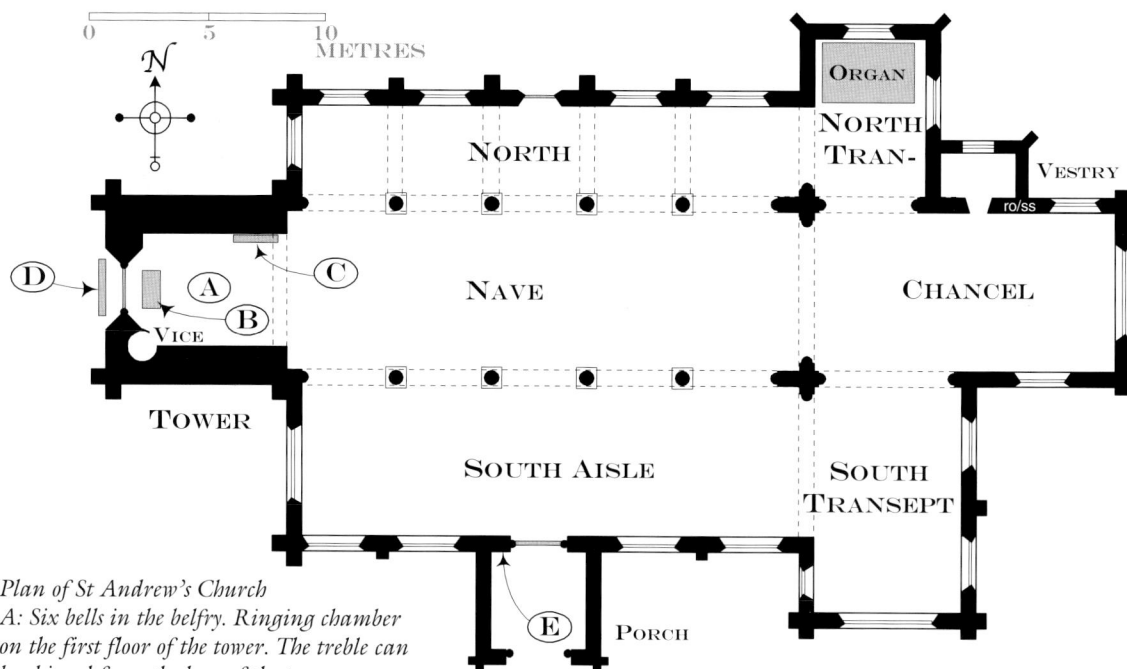

Plan of St Andrew's Church
A: Six bells in the belfry. Ringing chamber
on the first floor of the tower. The treble can
be chimed from the base of the tower
B: Clock on the third floor of the tower
C: Ellacombe chiming frame in the ringing chamber
D: Clock dial
E: Scratch dials 1 & 2
The vice provides access to the ringing chamber, clockroom
and belfry

possibly by Henry Penn of Peterborough, and recast again by John Taylor in 1872. The lack of any record of payments in the Churchwardens' Accounts suggests that the cost of recasting on both occasions was paid for by subscription or donation. By 1754 it seems that there were definitely four bells as a Jacob Fowler was paid 14s for '4 Bell Ropes' in that year.

The following item in the accounts probably refers to the recasting of the then treble by an unknown founder:

| 1777 | Pd the Bell Founders Bill | £16 0s 0d |
| | Pd Jno Fowler for Carage of ye Bell | 14s 0d |

Eight years later, a bell was taken to Edward Arnold's foundry in Leicester. It may be that the treble had to be recast again and this would be the present third inscribed EDW^D ARNOLD LEICESTER FECIT 1785:

| 1785 | Paid the Bell Founders Bill | £18 18s 10d |
| | the Caridge of the Bell to Leicester and home a gain | £1 1s 0d |

Other than a few payments at the beginning of the accounts there are no obvious payments for the repair of the bells. There are however numerous entries for the renewal of bellropes and the suppliers named include Jacob Fowler, John Cooke, Mr Wartnaby, and George

and Widow Toon during the second half of the eighteenth century, and Barsby and Mr Brasbridge into the first decade of the 1800s.

The following are some examples from the accounts relating to the bells:

1695	for a plate for ye bell wheel	6d
	for ye stays mending	6d
	for 4 gugins [gudgeons] for ye bells	4s 0d
1696	paide to John Sturges for makeing ye bell wheel	1s 6d
	paide for the bell rps	10s 0d
1746	Pd for Oyle and Lamp Black [graphite powder for lubricating sliders] for ye Bell frames	7s 11d

In 1872 when the present fifth was recast the other bells were put in good order. The Rector of that time recorded the costs (North 1880, 163):

New 3rd bell	£72 18s 3d
New hangings, clapper, time & carriage	£13 16s 5d
New hangings and clappers to 1st, 2nd, and tenor bells, time and carriage	£31 16s 0d
	£118 10s 8d
Allowed for old bell	£45 3s 6d
Total	£73 7s 2d

The Rev E L Horne requested quotations for the restoration and augmentation of the bells in 1897. John Taylor's report of 28 August included the recommendation that the old and decaying wooden frame be replaced by a new metal frame for the four existing and two new bells, all on one level. An alternative quotation was obtained from John Warner and in the event an order was placed with this founder for only one bell, the present second, which was installed in 1898. The cost of the new bell was £96 6s 3d. The existing frame was retained and some strengthening work was carried out by a local carpenter (Smith 2000, 83-4). The present treble was added eight years later by John Taylor.

These events, together with the installation of a new clock, are recorded on a brass plaque in the tower:

**A FIFTH BELL WAS ADDED TO THE
ORIGINAL PEAL IN COMMEMORATING THE
SIXTIETH YEAR OF QUEEN VICTORIA'S REIGN.
A SIXTH BELL WAS ADDED IN 1906.
THE CLOCK WAS ERECTED IN MEMORY OF
KING EDWARD VII AND TO COMMEMORATE
THE CORONATION OF KING GEORGE V IN 1911.
THESE ADDITIONS WERE MADE AT THE
COST OF THE PARISHIONERS OF WHISSENDINE.**

An adjacent brass plaque records that the bells were rehung in 1919. The work was carried out by Day of Eye who also reconstructed the oak frame (Powell 1938, 24):

**THE BELLS WERE RE-HUNG IN 1919:
THE CHURCH RE-POINTED AND THE
TOWER AND PORCH RESTORED BY
COLONEL JOHN GRETTON M.P.
OF STAPLEFORD PARK IN 1920.
THIS TABLET IS PLACED HERE
BY THE GRATEFUL PARISHIONERS.**

In 1956 a survey of the bells was carried out by John Taylor and the subsequent report was discussed at a meeting of the PCC held on 20 November. It stated that the headstocks and fittings were in a poor state, and to avoid complete renewal and rehanging they suggested that the bells be 'fixed immovable' and that an Ellacombe chime apparatus be installed in the ringing chamber. Taylor's recommendation was accepted and the apparatus was installed at a cost of £169.

In 1963 sufficient funds became available to rehang the bells with new fittings in the existing wooden frame. This work was carried out by John Taylor and from then on the bells could again be rung full circle.

BELL DETAILS

1 1906. Treble. Diameter 794mm (31¼in). Weight 329kg (6cwt 1qr 25lb). Cast without canons. Cast-iron headstock. Cast by John Taylor & Co, Loughborough.

JOHN TAYLOR & CO. [65] FOUNDERS [65]
~ LOUGHBOROUGH [65] 1906 [65]

On the waist:

E. L. HORNE VICAR
W. S. FOWLER } CHURCHWARDENS
M. CUNNINGHAM

The lettering is like [109]. Below the inscription band is decoration [54].

2 1897. Diameter 864mm (34in). Weight 358kg (7cwt 0qr 6lb). Canons removed. Cast-iron headstock.

The six bells await collection outside Whissendine Church in 1963 (Paul Phillips)

JOHN WARNER & SONS L^TD

An example of the lettering on the second bell at Whissendine

Cast by John Warner & Sons of London.
**CAST BY JOHN WARNER & SONS L^TD
~ LONDON 1897**

3 1785. Diameter 924mm (36⅜in). Weight 412kg (8cwt 0qr 12lb). Canons removed. Cast-iron headstock. Cast by Edward Arnold of Leicester.

**RICH^D. FLOAR CHURCHWARDEN [45]
~ EDW^D. ARNOLD LEICESTER FECIT 1785
~ [45] [45] [45]**

There were still many members of the Floar family in the village in 1846 (White, 618).

4 1609. Diameter 943mm (37⅛in). Weight 411kg (8cwt 0qr 10lb). Canons retained. Cast-iron headstock. Cast by Henry Oldfield II of Nottingham.

[34] GOD SAVE HIS CHURCH 1609 [7]

The words are placed in blocks. The letters are like [116].

5 1872. Diameter 1016mm (40in). Weight 532kg (10cwt 1qr 24lb). Canons removed. Cast-iron head-

Plan of the low-sided timber bellframe at Whissendine Church which was reconstructed by Day of Eye in 1919. The headstock bearings of the third, fourth, fifth and tenor were raised onto short 'I' beams in the restoration of 1963

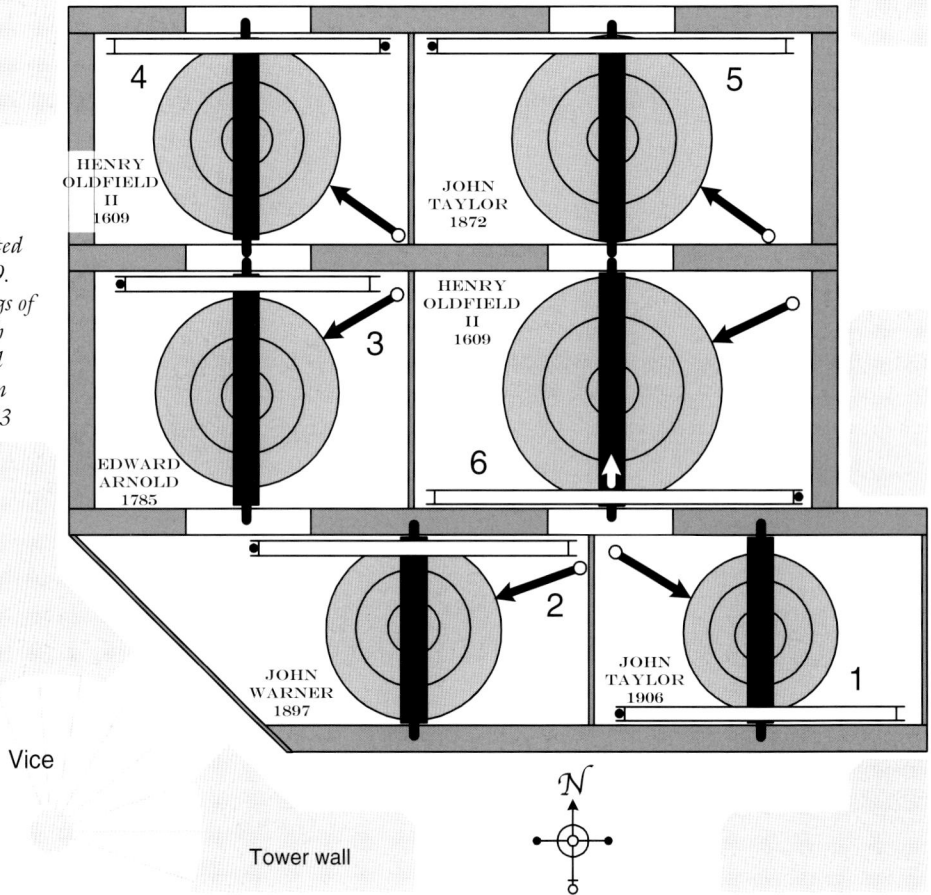

HENRY OLDFIELD II 1609

JOHN TAYLOR 1872

4

5

EDWARD ARNOLD 1785

3

HENRY OLDFIELD II 1609

6

JOHN WARNER 1897

2

JOHN TAYLOR 1906

1

Vice

Tower wall

N

stock. Recast by John Taylor & Co of Loughborough.

> ### SACRA CLANGO : GAUDIA PANGO :
> ### ~ FUNERA PLANGO

(Holy offices I proclaim: joys I spread: deaths I bewail)
On the waist:

> ### RECAST BY JOHN TAYLOR & Cº
> ### LOUGHBOROUGH 1872:

Former bell

1709. Ascribed to Henry Penn of Peterborough (Pearson 1989). Diameter 965mm (38in). Weight 504kg (9cwt 3qr 20lb).

> ### SACRA CLANGO : GAUDIA PANGO :
> ### ~ FUNERA PLANGO 1709
> ### E STAFFORD. J GREENFIELD WARDENS.
> ### ~ W CUMINGE VICKAR

6 1609. Tenor. Note E. Diameter 1108mm (43⅝in). Weight 647kg (12cwt 2qr 26lb). Canons removed.

warning geve that

An example of the lettering on the tenor at Whissendine

Cast-iron headstock. Cast by Henry Oldfield II of Nottingham.

> my roaringe sounde doth warning geve
> ~ that men cannot heare always lyve 1609 [7]

The words are placed in blocks. The lettering is like [114].

BELLRINGING CUSTOMS

Confirmation that a Curfew Bell and a Day Bell were rung in the early seventeenth century is provided by this extract from Irons' Notes (MS 80/5/24):

1614 Oct 11

Thomas White he doth not his office to attend on the

St. Andrews Church
Whissendine

A set of bellropes was
presented in memory of
Samuel Robert (Bob) Hoy
1917-1993
by his family

A resident of Oakham, he
taught a new band of ringers in
Whissendine, and often rang here.

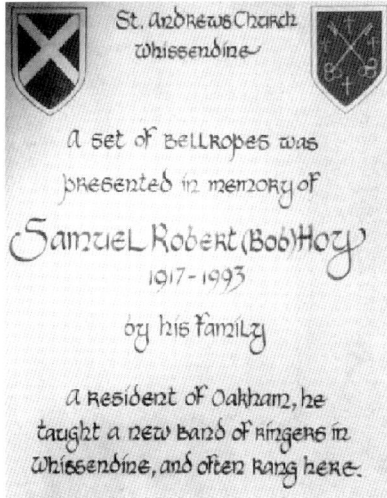

Whissendine Church.
Certificate in memory of
Bob Hoy of Oakham, who
was devoted to bellringing

A framed peal certificate
in the ringing chamber
at Whissendine

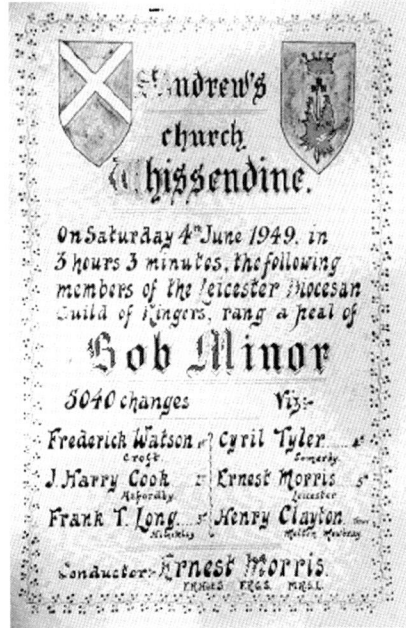

St Andrew's
church
Whissendine.

On Saturday 4th June 1949, in
3 hours 3 minutes, the following
members of the Leicester Diocesan
Guild of Ringers, rang a peal of

Bob Minor

5040 changes   Viz:-

Frederick Watson   Cyril Tyler
        Croft   formerly
J. Harry Cook   Ernest Morris
        Ashfordby
Frank T. Long   Henry Clayton
        Wilksby   Melton Mowbray

Conductor: Ernest Morris.
        Kinges Frea Mral

minister upon Woden's dayes [Wednesdays] and ffrydayes [Fridays] according to his duty. For refusing to ringe day bell and Curfey bell as it hath beene runge.

TW [Thomas White] says he beinge both Clarcke and Sexton is not to performe the office of the sexton for the wages they give him but that wages which he hath is given him as he is the Clarke.

On Sundays the treble was rung at 8am, the second at 9am, and all the bells were chimed for Divine Service. At the Death Knell four tolls were given for a male and three tolls for a female. The Pancake Bell was rung at 11am on Shrove Tuesday and the Gleaning Bell was rung at 8am and 6pm during harvest (North 1880, 164).

There are few payments in the accounts for bell-ringing other than the annual amounts paid on 5 November from 1754 to 1811.

At the time of writing St Andrew's has a band of bellringers which practises every Thursday evening from 7.30pm to 9pm. The bells are rung for half an hour

before the 11am Sunday Service, for the Easter Service and for all Christmas Services. They are also rung for the midnight services on Christmas Eve and New Year's Eve. Other occasions include the Armistice Day Service, when they are rung half muffled, and by request for weddings and funerals. Occasionally peals and quarter peals are rung by visiting bands.

Hourglass

References to hourglasses are almost non-existent in Rutland but a carving on a headstone to the north of the church gives the ominous message that life is as transient as the 'sands of time'.

Scratch Dials

There are faint remains of two scratch dials on the thirteenth-century jamb to the west of the south door. This doorway was apparently moved to its present position when the south aisle and south porch were added in the fourteenth century.

Clock History

The earliest reference to a clock may be in Irons' Notes (MS 80/5):

1635 June 25 Par clk [parish clock?] not allowed of by the bp [bishop]

This extract seems to imply that the Bishop would not allow them to have a parish clock. Alternatively, 'Par clk' may be a reference to the parish clerk.

The following items in the Churchwardens' Accounts record the installation of a clock by Thomas Eayre II,

An hourglass on a headstone in Whissendine churchyard

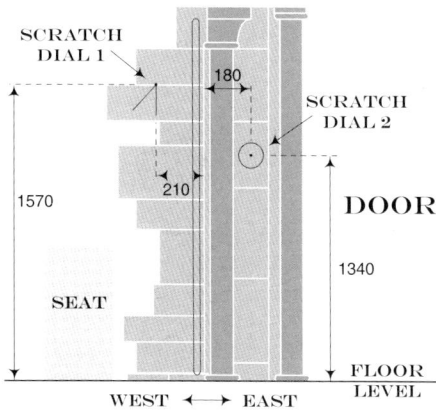

Location of the scratch dials inside the porch at Whissendine Church

WHISSENDINE 1		
Location	West of south door, inside porch	
Condition	Poor	
Gnomon Hole Diameter		10
Gnomon Hole Depth		Filled in
Height above ground level		1570
Line Ref	a	b
Length	95	135
Angle (°)	180	235

Scratch dial 1 details

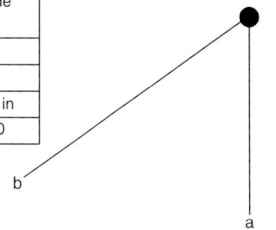

WHISSENDINE 2		
Location	West of south door, inside porch	
Condition	Poor	
Gnomon Hole Diameter		12
Gnomon Hole Depth		Filled in
Height above ground level		1340
Circle Diameter		110
Line Ref	No lines visible	

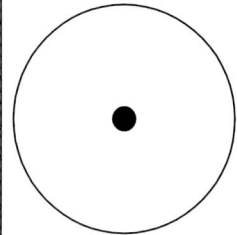

Scratch dial 2 details

clockmaker and bellfounder of Kettering:

1746	pd Mr Eayrs his Bill	£20 2s 11d
	pd Chris Hack for 2 poles for ye Clock	1s 0d
	pd George Snodin for going to 3 times to Kettering for Mr Eayr	12s 0d
	pd John Edgson his Bill	£19 6s 2d

John Edgson of Teigh was a carpenter and his bill may have been for a floor for the clockroom, a stand for the clock and a new clock case. The poles supplied by Chris Hack would have been for scaffolding.

There may however have been another clock prior to this. A Stephen Blackburn was a churchwarden at Whissendine in 1695 and it is believed that he was a blacksmith and a member of the family of clockmakers working at Oakham during the next century. Before the installation of the Eayre clock in 1746 Stephen Blackburn, the Oakham clockmaker, was paid 5s in 1742 and 1743. This is commensurate with amounts paid after this date for work on the clock:

| 1747 | pd Mr Blackburn for ye Clock for 2 years | 10s 0d |
| 1748 | paid Mr blackborn for the Clock Doing | 5s 0d |

Stephen Blackburn maintained and repaired the clock until 1771. During this time the dial was replaced and a considerable sum of money was spent. This may be when the clock was converted to two hands:

1761	pd for ye New Dyall	£4 4s 0d
	pd Mr Blackburn for ye Clock	5s 0d
	pd John Orton for helping down with ye dyal	1s 0d
1770	pd Mr Blackburns man for Raiseing ye dyal Board and makeing ye hand to go	3s 6d
1771	pd Mr Blackbourn for ye town Clock	£12 0s 0d

The 1770 entry indicates that the clock had a single hand and the use of the word 'Board' implies that it was a timber dial. Fifteen years later the dial required attention again:

| 1786 | paid for alle for the Dial Board puting up | 1s 3d |

There are no further entries regarding the clock until 1798 when Mr Boyfield, a clockmaker from Melton Mowbray, was paid for work for three consecutive years. Nothing more is known of the Eayre clock.

It was replaced in 1911 by a new clock from John Smith & Sons of Derby, purchased by the parishioners in memory of King Edward VII and to commemorate the coronation of King George V.

CLOCK DETAILS

Details of the present clock:

Maker:	John Smith & Sons of Derby
Signed:	On the setting dial
Installed:	1911
Donated by:	Whissendine parishioners
Frame:	Cast-iron flatbed
Trains:	Going and striking
Escapement:	Pinwheel
Pendulum:	Wooden rod and cast-iron cylindrical bob
Rate:	52 beats a minute
Striking:	Countwheel hour-striking
Weights:	Cast-iron
Winding:	Hand wound weekly
Dial:	Circular metal dial with blue background and gilded roman numerals and hands
Location:	West face of tower
Note:	A photograph taken in 1913 shows a circular dial (Henton 1163)

WHISSENDINE RECTORY

A table sundial is shown in the grounds of Whissendine Rectory on the early 1900s OS Second Edition 25 inch map.

WHITWELL

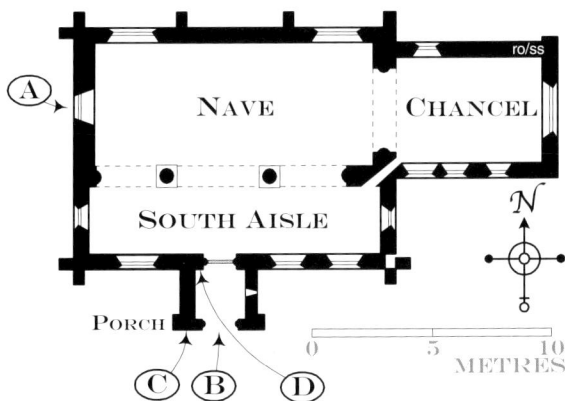

Plan of St Michael & All Angels' Church
A: Two bells in a bellcote rung from the west end of the nave
B: Sundial
C: Scratch dials 1 & 2
D: Scratch dials 3 & 4

Plan of the bellcote at Whitwell Church

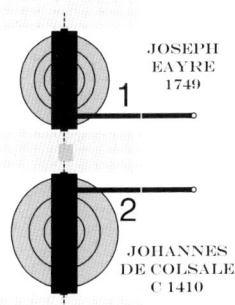

ST MICHAEL & ALL ANGELS Grid Ref: SK 924088

PARISH RECORDS

The following records have been searched:

Churchwardens' Accounts 1809-70 and 1870-1944 (DE 3013/61-62).
Vestry Minute Book 1894-1943 (DE 3013/68).
Parish Meeting Minute Book 1894-1945 (DE 3013/69).

BELL HISTORY

The thirteenth-century double bellcote at Whitwell is the earliest of the Rutland bellcotes and it supports one of the earliest bells in the county. This bell, the second, is thought to have been cast *circa* 1410 by Johannes de Colsale, an itinerant founder, probably from the Nottingham area. Joseph Eayre of St Neots cast the present treble in 1749, but nothing is known of the bell or bells that previously hung in this position.

There are few details regarding the present bells in surviving documents. They do, however, give the names of some bellrope suppliers along with the names of those who carried out repairs to the bells, and the following are typical examples. It is interesting to note that one of the bells had to be taken to an Exton blacksmith for repairs:

1812	Nov 9	Carriage of the Bell to Exton and bringing it back	5s 0d
	Dec 28	Paid Jacksons Bill for repairing the Bell	£1 0s 9½d
1824	Feb 23	Dr Mr Ogdens Bill Each for repairing the Bells & Steeple	£3 12s 0d
1826	Nov 24	Pd Wm Royce for lining the Bell clappers	6s 0d
1827	Nov 14	Mr Lowe for Bellropes	7s 0d
1837	Nov 4	Dr for Piper for bellropes & repairing the roof	£1 12s 2d

The church leaflet of 1975, prepared by the then Rector, itemises the restoration work carried out at the church in that year. It includes: 'the repair of the bell-cote the re-hanging of the two bells so that they could both be used again and before further weathering caused them to fall.'

Today both bells still have wooden headstocks and both are hung for chime ringing. Wires from headstock levers pass through tubes in the nave roof. These are attached to the bellropes which hang at the west end of the nave.

BELL DETAILS

1 1749. Treble. Diameter 438mm (17¼in). Weight approximately 63kg (1cwt 1qr). Canons retained. Timber headstock. Cast by Joseph Eayre of St Neots.

J EAYRE ST NEOTS 1749 [49] [49]

2 *Circa* 1410. Diameter 540mm (21¼in). Weight approximately 114kg (2cwt 1qr). Canons retained. Timber headstock. Ascribed to Johannes de Colsale, an itinerant founder (George Dawson), probably from the Nottingham area.

[28] IN ḢONORE SANCTI EIGDII
(In honour of St Ægidius)
[St Ægidius is the Latin form of St Giles]
Scheduled for preservation (Council for the Care of Churches).

BELLRINGING CUSTOMS

On Sundays the treble was rung at 8am and again at 8.30am if a sermon was to be preached at the Morning Service. During harvest the treble was rung as the Gleaning Bell at 8am. At the Death Knell thrice three tolls were given for a male and thrice two tolls for a female, both before and after the knell. At funerals both bells were rung (North 1880, 164).

In many Rutland parishes it was customary for the parish clerk to ring a bell at agreed times. Although the surviving accounts include occasional payments for bell-

ENTRANCE
TO
SOUTH
PORCH

SCRATCH DIAL 1

SCRATCH DIAL 2

WEST ←→ EAST

FLOOR LEVEL

Location of scratch dials 1 & 2 at Whitwell Church

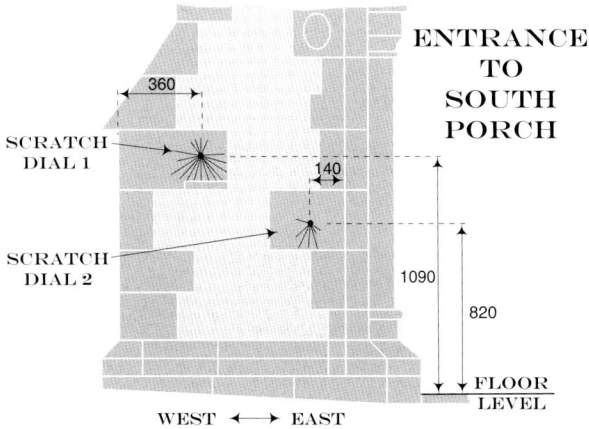

WHITWELL 2							
Location	Front face of porch, west side						
Condition	Very poor						
Gnomon Hole Diameter		15					
Gnomon Hole Depth		25					
Height above ground level		820					
Line Ref	a	b	c	d	e	f	g
Length	40	40	100	90	100	70	55
Angle (°)	125	156	170	185	205	225	292

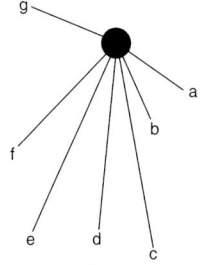

Scratch dial 1 details

WHITWELL 1									
Location	Front face of porch, west side								
Condition	Average								
Gnomon Hole Diameter		18							
Gnomon Hole Depth		25							
Height above ground level		1090							
Line Ref	a	b	c	d	e	f	g	h	i
Length	94	80	110	120	125	140	120	110	105
Angle (°)	0	34	85	104	117	133	149	160	178

Line Ref	j	k	l	m	n	o	p	q
Length	100	105	120	130	120	120	85	80
Angle (°)	193	207	222	238	254	276	320	340

Scratch dial 2 details

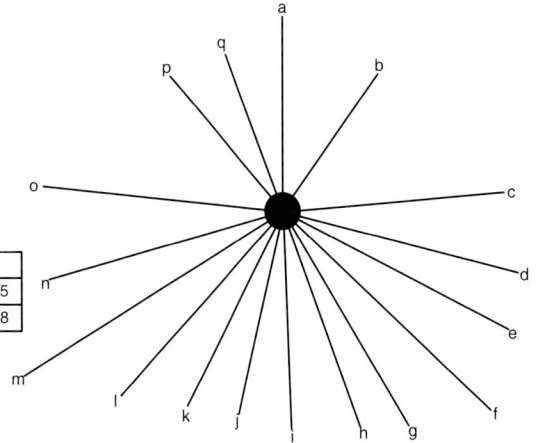

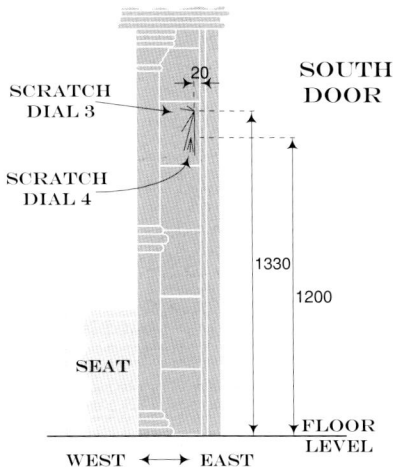

SOUTH
DOOR

SCRATCH DIAL 3

SCRATCH DIAL 4

SEAT

WEST ←→ EAST

FLOOR LEVEL

Location of scratch dials 3 & 4 at Whitwell Church

WHITWELL 3					
Location	West side of south door				
Condition	Poor				
Gnomon Hole Diameter		-			
Gnomon Hole Depth		-			
Height above ground level		1330			
Line Ref	a	b	c	d	e
Length	200	185	90	115	80
Angle (°)	177	196	215	278	310

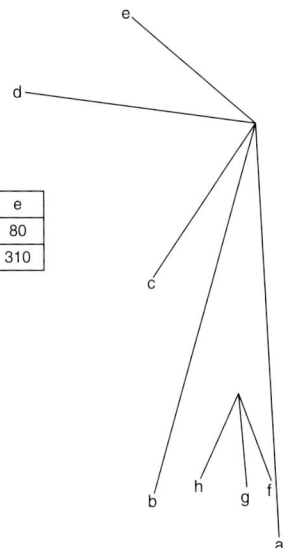

WHITWELL 4			
Location	West side of south door		
Condition	Poor		
Gnomon Hole Diameter		-	
Gnomon Hole Depth		-	
Height above ground level		1200	
Line Ref	f	g	h
Length	45	45	45
Angle (°)	160	175	195

Scratch dial 3 & 4 details

ringing, the following is the only entry indicating that it was the clerk's duty at Whitwell:

1936 Mr Springthorpe Bells etc £4 19s 0d

Elijah Needham Springthorpe was sub-postmaster, bootmaker and parish clerk. He lived at the Post Office (Kelly 1925, 747).

Today the tenor is rung for five minutes before each service and on the occasions of Easter and Christmas.

SCRATCH DIALS

There are two scratch dials on the south face of the thirteenth-century porch to the west of the entrance and a further two inside the porch on a stone to the west of the south door which is possibly of twelfth-century origin.

SUNDIAL

There are the remains of a sundial in the porch gable of Whitwell Church. No markings remain and the gnomon is missing. The dial stone is approximately 400mm (16in) high and 450mm (18in) wide. A drawing of *circa* 1793 (RCM F10/1984/66) shows that it then had a gno-

Whitwell church sundial from a drawing of circa 1793
(taken from RCM F10/1984/66)

mon, but the sundial is not included on a drawing of *circa* 1839 in Uppingham School Archives.

WHITWELL RECTORY

A table sundial is shown in the garden of the Rectory on the early 1900s OS Second Edition 25 inch map.

WING

ST PETER & ST PAUL   Grid Ref: SK 894030

PARISH RECORDS

The following parish records have been searched:

Churchwardens' Accounts 1898-1921 (DE 1846/23).
Vestry Minute Books 1833-91 (DE 1846/24-5).
Church Accounts 1932-46 (DE 5003/12).
PCC Minute Books 1920-84 (DE 5003/14-16).
Faculties for the removal of the spire (11 May 1841), alteration and restoration work in the church (31 August 1883), repairing and improving the church (23 April 1875), and restoring the church and rehanging the bells (13 May 1903) (DE 1846/15-18).
Records of Wing — an unpublished record of memories of parishioners from 1935 until about 1960 (held in the parish).

BELL HISTORY

Little is known of the early bells at Wing except that there was a bellcote over the western end of the nave until the tower was built at the end of the fourteenth century (*VCH* II, 105). If the bellcote was a typical Rutland structure it would have had two small bells in a double bellcote. It is known that there were two large bells in the tower in the sixteenth century and that the

Newcombe foundry of Leicester possibly supplied both. Irons' Notes (MS 80/1/3) confirm that the church also had a Sanctus Bell and handbells at about this time:

1605 The Santi bell about 16 yrs since & 2 hand bells were put into one of their bells then it was cast so that now they have not a Saints bell and it was supposed that they had a bigger & great bell which they sold.
1616 June 28 Gard [guardian or churchwarden] to provide a Sts [Saint's or Sanctus] bell

By the end of the eighteenth century, and probably well before this, the belfry had reached its present complement of five bells. The recasting of the fourth bell in 1903 has been the only change since then.

In 1789, the treble, second and third bells were recast by Robert Taylor of St Neots. This was, as Thomas North records, after an 'enterprising Churchwarden' by the name of George Paddy had made an unsuccessful attempt to recast them himself (North 1880, 165). Obviously the Churchwardens' Accounts for this period were still available at the time of North's survey as he states that the 'Church records give no reference to the recasting of these bells', implying that the cost was covered by donation. George Paddy's name is inscribed upon all three bells and he may have been the benefactor.

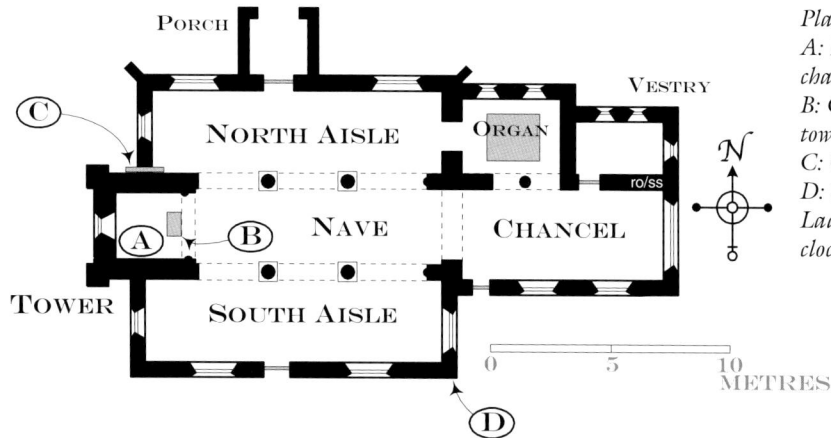

Plan of St Peter & St Paul's Church
A: Five bells in the belfry. Ringing chamber at the base of the tower
B: Clock on the first floor of the tower
C: Clock dial
D: Scratch dial
Ladders provide access to both the clockroom and the belfry

In 1841, Mr Flint, an architect of Leicester, was asked to carry out a survey on the tower and spire, the condition of which were causing some concern at the time. In his report, which was considered at a Vestry Meeting on 15 April of that year, he stated that the tower had been 'weakened by being partially cut away to place the Bell-frames'. As a result, a faculty dated 11 May 1841 was obtained and the spire removed soon after. The stone was used in the building of the Wesleyan Chapel which was erected in the same year.

Despite the removal of the spire, there was still some concern regarding the condition of the tower. A special Vestry Meeting was called in 1902 'to consider the condition of the Church Tower & Bells'. It was unanimously agreed that 'the restoration be undertaken & an appeal for funds be made for the purpose'. A faculty dated 13 May 1903 stated that the upper part of the tower 'was in such a dilapidated condition' that taking the tower down and rebuilding it was a necessity and 'the Bells were to be rehung and the fourth Bell recast'. The estimated cost of the work on the bells was £142 18s. This work was carried out by John Taylor & Co in the same year, the bells being hung in a new metal high-sided frame with the treble set above in a low-sided frame.

The Churchwardens' Accounts show virtually no expenditure on the maintenance of the bells and no further details have been found.

Wing Church just before the spire was removed in 1841
(Canon John R H Prophet). *This is based on a circa 1839 drawing of the church* (Uppingham School Archives)

BELL DETAILS

1 1789. Treble. Diameter 695mm (27⅜in). Weight 240kg (4cwt 2qr 26lb). Canons removed. Cast-iron headstock. Cast by Robert Taylor of St Neots.

[67] GEO: PADDY C ÷ WARDEN. 1789. [67]
There is a block of decoration [67] on the inscription band on the opposite side of the bell.

2 1789. Diameter 727mm (28⅝in). Weight 239kg (4cwt 2qr 23lb). Canons removed. Cast-iron headstock. Cast by Robert Taylor of St Neots.

GEORGE PADDY CHURCHWARDEN. [67]
~ ROBᵀ. TAYLOR Sᵀ. NEOTS FECIT 1789 [67]

3 1789. Diameter 756mm (29¾in). Weight 255kg (5cwt 0qr 3lb). Canons removed. Cast-iron headstock. Cast by Robert Taylor of St Neots.

[67] GEO: PADDY C ÷ WARDEN [67]
~ TAYLOR Sᵀ. NEOTS FECIT. 1789. [67]

4 1903. Diameter 775mm (30½in). Weight 266kg (5cwt 0qr 26lb). Cast without canons. Cast-iron headstock. Recast by John Taylor & Co of Loughborough.

[30] Gloria in Excelsis DEO
(Glory to God in the highest)

Plan of the bellframe at Wing Church

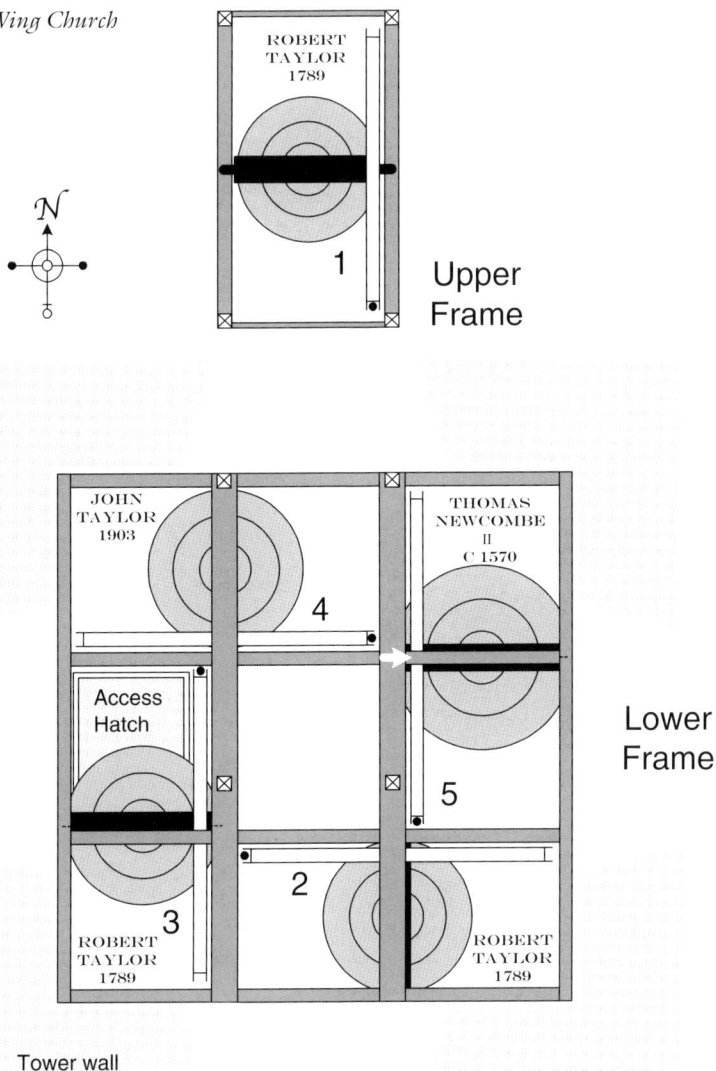

Upper Frame

Lower Frame

Tower wall

On the waist:

RECAST 1903

[65] [65]

F. J. BERRY [65] CHURCHWARDENS

E. WORRALL [65]

On the waist opposite:

[107]

The capital letters in the inscription are like [131], the lower case like [113]. The lettering on the waist is like [109]. There is a band of decoration [127] below the inscription.

Former bell

Cast in the sixteenth century. Ascribed to Newcombe of Leicester. Diameter 787mm (31in).

[30] Gloria In excelsis Deo

Note that North (1880, 165) did not record the c in 'excelsis'.

5 *Circa* **1570**. Tenor. Note B flat. Diameter 889mm (35in). Weight 353kg (6cwt 3qr 22lb). Canons removed. Cast-iron headstock. Ascribed to Thomas Newcombe II of Leicester.

[2] [27] [2] S THADDEE

(St Thaddæus)

Letters are like [108]. Letters and marks are spaced evenly around the inscription band. Scheduled for preservation (Council for the Care of Churches).

SANCTUS BELL

Evidently an early Sanctus bell and two handbells were melted down when one of the larger bells was recast about 1590. Although the churchwardens were later ordered by the Archdeacon to provide a replacement, it is not known if this order was complied with (see Bell History).

HANDBELLS

It has already been noted that Wing had at least two handbells prior to 1590. *Wing Parish Magazine* of February 1893 records that the Church held a set of handbells which at this date were used to provide entertainment at special events held at the National School and the Sunday School (information from Mr & Mrs Birch). The village still has a set of handbells today.

BELLRINGING CUSTOMS

On Sundays the first and second bells were rung at 9am. All the bells were chimed for Divine Service and after the Morning Service a bell was rung to announce that Evening Prayer was to be said. At the Death Knell thrice three tolls were given for a male and thrice two tolls for a female, both before and after the knell. The tenor was usually tolled at funerals (North 1880, 165).

The following extract is taken from *Wing Parish Magazine* dated February 1892 (information from Mr & Mrs Birch):

On Sunday, Jan. 17th, a Funeral March was played at the close of each Service in recognition of the death of the Duke of Clarence and Avondale .... After Service a muffled peal was rung to mark the sad event.

On Wednesday, Jan. 20th, the Church bell tolled at 3

Mrs Bagley, on the left, who at 77 years was still gleaning in the fields near Wing in 1935 (Records of Wing)

o'clock, the time fixed for the Prince's Funeral, and in the evening a muffled peal was rung.

According to *Records of Wing* gleaning in the fields by 1935 'was practically a thing of the past' although at that date a Mrs Bagley, aged 77 years, still continued with this tradition. When she was a child in the 1860s it was quite a common sight to see women working in the fields and she could earn 4d a day pulling weeds or picking stones. At the age of fourteen years she went into domestic service on a local farm. Her working day began when the Sexton rang a church bell at 6am. Another bell at 6pm signalled the end of the working day for most,

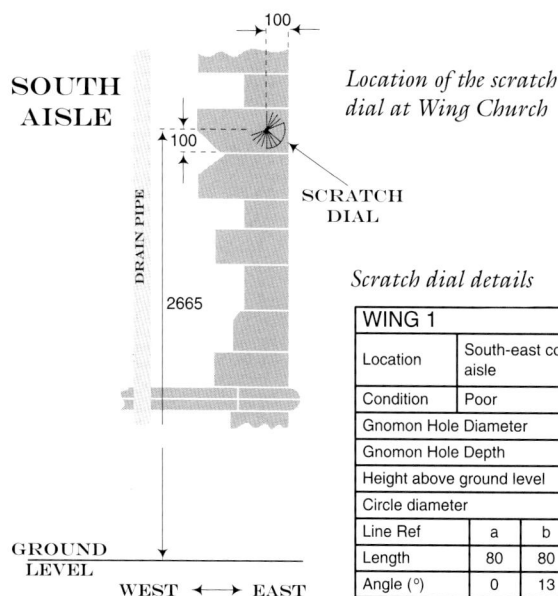

SOUTH AISLE

Location of the scratch dial at Wing Church

SCRATCH DIAL

DRAIN PIPE

100

100

2665

GROUND LEVEL

WEST ←——→ EAST

Scratch dial details

WING 1													
Location	South-east corner of south aisle												
Condition	Poor												
Gnomon Hole Diameter			23										
Gnomon Hole Depth			Filled in										
Height above ground level			2665										
Circle diameter			160										
Line Ref	a	b	c	d	e	f	g	h	i	j	k	l	m
Length	80	80	80	80	80	80	80	80	80	80	80	80	80
Angle (°)	0	13	28	44	55	120	138	160	180	193	210	228	238

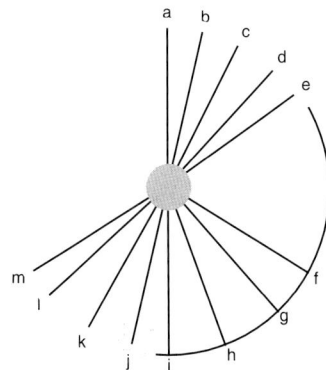

but she then had to milk the cows.

At the time of writing, Wing has a regular band of bellringers. One of the bells is chimed for approximately fifteen minutes before each service and all the bells are rung for special services including the Advent Songs of Praise and the Christmas Service. At the Millennium, the bells were rung at midnight and at noon on New Year's Day. They are rung for weddings if requested.

SCRATCH DIAL

There is a scratch dial on a quoin at the south-east angle of the south aisle. The original south aisle built *circa* 1140 was rebuilt and extended over the next two centuries. The church was extensively restored between 1875 and 1885 and this included the rebuilding of the chancel and the south aisle as well as the removal of the south porch. Although it is possible that the scratch dial quoin is somewhere near its original position, it may have been relocated from elsewhere during the restoration.

CLOCK HISTORY

Irons' Notes (MS 80/1/3) confirm that there was a clock at Wing before 1602:

> **1605** There was a clock in the church on the death of Mr Cooke the late parson & now taken away by Mrs Cooke the relict of the said Mr Cooke.

Robert Cooke was the rector of Wing from 3 February 1577. On his death he was succeeded by Walterus Baker on 28 April 1602. The widowed Mrs Cooke probably remarried to become the Isabella Simon referred to in the following notes:

> **1618 Oct 30** Isabella Simon of Glason for takinge of a clocke belonging to the churche — deferred to the next court at Uppingham.
> **1619 May 29** John Simon & Isabel his wife: to provide a clock for to be in ye Churche of Winge.
> **1619 July 9** J S & I S to restore the clocke againste the church at Winge which his wife caused to be taken from thence by Mich. next.

There are many similar notes through to 1621 demanding that the clock be returned but unfortunately the outcome is not revealed. No further details regarding this clock have survived.

The current clock at St Peter and St Paul's Church is a two-train flatbed movement by John Smith & Sons of Derby. It was installed in 1920 at a cost of £130 and dedicated to the memory of men from Wing who lost their lives in the First World War. The Rev Canon Ashby MC, Rector of Market Deeping, Dean of Stamford, and a former Chaplain of the Forces conducted the dedication service on Friday 13 February 1920. A memorial in the church records the names of those who served and died in the Great War and a replica was unveiled on 5 March 1920 in the Wesleyan Chapel in Middle Street by Mr A W Hickling of Wing Old Hall (Phillips 1920, 244).

The World War I memorial tablet adjacent to the churchyard gate at Wing

On the closure of the chapel in 1993 this memorial was transferred to the Village Hall. A tablet on the churchyard wall, adjacent to the entrance gate, records the dedication.

Automatic winding units were installed in 1997 by John Smith & Sons of Derby.

CLOCK DETAILS

Details of the present clock:

Maker:	John Smith & Sons, Midland Clock Works of Derby
Signed:	On the setting dial and on the frame. Dated 1919
Installed:	1920
Cost:	£130
Frame:	Cast-iron flatbed
Trains:	Going and striking
Escapement:	Pinwheel
Pendulum:	Wooden rod with a cylindrical cast-iron bob
Rate:	53 beats per minute
Striking:	Countwheel striking on the hour
Weights:	Cast-iron weights, now removed
Winding:	Wound weekly until automatic winding units were installed on both trains in 1997
Dial:	Skeleton dial with a black background and

Location:	gilded roman numerals and hands
	North face of the tower
Note:	The clock is housed in a wooden case

RAILWAY MISSION CHAPEL

Wing Mission Chapel was one of three chapels erected along the Rutland section of the Manton to Kettering railway line when it was being constructed in the mid 1870s. The others were at Glaston and Seaton (see under Glaston and Seaton). The Rev D W Barrett MA was the curate in charge of the Bishop of Peterborough's Railway Mission, and his *Life and Work among the Navvies* (1880, 112) includes the following:

> ... in the autumn of 1877 we had the satisfaction of seeing an old hut from Corby Wood being drawn up the hill towards Wing to be erected as a temporary church. Before the hill top was reached, the waggon which bore the load broke down, and it seemed as if we were to be doomed to disappointment .... But orders had been given at headquarters that a chapel was to be erected, and soon it was standing with its little turret [which had a small bell] and open porch under shelter of the hill at Wing cross-roads .... We held our first service there the last Sunday in September, making it a kind of Harvest Festival, and from that day till the end of 1878 regular services were held there.

SUNDIAL HOUSE

An engraved sundial is set into one of the dormers of the early seventeenth-century Sundial House, originally three almshouses, in Church Street. When the dial was restored in 1995, in memory of the late Mr Harris who lived here, an 'H' was included in the gnomon design and the motto **UT AVES SIC HORAE** (As birds fly so does time) was added. Note the intended connection between birds and Wing (Walter Wells).

The restored sundial at Sundial House in Church Street, Wing

Wing Railway Mission Chapel in 1877. It was located at Wing crossroads (Barrett 1880, 30)

CITY YARD HOUSE

There is a direct south vertical sundial on the south elevation of City Yard House. The arabic numerals and lines are engraved on the limestone dial which is complete with its gnomon. There appears to be a date and initials above the gnomon but they are illegible. City Yard House has datestones of 1622 and 1694. The sundial can be seen from the public footpath which passes through City Yard.

The seventeenth-century sundial on the south elevation of City Yard House, Wing

Select Glossary

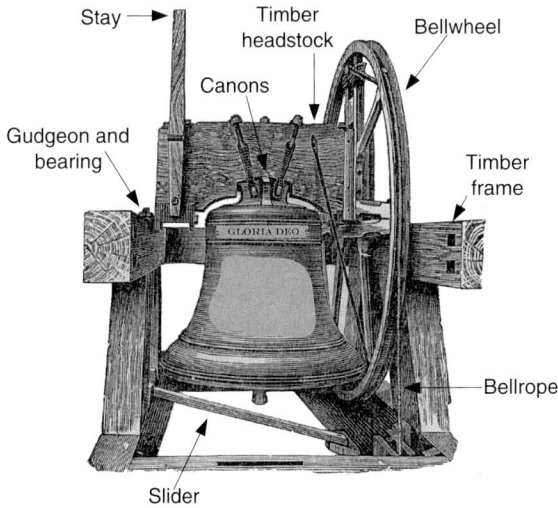

The parts of an early church bell

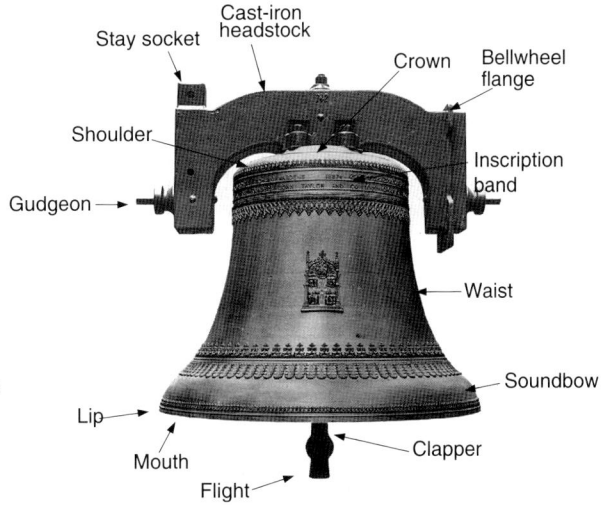

The parts of a modern church bell (John Taylor Bell Foundry Museum)

The parts of a circa 1890 three-train flatbed turret clock (Chris McKay)

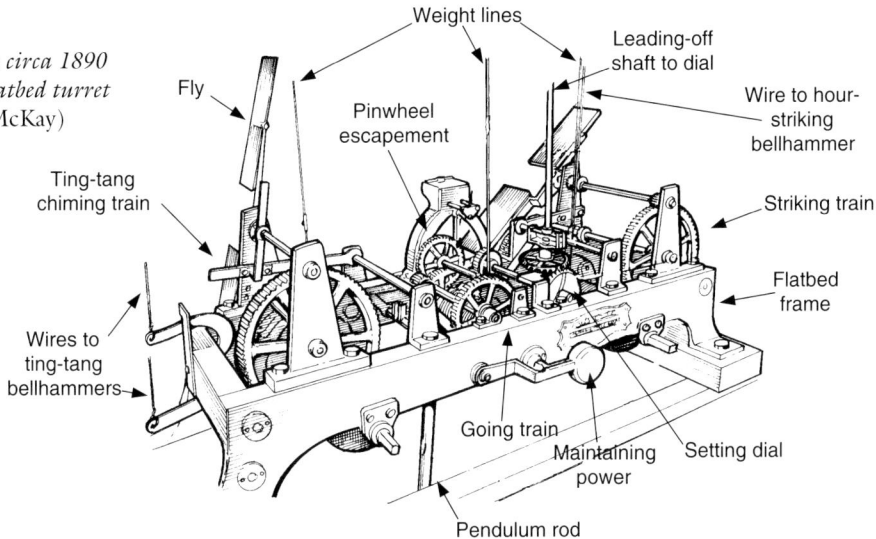

Words in bold are defined elsewhere in the Glossary.

'A' frame A **turret clock** frame shaped like the letter 'A'. Used mainly for small nineteenth-century single-train **movements** (see **endless chain automatic winder**).

anchor escapement The **pallets** resemble an anchor in shape. Also known as the recoil escapement. Invented *circa* 1670 by Joseph Knibb, Robert Hooke or William Clement for a long **pendulum** with a narrow arc. This **escapement** is easily identified because the **escape wheel** recoils (turns backwards) a fraction of a turn immediately after an **escape wheel** tooth has dropped onto a locking face of a **pallet**.

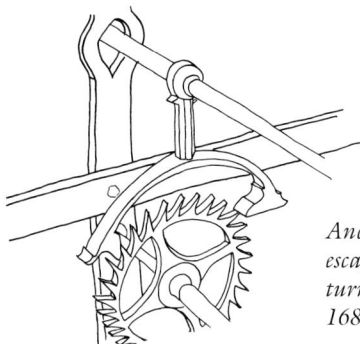

Anchor or recoil escapement from a turret clock of circa 1680 (Chris McKay)

arbor A horological term for the shafts or axles which carry the **wheels**, **pinions** and **pallets** in a clock or watch.

argent The central boss on the **crown** of an early bell around which the **canons** are grouped.

armillary sphere Originally a sphere made of metal rings which represented the great circles of the heavens. Adapted later to include a sundial.

augmentation The addition of a bell or bells to an existing **ring**.

automatic winding The use of an electric motor to wind a clock instead of manual winding. In most modern systems a low voltage motor is used to wind a small weight at frequent intervals. The energy source is a rechargeable battery and the clock will continue to run normally for several hours during a power-cut. One automatic winding unit is usually required for each **train** of a clock. The original heavy weights are not used, thus removing a safety hazard (see **endless chain automatic winder** and **epicyclic automatic winder**).

baldrick The leather-lined metal strap which was used on early bells to suspend the **clapper** from the **crown staple**.

bearing Part of a machine that supports a rotating or other moving part.

beat The sound made when an **escape wheel** tooth drops onto a locking face of a **pallet**. A slightly different sound is made by each of the two pallets and this produces the familiar 'tick' and 'tock' of a clock. A clock is said to be 'in beat' when the time intervals between succeeding beats are equal.

bell chamber see **belfry**.

belfry The space or room where bells are hung in a church tower. Also referred to as a **bell chamber**.

bellcote An open structure above a building where one or more bells are hung.

bell hammer The part of a **chime barrel** system which strikes on the **soundbow** of a bell.

bellrope The rope attached to the **bellwheel** which hangs down to the **ringing chamber**.

bell turret see **bellcote**.

bellwheel A large spoked wooden wheel which is attached to the **headstock** of a bell.

bezel A ring or frame which surrounds or holds a clock dial. In watches the ring which holds the glass.

birdcage frame A **turret clock** frame in the shape of a cage containing one or more clock **trains**. They were usually made of wrought iron. The end-to-end birdcage frame was in use in the thirteenth century and lasted until *circa* 1680 when side-by-side frames were found to be more suitable for the recently introduced long **pendulum**. These lasted until cast iron superseded wrought iron at the end of the eighteenth century and when **posted frames** were introduced. Some early side-by-side frames were made of timber and the clocks made by John Watts of Stamford are good examples.

End-to-end birdcage clock movement of circa 1680 (Chris McKay)

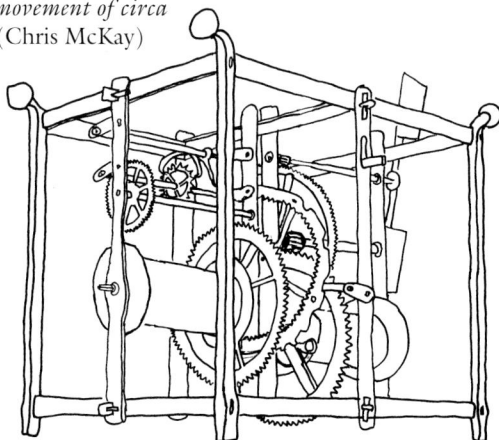

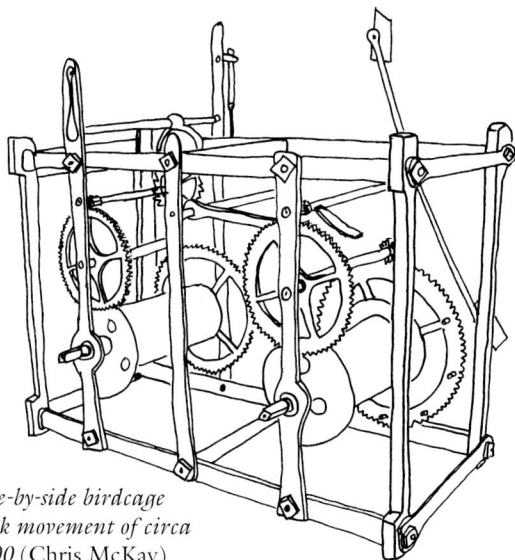

Side-by-side birdcage clock movement of circa 1700 (Chris McKay)

black-letter   Gothic lower case lettering. An old heavy style of typeface.

bob   The heavy weight at the lower end of a **pendulum** rod. Usually lenticular or cylindrical in form.

buttress   A projecting support built against a wall.

campanology   The study of bells, and the art or practice of bellringing.

canons   Loops cast in the **crown** of older bells by which the bell is suspended from the **headstock**.

change-ringing   Ringing a set of bells in a constantly varying order.

chapter ring   The ring on a clock or watch dial bearing the hour numerals and quarter hour or minute divisions.

chime   The sounding of a succession of notes on a bell or set of bells.

chime barrel   A wood or metal drum with protruding pins and a system of levers connected to **bell hammers** by wires. As the drum rotates the pins lift and release the levers which in turn cause the hammers to strike the bells. The relative circumferential positions of the pins determines the order in which the bells are struck, and hence the tune. Other tunes can be selected by moving the barrel along its horizontal axis so that the levers engage with different rows of pins. By this means the chime barrel can play a number of different tunes. Also known as a tune barrel.

chime hammer   A **bell hammer** used in conjunction with the **Ellacombe chiming apparatus**, a **clavier** or a **chime barrel**.

chiming train   The separate **train** of a clock which **chimes** every fifteen or thirty minutes.

clapper   The striking mechanism of a bell. It hangs from the **crown staple** inside a bell and consists of a head, shank, ball and **flight**.

clavier   The keyboard of a musical instrument. A simple clavier for playing church bells consists of wooden levers in a frame, each lever being connected by rope to a different **chime hammer** via a series of pulleys. Pressing a lever causes a hammer to strike the **soundbow** of a bell.

click   That part of a ratchet mechanism which allows the ratchet wheel to turn in one direction only.

clock hammer   The part of a clock striking system which strikes on the **soundbow** of a bell.

clocking   The **clapper** is 'clocked' against the bell by pulling on a rope tied round the **flight** of the **clapper** (*see* Chapter 4 — Gazetteer, Tixover). Some early clappers had an eye below the **flight** for this purpose, but this is an illegitimate method of chiming which can result in a cracked bell.

clocksmith   A blacksmith-cum-clockmaker.

cope   The outer mould of a bell when it is being cast. The inner mould is known as the core.

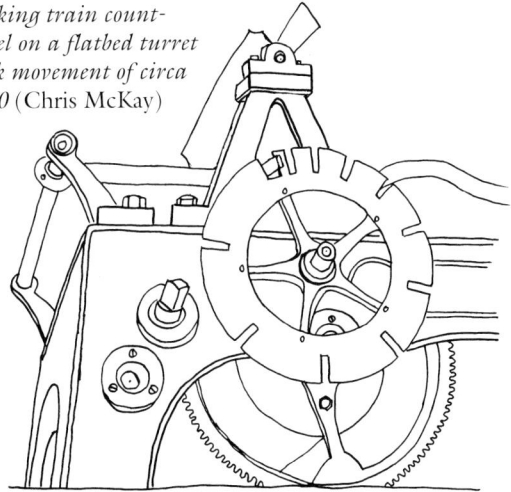

Striking train countwheel on a flatbed turret clock movement of circa 1890 (Chris McKay)

countwheel striking   A type of striking used in thirty-hour longcase clocks and most turret clocks in which a countwheel is used to determine the number of blows to be struck by the **clock hammer** of the **striking train**. A countwheel is a disk with notches at increasing intervals round its periphery. Some later clocks have a wheel with pins at increasing intervals. The notches or pins are spaced so that the striking train will first run for one blow of the clock hammer and then stop. At the next hour it will run for two blows and stop; and so on until twelve o'clock, after which the cycle is repeated. A countwheel is also used to determine the number of quarters to be chimed by the **chiming train** of many three-train turret clocks. Countwheel striking is also known as locking plate striking.

crown   The upper part of a bell above the **inscription band** and **shoulder**.

crown wheel   The **escape wheel** of a **verge and foliot escapement** so called because it resembles a crown.

crown staple   Cast into or bolted through the **crown** of a bell to provide a means of attaching the **clapper**.

crutch   That part of a clock mechanism which transmits the power from the **escapement** to the **pendulum**. It also transfers the movement of the pendulum back to the escapement to ensure that the **escape wheel** is released at regular intervals. It is the basis of the invention by Christiaan Huygens for which he was granted a patent in 1657.

cupola   A small rounded dome above a roof or tower, often housing a clock bell.

deadbeat escapement   Similar to the **anchor escapement** except that there is no recoil. Used in good quality **turret clocks** and regulator clocks. Invented by George Graham in 1715.

door frame   An early type of vertical wooden **turret**

*Turret clock deadbeat
escapement of circa 1740*
(Chris McKay)

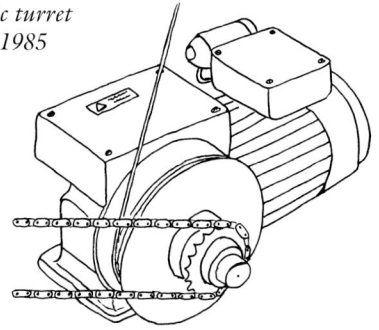

*Epicyclic automatic turret
clock winder circa 1985*
(Chris McKay)

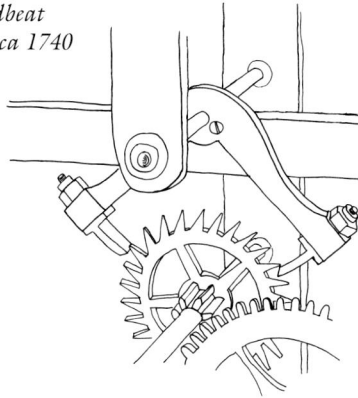

clock frame in which the **trains** are mounted one above the other.

electric master clock The main clock of an electrical impulse time-keeping system.

Ellacombe chiming apparatus A frame for a set of bellropes which by means of pulleys and hammers enables bells to be chimed without swinging them. One person can chime any number of bells using this apparatus.

endless chain automatic winder Christiaan Huygens used an endless rope system to provide **maintaining power** on his first pendulum clock of *circa* 1657. This system was used later on thirty-hour longcase clocks so that one weight drives both the **going train** and the **striking train**. On endless chain automatic winders the rope is replaced by a roller chain and an electric motor lifts the weight.

end-to-end frame see **birdcage frame**.

epicyclic automatic winder A type of **automatic winder** used on **turret clocks** which utilises an **epicyclic gear-**

box. The main advantage is that it maintains the drive to the clock **train** during winding.

epicyclic gearbox A gearbox in which a small gear wheel moves round the circumference of a larger one.

equation of time The difference between apparent solar time (the time indicated by a sundial) and mean solar time (the local time indicated by a clock). A table showing the correction for each day of the year used to be included in almanacks (*see* Chapter 1.5 — Introduction, Scientific Sundials).

escape wheel The last wheel in a clock or watch **movement**. It maintains the oscillations of the **pendulum** or balance wheel.

escapement That part of a clock or watch **movement** which allows the motive power to escape.

extended barrel frame Sometimes incorrectly referred to as a chair frame. It is a double-framed construction with a short frame for the **trains** and a long frame for the barrels. It therefore achieves maximum strength for the trains and a long going period for the barrels. It was used by manufacturers from 1730 to *circa* 1840.

faculty Authorisation given by a church authority for work to be undertaken.

fieldgate frame An early wooden **turret clock** frame in which the **trains** were mounted end to end, and which resembled a field gate.

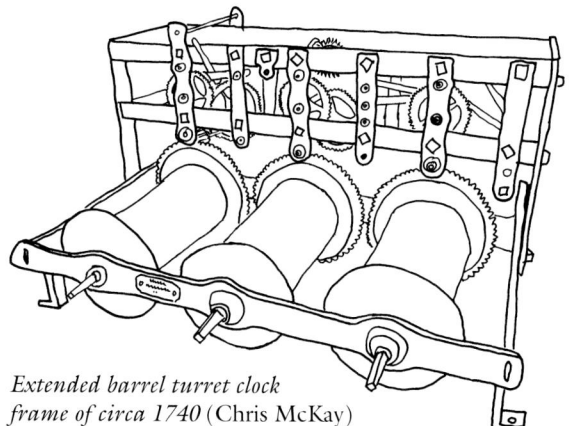

*Huygens' endless chain automatic winder on an 'A'
frame turret clock movement* (Chris McKay)

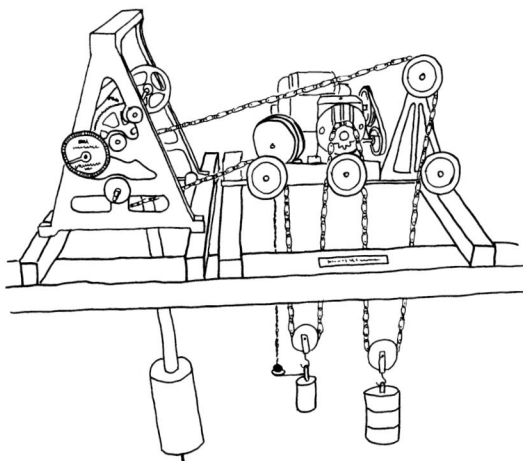

*Extended barrel turret clock
frame of circa 1740* (Chris McKay)

Flatbed turret clock movement of 1874 by W F Evans & Son of Handsworth, Birmingham (Chris McKay)

flatbed frame A flat horizontal cast-iron **turret clock** frame on the upper surface of which the **arbor** bearings, or frames holding the bearings, are bolted. Ideally it was possible to remove any arbor without disturbing another, but in reality this was seldom achieved. First introduced *circa* 1850, it brought a major change to British turret clock making.

flight The lower end of a bell **clapper**, immediately below the ball.

fly Two or more vanes mounted on the last **arbor** of clock **striking** and **chiming trains**. It governs the speed of the train by air resistance. A fly is also used in conjunction with **gravity escapements** to control the fall of the escape tooth onto the **pallet**.

foliot A pivoted bar fixed to the **verge** in a **verge and foliot escapement**. It oscillates in a horizontal plane and an adjustable weight on each arm determines the period of oscillation.

gilding Covering with gold leaf.

gleaning Gathering the remaining ears of corn after the harvest.

Glebe Terrier An inventory of church land. Sometimes includes a record of items held within the church.

gnomon That part of a sundial which casts a shadow to indicate solar time. Also known as a **style**.

going train The timekeeping **train** of a clock.

gravity escapement An escapement in which gravity provides the impulse to the **pendulum**. A weight is lifted a set distance and then released to give the constant force impulse. Gravity escapements are normally found on better quality **turret clocks**, the major benefit being that the **escapement** is isolated from the effects caused by wind, ice and snow on the clock hands. This type of escapement was developed by Edmund Beckett Denison (Lord Grimthorpe) for the clock at the Palace of Westminster. His final version, the double three-legged gravity escapement, was ultimately fitted to the Westminster clock, and this became the standard escapement for all better quality turret clocks.

gudgeon The pivot of a bell **headstock**.

headstock A wooden, cast-iron or fabricated steel structure to which a bell is attached by straps or bolts. It has a **gudgeon** at each end which is mounted in a **bearing** on the frame so the bell can be swung.

horology The art of measuring time, the making of clocks, watches and other time telling, measuring and recording devices, and the study of these.

inscription band A narrow band between the **moulding wires** just below the **shoulder** of a bell. The traditional location for the bell inscription.

itinerant founder A bellfounder who travelled the country setting up temporary foundries near to where bells were required. This was often in churchyards, and sometimes even in churches.

journeyman A qualified craftsman who works for another.

key A brass wedge used to locate and adjust the **headstock** bearings. These were sometimes referred to as 'brasses'.

lantern clock An early form of domestic clock, developed from, and resembling a **turret clock**. They were sometimes referred to as 'sheep's head' clocks from the appearance of the dial.

leading-off work The shaft or shafts which connect a **turret clock** to its dial or dials.

lip The lower part of a bell, below the **soundbow**.

local time Mean solar time at a particular location. It differs from standard time, known as Greenwich Mean Time (GMT), according to the number of degrees east or west the location is from the Greenwich Meridian. For example, the local time of a point 2 degrees west of this meridian would be eight minutes behind standard time.

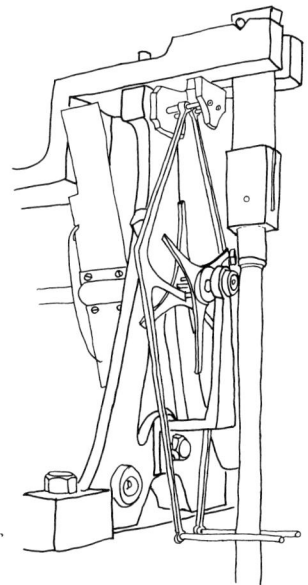

Double three-legged gravity escape-ment of 1878 (Chris McKay)

locking plate striking   see **countwheel striking**.

maintaining power   A mechanism which maintains the drive to the **going train** of a clock when it is being wound. Without such a device the clock will lose time. It also prevents damage to the escape wheel.

marquetry   From the French 'marqueter' meaning to inlay. A form of inlay work on veneered clock cases and other furniture using contrasting coloured woods in a profuse pattern.

motionwork   The gears, usually just behind a dial, which provide concentric outputs of one revolution in one hour and one revolution in twelve hours.

motto   The legend on a sundial.

moulding wires   Lines cast around the shape of a bell, particularly above and below the inscription, at the lower **waist** and near the **lip** of the bell. Originates from the need to temporarily hold the 'model' bell to the outer mould (**cope**) during the moulding of medieval bells.

mouth   The open area of a bell, below the **lip**.

movement   That part of a clock which includes the frame and the **trains** of gears. It does not include the dial, case, remote **motionwork**, or bells.

muffled   Bells rung muffled have a leather cover over the ball of the **clapper** to produce a more mellow tone.

nocturnal   An instrument for determining the time at night by the position of certain fixed stars. Also known as a night dial.

noctuary   A watchman's clock.

offset pendulum   A **pendulum** is sometimes offset from the centre line of the **going train** in order to avoid the **leading-off** shaft. One way of achieving this is by inserting a horizontal link between the **crutch** and an offset **pendulum**.

pallet   That part of the **escapement** on which the **escape wheel** teeth lock and provide an impulse. Most escapements have two **pallets**.

pallet arbor   The **arbor** on which the **pallets** of a **pendulum** clock are mounted. It is sometimes referred to as a **verge**.

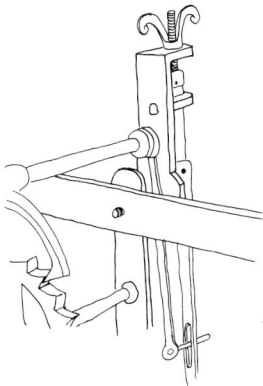

A top pendulum adjuster on a turret clock of circa 1760 (Chris McKay)

Pinwheel escapement on a turret clock of 1918 (Chris McKay)

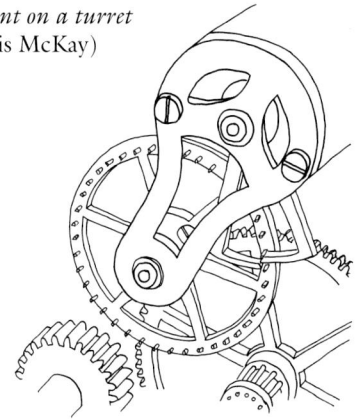

peal   The ringing of the maximum number of changes for a particular ring of bells. For a ring of seven bells it is 5040 changes.

pendulum   A rod with a weighted end (**bob**) which swings in a small arc and controls the rate of the clock through the **escape wheel**.

pendulum adjuster   A nut at the top or bottom of a pendulum by which the pendulum length can be adjusted by small increments. This is the normal way of **regulating** a pendulum clock.

pinbarrel   A wood or metal drum with protruding pins. As the drum rotates the pins lift and release levers which in turn cause hammers to strike bells. See also **chime barrel**.

pinion   A small gear wheel in a clock or watch train with up to twelve teeth or leaves.

pinwheel escapement   An **escapement** in which the teeth of the **escape wheel** are in the form of pins projecting at right angles to the face of the wheel. Used in good quality French clocks and some **turret clocks**.

plate and spacer frame   Consists of cast-iron front and rear frames, held apart by four pillars. This type of frame was popular from 1800 to 1850, but continued to be used on small single-train **turret clocks**.

pocket chronometer   A pocket watch with a chronometer **escapement**, but often applied to any high grade precision pocket watch.

posted frame   A development of the **birdcage frame** for **turret clocks**. The posted frame was popular from *circa* 1790 when cast iron became more available. All the frame and **train** bars were made of cast iron, with the corner posts of round or square section. On smaller clocks the end frames were often cast as one unit. The cast-iron posted frame lasted until the 1850s when **flatbed frames** were introduced.

quarter chiming   Part of the description of a clock which **chimes** every fifteen minutes.

rack striking   A type of striking used in most eight-day domestic clocks and some **turret clocks**. A snail-shaped cam, which turns once in twelve hours, governs the

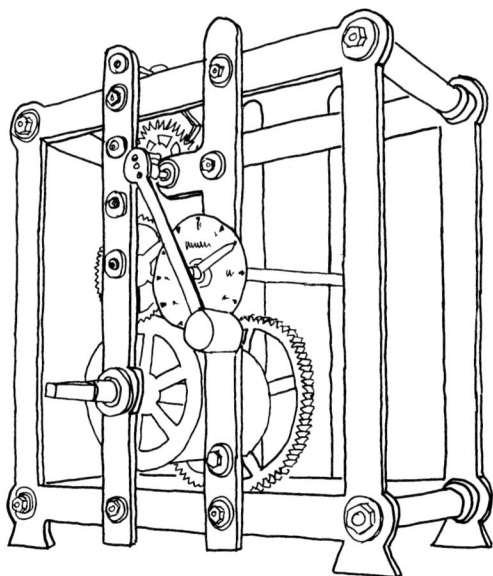

A single train plate and spacer turret clock movement of circa 1850 (Chris McKay)

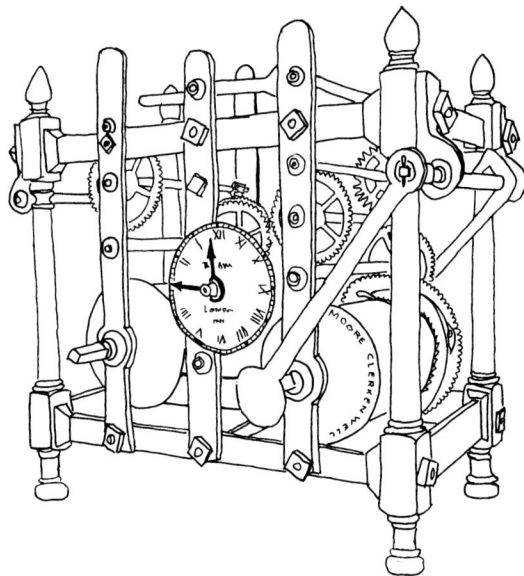

Posted frame turret clock movement of circa 1830 (Chris McKay)

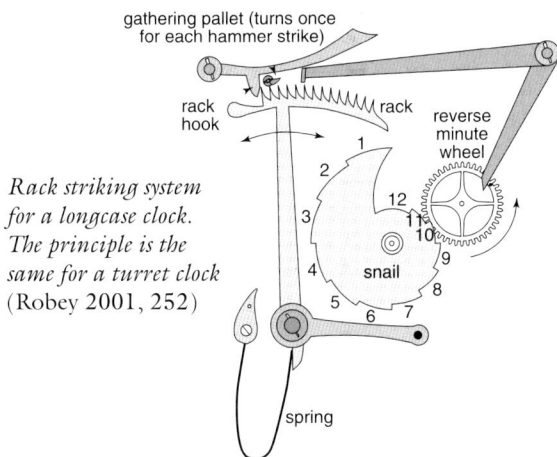

Rack striking system for a longcase clock. The principle is the same for a turret clock (Robey 2001, 252)

distance by which a rack is allowed to fall back just before striking commences. During striking, a gathering pallet gathers in the rack by one tooth for each blow of the clock hammer. Just before the next hour the rack falls back by one more tooth and this cycle is repeated until twelve o'clock has been struck.

recast bell   A bell that has been cast using the metal of a former bell. The inscription and other markings are usually, but not always, repeated on the recast bell.

regulate   To adjust a clock or watch in order to achieve more accurate timekeeping.

remontoire   A device in which the main source of power is used to wind up a small weight or auxiliary spring at regular intervals to drive the mechanism. Where a remontoire is used to drive the hands of a **turret clock** it isolates the varying forces due to wind, ice and snow from the **escapement**. There is therefore a constant torque at the **escape wheel** and consequently the **pendulum** receives impulses of constant magnitude. This results in improved timekeeping.

ring (of bells)   The collective term used to indicate the full complement of bells in a tower, for example a ring of six bells.

ringing chamber   The room or area where the bellringers stand to ring church bells.

round   The ringing of bells in order from the **treble** to the **tenor**.

scratch dial   A primitive form of sundial, usually scratched onto an external south-facing church wall. Sometimes referred to as a mass dial or a mass clock.

side-by-side frame   see **birdcage frame**.

shooting a bellrope   Repairing a **bellrope** by splicing. This was a means of extending the life of a rope which was broken or frayed.

shoulder   The upper part of a bell above the **inscription band** and below the **crown**.

slider   The slider engages with the **stay** on a bell which is intended to be rung full circle. It holds the bell in the inverted position (**mouth** up) on both forward and backward swings. The bell is then said to be set.

soundbow   The lower part of a bell, just above the **lip**, which when struck internally by a **clapper** or externally by a **bell hammer** causes the bell to resonate and hence produce the characteristic sound.

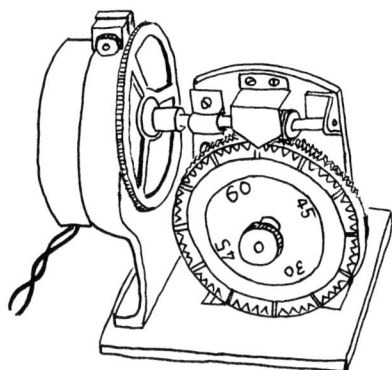

Synchronous electric turret clock movement of circa 1940 (Chris McKay)

smalt A pigment made by grinding blue glass to a powder.

spandrels The cast brass triangular decorations placed at the four corners of late seventeenth-century and eighteenth-century brass dial clocks.

stay The stay is attached to and projects upwards from the **headstock** of a bell which is intended to be rung full circle. When the bell is swung round until it is inverted (**mouth** up) the stay engages with the **slider** so that the bell will remain in this position. The bell is then said to be set.

stook Sheaves of corn collected together and stood on end in the harvest field.

striking train The separate **train** of a clock which strikes the hours on a bell or gong.

style see **gnomon**.

suspension block The structure on a clock frame from which a **pendulum** is suspended by its suspension spring.

synchronous electric motor An electric motor in which the speed of revolution is determined by the frequency of the alternating current supplied to it. In the United Kingdom the frequency of the public electricity supply is 50 hertz (cycles per second); this is maintained to a very high degree of accuracy, thus providing an excellent time source.

tenor The largest and lowest pitched bell in a **ring** of bells.

thirty-hour clock A clock which is designed to run for a maximum of thirty hours before rewinding is required. In practice such clocks should be wound every twenty-four hours.

ting-tang quarters The **chiming train** chimes on two bells at each quarter, once at the first quarter, twice at the second quarter and three times at the third quarter.

tolling Sounding a bell with a slow uniform succession of strokes.

train A group of **wheels** and **pinions** which are geared together in a clock or watch.

treble The smallest and highest pitched bell in a **ring** of bells.

tune barrel see **chime barrel**.

turret clock A clock in a tower which makes the time known publicly either by striking on a bell or showing the time on an external dial, or both. Some turret clocks have multiple bells and dials.

verge The **pallet arbor** of a **verge escapement**.

verge and foliot escapement The earliest type of **escapement** known in **horology**. It consists of an **escape wheel**, known from its shape as a **crown wheel**, and a **verge** with two **pallets** which engage alternately with the teeth of the **crown wheel**. Control is by a **foliot** which has an adjustable weight on each arm which determine the period of oscillation (*see* Chapter 1.6 — Introduction, Clocks and Watches).

Vestry Formerly the governing body of a parish. The name was derived from the room in the church in which its meetings were held. In 1894 the civil functions of the parish were transferred to parish councils and parish meetings.

vice The winding stairs in a church tower.

Visitation An official visit of inspection by a bishop to a church in his diocese.

waist The part of a bell between the **inscription band** and the **soundbow**.

wheel A gear wheel in a clock or watch **train** with more than twelve teeth.

yoke An old term for **headstock**.

Stamp used by Robert Mot,
a London Founder (North 1880, 112)

Bibliography

Baillie, G H, *Watchmakers & Clockmakers of the World* **1** (2nd ed, N A G Press, London, 1947, reprinted 1976).

Barrett, the Rev D W, *Life and Work among the Navvies* (2nd ed, Wells Gardner, Darton, London, 1880).

Barker, S, *Directory of the Counties of Leicester, Rutland &c* (Leicester, 1875).

Beckett, Edmund, Lord Grimthorpe, *A Rudimentary Treatise on Clocks, Watches, & Bells for Public Purposes* (Crosby Lockwood, London, 1903, reprinted 1974).

Beeson, Dr C F C, *English Church Clocks 1250-1850* (Brant Wright, Ashford, 1977).

Bell, Alice M, Vanishing Uppingham, *Rutland Magazine and County Historical Record* **V**, No 34 (Oakham, 1911-12), 46-9.

Benson, James W, *Time and Time-Tellers* (Robert Hardwicke, London, 1875).

Bird, Clifford, & Bird, Yvonne (eds), *Norfolk & Norwich Clocks & Clockmakers* (Phillimore, Chichester, 1996).

Blagg, T F C, The development of bell-casting, and some Nottinghamshire bell-founders, *British Archaeological Association Conference Transactions* **XXI** (Leeds, 1995), 126-35.

Blore, Thomas, *The History and Antiquities of the County of Rutland* **I**, Pt 2 (Blore, Stamford, 1811).

Bourne [*recte* Bourn], Jill, & Goode, Amanda (eds), *Rutland Hearth Tax 1665* (Rutland Record Society, Oakham, 1991).

Brandwood, Geoffrey K, Some Early Drawings of Rutland Churches, *Rutland Record* **9** (Rutland Record Society, Oakham, 1989), 316-19.

Brewer, J N, *Beauties of England & Wales* (Longman, London, 1813).

Britten, F J, *Old Clocks and Watches & their Makers* (6th ed, Spon, London, 1932. Republished, S R Publishers, Wakefield, 1971).

Buchan, James, *Thatched Village* (Hodder & Stoughton, Sevenoaks, 1983).

Burke's Genealogical & Heraldic History of the Peerage, Baronetage and Knightage (104th ed, London, 1967).

Butterworth, Andrew, St James Church Clocks - The Most Familiar Face in Gretton, *Taking Stock* **3** (Gretton Local History Society, May 1994).

Camp, John, *Discovering Bells & Bellringing* (4th ed, Shire Publications, Princes Risborough, 1997).

Clough, T H McK, Dornier, A, & Rutland, R A, *Anglo-Saxon and Viking Leicestershire including Rutland* (Leicester, 1975).

Clough, T H McK, An Early Seventeenth Century Dial from Ridlington, Rutland, *Transactions of Leicestershire Archaeological & Historical Society* **LII** (Leicester, 1978), 69-72.

Clough, T H McK, The Documentation of the Oakham Workhouse Clock, *Rutland Record* **2** (Rutland Record Society, Oakham, 1981), 82-3.

Clough, T H McK, *Rutland in Old Photographs* (Alan Sutton, Stroud, 1993).

Clutton, Cecil (ed), *Britten's Old Clocks & Watches & Their Makers* (9th ed, Methuen, London, 1982).

Cocks, A H, *The Church Bells of Buckinghamshire* (London, 1897).

Colchester, W E, *Hampshire Church Bells* (1920, reprinted 1979).

Cornwall, Julian (ed), *Tudor Rutland - The County Community under Henry VIII* (Rutland Record Society, Oakham, 1980).

Corrie, G E, *Sermons by Hugh Latimer* **I** (1844).

Council for the Care of Churches, *Schedule of Bells for Preservation: Diocese of Peterborough* (1977).

Coyne, Patrick, For the Love of Rutland: The Life and Times of George Phillips and his Family, *Rutland Record* **20** (Rutland Local History & Record Society, Oakham, 2000), 437-44.

Cox, B, *The Place-Names of Rutland* (English Place-Name Society, Nottingham, 1994).

Craven, Maxwell, *John Whitehurst of Derby Clockmaker & Scientist 1713-88* (Mayfield Books, Ashbourne, 1996).

Cutmore, Maxwell, *The Pocket Watch Handbook* (Bracken Books, London, 1985).

Daniel, Christopher St J H, *Sundials* (Shire Publications, Princes Risborough, 1986).

Daniell, John, *Leicestershire Clockmakers* (Leicester, 1975).

Dawin, John, *The Triumphs of Big Ben* (Hale, London, 1986).

Dawson, G A, *The Church Bells of Nottinghamshire* (privately published, Willoughby-on-the-Wolds), Pt 1 (1994), Pt 2 (1995), Pt 3 (1995).

de Carle, Donald, *Watch & Clock Encyclopedia* (NAG Press, London, 1975).

Dimock Fletcher, the Rev W G, The Late Thomas North FSA, *Transactions of the Leicestershire Architectural and Archaeological Society*, **VI** (LAAS, Leicester, 1889) 91-93 & 160

Drake, E S, *Directory of Leicestershire* (E S Drake & Co, Sheffield, 1861)

Dryden, Sir Henry, Squints and Dials, *Transactions of the Worcestershire Society* **XXXIII**, Pt 2 (Worcester 1897).

Dickinson, Gillian (ed), *Rutland Churches before Restoration* (Barrowden Books, London, 1983).

Eisel, John C, Herr Ohlsson of Lubeck, *Ringing World* No 4715 (The Ringing World Ltd, Woodbridge, 2001), 909.

Ellacombe, Henry Thomas, *The Church Bells of Devon* and *Bells of the Church* (single publication, Ellacombe, Exeter, 1872).

Elphick, George P, *Sussex Bells and Belfries* (Phillimore, London, 1970).

Ennis, P, *Rutland Rides* **II** (Spiegl Press, Stamford, 1979).

Fernyhough, Bernard (ed), *Recollections of a Rutland Schoolmaster* (Spiegl, Stamford, 2002)

Field, the Rev Lawrence P, *Isaac Johnson and the Bells of North Luffenham* [appeal leaflet] (North Luffenham, *circa* 1950).

Finch, Edith, *A short description of the Church of St Mary's Greetham, Rutlandshire, and an account of the recent repairs* (privately published, Burley on the Hill, 1897).

Finch, Pearl, *History of Burley on the Hill, Rutland* **I** (John Bale, Sons & Danielsson, London, 1901).

Fraser, the Rev A E, *The Story of Oakham Church, School & Castle* (Oakham, 1932).

Gatty, Mrs Alfred, *The Book of Sun-Dials* (George Bell, London,

1890).

Gentleman's Magazine Library **IX**, No 9 (Elliot Stock, London, 1897), 264-7 (re Caldecote).

Greenlaw, Joanna, *Longcase Clocks* (Shire Publications, Princes Risborough, 1999).

Haddelsey, the Rev Stephen, *Oakham Parish Church* (Oakham, 1972).

Harding, Carol, *The Harringworth Millennium Book* (Spiegl Press, Stamford, 2001).

Harrod, J G, *Postal and Commercial Directory of Leicestershire and Rutland* (1870).

Hartopp, Henry, *Register of the Freemen of Leicester, 1196-1770* **I** (Edgar Backus, Leicester, 1927).

Hartopp, Henry, *Roll of the Mayors of the Borough and Lord Mayors of the City of Leicester, 1209 to 1935* (Edgar Backus, Leicester, 1936).

Hatherley, A A, Restoration of a Commonwealth Clock, *Horological Journal* **109** (British Horological Institute, London, 1967).

Healey, John M C, *The Last Days of Steam in Leicestershire & Rutland* (Alan Sutton, Gloucester, 1989).

Hewitt, P A, *Leicestershire & Rutland Clockmakers* (Antiquarian Horological Society, Ticehurst, 1992).

Hewitt, P A, *Turret Clocks in Leicestershire & Rutland* (Leicestershire Museums Arts & Records Service, Leicester, 1994).

Higgs, William, My Connection with the Grand Old Church of All Saints', Oakham (unpublished, nd).

Hill, R M T (ed), *The Rolls and Registers of Bishop Sutton 1280-1299*, 4 (Lincoln Record Society **52**, Lincoln, 1958).

Horne, Dom Ethelbert, *Primitive Sun Dials or Scratch Dials* (Barnicott & Pearce, Taunton, 1917).

Horne, Dom Ethelbert, *Scratch Dials. Their Description and History* (Simpkin Marshall, London, 1929).

Howlett, Sue, *St Nicholas' Church, Stretton, Rutland: Visitor's Guide* (Stretton, 1998).

Hughes, Roy G, & Craven, Maxwell, *Clockmakers & Watchmakers of Derbyshire* (Mayfield Books, Ashbourne, 1998).

Irons, the Ven E A, Notes on Rutland (unpublished; University of Leicester Library Special Collections ref MS 80).

Jagger, Cedric, *Royal Clocks* (Hale, London, 1983).

Jennings, Trevor S, *Master of my Art: The Taylor Bell Foundries 1784-1987* (John Taylor & Co (Bellfounders) Ltd, Loughborough, 1987).

Jennings, Trevor S, *Handbells* (Shire Publications, Princes Risborough, 1989).

Jennings, Trevor S, *The Development of British Bell Fittings* (privately published, Loughborough, 1991).

Kelly, A Lindsay, *Directory of Leicestershire and Rutland* (London, 1900).

Kelly, A Lindsay, *Directory of the Counties of Leicester and Rutland* (London 1891, 1912, 1925, 1928).

Ketteringham, John R (ed), *Lincolnshire Bells & Bellfounders* (Lincoln, 2000).

Kieve, Jeffrey L, *Electric Telegraph - A Social and Economic History* (David & Charles, Newton Abbot, 1973).

Kightly, Charles, *The Customs and Ceremonies of Britain* (Thames & Hudson, London, 1986).

King, Patrick I, Thomas Eayre of Kettering and other Members of his Family, *Northamptonshire Past & Present* **I**, No 5 (1952), 11-23 and No 6 (1953), 10.

Lane, Peter (ed), Notes on the History of Uppingham (unpub-

lished, Uppingham Local History Group, Uppingham, 1999).

Lee, Michael, *A Clock At Browne's Hospital Stamford* (M Lee, Wansford, 2001).

Lee, Michael, *Henry Penn Bell Founder* (Link Publications, Peterborough, *circa* 1988).

Lee, Michael, *Henry Penn Bell Founder* (M Lee, Wansford, 1999).

Lee, Michael, *John Watts Stamford Clockmaker 1686-1704* (M Lee, Wansford, 2000).

Leicestershire & Rutland Federation of Women's Institutes, *Leicestershire & Rutland within Living Memory* (Countryside Books, Newbury & LRFWI, Leicester, 1994).

Leicestershire & Rutland Notes & Queries **I**, April 1889 - January 1891 (Elliot Stock, London, 1891).

Llewellin, John, *Bells and Bellfounding* (Bristol, 1879).

Lock, Lucy Adela, Memories of a Villager (unpublished, Yarwell, 1957).

Loomes, Brian, *Country Clocks and their London Origins* (David & Charles, Newton Abbot, 1976a).

Loomes, Brian, *Watchmakers & Clockmakers of the World*, **2** (N A G Press, London, 1976b).

Loomes, Brian, *Grandfather Clocks and their Cases* (David & Charles, Newton Abbot, 1985).

Loomes, Brian, Another Hybrid Lantern, *Clocks* **25**, No 3 (Splat Publishing, Edinburgh, 2002) 16-19.

McKay, Chris (ed), *Guide to Turret Clock Research* (Turret Clock Group - Antiquarian Horological Society, Ticehurst, 1991).

McKay, Chris (ed), *J. Smith & Sons of Clerkenwell Facsimilies* (Pierhead Publications, Herne Bay, 2001).

McKay, Chris, *The Turret Clock Keeper's Handbook* (Turret Clock Group - Antiquarian Horological Society, Ticehurst, 1998).

McKenna, Joseph, *Clockmakers & Watchmakers of Central England* (Mayfield Books, Ashbourne, pending 2002).

McKenna, Joseph, *Watch & Clockmakers of the British Isles - Warwickshire* (Pendulum Press, Birmingham, *circa* 1980).

Mather, Harold H, *Clock and Watch Makers of Nottinghamshire* (Friends of Nottingham Museum, Nottingham, 1979).

Matthews, Bryan, *The Book of Rutland* (Barracuda Books, Buckingham, 1978).

Matthews, Bryan, *By God's Grace ... A History of Uppingham School* (Whitehall Press, Maidstone, 1984).

Matkin, C, *Oakham Almanack* (Oakham, 1881-1941).

Meadwell, Nicholas, Church Bells of Rutland, in Waites, B (ed), *A Celebration of Rutland* (Oakham, 1994), 64-5.

Mercer, Vaudrey, *The Life and Letters of Edward John Dent Chronometer Maker and Some Account of his Successors* (Antiquarian Horological Society, Ticehurst, 1977).

Mills, Alan A, Isaac Newton's Sundials, *Antiquarian Horology* **XX**, No 2 (Antiquarian Horological Society, Ticehurst, 1992a) 126-39.

Mills, Alan A, The Scratch Dial and its Function, *Bulletin of the British Sundial Society* No 3 (1992b), 5-8.

Mills, Ernest, *Empingham Remembered* (Mills, Empingham, 1984).

Nichols, John, *The History & Antiquities of the County of Leicester* (Nichols, 1800. Republished, S R Publishers, Wakefield, 1971).

North, Thomas, *The Church Bells of Leicestershire* (Samuel Clarke, Leicester, 1876).

North, Thomas, *The Church Bells of Northamptonshire* (Samuel Clarke, Leicester, 1878).

North, Thomas, *The Church Bells of Rutland* (Samuel Clarke, Leicester, 1880).

North, Thomas, *The Church Bells of the County and City of Lincoln* (Samuel Clarke, Leicester, 1882).

North, Thomas, *The Church Bells of Bedfordshire* (Elliot Stock, London, 1883).

Ovens, Robert, & Sleath, Sheila, Keeping Time in Rutland, in Waites, B (ed), *A Celebration of Rutland* (Oakham, 1994), 57-63.

Palmer, Arnold (ed), *Recording Britain* II (Oxford University Press, London, 1947).

Palmer, Roy, *The Folklore of Leicestershire and Rutland* (Sycamore Press, Wymondham, 1985).

Parkes, Gill, Rutland Records in the Leicestershire Record Office, *Rutland Record* 4 (Rutland Record Society, Oakham, 1984), 155-6.

Parkin, David, *The History of the Hospital of Saint John the Evangelist and of Saint Anne in Okeham* (Rutland Local History & Record Society, Oakham, 2000).

Parr, D R, British Railway Clocks, Pt II, *Antiquarian Horology* XXIII, No 4 (Antiquarian Horological Society, Ticehurst, 1997), 322-40.

Parsons, David, *A Bibliography of Leicestershire Churches*, Pt 3 (Leicester, 1983).

Pearson, Brian, The Burley Clock, *Antiquarian Horology* IX, No 5 (Antiquarian Horological Society, Ticehurst, 1975) 580-2.

Pearson, Denis, *An Inventory of the Church Bells of the Diocese of Peterborough* (Peterborough Diocesan Guild of Church Bell Ringers, Burton Latimer, 1989).

Peterborough Diocesan Directory 2000, No 131 (Diocese of Peterborough, 2000).

Peterborough Diocesan Records, *Church Survey* 3 & 7 (Northamptonshire Record Office, 1619 & 1681).

Pevsner, Nikolaus, *The Buildings of England: Leicestershire and Rutland* (2nd ed, Penguin Books, Hardmondsworth, 1984).

Phillips, George, Langham, *Rutland Magazine and County Historical Record* I (Oakham, 1903-04), No 5, 137-51.

Phillips, George, Egleton, *Rutland Magazine and County Historical Record* I (Oakham, 1903-04), No 6, 169-76.

Phillips, George, Ridlington, *Rutland Magazine and County Historical Record* II (Oakham, 1905-06), No 12, 96-101, No 13, 129-36.

Phillips, George, Parish Officers and their Books, *Rutland Magazine and County Historical Record* III (Oakham, 1907-08), No 17, 29-32.

Phillips, George, North Luffenham, *Rutland Magazine and County Historical Record* III (Oakham, 1907-08), No 17, 1-8, No 18, 33-40, No 19, 66-73.

Phillips, George, Courts Leet, *Rutland Magazine and County Historical Record* IV (Oakham, 1909-10), No 25, 24-31.

Phillips, George, Market Overton, *Rutland Magazine and County Historical Record* IV (Oakham, 1909-10), No 27, 65-72, No 28, 97-101, No 29, 129-36, No 30, 161-7; V (1911-12), No 38, 186-90.

Phillips, George, Oakham Church Bells, *Rutland Magazine and County Historical Record* V (Oakham, 1911-12), No 34, 41-5.

Phillips, George, Caldecott, *Rutland Magazine and County Historical Record* V (Oakham, 1911-12), No 35, 65-72.

Phillips, George, Curious Items from the County Accounts, *Rutland Magazine and County Historical Record* V (Oakham, 1911-12), No 35, 94.

Phillips, George, Manton and Martinsthorpe, *Rutland Magazine and County Historical Record* V (Oakham, 1911-12), No 40, 225-34.

Phillips, George, *Rutland and the Great War* (Padfield, Manchester, 1920).

Pickford, Christopher J, *The Steeple, Bells, and Ringers of Coventry Cathedral* (C J Pickford, Bedford, 1987).

Pickford, Christopher J, *Bellframes* (C J Pickford, Bedford, 1993).

Pickford, Christopher J, *Whitechapel ... where the tradition of British bellfounding lives on* (Whitechapel Bell Foundry Ltd, London, 1994).

Pickford, Christopher J, *Turret Clocks: Lists of clocks from catalogues and publicity materials* (Turret Clock Group - Antiquarian Horological Society, Ticehurst, 1995).

Pickford, Christopher J, Notes on Bells in the Stamford Churches (unpublished, Bedford, 1998).

Pigot, J & Co, *Directory of Leicestershire and Rutland* (1835, 1841).

Post Office Directory of Leicestershire (London, 1855).

Powell, the Rev E S, *Inventory of the Church Bells of Northamptonshire and Rutland in 1938* (Peterborough Diocesan Guild of Church Bell Ringers, Geddington, 1938).

Price, Laurence N, *Scratch Dials* (Sun and Harvest Publications, Weston-super-Mare, 1991).

Price, Percival, *Bells & Man* (Oxford University Press, Oxford, 1983).

Rawlings, A L, *The Science of Clocks & Watches* (3rd ed, British Horological Institute, Upton, 1993).

Rawnsley, W F ('An Old Boy'), *Early Days at Uppingham under Edward Thring* (Macmillan, London, 1904).

Redburn, H A, *A History of the Church of St Peter and St Paul, Market Overton, Rutland* (Market Overton, 1976).

Redlich, the Rev E B, *The History of Teigh* (Kings Stone Press, Shipston-on-Stour, 1926).

Robey, John, *The Longcase Clock Reference Book* 1 & 2 (Mayfield Books, Ashbourne, 2001).

Robey, John, Sprung Impulse, *Horological Journal* 142 (British Horological Institute, Upton, 2000), 403.

Rooksby, John, The Restoration of Oakham Church 1857-1858 (unpublished notes, 1995).

Seaby, Wilfred A, *Clockmakers of Warwick and Leamington to 1850* (Warwickshire Museum, Warwick, 1981).

Sharman, Charles A, The Sharmans of Greetham 1538-1863, *Rutland Record* 4 (Rutland Record Society, Oakham, 1984), 131-4.

Slater, *Directory of Leicestershire & Rutland* (1850, 1858).

Smith, Martin, *Stamford Then & Now* (Paul Watkins, Stamford, 1992).

Smith, Stan, *Bell Tales* (J H Hall, Nottingham, 1994).

Standage, Tom, *The Victorian Internet* (Phoenix, London, 1999).

Stevens, the Rev C A, Notes on Oakham Church Restoration, *Rutland Magazine and County Historical Record* II (Oakham, 1905-6), No 9, 25-30.

Swaby, J E, *A History of Empingham AD 500 - AD 1900* (privately published, 1988).

Tebbutt, Laurence, *Stamford Clocks & Watches* (Stamford, 1975).

Tew, David, A History of the Parish and Church of Wing, Rutland (unpublished, Oakham, *circa* 1975).

Till, E C, *A Family Affair. Stamford and the Cecils 1650-1900* (Jolly & Barber, Rugby, 1990).

Tilley, the Rev H T, & Walters, H B, *The Church Bells of Warwickshire* (Cornish Bros, Birmingham, 1910).

Thompson, A, The Use of Scratch Dials by the Rural Church and Laity as evidenced from their occurrences on Churches in the East Midlands (unpublished MA Thesis, Nottingham University, 1995).

Traylen, A R, *The Villages of Rutland*, Pts 1 & 2 (In Rutland Series, Spiegl Press, Stamford, nd).

Traylen, A R, *Oakham in Rutland*, (In Rutland Series, Spiegl Press, Stamford, 1982a).

Traylen, A R, *Turnpikes & Royal Mail of Rutland* (In Rutland Series, Spiegl Press, Stamford, 1982b).

Traylen, A R, *Uppingham in Rutland* (In Rutland Series, Spiegl Press, Stamford, 1982c).

Traylen, A R, *Churches of Rutland* (In Rutland Series, Spiegl Press, Stamford, 1988).

Traylen, A R, *Dialect, Customs & Derivations in Rutland* (In Rutland Series, Spiegl Press, Stamford, 1989).

Traylen, A R, *Old Village Schools of Rutland* (In Rutland Series, Spiegl Press, Stamford, 1999).

Universal Directory: Leicestershire & Rutland (1791).

Uppingham Local History Studies Group, *Uppingham in 1851: A Night in the Life of a Thriving Town* (Uppingham, 2001).

Uppingham School Magazine (Uppingham, 1893).

Victoria County History: Rutland **I** (1908), **II** (1935).

Walters, H B, *Church Bells of England* (Oxford University Press, Oxford, 1912, reprinted E P Publishing, Wakefield, 1977).

Ward, Dr F A B, *Time Measurement: Pt 1 - Historical Review* (Science Museum, London, 1947).

Waites, Bryan (ed), Who was Who in Rutland, *Rutland Record* 8 (Rutland Record Society, Oakham, 1987).

Waugh, Albert E, *Sundials - Their Theory and Construction* (Dover, New York, 1973).

White, A H (ed), *William Stukeley: Memoirs of Sir Isaac Newton's Life* (Taylor and Francis, London, 1936).

White, William, *History, Gazetteer, and Directory of Leicestershire and Rutland* (Sheffield, 1846, 1863 and 1877).

Wilbourn, A S H, and Ellis, R, *Lincolnshire Clock Watch and Barometer Makers* (Hansord, Ellis and Wilbourn, Lincoln, 2001).

Wilton Hall, H R, Some Recollections of the Parish of Exton in the County of Rutland from AD 1875-1877 (unpublished, St Albans, 1913).

Wright, C N, *Commercial and General Directory of Leicestershire & Rutland* (Leicester, 1880).

Woodburn, Anthony, A Comparison of the Lives and Work of Thomas Tompion and Joseph Knibb, *Proceedings of the Oxford 2000 Convention* (Antiquarian Horological Society, Ticehurst, 2000), 35-40.

Wordsworth, the Rev Christopher, Glaston Parish Charities and Memorials (unpublished, *circa* 1889 with later additions).

Wright, James, *History and Antiquities of the County of Rutland* (London 1684 and later Additions, reprinted E P Publishing, Wakefield, 1973).

King David playing a 'carillon of five bells'
(fourteenth century) (North 1880, 81)

Appendix 1
Bellfounders' Marks, Devices, Decoration & Lettering on Rutland Bells

The founders' marks, devices and examples of decoration and lettering illustrated within the main text are repeated here in numerical order. They are accompanied by the names of the founders and the Rutland bells on which they occur.

[1] to [40] are copies of woodcuts from *The Church Bells of Rutland* (North 1880). [20], [37] & [38] are not on any current Rutland bell. [118] on the third bell at Teigh was incorrectly recorded by Thomas North as [20]. He also recorded [13] on the South Luffenham treble as [37]. [38] is an example of gothic lettering occasionally used by Francis and Hugh Watts I but there are no known examples in Rutland. It is included here to preserve Thomas North's numbering system.

[72] & [78] are from *The Church Bells of Leicestershire* (North 1876), [55], [73], [106], [130] & [131] are from *The Church Bells of the County and City of Lincoln* (North 1882), [133] is from *The Church Bells of Warwickshire* (Tilley & Walters 1910) and [87], [88], [110] & [111] are from *The Church Bells of Buckinghamshire* (Cocks 1897). The remainder were drawn by the authors from castings and rubbings.

Note that a bellfounder's name in parentheses, for example (**JOHN TAYLOR**) indicates that this founder transferred the mark, device, decoration or lettering from a former bell when it was recast.

[1]
LONDON FOUNDER
circa 1480
Langham 4

[2]
THOMAS BETT
Wardley 2
ROBERT NEWCOMBE I
North Luffenham —
The Pastures clock bell
THOMAS NEWCOMBE II
Barrowden 5, Braunston 3
Wing 5
(**JOHN TAYLOR**)
Manton 2

[3]
NEWCOMBE & WATTS
Glaston 6, Ketton 4, Preston 6
FRANCIS WATTS
Barrowden 4
HUGH WATTS I
Braunston 5, Ketton 5
South Luffenham 1
(**JOHN TAYLOR**)
Barrowden 3, Ridlington 4

[4]
JOHN RUFFORD
Ayston 2

[5]
TOBY NORRIS I
Edith Weston 5
Glaston 4
North Luffenham 4

[6]
ROBERT MELLOUR
South Luffenham 4
(**JOHN TAYLOR**)
Ridlington 2

[7]
HENRY OLDFIELD II
Cottesmore 3
Ketton 2
Whissendine 4, 6
(JOHN TAYLOR)
North Luffenham 6

[8]
RICHARD HOLDFIELD
possibly with MATTHEW
NORRIS
Seaton 4

[9]
ROBERT MELLOUR
South Luffenham 4
NEWCOMBE 16th century
Morcott 3
(JOHN TAYLOR)
Ridlington 2

[10]
THOMAS NORRIS
Clipsham 1
Cottesmore 4
Langham 6
Stretton 2
Tinwell 3
TOBY NORRIS III
Caldecott 2, 3, 5, 6
Lyddington 2, 3, 5, 6
Lyndon 4
(JOHN TAYLOR)
Ridlington 3

[11]
ROBERT
NEWCOMBE I
Ayston 1
North Luffenham —
The Pastures clock bell
(JOHN TAYLOR)
Manton 2

[12]
RICHARD HILLE
Tixover 1

[13]
NEWCOMBE 16th century
Morcott 3
FRANCIS WATTS
Barrowden 4
HUGH WATTS I
South Luffenham 1
(JOHN TAYLOR)
Barrowden 3

[14]
TOBY NORRIS I
Brooke 2, Hambleton 2
Little Casterton 2
THOMAS NORRIS
North Luffenham 2
TOBY NORRIS III
Exton 5
ALEXANDER RIGBY
Burley 1
(JOHN TAYLOR)
Barrowden 2, 6

[15]
JOHN BARBER
Preston Sanctus Bell

[16]
NEWCOMBE & WATTS
Glaston 6, Ketton 4, Preston 6
FRANCIS WATTS
Morcott 2
HUGH WATTS I
Braunston 5, Ketton 5
(JOHN TAYLOR)
Ridlington 4

[17]
NEWCOMBE & WATTS
Glaston 6, Ketton 4, Preston 6
FRANCIS WATTS
Morcott 2
HUGH WATTS I
Braunston 5
Ketton 5
(JOHN TAYLOR)
Ridlington 4

[18]
HENRY OLDFIELD I
Teigh 3

[19]
HENRY OLDFIELD I
Teigh 3

[20]
JOHN WOOLLEY
No known bell in Rutland
with this founder's mark

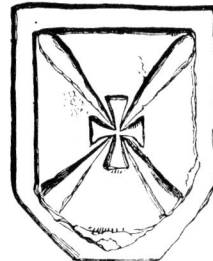

[21]
ROBERT MELLOUR
South Luffenham 4
(JOHN TAYLOR)
Ridlington 2

[22]
THOMAS NEWCOMBE II
Barrowden 5

[23]
NEWCOMBE & WATTS
Glaston 6
Ketton 4
Preston 6

[24]
ROBERT NEWCOMBE I
Ayston 1

[25]
RICHARD BENTLEY
Seaton 3

[26]
TOBY NORRIS I
Ayston 3, Edith Weston 5
Glaston 4, 5, Hambleton 3
North Luffenham 4
THOMAS NORRIS
Braunston 6, Clipsham 3
Cottesmore 2, Langham 3
Morcott 1, Ryhall 2, 3
Tinwell 1
TOBY NORRIS III
Clipsham 2, Exton 2, 3, 4, Seaton 5

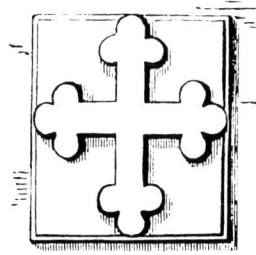

[27]
THOMAS NEWCOMBE II
Braunston 3
Wing 5

[28]
JOHANNES DE COLSALE
Whitwell 2

[29]
RICHARD HILLE
Tixover 1

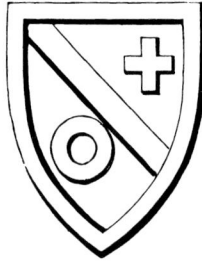

[30]
(JOHN TAYLOR)
Wing 4
Copied from a former
bell by the Newcombe
foundry, Leicester

[31]
EDWARD NEWCOMBE
Ketton 6

[32]
RICHARD HILLE
Tixover 1

[33]
LEICESTER FOUNDER
circa **1400**
Preston 5

[34]
HENRY OLDFIELD II
Ketton 2
Whissendine 4
(JOHN TAYLOR)
North Luffenham 6

[35]
LONDON FOUNDER
circa **1480**
Langham 4

[36]
LONDON FOUNDER
circa **1480**
Langham 4

[37]
WATTS
There is no known bell in
Rutland with this mark.
Note that Thomas North
recorded [37] on South
Luffenham 1 instead of [13]

[38]
WATTS
An example of
the gothic letter-
ing used by Watts,
but not on a
current
Rutland bell

[39]
TOBY NORRIS I
Brooke 2, Hambleton 2, 3

[40]
NEWCOMBE & WATTS Ketton 4
WILLIAM NOONE Ashwell 3, 4, 6, Braunston 4

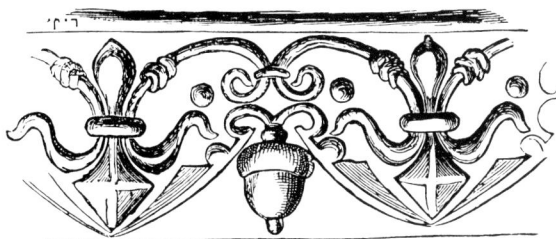

[41]
WILLIAM NOONE
Ashwell 3, 4

[42]
WILLIAM NOONE Ashwell 4, 6, Braunston 4
THOMAS HEDDERLY I Ashwell 2, Langham 5
(JOHN TAYLOR) North Luffenham 6

[43]
THOMAS HEDDERLY I
Ashwell 2

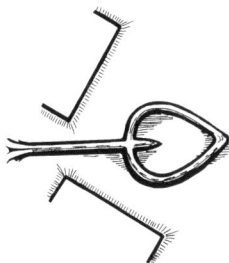

[44]
WHITECHAPEL —
C & G MEARS
Ashwell 1

[45]
JOSEPH EAYRE Former Normanton 1 removed to
St Jude's Church, Peterborough
EDWARD ARNOLD
Ashwell 5, Brooke 3, Whissendine 3

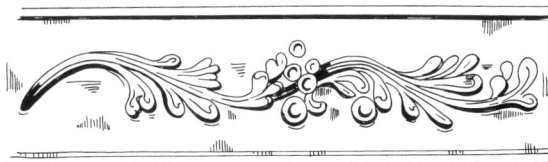

[46]
EDWARD
ARNOLD
Ashwell 5
Ryhall 1

[47]
JOHN TAYLOR
Cottesmore 1
Glaston 2
Lyddington 1
Tickencote 2

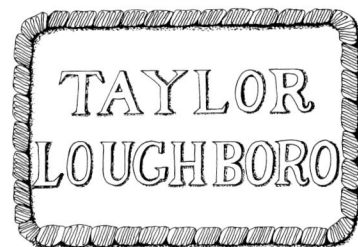

[48]
ROBERT
NEWCOMBE I
Ayston 1

[49]
THOMAS EAYRE II
Morcott 4
JOSEPH EAYRE
Whitwell 1
EDWARD ARNOLD
Barrowden Priest's Bell
(JOHN TAYLOR)
Langham 1

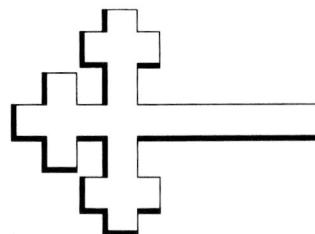

[50]
JOHN RUFFORD
Ayston 2
HENRY PENN
Edith Weston 6
(JOHN TAYLOR)
Ketton 3

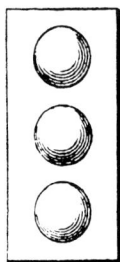

[51]
THOMAS EAYRE II
Market Overton 2, 4
North Luffenham 5
Oakham, Chapel of St
John & St Anne

[52]
THOMAS NORRIS
Braunston 6, Clipsham 1, Cottesmore 2, 4, Langham 6
Morcott 1, Stretton 2, Tinwell 1, 3
TOBY NORRIS III
Exton 2, 3, 4, 5, Seaton 5, Wardley 1
(JOHN TAYLOR)
Exton 1, Ridlington 3

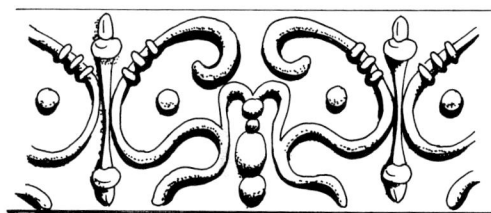

[53]
TOBY NORRIS I
Ayston 3
THOMAS NORRIS
Ryhall 2, 3, North Luffenham 2, Tickencote 1

[54]
JOHN TAYLOR
Barrowden 1, Brooke 1, Empingham 1, 2, 3, 4, 5, 6
Exton 1, Ketton 3, Langham 1, 2, North Luffenham 1
Preston 2, 3, Ridlington 1, Whissendine 1

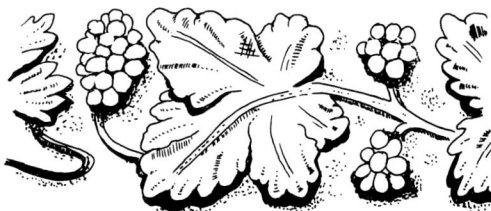

[55]
JOHN TAYLOR
Barrowden 1

[56]
JOHN TAYLOR
Barrowden 1, 2, 3, 6
Brooke 1, 6
Caldecott 1, 4
Manton 1, 2
North Luffenham 1, 6
Uppingham Angelus Bell

[57]
ALEXANDER RIGBY Burley 1
(JOHN TAYLOR) Barrowden 2, 6

[58]
GILLETT & JOHNSTON
Belton in Rutland
1, 2, 3, 4, 5, 6
Oakham 1, 2, 3, 4,
5, 6, 7, 8
Uppingham School
Bell
Uppingham School
Chapel 1, 2, 3

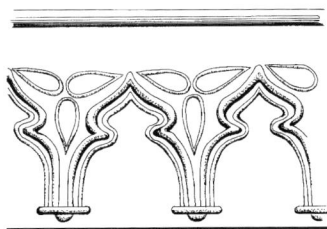

[59]
**WHITECHAPEL —
MEARS**
Braunston 1, 2

[60]
JOHN TAYLOR
Brooke 6

[61]
JOHN TAYLOR
Preston 2, 3

[62]
TOBY NORRIS I
Brooke 2
Glaston 5
Hambleton 2

[63]
**THOMAS
NORRIS**
Brooke 4

[64]
JOHN TAYLOR
Brooke 6

[65]
JOHN TAYLOR
Brooke 6
Edith Weston 1, 2, 3
Empingham 1 to 6
Exton 6, Ketton 3
Langham 1, 2
Preston 1, 2, 3
Tickencote 2
Uppingham 5
Whissendine 1, Wing 4

[66]
JOHN TAYLOR Brooke 1

[67]
ROBERT TAYLOR
Brooke 5, Uppingham 4, Wing 1, 2, 3

[68]
HENRY
OLDFIELD II
Cottesmore 3

DIEV ET MON DROIT

[69]
THOMAS NORRIS Langham 3, Ryhall 5
TOBY NORRIS III Clipsham 2, Lyndon 4

[70]
TOBY NORRIS III
Caldecott 2, 3, 5, 6, Lyddington 2, 3, 5, 6

[71]
JOHN TAYLOR
Cottesmore 1, Glaston 2, 3, Tickencote 2

[73]
HENRY OLDFIELD II
Cottesmore 3

[72]
HENRY OLDFIELD II
Cottesmore 3

[74]
GILLETT & JOHNSTON
Greetham 1

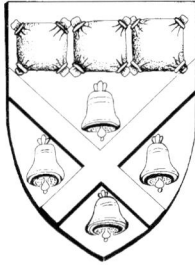

[75]
GILLETT & JOHNSTON
Greetham 2, 3, 4, 5, 6

[76]
GILLETT & JOHNSTON
Greetham 2, 3, 4, 5, 6, Oakham 7

[77]
THOMAS BETT
Wardley 2

[78]
ROBERT NEWCOMBE I
Ayston 1

[79]
TOBY NORRIS I Edith Weston 5

[80]
HENRY PENN Edith Weston 6

[81]
JOHN TAYLOR
Empingham 1, 2, 3, 4
Exton 1, 6
Ketton 3

[82]
JOHN TAYLOR
Preston 2, 3

[83]
TOBY NORRIS I
Glaston 4

[84]
HENRY OLDFIELD II
Ketton 2

[85]
JOHN TAYLOR
Exton 1,
Lyndon 2
Market Overton 5

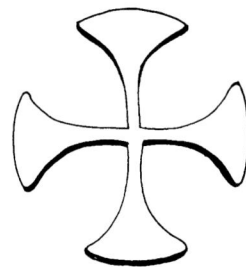

[86]
JOHN TAYLOR
Exton 6, Uppingham 5

[88]
RICHARD BENTLEY Seaton 3

[87]
RICHARD BENTLEY Seaton 3

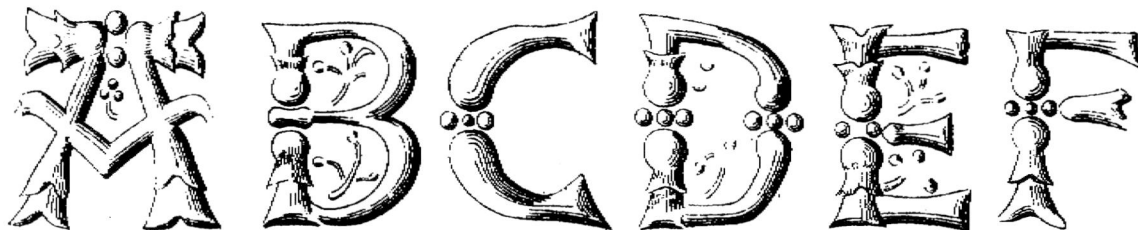

[89]
LEICESTER FOUNDER *circa* **1400**
Preston 5

[90]
TOBY NORRIS I Ayston 3
THOMAS NORRIS Ryhall 3

[91]
JOHN TAYLOR
Hambleton 4

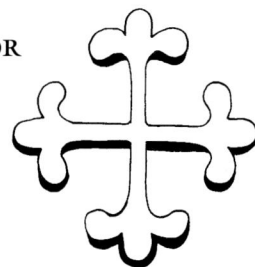

[92]
JOHN TAYLOR
Hambleton 4

[93]
WHITECHAPEL
Great Casterton 1

[94]
JOHN TAYLOR
Preston 2, 3

[95]
JOHN TAYLOR
Glaston 1

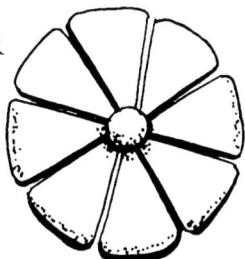

[96]
HENRY PENN
Stretton 1

[97]
WHITECHAPEL — PACK & CHAPMAN
Uppingham 1, 2, 3, 6, 7

[98]
JOHN TAYLOR Uppingham Angelus Bell

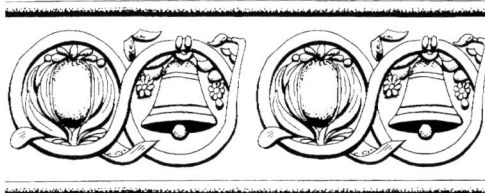

[99]
THOMAS EAYRE II
Teigh 1, 2
(JOHN TAYLOR)
Langham 1

[100]
HENRY BAGLEY II
Seaton 1

[101]
THOMAS EAYRE III
Stoke Dry 1

[102]
WHITECHAPEL
Great Casterton 1

[103]
NEWCOMBE 16th century
North
Luffenham 3

[104]
(JOHN TAYLOR)
Langham 1
Possibly copied from
a former bell cast by
Thomas Eayre II

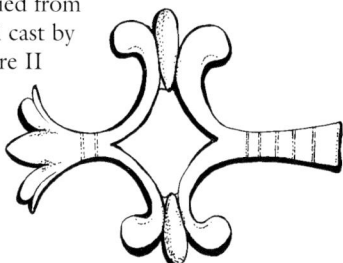

[105]
JOHN TAYLOR
Ridlington 2, 3, 4

[106]
NEWCOMBE 16th century
North Luffenham 3

[107]
JOHN TAYLOR
Langham 1, 2
Wing 4

[108]
THOMAS BETT Wardley 2
ROBERT NEWCOMBE I
Ayston 1, North Luffenham — The Pastures clock bell
THOMAS NEWCOMBE II
Barrowden 5, Braunston 3, Wing 5
NEWCOMBE 16th century Morcott 3
(JOHN TAYLOR) Manton 2

[109]
JOHN TAYLOR
Ridlington 1, Whissendine 1, Wing 4

[110]
NEWCOMBE & WATTS
Glaston 6, Ketton 4

[111]
NEWCOMBE & WATTS
Preston 6

[112]
JOHN RUFFORD Ayston 2

[113]
JOHN TAYLOR
Preston 2, 3, Wing 4

[114]
HENRY OLDFIELD II
Ketton 2, Whissendine 6

[115]
JOHN BARBER
Preston Sanctus Bell

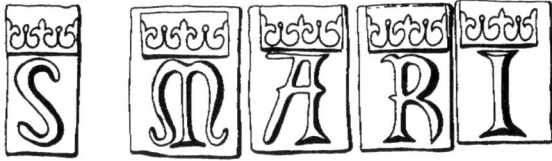

[116]
HENRY OLDFIELD II
Cottesmore 3, Whissendine 4
(JOHN TAYLOR)
North Luffenham 6

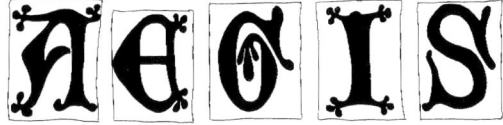

[117]
THOMAS EAYRE II
Teigh 1

[118]
HENRY OLDFIELD I
Teigh 3

[119]
NEWCOMBE & WATTS
Ketton 4

[120]
JOHN TAYLOR
Hambleton 1, 5, Market Overton 1, 5

[121]
JOHN TAYLOR
Lyndon 1, 2

[122]
JOHN TAYLOR
Ridlington 1

[123]
UNKNOWN FOUNDER *circa* 1620
Morcott clock bell

[124]
ROBERT MELLOUR
South Luffenham 4
(JOHN TAYLOR)
Ridlington 2

[125]
**RICHARD HOLDFIELD possibly with
MATTHEW NORRIS**
Seaton 4

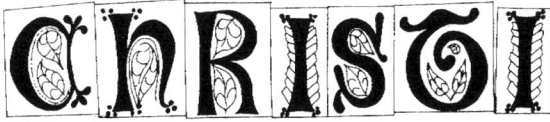

[126]
JOHN TAYLOR Preston 2, 3

[127]
JOHN TAYLOR Wing 4

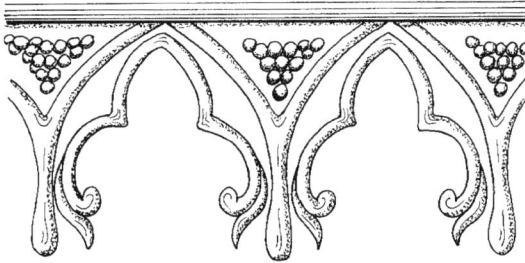

[128]
(JOHN TAYLOR) Langham 1
Possibly copied from a former bell cast by Thomas
Eayre II

[129]
THOMAS HEDDERLY I
Langham 5

[130]
JOHN TAYLOR
Barrowden 1, Preston 2, 3, Ridlington 2, 3, 4

[131]
JOHN TAYLOR
Barrowden 1, Wing 4

[132]
JOHN BARBER
Preston Sanctus Bell

[133]
THOMAS HEDDERLY I
Langham 5

[135]
JOHN WARNER
Stretton — former school bell

[134]
JOHN WARNER
Stocken Hall Coach House clock bell

[136]
HENRY OLDFIELD I
Teigh 3

[137]
JOHN TAYLOR
Empingham 1, 2, 3, 4, 5, 6, North Luffenham 1, 6

AND EVERY TONGUE

[139]
JOHN WARNER
Stocken Hall Coach
House clock bell

[138]
**RICHARD HOLDFIELD
& MATTHEW NORRIS**
Terrington St Clements,
Norfolk (not found on a
Rutland bell)

[140]
JOHN WARNER
Stocken Hall Coach House clock bell

[141]
JOHN TAYLOR Cottesmore 1

[142]
HUGH WATTS II
Oakham former 6

Appendix 2
Rutland Bells Scheduled For Presevation

The following bells, by virtue of their age (cast before 1600) or because of their historical importance, have been recommended for preservation by the Council for the Care of Churches. The list is based on detailed inspection of the individual bells by Robert Ovens and Sheila Sleath, George Dawson and others. Full details of these bells are given in Chapter 4 — Gazetteer.

PARISH CHURCH	BELL	DATE	FOUNDER
Ayston	1 of 4	c1550	Robert Newcombe I of Leicester
Ayston	2 of 4	c1365	John Rufford of Toddington
Barrowden	4 of 6	1595	Watts (possibly Francis) of Leicester
Barrowden	5 of 6	c1570	Thomas Newcombe II of Leicester
Braunston	3 of 6	c1570	Thomas Newcombe II of Leicester
Burley	1 of 1	1705	Alexander Rigby of Stamford
Cottesmore	3 of 6	1598	Henry Oldfield II of Nottingham
Edith Weston	4 of 6	1597	Matthew Norris*
Glaston	6 of 6	1598	Newcombe and Watts of Leicester
Ketton	4 of 6	1598	Newcombe and Watts of Leicester
Langham	4 of 6	c1480	Unknown London founder (Brede mark)
Morcott	2 of 4	16th C	Watts (possibly Francis) of Leicester
Morcott	3 of 4	16th C	Newcombe of Leicester
Morcott	Disused clock bell	c1620	Unknown founder
North Luffenham	3 of 6	16th C	Newcombe of Leicester
Preston	5 of 6	c1400	Unknown Leicester founder
Preston	6 of 6	c1598	Newcombe and Watts of Leicester
Preston	Sanctus Bell	c1400	John Barber of Salisbury
Seaton	2 of 5	1597	Matthew Norris, possibly with Richard Holdfield*
Seaton	3 of 5	c1585	Richard Bentley, possibly of Buckingham
Seaton	4 of 5	1597	Richard Holdfield, possibly with Matthew Norris*
South Luffenham	1 of 4	1593	Hugh Watts I of Leicester
South Luffenham	4 of 4	c1510	Robert Mellour of Nottingham
Teigh	3 of 3	c1540	Henry Oldfield I of Nottingham
Tickencote	Disused	c1500	Unknown London founder
Tixover	1 of 1	c1430	Richard Hille of London
Wardley	2 of 2	c1529	Thomas Bett of Leicester
Whitwell	2 of 2	c1410	Johannes de Colsale, itinerant founder
Wing	5 of 5	c1570	Thomas Newcombe II of Leicester

SECULAR BUILDING
North Luffenham

The Pastures	Clock bell	c1550	Robert Newcombe I of Leicester

* Richard Holdfield and Matthew Norris were possibly itinerants at this time (*see* Chapter 2 — Bellfounders, Cambridge, and Stamford).

Appendix 3
Rutland Ringing Customs

This Appendix provides further information on the ringing customs noted in Chapter 4 — Gazetteer. These customs, both religious and secular, varied slightly between communities, and often depended on the number of bells in the church tower. Even in the same church customs varied over the years, times and methods of ringing differed, and some customs were dispensed with whilst others were reinstated. Names given to bells, used originally for a specific custom, were often altered or used in a different context. Examples of customs have been taken from Churchwardens' Accounts to illustrate this Appendix, but the lack of such documents for many villages means that it is impossible to give a complete picture. Of those accounts accessed, many entries were found to be irregular and lacking in detail.

CURFEW BELL AND ANGELUS BELL

The use of the **Curfew Bell** was enforced throughout the country by William I who was familiar with its use in Normandy. It was originally introduced as a fire prevention measure and the sound of this bell was a signal for all lights and fires to be covered. Unattended hearths and candles in closely-packed wooden dwellings with thatched roofs were a real fire risk. 'Curfew' is from the French '*couvre-feu*' meaning 'cover the fire'. The bell was usually rung from Old Michaelmas Day to Old Lady Day (29 September to 25 March).

Although curfew was abolished by Henry I in 1100 it appears that the Curfew Bell continued to be rung in many places for its original purpose. Up to the middle of the sixteenth century it acted as a signal for all taverns and ale houses to close. Although the ringing of this bell was generally 9pm in London and the larger towns, it was somewhat more flexible in the countryside.

In some places, the continuance of the Curfew Bell over the next few centuries can probably be attributed to a religious purpose rather than a civil one. The ringing of a bell, referred to as and eventually called the 'last Angelus', was closely associated with the ringing of the Curfew Bell. In the fourteenth century an evening 'Hail Mary' was ordered by Pope John XXII to be said at the sound of a bell, and this practice of saying an '*Ave*' to the Blessed Virgin at nightfall coincided with the ringing of the Curfew Bell. Although a separate bell was intended for the **Angelus Bell**, invariably one bell sufficed.

It is likely that the custom of ringing this bell, still referred to as the Curfew Bell, prevailed in many parts of Rutland, but unfortunately reference has only been found in six of the county's parishes. It was rung in the early 1600s at Whissendine and Uppingham and in the latter town at the end of the eighteenth century. In the 1880s it was recorded at Langham, Oakham and South Luffenham, and in the first decades of the twentieth century at Langham, Oakham, South Luffenham, Teigh and Uppingham. In 1880 the treble was used for the Curfew Bell at South Luffenham, and after ringing this bell at Langham, the day of the month was tolled. At both of these villages this bell was rung at 8pm.

Sometimes the continuance of the Curfew Bell was secured by an endowment, provided by persons having been guided home by the sound of the church bells on a dark night. Sometimes a condition was attached to the endowment, stating that another bell had to be rung at 5am and that both were to be rung daily over the winter months. This was the case at Langham and South Luffenham in 1880, but in the latter village the bells were not rung during a fortnight at Christmas.

DAILY BELL AND DINNER BELL

'A mid-day *Angelus* was rung in France in the fifteenth century, but the practice does not appear to have been introduced into Rutland' (North 1880, 102). However Thomas North did record a few villages in Rutland which traditionally rang a **Daily Bell**. In 1880 a 1pm bell was rung at Belton, Thistleton and Lyddington, its purpose to let the villagers and particularly the workers in the fields know that it was dinnertime. Although all three villages had a clock at this time, the one stroke of the hour may have been insufficient to ensure that everyone heard it. It was therefore essential that the church bell was rung repeatedly in an accustomed manner to get the message over loud and clear. By 1891 this bell at Belton was called the **Dinner Bell** and it was the duty of the sexton to combine the daily winding of the clock with the ringing of this bell.

In 1880 an 8am Daily Bell was still being rung at Thistleton and prior to this date Daily Bells had been rung at 5am at Belton and 8am at Lyddington.

Although the earlier bells may originally have been early Morning or early Angelus Bells, the 1pm bell seems to have been used solely as a call for dinner.

As previously noted, customs over the ages changed and this was sometimes due to personal idiosyncracies. A recent example was when Father Brian Scott moved to Barrowden in 1983. He rang the mid-day Angelus thrice three times on the treble every day. When the Priest's Bell was rehung in 1990, he chimed that instead, continuing to do so until he left the village in 1998.

Funeral Bells: Invitation, Warning, Carrying or Call Bell

The sounding of bells at funerals was an important and essential custom which prevailed in pre-Reformation times. However the practice became excessive and in certain parts of the country the custom needed to be curbed. In order to do this the Canons of 1603 instructed that, in addition to any Passing Bell or Death Knell, there should only be one short peal before and after burial. By 1880 in Rutland, the custom of chiming or ringing the bells at funerals was on the wane and it was customary just to toll the tenor.

The **Invitation Bell** is known to have been tolled during the late nineteenth century at Braunston, Empingham, Exton, Manton, South Luffenham and Tickencote, and as late as the 1920s at Langham and Teigh. According to the custom of the place, this bell was also referred to as the **Warning, Carrying or Call Bell**. These bells warned the bearers to assemble in preparation for the funeral cortège and were timely reminders to neighbours of the forthcoming funeral service.

Customs varied in each village and the following are additional funereal customs noted. At Braunston there were three tollings at intervals of an hour: at 1pm to give warning, at 2pm to call the bearers together and at 3pm for the funeral. At South Luffenham the tenor would toll the age of the deceased an hour before the funeral and it would toll continuously as the cortège processed to the church. At Tickencote it was customary to toll the tenor for an hour while preparations were being made to convey the corpse to the church and at Manton the tenor was tolled as interment took place. At Burley the 'tellers' (see under Passing Bell) were given before and after tolling for a funeral.

Gleaning Bell

The two earliest records found of a **Gleaning Bell** being rung in Rutland are 1748 at Oakham and 1760 at Empingham. This bell is known to have been rung in five other villages prior to 1880; Bisbrooke, Manton,

Ridlington, Seaton and Tickencote. It was still being rung in the following parishes in 1880:

Ashwell	Braunston	Clipsham
Cottesmore	Egleton	Empingham
Exton	Great Casterton	Greetham
Hambleton	Ketton	Langham
Lyddington	Market Overton	Morcott
Oakham	Teigh	Thistleton
Tinwell	Whissendine	Whitwell

The treble acted as the Gleaning Bell at Whitwell. It is not possible to state when this custom ceased in each of these villages but it continued at Empingham, Ketton and Teigh into the early twentieth century and as late as the 1930s at Ketton.

Records of Wing (*see* Chapter 4 — Gazetteer, Wing Parish Records) reveal that gleaning took place in the fields near the village in the 1860s and that a Mrs Bagley was still gleaning until *circa* 1935. It is likely that a Gleaning Bell was rung at Wing but no such record has been located.

The usual hours for the Gleaning Bell to be rung during harvest were 8am and 6pm, signalling the time when gleaning could begin and when it should terminate. This was done in order to give the old and young, the active and the weak a fair start and an equal chance. Individual payments made to the clerk for ringing this bell at the beginning of the twentieth century were one penny a week at Lyddington, two pence at Ketton and three pence at Empingham. A list of those who paid the clerk at various times is given in the Ketton Memoranda Book (DE 1944/12). Women and children made up the majority of the gleaners and an entry in Thistleton School Manager's Book for 1888 notes 'the annual absence as all go out to the fields, gleaning' (Traylen 1999, 200). This may have been the situation in many Rutland schools at that time.

Morning Bell, Maria Bell and Day Bell

The early **Morning Bell** had close associations with the custom of ringing the Angelus at nightfall, for in 1399 Archbishop Arundel issued a mandate that the practice of saying an '*Ave Maria*' in the evening should also apply to early dawn. The Angelus was rung as a reminder to all of this duty.

The bell used for the early morning Angelus (and for the evening Angelus) was often called 'Gabriel' and the fifth bell at Preston, cast *circa* 1400 and appropriately inscribed, was probably rung for this purpose. The *Maria* bells may also have been used for this function. The third bell at Morcott dedicated to **S MARIA** and three other sixteenth-century bells in Rutland with inscriptions that include the name of Mary may have been rung for the early morning 'Ave' or Angelus. They are the second at Seaton, the third at Teigh and the fourth at Edith Weston.

The Ringers of Launcells *by Frederick Smallfield ARWS (1829-1915)* (The Royal Institution of Cornwall, Royal Cornwall Museum, Truro). *This painting is probably typical of the scene in many Rutland ringing chambers during the eighteenth and nineteenth centuries. Rutland Churchwardens' Accounts frequently record payments to bellringers and, as ringing was thirsty work, such payments would no doubt have been quickly exchanged for liquid refreshment at the local inn. Some receipts reveal that they were also paid in kind, usually in the form of ale, tobacco and cheese. Successful peals were often recorded on a board in the ringing chamber; the one shown here commemorates the accession of George III in 1760. Other boards listed rules for ringing and for general behaviour. Unfortunately many of these old boards were discarded in Victorian times and only a few survive in Rutland. The following poem on a tablet at Great Bowden Church, Leicestershire, was recorded by Thomas North (1876, 152). The words add atmosphere to this early bellringing scene:*

If you get Drunk and hither Reel,
Or with your Brawl Disturb the Peal,
Or with mumlungeous* horrid Smoak
You cloud the Room, and Ringers Choak;
Or if you dare prophane this Place
By Oath, or Curse, or Language Base,
Or if you shall presume in Peal
With Hatt or Coat, or armed Heel:
Or turn your Bell in careless way,
For each Offence shall Two Pence pay;
To break these Laws if any hope
May leave the Bell, and take the Rope.

Edward Englehern churchwarden.

N.B. *He who plucks his Bell over when turned shall pay Six Pence.*

*[North interprets this word as 'stinking tobacco']

It is possible that all three could have been recasts of much earlier bells and their inscriptions retained. If so their primary function had ceased before North's survey of 1880.

Over the years, particularly after the Reformation, the religious significance of this bell became obscure. In an agricultural county like Rutland this early bell must have sounded in many parishes but it would have been rung purely as a call to work. It appears that the custom had almost ceased by 1880, but it continued into the twentieth century at South Luffenham and Preston. At both of these villages the bell was rung at 5am and at Preston the treble was used. This bell was known in both villages as the **Maria Bell** (Traylen 1989, 34).

A **Day Bell**, presumably the early Morning Bell, is recorded at Whissendine and Uppingham in the seventeenth century. At Uppingham it was rung at 4am but in 1790 at 5am. At both parishes this bell was rung in conjunction with the Curfew Bell and only rung in the winter months. At Wing *circa* 1872 a bell was rung at 6am and 6pm to mark the beginning and end of the working day.

Mote Bell, Meeting Bell, Rent Bell and Vestry Bell

A **Mote Bell** was ordered to be rung by Edward the Confessor in order to convene the people in times of danger, and this may have been the forerunner of bells being rung to announce public meetings. Bells were rung in connection with the manorial courts but the only reference found to such an event in Rutland was at Bisbrooke when a bell was being rung *circa* 1880 to announce the 'Duke's Court'. A **Rent Bell** was traditionally rung at Langham but ceased in 1914. It announced the arrival of the steward to collect the half-yearly rents.

During the nineteenth century, a **Vestry Bell** was tolled at Oakham and Bisbrooke, and a **Meeting Bell** (the seventh bell) was rung at Oakham as a summons to town meetings. By 1910 the Meeting Bell at Oakham was no longer rung for this purpose, but was still rung for Vestry Meetings.

Obit Bell

In addition to ringing after death and at a funeral, it was the custom in pre-Reformation times, for those who could afford it, to keep 'mind days'. The 'month's mind' was a monthly commemoration of the dead. Occasionally bellmen were employed to ring their handbells exhorting the parishioners to pray for all Christian souls.

The **Obit Bell** or 'year mind' was the anniversary of the death of a particular person, and it was upon that day that bells were rung and masses said for the dead. For the Obit, provision was usually made by the deceased person in his will, or by his relatives or friends. Although the

'Obits' were abolished at the Reformation 'commemorative peals of a similar type survived in a more secular guise' (Walters 1977, 164). Charitable gifts were often dispensed on these days. The only reference found to such a provision being made in Rutland was by Thomas Drake of Uppingham in the late sixteenth century (Irons' Notes MS 80/1/3). Whether one bell was tolled, a peal rung or bells chimed is not known.

Passing Bell, Death Knell and Death Bell

In medieval times the **Passing Bell** was rung to summon the priest to the bedside of a dying person and in order that those who heard it might speed and assist the departing soul with their prayers. It was a popular belief that the sound of the bells would scare off any demons lying in wait for the 'departing spirit'. A certain number of tolls denoted the sex of the sick person. The use of this bell meant that it could be rung at any time of the day or night. The custom continued well after the Reformation, the Canons of 1603 confirming the use of the Passing Bell and the ringing of no more than one short peal (the Death Knell) after death. This practice seems to have lasted in this form until *circa* 1755, but after this date until the end of the nineteenth century, the tolling of the **Death Knell** generally superseded the use of the Passing Bell.

An early reference to the Passing Bell at Teigh in 1707 is found in the Churchwardens' Accounts. In the North Luffenham accounts for 1820, payment for the Passing Bell actually announced a death (that of George III) and therefore more properly referred to the Death Knell. This seemed to be the case in several Rutland villages. For instance, at Teigh in the 1920s when villagers remembered a bell being rung on the occasion of death, they called it the Passing Bell. It is therefore understandable that the names of these two bells became interchangeable in different communities over the ages.

According to Thomas North in 1880 the procedure for ringing at death did not vary much in the parishes of Rutland. His survey reveals that a combination of strokes was made on the tenor, either before or after the Death Knell, to denote whether the deceased was male or female. In some villages these strokes were made before and after the knell. Three times three strokes were given to indicate that a male parishioner had died, and twice by two strokes given for a female. These strokes were often called tellers. 'Tellers' may be a word in its own right or it may be a corruption of 'tailers', the word referring to the bells being rung at the tail or end of the knell. It is from this source that the saying 'nine tailors make a man' originates.

At least three villages, Market Overton, North Luffenham and Tinwell, tolled a different combination of strokes to denote whether the death was of a young man or woman, or of a boy or girl. Generally the treble was tolled to indicate the death of a child and this was used at Tinwell.

Occasionally the age of the departed was tolled but only four instances of this custom have been recorded in Rutland. Thomas North (1880, 123 & 153) stated that although the custom prevailed at Teigh in 1880 the isolated occurrence at Braunston at that date was 'a custom … quite unknown to the villagers generally'. This custom was still adhered to at Barrowden and Empingham well into the twentieth century.

Although the old customs fell into disuse the 'regular' tolling of a **Death Bell** or Death Knell continued, but the custom was generally discontinued by *circa* 1950. Still known as the Passing Bell it was being rung at Caldecott in the 1960s. Today the solitary bell at Stoke Dry continues to toll at the death of a parishioner.

In 1880 the then third at Hambleton and the tenors at Hambleton, Tinwell and Whissendine had inscriptions concerning death. Of these the tenor at Whissendine, cast in 1609 by Henry Oldfield II, remains, and the inscription reads my roaringe sounde doth warning geue that men cannot heare always lyue. The other bells have been recast and with the exception of the third at Hambleton retain their Latin inscriptions. These translate as: **I sound not for the souls of the dead but for the ears of the living** and **Holy offices I proclaim: joys I spread: deaths I bewail**.

Peals rung for Oak Apple Day and Gunpowder Plot

Two of the most widespread and flourishing of the celebrations held were those of **Gunpowder Plot** on 5 November (when Parliament was saved from being destroyed in 1605) and **Oak Apple Day** on 29 May. It was on this latter day, his birthday, that the future Charles II made his triumphal entry into London in 1660, thus ending his long exile and so announcing the Restoration of the Monarchy. An essential element of the celebrations on this day was the use of oak leaves for decorative purpose, thus recalling the oak tree in which Charles had hidden after his defeat at the Battle of Worcester in 1651.

Parliament ordered that both of these days were to be commemorated annually with church services, bellringing and bonfires (Kightly 1986, 130 & 176).

Exactly how many villages commemorated these two events is not known, but payments made at dates near to their anniversaries indicate that there were more than those listed here. Ashwell, Edith Weston, Empingham, Teigh and Whissendine paid ringers in celebration of Guy Fawkes Night during the eighteenth century. Payments were made to ringers during the twentieth century to celebrate this anniversary at Braunston, Langham,

Morcott, Oakham, South Luffenham and Uppingham. Only Oakham and Uppingham Churchwardens' Accounts record specific payments for ringing on 29 May during the nineteenth century.

Pudding Bell

A **Pudding Bell** still rang out in many Rutland villages until the outbreak of the Second World War. This bell usually sounded after a particular Sunday service bell, indicating to the villagers that it was time to take the Sunday dinner to the bakehouse. The baker made a charge of one or two pence for cooking the roast and Yorkshire pudding. This bell was rung at 11am at Egleton in 1880. Sometimes the Pudding Bell was rung at the end of a midday service to alert housewives to prepare the dinner for the table.

Sanctus Bell and Priest's Bell

In medieval times a 'little bell', the **Sanctus Bell** (Sancte or Saunce Bell), was owned by most churches. Although it was usually hung in a special bellcote on the eastern gable of the nave it was sometimes hung in a belfry window. The bellrope would have passed down into the church to be within easy reach of the server at the altar. The original use of this bell was during the celebration of Mass. When the priest said the Sanctus, three strokes were made on the Sanctus Bell and a little later in the service the bell was rung again at the Elevation of the Host. The sound of the bell gave the opportunity to those not attending church to bow their heads in prayer.

The only medieval Sanctus Bell remaining in Rutland is at Preston dedicated to S MARI (St Mary). The following parish churches are also known to have owned a Sanctus Bell: Caldecott, Langham, Lyndon, Manton, Market Overton, Stretton, Uppingham, Wardley and Wing.

The tenor at South Luffenham cast *circa* 1510 by Robert Mellour of Nottingham has two capital Ss for its inscription and this possibly denotes that it was used as a Sanctus Bell. Two former bells in Rutland, now recast, are known to have had three Ss for their inscription and both were cast at the same time by the same founder. One of these, the second at Ridlington, retains its old inscription. Although not 'little bells' it is understandable that some parishes could not always afford separate bells, distinct in tone, for each function. In such cases it was prudent to have a larger bell that doubled up as the Sanctus Bell.

The Reformation basically ended the original function of the Sanctus Bell but another use was soon found, and it became the **Priest's Bell**. This bell was sometimes referred to as the 'ting-tang' and was rung just before a service telling the priest that the congregation was waiting for him. The Priest's Bells at Barrowden, Egleton,

Little Casterton and Oakham (and a former Priest's Bell at Hambleton) may originally have been former Sanctus Bells before being recast.

At Belton today a bell is rung as the Priest's Bell, three minutes prior to the start of the Sunday service.

At Uppingham the small bell currently used as a Sanctus Bell has for its inscription **ANGELUS AD VIRGINEM** (Angelus to the Virgin). It is chimed for five minutes to call people to Holy Communion and although it is no longer used at the Elevation of the Host, because the communicants could not hear it within the church, it was originally intended for this purpose. Instead the second bell is used for this part of the service. It is interesting that this bell is also referred to as the Angelus Bell, the Service Bell and the Priest's Bell, which is indicative of the difficulty in trying to ascribe a definite name and function to a particular bell at any one time throughout the ages.

Sermon Bell

Evidence suggests that a bell was sometimes rung in medieval times at midday to give notice that a sermon was to be preached in the afternoon. The Royal Injunctions of 1547 ordered that the ringing and knelling of bells, such as during Litany and Mass, should be abolished except for one bell being rung at a convenient time before the sermon. In 1880 it is recorded that a quarter of the churches in Rutland rang the Sermon Bell directly after the bells were chimed for Divine Service. Generally the bell used for the **Sermon Bell** was the tenor. The exception seems to have been at Whitwell when the treble was rung at 8am and then again at 8.30am if a sermon was to be preached.

Shrive Bell, Pancake Bell and Lenten Bell

A bell was originally rung in medieval times to give parishioners their last chance of getting 'shriven' before the start of Lent. The week preceding Lent was called Shrove-tide, and the Tuesday within it 'Shrove Tuesday' ('Shrive Tuesday' or 'Confession Tuesday') and it was upon this day that a **Shrive Bell** was rung to remind people that the priest was ready to receive them in church so that they could make their confessions.

Another custom naturally followed, for at that time there was strict abstinence from eating fresh meat within the period of Lent. On Shrove Tuesday in order to comply with this practice, the housewives would use up all of the fats within the home and make pancakes. At the summoning of the **Pancake Bell**, the apprentices (as recorded at Teigh) and others about the house would gather to eat the meal. In general the Pancake Bell was either rung at 11am as at Belton, Lyddington and Thistleton or noon as at Ashwell, Ayston and Market Overton. In 1880 Thomas North recorded that almost half of the villages in Rutland still observed this custom. At Glaston, the second bell was rung as the Pancake Bell. Although the custom still continued in Oakham in 1902 it appears that it had died out by this date in all other Rutland parishes. An exception may have been at Caldecott where a Pancake Bell was rung in the 1960s, but this may have been a reinstated custom.

Although it was a tradition of the Church not to ring the bells during Lent a **Lenten Bell** was rung at Oakham in 1861 and at Caldecott just prior to 1880. At Caldecott it was rung daily at 11am.

Spur Bell or Banns Peal

Some churches used to ring a bell or a peal after the Sunday service when the banns of a marriage were first announced. This custom was still kept at Pickworth and North Luffenham in 1880, at Lyndon in 1890, and survived at Ridlington until the 1930s. At North Luffenham this bell was also rung when the banns were read for the third time.

Ancient bell-tile found at Repton,
Derbyshire (North 1880, 39)

Appendix 4
Rutland Sundials

The sundials listed here, with the exception of a few notable modern examples, are all known to have existed in the county of Rutland prior to the twentieth century.

Little dating evidence is available but some guidance is given based on characteristics of the sundials, and relevant architectural features of the buildings.

KEY

ACC = ACCESS
Op = open
Pr = private property
Hi = can be seen from public highway
na = not applicable

MATL = MATERIAL
Ls = limestone or similar
Sa = sandstone
B = brass, bronze or gunmetal
Fe = iron or steel
W = wood

COND = CONDITION

FURNITURE
H = hour lines
30, 15, 10, 7.5, 5, 2 = minute divisions
A = arabic numerals
R = roman numerals
S = sunburst decoration
C = compass points
EOT = equation of time
Pa = painted
E = engraved / incised
N = no visible markings
L = legend
G = gilded

TYPE
dec = declining

SIZE
approx width x height in mm, or approx diameter in mm

NOTES

1839 drg	= _circa_ 1839 drawing in Uppingham School Archives
1793 drg	= _circa_ 1793 drawing in Rutland County Museum
Wright 1684	= James Wright's _History of Rutland_ of 1684
CW A/C	= Churchwardens' Accounts
Inscr	= inscription
OS 2nd ed 25″ map	= early 1900s Ordnance Survey 2nd edition 25″ map

MOTTO
M1 TEMPUS FUGIT
M2 A shadow round about my face. The sunny hours of day will grace
M3 Your sunny hours alone I tell
M4 NUNC HORA BIBENDI le temps passe l'amitie reste c'est l'heure de bien faire
M5 MORNING EVENING
M6 IMPROVE THE TIME
M7 NON REGO NISI REGAR
M8 Make time, save time. While time lasts. All time is no time. When time is past
M9 memento mori
M10 UT AVES SIC HORAE

INSCRIPTION
Inscr 1 Isaack 1614 Symmes; the gift of Sir Willyam Bulstrode

PARISH	BUILDING	GRID REF	ACC	DATE	TYPE
Ashwell	church tower	SK 866137	na	*c* 1692	vertical
Ayston	church porch	SK 859009	Op	18th C?	vertical S
Ayston	Ayston Hall	SK 859011	na	19th C?	horizontal
Barleythorpe	Barleythorpe Hall	SK 847098	na	19th C?	horizontal
Barrow	Barrow House	SK 894152	Pr	19th C?	horizontal
Barrowden	church tower	SK 945999	Op	18th C?	vertical S
Belton in Rutland	church porch	SK 816014	Op	*c* 1841	vertical dec SE
Belton in Rutland	Westbourne House	SK 816013	na	19th C?	horizontal
Belton in Rutland	Southview Cottage, Main St	SK 817012	Hi	18th C?	vertical (S?)
Belton in Rutland	Gorse View, Nether St	SK 817013	Pr	18th C?	vertical (S?)
Belton in Rutland	former Vicarage	SK 815011	na	19th C?	horizontal
Belton in Rutland	The Cottage, Nether St	SK 818013	Pr	19th C?	horizontal
Belton in Rutland	1 Oakham Rd	SK 817015	Hi	19th C?	horizontal
Braunston in Rutland	church nave	SK 832066	na	18th C?	vertical (S?)
Burley	church	SK 883102	na	*c* 1700	vertical
Caldecott	church porch	SK 868937	Op	1935	vertical S
Caldecott	former Sun Inn	SP 867934	Hi	18th C?	vertical dec SE
Clipsham	church porch	SK 971164	na	18th C?	vertical S
Clipsham	Clipsham Hall, east garden	SK 970164	Pr	19th C?	horizontal
Clipsham	Clipsham Hall, west garden	SK 970164	Pr	19th C?	horizontal
Edith Weston	Sundial House, Weston Rd	SK 926055	Hi	1773?	vertical dec SE
Edith Weston	former Edith Weston Hall	SK 927055	na	19th C?	horizontal
Egleton	church porch	SK 876075	Op	19th C?	vertical S
Exton	Exton Hall nursery garden	SK 917117	na	19th C?	horizontal
Exton	Old Hall	SK 922112	na	17th C?	2 x vertical S
Glaston	cottage in Main St	SK 897004	Hi	18th C?	vertical S
Glaston	church porch	SK 896005	Op	18th C?	vertical S
Great Casterton	church porch	TF 001088	Op	18th C?	vertical dec SE
Greetham	church porch	SK 924147	Op	1673?	vertical dec SW
Greetham	Chapman's Cottage, Main St	SK 925145	na	18th C?	vertical dec SE
Greetham	Greetham House	SK 925146	na	19th C?	horizontal
Greetham	Greetham Vicarage	SK 926146	Pr	18th C?	vertical dec SW
Greetham	Ram Jam Inn, Gt North Rd	SK 945160	Pr	Saxon?	vertical
Hambleton	church porch	SK 900076	na	18th C?	vertical (S?)
Hambleton	Hambleton Hall	SK 902075	Pr	1897	vertical dec SE
Hambleton	Manor House	SK 901076	na	19th C?	horizontal
Hambleton	Old Hall	SK 899069	na	19th C?	horizontal
Hambleton	Home Farmhouse, Ketton Rd	SK 903076	Hi	19th C?	vertical
Hambleton	Hilltop, Oakham Rd	SK 900077	Hi	18th C?	vertical S
Ketton	Grange Cottage	SK 985051	Pr	17th C?	vertical S
Ketton	The Firs, Stamford Rd	SK 983051	na	19th C?	horizontal
Ketton	The Priory	SK 982043	Pr	19th C?	horizontal
Ketton	Geeston House	SK 986040	Pr	19th C?	horizontal
Ketton	The Grange, Stamford Rd	SK 985051	Pr	18th C?	vertical
Ketton	Ketton Hall	SK 980042	Pr	19th C?	horizontal
Langham	Manor House, Church St	SK 844114	Hi	18th C?	vertical dec SE
Langham	Ranksborough Hall	SK 836110	na	19th C?	horizontal
Lyddington	Manor House, Main St	SP 875971	Pr	17th C?	vertical S
Lyndon	Lyndon Hall	SK 906044	Pr	18th C?	horizontal
Market Overton	church tower, S face	SK 886165	Op	17th C?	vertical dec SE

SIZE	GNOMON	FURNITURE	MATL	COND	NOTES
				lost	1692 CW A/C — 'drawing the dial'
575 x 650	Fe	H, 15, A, S, E	Ls	average	from demolished cottage
				lost	OS 2nd ed 25" map
300 dia	B	H, 7.5, R, S, E, M1	B & Ls	relocated	moved to Westbourne House, Belton
				lost?	OS 2nd ed 25" map
600 x 600	Fe	H, 30, R, E	Ls	fair	
600 x 600	Fe	H, 30, R, E, G	Ls	good	recut and gnomon gilded in 2001
300 dia	B	H, 7.5, R, S, E, M1	B & Ls	lost	originally at Barleythorpe Hall
300 x 375	missing	N	Ls	very poor	covered by modern table dial
	missing	N	Ls	very poor	
				lost	OS 2nd ed 25" map
250 dia?	missing	H, 10, R, S, E, M2	B & Ls	average	
200 dia?	B	H, E	B & Ls	lost	
				lost	1839 drg; painted timber dial?
				lost	1703 CW A/C — 'Dyall' whitewashed
550 x 550	B	H, 5, R, S, E, M3	B	excellent	commemorates Silver Jubilee of 1935
700 x 700	missing	N	Ls	very poor	building now part of garage premises
				lost	1839 drg
	B	H, R	B & Ls	vandalised	OS 2nd ed 25" map; new dial
175 x 175	B	H, 5, R, S, C, EOT, E	B & Ls	good	signed W Deane
550 x 550	Fe	H, 15, A, R, S, Pa	Ls	very good	date stone 1773
				lost	OS 2nd ed 25" map; house demolished
500 x 500	Fe	H, 15, R, S, Pa	Ls	poor	furniture can be seen in oblique sunlight
				lost	OS 2nd ed 25" map
				lost	Wright 1684; building now in ruins
500 x 500		A (faint)	Ls	very poor	
375 x 375	Fe	R (faint)	Ls	very poor	1793 drg
550 x 550	Fe	H, R (both faint)	Ls	very poor	1839 drg; to be refurbished in 2002
500 x 500		R	Ls	very poor	1839 drg; porch rebuilt in 1673
			Ls	lost	cottage demolished in 1990s
				relocated	OS 2nd ed 25" map
650 x 600	Fe	R, Pa	Ls	average	from former vicarage
350 x 350	missing		Ls		found in 1929 excavations
				lost	1751 CW A/C; 1793 drg
1000 x 775	B	H, 15, A, Pa, M4	B	very good	commissioned by Walter Gore Marshall
				lost?	OS 2nd ed 25" map
				lost?	OS 2nd ed 25" map
300 x 300	missing	N	Ls	very poor	gnomon holes visible
300 x 300	missing	H, R (both faint)	Ls	very poor	traces of painted numerals and lines
400 x 500	Fe	H (faint)	Ls	poor	cottage dated 1689
				lost?	OS 2nd ed 25" map
				not known	OS 2nd ed 25" map
				not known	OS 2nd ed 25" map
				not known	
				not known	OS 2nd ed 25" map
800 x 800	Fe	H, 15, R, E, M5	Ls	good	
				lost?	OS 2nd ed 25" map
200 x 300	missing	H, R (faint), E	Ls	very poor	on thrall in cellar; has initials NH
230 dia	B	H, 2, R, S, C, E	B & Ls	good	
900 x 900	Fe	N	Ls	very poor	1839 drg; originally painted

PARISH	BUILDING	GRID REF	ACC	DATE	TYPE
Market Overton	church tower, W face	SK 886165	Op	17th C?	vertical W dec SW
Market Overton	Manor House	SK 888163	na	17th C	ceiling
Martinsthorpe	Martinsthorpe House	SK 868046	na	c 1620	vertical
Morcott	church porch	SK 925008	Op	c 1695	vertical dec SE
Morcott	Sundial House, Church Lane	SK 924008	Hi	1627	vertical dec SE
Normanton	Normanton Hall	SK 935063	na	18th C?	horizontal
North Luffenham	Manor Farm	SK 932035	Hi	18th C?	vertical S
North Luffenham	Sundial Cottage, Digby Drive	SK 937032	Hi	18th C?	vertical S
North Luffenham	Sundial Cottage (to NE of)	SK 937032	na	19th C?	horizontal
North Luffenham	Luffenham House	SK 934033	na	17th C?	vertical (dec SE?)
North Luffenham	North Luffenham Hall	SK 935032	na	19th C?	horizontal
North Luffenham	The Pastures, Glebe Road	SK 931034	na	1901	horizontal
North Luffenham	barn near Dovecote House	SK 934034	na	?	vertical
North Luffenham	church, chancel buttress	SK 934033	Op	18th C?	vertical dec SW
Oakham	Buttercross, Market Place	SK 861088	Op	18th C?	4 x vertical N S E W
Oakham	Catmose House	SK 863086	na	19th C?	horizontal
Oakham	The Cottage, Uppingham Rd	SK 862085	na	19th C?	horizontal
Oakham	13 High Street	SK 861088	Op	19th C?	vertical S
Oakham	St John & St Anne's Chapel	SK 857087	Pr	17th C?	vertical S
Oakham	Oaklands, Peterborough Ave	SK 865099	Hi	modern	armillary sphere
Pickworth	Sundial Cottage	SK 994138	Pr	19th C?	vertical S
Preston	Manor House	SK 872024	Hi	18th C?	vertical S
Preston	10 Main Street	SK 872023	Pr	17th C?	vertical dec SE
Preston	church, over south porch	SK 870024	Op	17th C?	vertical S
Preston	Wings House, Main Street	SK 872027	Hi	18th C?	vertical S
Ridlington	Ridlington House	SK 848027	Pr	1614	horizontal
Ryhall	church, over south porch	TF 036108	na	17th C?	vertical S
Ryhall	Vicarage	TF 036108	Pr	19th C?	horizontal
Seaton	church, south porch gable	SP 904983	Op	17th C?	vertical S
South Luffenham	South Luffenham Hall	SK 942018	na	19th C?	horizontal
Stretton	church, south porch gable	SK 949158	Op	16th C?	vertical S
Thorpe by Water	1 Main Street	SP 892965	Hi	17th C?	vertical dec SE
Tixover	church, south porch gable	SP 971998	na	18th C?	vertical (S?)
Tixover	Tixover Grange, Magnolia Cottage	SK 980017	Pr	18th C?	vertical S
Uppingham	former Unicorn Inn, 11 High St E	SP 867997	Hi	1765	vertical dec SE
Uppingham	Crown Hotel, 19 High St E	SP 867997	Hi	18th C?	vertical dec SE
Uppingham	Sundial House, 45 High St E	SP 868997	Hi	18th C?	vertical dec SW
Uppingham	Thring Centre, High St W	SP 865997	Pr	18th C?	vertical dec SW
Uppingham	Archdeacon Johnson's School	SP 867996	na	1741	vertical
Uppingham	Pleasant Terrace, Adderley St	SP 870996	Pr	1661	vertical S
Uppingham	Uppingham Hall	SP 869996	Pr	19th C?	horizontal
Uppingham	Masons' Lawn, Uppingham School	SP 866996	Pr	2001	horizontal
Uppingham	Brooklands, London Rd	SP 867991	na	19th C?	horizontal
Uppingham	Sundial Cottage, 1 Stockerston Rd	SP 863997	na	?	(vertical?)
Wardley	church, over south porch	SK 832002	Op	1694	3 x vertical E S W
Whissendine	Rectory	SK 833143	na	19th C?	horizontal
Whitwell	church, south porch gable	SK 924088	Op	17th C?	vertical S
Whitwell	Rectory	SK 925088	na	19th C?	horizontal
Wing	Sundial House, Church St	SK 893032	Hi	17th C?	vertical dec SW
Wing	City Yard House	SK 892029	Hi	17th C?	vertical S

SIZE	GNOMON	FURNITURE	MATL	COND	NOTES
900 x 900	missing	N	Ls	very poor	1839 drg; originally painted
				lost	constructed by Isaac Newton
				lost	Wright 1684; house demolished 1755
400 x 400	missing	N	Ls	very poor	1695/6 CW A/C — painted for 6s 6d
400 x 400	Fe	N	Ls	poor	datestone (WD 1627) with gnomon
				lost	OS 2nd ed 25" map demolished 1926
450 x 450	Fe	N	Ls	poor	
350 x 350	Fe	N	Ls	very poor	renovated 1894
				lost?	OS 2nd ed 25" map
				lost	Wright 1684; house demolished
				lost?	OS 2nd ed 25" map
				lost	by C F A Voysey; base remains
				lost	
425 x 450	Fe	R (faint traces)	Ls	poor	originally painted
300 x 400		N		very poor	originally painted
				lost	OS 2nd ed 25" map
				not known	OS 2nd ed 25" map
1200 x 1200	B	H, R, Pa, M1	Ls	excellent	former Matkin's building
270 dia	Fe	H, R, E	Ls	excellent	restored 1983
450 dia				excellent	
500 x 450	Fe	N	Ls	very poor	formerly the Bluebell Inn
700 x 700	Fe	H, 15, R, E	Ls	good	
750 x 750	Fe	H, R, E	Ls	good	formerly the Chaise public house
11 x 711 x 203	Fe	A, E	Ls	excellent	restored mid 1990s; new gnomon
350 x 400	Fe	N	Sa	poor	new gnomon
159 x 159	B	H, 15, R, S, C, E	B & Ls	excellent	Inscr 1 — former church sundial?
500 x 500?	Fe	N	Ls	lost	1793 drg; removed in 1980s
				lost	OS 2nd ed 25" map; stolen
550 x 400	missing	N	Ls	very poor	1793 drg
				not known	OS 2nd ed 25" map
300 x 200	missing	H, R, E	Ls	poor	transitional scratch dial?
500 x 600	missing	N	Ls	very poor	
				lost	1793 drg
350 x 350	Fe	H, R, E	Ls	average	across corner of SE and SW walls
900 x 900	Fe	H, 15, A, M6		poor	furniture painted out
1000 x 1000	Fe	H, R, Pa, M7, G	Ls	excellent	
450 x 450	B	H, R, E	B	good	
450 x 450	Fe	H, 30, A, E	Ls	average	relocated to garden wall in mid 1900s
	missing	N	Ls	removed	dismantled — in garden at rear
210 x 240	missing	H, A, E	Ls	poor	dated 1661 and 'RPA'
				not known	OS 2nd ed 25" map; now on south side
400 dia	B	H, 15, 10, R, E, M8	B & Ls	excellent	
				not known	OS 2nd ed 25" map
				lost	
00 x 600 x 150	Fe	R, E, M9	Ls	average	east and west faces have no markings
				not known	OS 2nd ed 25" map
450 x 400	missing	N	Ls	very poor	1793 drg
				not known	OS 2nd ed 25" map
400 x 500	Fe	H, 30, R, E, M10	Ls	excellent	restored 1995 and motto added
400 x 500	Fe	H, 30, A, E	Ls	good	

William Potts & Sons' Specification and Tender of 1919 for Seaton Church Clock

[Tender No 1.]

<u>Specification & Tender</u> for a new Turret Clock. To show the time upon one solid cast iron dial four feet to four feet six in diameter & to strike the hours on the largest bell at Seaton Church Uppingham.

Hour Striking Clock.

Dial & hands	<u>The Clock</u> to show the time on one solid cast iron dial, also to have one strong set of motion wheels of gun metal cut & polished on the engine from the solid, double & single universal joints of hard brass & connections. The Clock hands to be of stout copper & dished also backed with copper to better resist the weather & be stronger & correctly ballanced or counterpoised within. The whole of the clock to be constructed from the designs & plans of Lord Grimthorpe, the greatest authority on clocks, & to have all the latest improvements inserted.
Frame, bushes, pinions, pivots, wheels & barrel.	<u>The Clock</u> frame to be on the horizontal plan cast in one piece from the solid & planed perfectly flat on the top & bottom surfaces, so that all the necessary wheels & fittings can be accurately adjusted upon it. All the bushes & bearings to be of gun metal, screwed separately into the frame, so that each or any wheel may be moved separately without disturbing the others, in case of alteration or accident. The barrel to be long to keep the cord from over-lapping, the pinions arbors pivots of steel hardened tempered & polished & pinions cut on the engine from the solid where not lantern. All the wheels to be of hard gun metal, cut & polished on the engine from the solid. Lantern pinions to be used wherever possible. The wheels & pinions to be cut by Brown Sharpes machine.
Going train & maintaining power	<u>The Going mainwheel</u> to be 13½ inches in diameter with maintaining power attached of the most approved & latest design & go 8 days with once winding.
Pendulum bob	The Pendulum bob to be of cylindrical form to be of solid cast iron.
Compensation pendulum 1¼ seconds	The above will be compensated by iron & zinc tubes & have screw at the bottom for regulation & collar conveniently placed to hold small weights for adjustment to beat 1¼ seconds. Provision to be made to prevent the pendulum from falling in the event of a spring breaking.
Escapement	<u>The Escapement</u> to be Lord Grimthorpes (Denisons) double three legged gravity & made to work as light as possible consistently with strength & to be carefully made so the clock will not trip, when set going without the fly.
Hand adjustment	<u>There will be provided</u> a small dial on frame, in clock room for minutes & one also for showing the seconds in order that the clock hands, can be accurately adjusted.
Hour Striking	<u>The Hour mainwheel</u> will be 15 inches in diameter, fly will be of large size to make the striking uniform & will be adjustable & not placed in front of the clock, with ratchets of brass squared & pinned on the arbor with 2 hardened and tempered click & springs. The First blow of the Hour hammer will strike at the correct second.
Apparatus for pulling off the hammer, shank, steel click springs crank & weights	<u>The Clock to have</u> an apparatus for pulling off the clock hammer when required. The Clock hammer will have wrought iron shank, or stem fixed on arbor & pivoted into iron frame with hardened steel click spring attached. The weights will be of cast iron in slips for easier adjustment, there will also be cranks levers, bevel wheels, fitted into gun metal bushed iron

frame & everything necessary & requisite in a First Rate Clock.

Pulleys & barrels
: The Pulleys to be of iron with turned grooves & steel pivots running in gun metal bushes & the barrels to be of iron, with 2 steel clicks & springs hardened & tempered to each barrel.

Steel cords & lines
: The cords to be of steel wire not to grind in running & not to overlap on barrel & not to be zinc coated.

Carriage & Fixing
: To include the carriage of the Clock & Material our mens time Railway fares board & lodgings during the time of fixing & leave all in good working order (this does not include joiners work masons hoisting or scaffolding if necessary).

Warranty
: The Clock Warranted for Twelve months, & after once regulated not to alter or vary in its timekeeping more than 5 seconds per month for the sum of £115.

One hundred & fifteen pounds

Extra dial & dial works & motion works
: £10 per dial extra = Clock as above with 2 dials £125

[Tender No 2.]
Similar to No 1. Tender but to strike the Hours on the largest bell & the Ding Dong ¾ Chimes on two of the smaller bells.

Ding dong Chimes
: The ding dong chimes great or mainwheel will be 15 inches in diameter with cams attached, or fixed on it for raising the hammers, for striking correct Quarters on the bells, the first stroke of the Quarters to be struck exactly at the second at 15, 30 & 45 minutes past the hour & at the hour the first blow of the hammer will strike at the correct second.

Warranty
: as No1 for the sum of £145. One hundred & forty five pounds.

Two dials £155 One hundred & fifty five pounds.

Tender No 3.

As No 1. but Cambridge or Westminster Chimes added.

Cambridge or Westminster Quarter Chimes
: The Cambridge or Westminster Quarter Chimes mainwheel will be 15 inches diameter with cams attached or fixed on it, for raising the hammers for striking the correct musical Quarters on the bells, the first stroke of the Quarters, to be struck exactly at the second, at 15, 30 & 45 minutes past the hour & at the hour they will be struck a little earlier, so that the first blow of the hour hammer, will strike at the correct second, as was first introduced by our Firm at Lincoln Cathedral, afterwards at Bolton, Rochdale, Dewsbury & Sunderland Town Halls, Leeds, S. Shields, Blackburn, Mossley Hill Liverpool, Holy Trinity Windsor & other Church Clocks, Newcastle Cathedral, Hexham & Bath Abbeys & Liverpool University Clocks.

Warranty
: As No 1 & 2 for the sum of £175. One hundred & seventy five pounds.

Two dials £185. One hundred & eighty five pounds.

To be paid for by instalments a sum down on completion of work & the balance within a reasonable time that time to be agreed upon by

{The Rector & Churchwardens } & Mr Thomas Edmund Potts Director.
{Representing The Church } & Messrs William Potts & Sons Ltd.

28th April 1919
Thomas Edmund Potts
Director

Wm Potts & Sons Ltd
Leeds & Newcastle-on-Tyne

Transcribed from the handwritten original (DE 1883/33/1). Spelling retained.

Appendix 6
Thomas North FSA

THOMAS NORTH AND HIS FAMILY

Thomas North was born on 24 January 1830, the son of Thomas and Mary North of Burton End, Melton Mowbray, Leicestershire.

His father, a butcher, was born at nearby Burton Lazars. He married Mary Raven at Melton Mowbray Parish Church in 1825 and they had two other children, Mary, baptised in 1827, and George, baptised in 1830.

On leaving school Thomas went to work for William Latham, a solicitor in Melton Mowbray, but in 1845, at the age of 15, he moved to Paget's Bank, in High Street, Leicester, where he was employed as a clerk (Drake 1861, 51). Whilst in Leicester he is known to have resided at 19 Princess Street (1860), 79 Regent Street (1861) and the Bank House in High Street.

On 23 May 1860 he married Fanny Luck, the only daughter of Richard and Anne Luck, at St Martin's Church, now Leicester Cathedral, where Thomas was at some time a churchwarden. He was also churchwarden at St Matthew's Church, Leicester, and the Rev George Venables, who had served a four-year ministry at this church, held him in high regard, referring to him as a model churchwarden (Dimock Fletcher 1888, 92).

Thomas and Fanny's only child, Herbert Luck North, was born in November 1871. He attended Uppingham School (Brooklands) from September 1885 until July 1890, followed by Jesus College, Cambridge. Herbert North BA FRIBA worked for some time as first assistant to Sir Edward Lutyens, the Victorian architect, and became well known in his own right as an 'Arts and Crafts'

Thomas North at the age of 19. This is from a tinted daguerreotype photograph taken on 24 January 1849. It appears to be his only surviving photograph (Pam Phillips)

Herbert Luck North BA FRIBA, the only son of Thomas & Fanny North (Pam Phillips)

Thomas North = **Mary Raven**
born 1804 Burton Lazars
married 1825 Melton Mowbray
butcher & grazier

Richard Luck = **Anne Tayler**
born 1812
solicitor in Leicester
moved to Llanfairfechan 1856/7
died 18-1-1898
buried Llanfairfechan

died 4-11-1890
buried Llanfairfechan

Mary North   **George North**

Thomas North = **Fanny Luck**
born 24-1-1830 Melton Mowbray
married 23-5-1869 Leicester
clerk at Paget's Bank, Leicester
retired 1872 due to ill health
died 27-2-1884
buried Llanfairfechan

born 25-10-1835 Leicester
baptised 11-11-1836 Leicester
died 10-2-1917
buried Llanfairfechan

Herbert Luck North = **Ida Maud Davies**
born 1872 Leicester
'Arts & Crafts' architect
died 9-2-1941
buried Llanfairfechan

born Dudley
died 25-2-1961
buried Llanfairfechan

Perceval Mitchell Padmore = **Ida Joan North**
born 1896 Melton Mowbray
died 1992
buried Llanfairfechan

born 1898 London
died 1990
buried Llanfairfechan

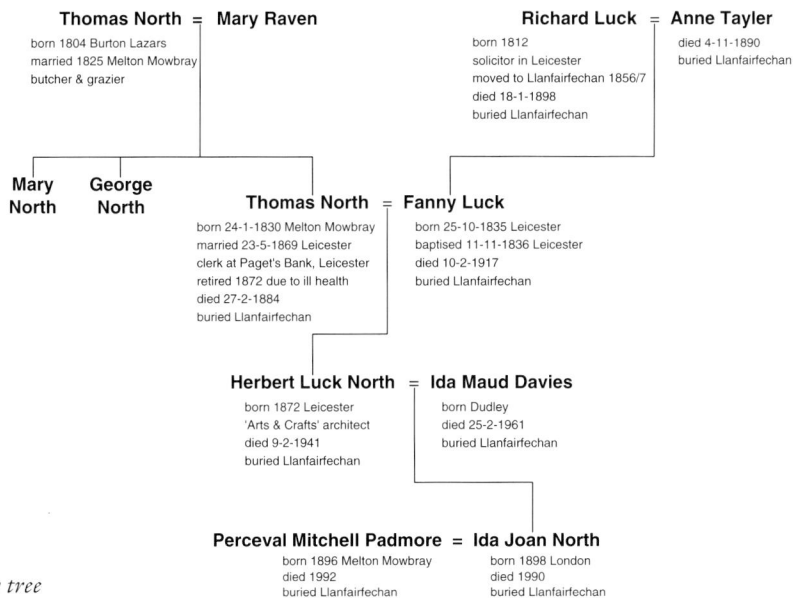

The North family tree

architect. In 1901 he built 'Wern Isaf' as his family home in Llanfairfechan and this Grade II* property is still owned by his descendants. He was also a historian of note and his publications included *The Old Cottages of Snowdonia* and *The Old Churches of Arllechwedd*. He died 9 February 1941 at the age of 69.

Eighteen months after Herbert's birth, Thomas North contracted tuberculosis, and his failing health compelled him to give up work altogether, and retire with his family to Ventnor, Isle of Wight. The Census of 1881 records his family as follows:

Dwelling: Zig Zag Road
Census Place: Ventnor, Hampshire, England

		Age	Birthplace	Occupation
Thomas NORTH	Head	51	Melton Mowbray	Income from Interests
Fanny NORTH	Wife	45	Leicester	
Herbert L NORTH	Son	9	Leicester	

Richard Luck, Thomas's father-in-law, was a partner in Harris & Luck, solicitors, at 63 High Street, Leicester. From 1852 he and his family were frequent visitors to North Wales, and in 1856 he purchased part of the Bulkeley estate at Lanfairfechan. He built his first house here in 1857, followed by a small mansion, which became known as Plâs Llanfair, in 1864.

After 1872, Thomas and his family were frequent visitors to Llanfairfechan, often spending whole summers at Plâs Llanfair. In 1881 they moved from Ventnor to live with Richard and Anne Luck. As a result of his illness, Thomas died at Plâs Llanfair on 27 February

1884 at the age of 54. Fanny, his widow, continued to live at the mansion until her death in 1917. After this the house was converted into a convalescent hospital. At the end of the First World War, Herbert North moved his architectural practice here and it also became his family home. In the early 1920s it was sold to St Winifred's School for Girls, but in the 1970s it was demolished and the land is now occupied by a housing estate.

Thomas was buried in the cemetery near the church of St Mary in Llanfairfechan where the family worshipped. The family memorial, in the shape of a Celtic cross on a cuboid stone base, was designed by Herbert North. It records the lives of Richard and Anne Luck, Thomas and Fanny North, Herbert and Ida North, and Herbert's daughter and husband.

In 1885 the north aisle of St Mary's Church was built to the memory of Thomas North and this is recorded on a brass plaque in the church. Corbels in the form of

Thomas and Fanny North are remembered on the family memorial in the cemetery at Llanfairfechan. Thomas was actually 54 when he died

IN·MEMORY·OF THOMAS·NORTH F·S·A·WHO·DIED·27 FEB·1884·AGED·53 AND·OF·FANNY·HIS WIFE·WHO·DIED·10 FEB·1917·AGED·82

The carved stone heads of Thomas and Fanny North which are placed either side of the north aisle west window at St Mary's Church, Llanfairfechan

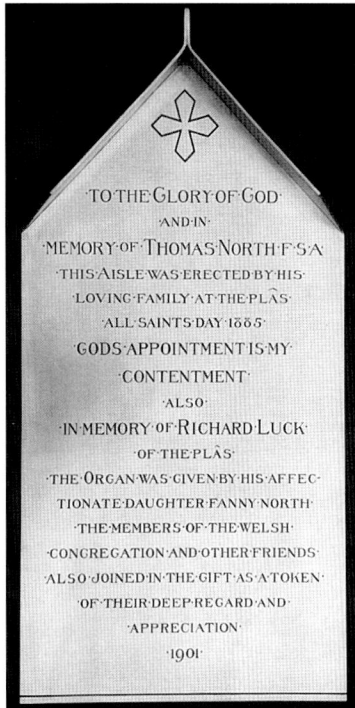

The brass memorial plaque in St Mary's Church, Llanfairfechan

carved stone heads of Thomas and Fanny are placed outside the church, either side of the north aisle west window.

Herbert North placed a stained glass window to his father's memory in the church. The dedication reads: IN LOVING MEMORY OF THO^S NORTH F.S.A. DIED FEB 27TH 1884. St Mary's is now a redundant church and unfortunately much of this window has been destroyed by vandals.

THOMAS NORTH'S INTERESTS

Most of Thomas North's spare time was spent in the study of archaeology and local history. In 1861 he was elected Secretary of the Leicestershire Architectural and Archaeological Society and later became an Honorary Member. He was also an Honorary Member of the Derbyshire Archaeological and Natural History Society and was elected a Fellow of the Society of Antiquaries of London on 2 March 1876.

Although ill health forced him to retire at the age of 42 to Ventnor, and nine years later to Llanfairfechan, he continued to pursue his interests in Leicestershire, collecting vast amounts of material to further his research.

From 1861 until his death in 1884 he was editor of the *Transactions of the Leicestershire Architectural and Archaeological Society*, contributing some thirty papers

The stained glass window to the memory of Thomas North at St Mary's Church, Llanfairfechan (Pam Phillips)

on various subjects in this time. Amongst these the most important were 'Tradesmen's Tokens issued in Leicestershire', 'The Mowbrays, Lords of Melton', 'Leicester Ancient Stained Glass' and 'The Letters of Alderman Robert Heyricke'. In later life he wrote a number of books on church bells, one of these being *The Church Bells of Rutland* which was published in 1880. It is a hardbound book of 176 pages in foolscap quarto with 40 woodcuts. 250 copies were printed by Samuel Clarke of 5 Gallowtree Gate, Leicester, and they were sold at a cost of 10s 6d. Although this book was complete in itself, it was the concluding volume of a *Description of the Church Bells of the Diocese of Peterborough*, the other parts being *The Church Bells of Leicestershire* which was published in 1876 and *The Church Bells of Northamptonshire* in 1878.

In order to complete this huge task, and because of his disability, he had to enlist local help. In his preface to *The Church Bells of Rutland* Thomas North acknowledges the help given by eighteen 'gentlemen', many being incumbents, for the 'efficient and valuable help in procuring rubbings or casts from bells in the parishes placed against their names'. They were:

The Rev John Baker — Exton
The Rev E Bradley — Stretton
The Rev T Lowick Cooper — Empingham
John Day Esq — Ashwell, Thistleton, Whissendine
The Rev Philip G Dennis — North Luffenham
The Rev Sir J Henry Fludyer Bart — Ayston
The Rev J Freeman — Manton
The Rev W Gay — Burley
Mr M Handscombe — South Luffenham
The Rev Charles T Hoskins — Clipsham
W H Jones Esq — Ayston, Belton, Bisbrooke, Braunston, Caldecott, Essendine, Glaston, Greetham, North Luffenham, South Luffenham, Lyddington, Normanton, Oakham, Pilton, Ridlington, Seaton, Stoke Dry, Tinwell, Uppingham School, Wardley, Whitwell
Mr Matthew Pearson — Barrowden, Brooke, Edith Weston, Egleton, Great Casterton, Hambleton, Ketton, Little Casterton, Lyndon, Morcott, Ryhall, Tickencote, Tinwell, Tixover, Wing
The Rev R Basset Rogers — Preston
J O Spreckley Esq — Langham
The Rev the Hon A J Stuart — Cottesmore
The Rev Chancellor Wales — Uppingham
The Rev H L Wingfield — Market Overton
Mr George Wood — Teigh

The following extract taken from *Some Recollections of the Parish of Exton in the County of Rutland from AD 1875 to AD 1877* (Wilton Hall 1913, 11) gives an insight into the help given by these 'gentlemen':

Shortly before my arrival Mr Baker, the Assistant Priest, had accomplished an important and valuable piece of work in connection with the Church Bells. He climbed up amongst the Bells, took their measurements, made rubbings of their inscriptions, detailed the local customs connected with their use, and the folk-lore concerning them,

which Mr North in his "*Church Bells of Rutland*" utilised. It was an awkward job as I know from my own experience, for upon my own account I climbed about there a little, and some years afterwards did the same kind of thing for the Bells of Essendon, Co Hertford, which Mr North used in his "*Church Bells of Hertfordshire*".

A further example is given by Thomas North in his *Church Bells of Northamptonshire* under Harringworth (1878, 297):

Priest's Bell:
The small bell, now without a clapper, hangs to a beam which projects from a spire light. It was used some years ago as a clock bell but probably was originally the sanctus bell. To obtain the inscription upon it was a work of some difficulty. The gentleman who undertook the task, and to whose intelligent perseverance and agility I am so much indebted for help amongst the bells of the county, informs me that he first pulled up a ladder outside the tower into a belfry window. With the ladder he got to the spire light from which the bell projected. He was there enabled to take a rubbing of a portion of the inscription, to read a portion he could not rub, and by use of a hand mirror to obtain a reflection of the remaining portion which he could neither rub nor see in the ordinary way.

The majority of Thomas North's collection of rubbings of bell inscriptions were presented to the British Museum and this is confirmed by a letter, dated 17 July 1903, from the Director and Principal Librarian to 'Mrs Thomas North':

I am directed by the Trustees of the British Museum to convey to you the expression of their best thanks for the Present mentioned on the other side, which you have been pleased to make to them jointly with Mr. Herbert L. North .... Rubbings of inscriptions on church bells chiefly in counties Bedford, Leicester & Lincoln, in 448 rolls (information from Pam Phillips).

These are now listed in the Catalogue of the British Library under 'Additional MSS':

36819. RUBBINGS of inscriptions on Church Bells, arranged topographically as follows: A. 1-75, co. Bedford; B. co. Hertford; C. 1-177, co. Leicester; D. 1-194, co. Lincoln; E. co. Warwick. The names of places are given in the Index.

There are 154 subscribers listed in North's *The Church Bells of Rutland*. Amongst these the following are of particular interest:

From Rutland —
  E N Conant Esq, Lyndon Hall (*see* Chapter 4 — Gazetteer, Lyndon)
  The Reverend Sir J H Fludyer Bart MA, Ayston Rectory (*see* Chapter 4 — Gazetteer, Ayston and Thistleton)
  Mr T C Halliday, Greetham [and Stamford] (*see* Chapter 4 — Gazetteer, North Luffenham)
  The Rev C Wordsworth MA, Glaston (*see* Chapter 4 — Gazetteer, Glaston)
From out of the county —
  J R Daniel-Tyssen Esq FSA, Brighton (see below)

William Latham Esq, Melton Mowbray (Thomas North's first employer: see above)

Richard Luck Esq, Plås, Llanfairfechan (Thomas North's father-in-law: see above)

Thomas Tertius Paget Esq, MP, Humberstone, Leicester (of Paget's Bank: see above)

Robert Stainbank Esq, Spring Lodge, Laurie Park, Sydenham (*see* Chapter 2 — Bellfounders, London)

The Rev Canon Sutton MA, West Tofts Rectory, Mundford, Norfolk (*see* Chapter 3 — Clockmakers & Chapter 4 — Gazetteer, Cottesmore and Greetham)

Messrs John Taylor and Co, Loughborough (*see* Chapter 2 — Bellfounders, Loughborough)

Messrs John Warner and Sons, the Crescent Foundry, Cripplegate, London, EC (*see* Chapter 2 — Bellfounders, London)

Thomas North, in *The Church Bells of Leicestershire* (1876, xviii), states: 'I have pleasure in adding that Mr. Utting has most carefully engraved for me, from casts taken direct from the bells, a large number of the woodcuts which illustrate the following pages.' Mr Utting's name can be seen on the following woodcuts included in *The Church Bells of Rutland*: **[2]**, **[13]**, **[16]**, **[23]** & **[33]** (*see* Chapter 2 — Bellfounders and Appendix 1).

Over 200 of the printer's blocks, many of them woodcuts, are still held by Thomas North's descendants. They represent the majority of the illustrations used in his various books on campanology, and thirty of the forty blocks used in *The Church Bells of Rutland* still survive.

Mr Utting's name and address is on the rear of one of the surviving woodcuts: 'R B Utting, 97 Saisford Street N. W.' The printer's block used to print the badge of the Leicestershire Architectural and Archaeological Society in their publications was also made by Mr Utting. The 1881 Census Return indicates that by then he had died and that his business had been taken over by his son, William:

Dwelling: 86 Park St
Census Place: St Pancras, London, Middlesex

		Age	Birthplace	Occupation
Susannah UTTING	Head — Widow	55	Yarmouth, Norfolk	
William S UTTING	Son	28	Yarmouth, Norfolk	Wood Engraver (Artist)
Arthur G UTTING	Son	26	St Pancras, Middlesex	Tailor
Alexander R UTTING	Son	20	St Pancras, Middlesex	Upholsterers Clerk

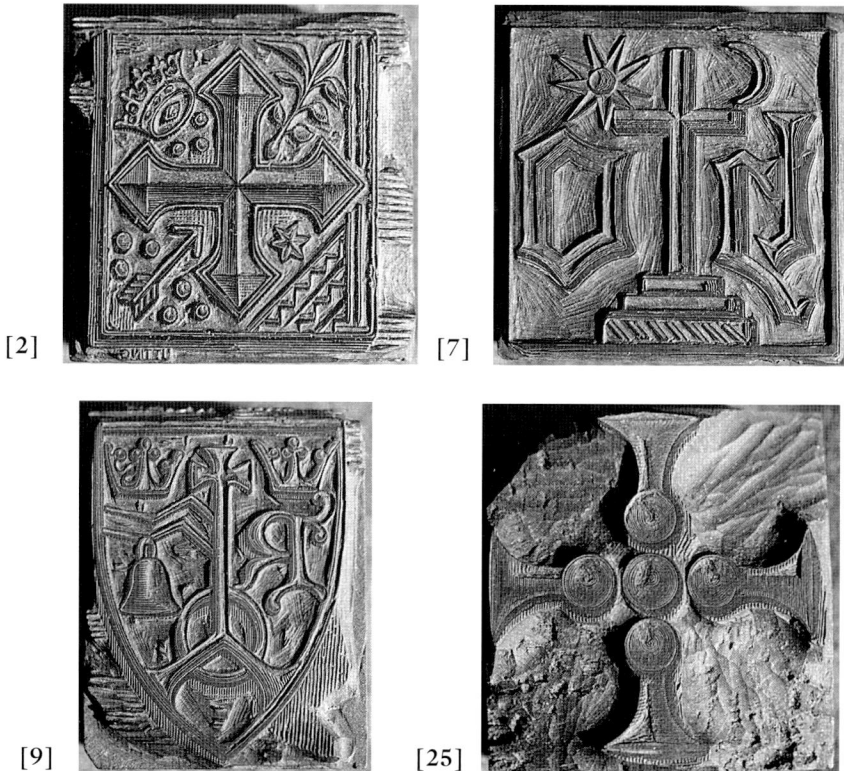

[2]

[7]

[9]

[25]

Four of the surviving printer's blocks (woodcuts) used to illustrate Thomas North's 'The Church Bells of Rutland'

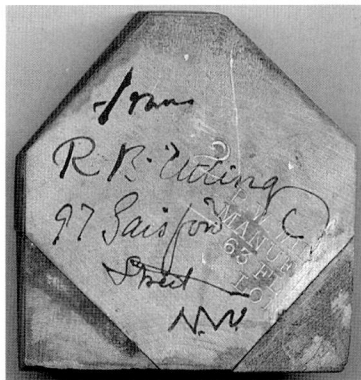

Mr Utting's name and address on the rear of one of the woodcuts which he made for Thomas North

Thomas North also borrowed printer's blocks from other campanologists to illustrate his books, as recorded in the following extracts:

My thanks are tendered to John Robert Daniel-Tyssen, Esq., F.S.A., to the Rev. H.T. Ellacombe, F.S.A., and to Llewellyn Jewitt, Esq., F.S.A. for the loan of several woodcuts ... (North 1876, xiii).

The moulds and casts of these letters and stamps [*see* Appendix 1, [55], [130] & [131]] ... were made by J.R. Daniel-Tyssen, Esq., F.S.A., which casts were engraved by the late Mr. Orlando Jewitt for W. A. Tyssen-Amherst, Esq., F.S.A., M.P., of Didlington Hall, Norfolk, by whose courteous permission they appear in this volume (North 1882, 79).

When Thomas North died on 27 February 1884 of tuberculosis, his death was noted in the President's anniversary address to the Society of Antiquaries in April 1884, and mention made of his interest in campanology. A brass tablet to his memory, erected by the Leicestershire Architectural and Archaeological Society, is in the south transept of St Martin's Church, now Leicester Cathedral:

THIS TABLET WAS ERECTED BY THE
LEICESTERSHIRE
ARCHITECTVRAL AND ARCHÆOLOGICAL
SOCIETY
IN MEMORY OF THOMAS NORTH FSA
FOR 23 YEARS HONORARY SECRETARY OF
THE SOCIETY
AND HISTORIAN OF THIS CHVRCH
BORN AT MELTON MOWBRAY IAN 24 1830
DIED AT LLANFAIRFECHAN FEB 27 1884.

The design and installation of this plaque is recorded in the *Transactions* of the Society (Dimock Fletcher 1888, 160):

Mr F. W. Ordish submitted a drawing, which he had kindly prepared, of the proposed brass to the memory of Mr. North, which was approved of by the Members present, and the Committee were directed to spend £35 for the erection of the same....

The plaque was engraved by Elgood Brothers, art metal workers, of Leicester, on an oblong brass tablet. The inscription was composed by the Rev W G Dimock Fletcher.

A list of those who attended Thomas North's funeral at St Mary's Church, Llanfairfechan, is recorded in the notes and diaries kept by Richard Luck, his father-in-law, some of which are still held by his descendants. The list includes the following from Leicestershire and Rutland who subscribed to Thomas's books. Those marked * also carried out bell surveys for Thomas:

Edwin Clephan Esq, Leicester
The Rev Robert Dalby*, Staunton Harold
John Day Esq*, Wymondham House
Elliott J Gill Esq*, Princess St, Leicester
Misses Gill, Leicester
The Rev Joseph Harris*, Westcotes, Leicester
Thomas Ingram*, Hawthorn Field, Wigston Magna
The Rev T Harry Jones MA, Ashwell Rectory
The Rev William Langley, Leicester
William Latham Esq, Melton Mowbray
Thomas Tertius Paget Esq, Humberstone, Leicester
Mrs Tayler, Rothwell

Also included in the list of mourners were:

George Henton Esq* [photographer and artist] (*see* Chapter 3 — Gazetteer Introduction)
Samuel Clarke [printer of Thomas North's books]

The Rev George Venables, Vicar of Great Yarmouth

... paid a touching tribute to his memory in "Church Bells", in the course of which he says: "A most humble-minded and unpretending man, he had thought, and read, and studied much, and was a man of unusually sound judgement and discretion. An honest, straightforward churchman, his influence for good was very much greater than it appeared to be, for no one could listen to his humble, kind, and withal uncompromising way of stating any matter without being affected by it ..." (Dimock Fletcher 1888, 92).

Thomas North's other publications include *English Bells and Bell Lore* (1880) and *The Church Bells of Bedfordshire* (1883). At the time of his death he was preparing a *History of Melton Mowbray*, *The Church Bells of Hertfordshire*, *The Church Bells of Essex* and *The Church Bells of Shropshire*. *The Accounts of the Churchwardens of S. Martin* (Leicester) was published in 1884.

As a final note to this Appendix the authors would like to record their own admiration for the research carried out by Thomas North. His publications were a major inspiration in producing *Time in Rutland*.

Index

This index is divided into four sections: Bellfounders, Clockmakers, Places, and Surnames. In the Places section the county is Rutland unless indicated otherwise. Current county status is shown for most places outside Rutland. The Surnames section excludes bellfounders and clockmakers. Original spelling has been retained.

Appendices 1 (Bellfounders' Marks, Devices, Decoration and Lettering on Rutland Bells), 2 (Rutland Bells Scheduled for Preservation) and 4 (Rutland Sundials) are not included.

Bellfounders

Appowell, George 27; John 27
Arnold, Edward 37, 49, 95-7, 111, 126, 129, 182, 253, 258, 261, 277-8, 334-6
Atton, Bartholomew 27; of Buckingham 28; Robert 27
Bagley, Henry I 28; Henry II 28, 48, 283-4; Matthew 28; William 28
Barber, John 13, 50, 269-70
Bentley, Richard 27-8, 283-4
Bett, Thomas 32-4, 332
Brasyers of Norwich 35
Briant, John 37
Burfords of London 38
Chapman, William 39
Clay, Thomas 30, 37, 318
Curtis, George 37
Daniel, John 38; John, successor 38
de Colsale, Johannes 57, 339
Eayre, of Kettering 30-1; Joseph 37, 49, 174, 176, 178, 192, 242, 339; Thomas I 49; Thomas II 30-1, 49, 114, 182, 203-4, 208-9, 231-2, 237-8, 243, 245-7, 263, 279, 303-4; Thomas III 31, 295, 297
Edmonds, Islip 37
Gillett & Johnston 29, 54, 112-14, 190-2, 254, 255-8, 326-8
Hedderly, George 47, 306; Thomas I 47, 95-7, 208, 210, 241, 242; Thomas II 47
Hille, Richard 38, 315
Holdfield, Richard 27-8, 52, 283-4
Hughes, Alan 41; Albert A 41; Douglas 41;

William A 41
Itinerant bellfounders 166
Johnston, Arthur 29; Cyril F 29
Lester, Thomas 39
Mears 121-2, 254, 257; & Stainbank 41, 154, 277-9; C & G 96, 97; Charles 39; George 38-9; George & Co 167-8; Thomas I 37, 39, 172-3; Thomas II 38-9, 107, 172-3, 266; William 39
Mellars, William 34
Mellour, Richard 45, 47; Robert 34, 44-5, 196, 198, 273, 275, 288-90
Mot, Robert 39
Newcombe, & Watts 179, 181, 203-4, 269-70; of Leicester 28, 32, 33-5, 47, 51, 237-8, 245-7, 341, 343; Edward 33-5, 203, 205; Robert I 33, 44, 102, 228, 251; Robert II 33-5; Thomas I 32-4; Thomas II 27, 33-4, 50, 108-10, 120-2, 343
Noone, William 46-7, 96-8, 120-2
Norris, bell foundry 27-8, 50-6, 166, 208-9, 223, 277, 310, 318; Matthew 27-8, 50-3, 153, 155, 224, 28-4; Thomas 51, 54-5 113-14, 120-2, 126, 129, 142-3, 148, 150-1, 168, 192, 204, 209-10, 231-2, 237-8, 246, 273, 275, 277-9, 297-9, 309-13, 318; Toby I 50, 51-4, 56, 102, 125-6, 129, 153, 155-6, 168, 180-1, 196-8, 215-16, 224, 228, 246-7, 289-90, 310, 313; Toby II 51, 54; Toby III 43, 51, 55-56, 114, 135, 137-8, 143, 148, 151, 167, 174-6, 217-18, 224,

254, 257-8, 283-5, 303, 332
Oldfield, George I 37; George III 46-7; Henry II 44-6, 148, 150-1, 203-4, 247, 303-5, 333, 335-6; Henry III 379; of Nottingham 28, 51-3; Richard 52
Osborn, Thomas 49
Pack, & Chapman 38, 318-21; Thomas 39
Penn, Henry 48-9, 153, 155-6, 187-8, 204, 223-4, 254, 257, 267-70, 298, 312, 334, 336; William 48
Rigby, Alexander 50, 56, 108-10, 131-3, 192, 303
Rog le Belleyetere 32
Rufford, John 33, 47, 56, 102; of Toddington 45; William 56
Smith, John, of Louth 44
Stainbank, Robert 41
Stephen le Belleyetere 32
Taylor, John 29, 31, 33-4, 36, 41-5, 49, 52, 102-343 passim, 392; Robert 49, 126, 129, 310-11, 319, 320-1, 341-3; William 49
Tho de Melton 32
Unknown bellfounders 56, 209-10, 216, 238, 267-8, 310
Warner, John 38-9, 302, 334-6, 392; Tomson 38
Watts, Francis 34-6, 108, 110, 237; Hugh I 36, 121-2, 203, 205, 288-90; Hugh II 28, 35-7, 254, 257; of Leicester 32, 34-6, 44, 47, 51, 108-9, 273, 275; William 36
Whitechapel Bell Foundry 38, 39-41, 96, 107-8, 121-2, 166-8, 186-8, 257-8, 266, 318-21

Clockmakers

Adams, Clement 59
Aris, Thomas 59-61, 117, 140, 157, 249, 272, 291, 324; William I 23, 60-1, 71-2, 184, 292; William II 59-61, 117, 157, 200, 249, 272, 324
Arnold, Edward 24-5, 37, 61-2, 66
Barker, John 76
Bates, John 62
Beaumont, Joseph 71
Billington, Everard 62, 220
Bird, Richard 62; William (jnr) 62, 28-2; William (snr) 280
Blackbond, Peter 62, 306
Blackburn, — 23, 58, 62, 99; Stephen 23-4, 57, 62-3, 199, 338
Bland, Charles 29, 71
Bloxham, Thomas 66

Bosworth of Nottingham 24
Boyfield, Richard 63, 338; Thomas 63, 338
Bradley, Simpson 63
Britton, John 63, 280
Brooks, Charles 63-4
Brumhead, John (jnr) 64; John (snr) 64, 74, 145, 280; Mary 64, 74; William 64, 280
Byron, Thomas 64
Bywater, Boniface 58, 64, 85
Clement, William 23
Cliff, George 324
Cooke, John 64, 65; Thomas 64-5, 212
Coombes, Cyril 29
Cope, Francis 65, G & F 25, 65, 74, 147, 157-8, 160, 314; George 65; William W 65
Corney, Mrs Mary Ann 65; Robert 26, 65, 74, 124, 201

Crane, — 65, 213
Cure, William 65
Denison, Sir Edmund Beckett 25, 65-6, 71, 75, 82, 262
Dent, E, & Co 66, 105; Edward 66, 91, 262; Frederick 25, 66, 261-3
Eayre, George 30; Joseph 30, 61-2, 64, 66, 261; Thomas I 30, 67; Thomas II 24, 59, 62-3, 66-70, 83, 118, 140-2, 183-5, 220, 243, 249-50, 337-8
Ecob, William 68
Esam, — 68
Evans, William Frederick 68, 101; W F, & Son 25, 68, 76
Flint, Mark 68-9, 74, 77-8, 80, 117, 249
Fox, George 69, 272; John 69, 140, 184, 220, 323; Robert 67, 69-70, 183, 220, 272, 291-2,

323; Thomas Henry 69; William 69, 70, 220, 239, 272, 323

Fromant, — 70, 157

Fromanteel, Ahasuerus 23, 75

Furniss, Joseph 70-1, 199, 291, 292, 324

Gent, — 265; & Co 71; James, & Son 25, 71, 234-5; John Thomas 71

Gillett, & Bland 25, 29, 71, 124-5; & Co 29, 308; & Johnston 29, 71, 213; William 29, 71

Goodman, William 71, 283

Grimadell, Peter 71; Samuel (jnr) 71; Samuel (snr) 71

Grimthorpe, Lord 25, 65-6, 71, 75, 82, 262

Hackett, Richard 58-61, 71-3, 77, 156-7, 159, 184, 291-2

Harding, William Pilkington 73

Hartshorn, John Stanley 73, 234-5

Haselwood, Samuel 59, 73

Haycock of Ashbourne 79, 250

Haynes, Robert Broughton 63, 73-4, 145, 314; Thomas 73-4, 92, 145, 314

Hedges, Frederick 74-6

Hedley, Amos 74

Hetterley, Charles 65, 74

Hickman, William 64, 74

Hicks, John 74, 249, 280, 292, 314

Hinds, Joseph 74

Holman, John 74, 140

Holt, — 213

Hooke, Robert 23

Horz, Phil, of Germany 65, 157-8, 160

Houghton, John 68, 74, 117, 220, 249

Houser, Matthias 75

Hubbard, Edward 75

Huygens, Christiaan 75

Jackson, John 75

Joyce, J B, & Co 75, 82, 308, 328; John Barnett 75; Thomas 75

King, William 22, 75, 323

Knibb, John 75; Joseph 11, 23, 75, 87, 132-4, 159; Samuel 75

Knight, John James 74, 76

Large, Thomas 68, 76, 101

Larratt, George 76

Line, John 76

Monck, Edmund 76, 92; Thomas 76, 92, 157

Newton, Sir Isaac 76-7, 114, 235

Nicholls, Henry 22, 77, 271-2

North, Samuel 69, 77

Norton, Henry 77; Robert 77

Nutt, Thomas Cornelius 77

Ogden, W 77

Parmiter, John F 77

Payne, Charles 77, 78, 124, 213, 262; William 78

Phillips, Richard 78

Phillpot 78, 306

Pinney, & Son 117, 324; Charles 78; F, & Sons 78; Francis 69, 78, 280; Richard Matthew 78

Potts, Anthony 79; Charles 79; Robert 79; Tom 79; William 64; William, & Sons 25-6, 79, 82, 196, 220-2, 250, 287, 292-3, 386-7

Rayment, Thomas (jnr) 79; Thomas (snr) 58, 79-80, 91, 323

Reed, George Jeremiah 83

Richardson, Ebenezer 80

Robinson, David 68-70, 80, 117

Rodely, Stephen 80, 152, 261-2

Roskell, William 91

Sharman, John 80

Sharpe, Hugh 80, 220

Simmons, Alexander Sadler 81; Charles Sadler 81; John Sadler 81

Simpson, John 23, 80-1, 99, 199, 261; Stephen 80-1, 152, 212, 261, 265

Smith, Frank Symon 82; John, & Sons, Clerkenwell 25, 81-2, 92, 227; John, & Sons, Derby 25, 63, 75, 78-9, 82, 133, 135, 152-3, 171-2, 179, 185-6, 194-5, 200, 213-14, 235, 240, 281, 286-7, 308, 324-5, 328-9, 338, 345; John Henry 82; William 79

Sparkes, James 69, 83, 184

Sutton, Rev Arthur F 83; Rev Augustus 83, 151, 194-5

Synchronome 29

Thwaites, & Reed 25, 83-4, 100-1; Aynsworth 83

Tilley, Samuel 84

Tiplee, William 60, 71

Tompion, Thomas 23, 75

Tucker, Elisha 25, 84, 212-13

Vines, John 84

Walker, John 24-5, 84, 300-1; Lewis 84

Warren, Henry 84-5, 280

Watts, Charles 85; John (jnr) 85, 88-90; John (snr) 12, 18, 22-4, 58-9, 64, 70, 73, 77, 85-90, 99, 144-6, 159-60, 170, 171, 194-5, 199, 220-1, 249, 260, 313; Robert 58, 85, 89-90, 91; William 90

Whitehurst, John I 24, 70, 90-1, 32-5; John II 82, 91; John III 91

Wilkins, John 23, 91, 199; Ralph 91

Wilkinson, Samuel 91

Wilson, James 73-4, 76, 91-2, 145, 157, 314; Joseph 24, 82, 92, 226-7; Ralph 58, 92, 156

Woodward, Thomas 91

Places

Apethorpe Church, Northants, clock 85, 145

Ashwell Church
bellringing customs 98, 376, 379, 380; bells 14, 39, 47, 95-9; clock 62, 78, 80, 84-5, 95, 99-101; clock bell 98, 131; cupola 99; Ringers' Rules 98-9; sundial 95, 99

Ashwell Hall, clock 101

Ayston Church
bellringing customs 103, 380; bells 18, 33, 42, 53, 102-3; scratch dial 18, 102, 103-4; sundial 102, 104

Ayston Hall, sundial 104

Barkby Church, Leics, clock 24, 62-3, 67

Barleythorpe
Clock House Court sundial 20; Hall sundial 118; Hall stables clock 66, 105-6

Barnack Church, Cambs, Saxon sundial 15, 16

Barrow Chapel, bells 38, 107

Barrow House, sundial 107

Barrowden Church
Angelus Bell 376; bellringing customs 111, 376, 379; bells, 33, 36-7, 42-4, 56, 96, 108-11; Priest's Bell 96, 108-111, 253, 376, 379; Sanctus Bell 111; sundial 108, 111

Beckingham Church, Notts, bells 57

Belmesthorpe
Blue Bell public house 111; Chapel 111

Belton in Rutland
Vicarage sundial 118; Gorse View sundial 118; Southview Cottage sundial 118; The Cottage sundial 20; Westbourne House sundial 105, 118

Belton in Rutland Church
bellringing customs 115-16; 375-76, 380; bells 15, 22, 29, 55-6, 112-16; clock 60-1, 69, 74, 78, 80, 82, 85, 112, 116-18; clock weights 117; hour glass 116; Priest's Bell 116, 380; scratch dial 112, 116; sundial 20, 112, 116-17; handbells 115

Bewcastle Cross, Cumbria, Saxon sundial 15

Big Ben, London 26, 39, 66, 91, 233, 262

Bisbooke Church
bell 42, 119-20; bellcote 12; bellringing customs 120, 376, 378

Braunston in Rutland Church
bellringing customs 122-3, 376, 379; bells 33, 36, 41, 47, 54, 120-3; clock 65, 71, 121, 123-5; coins on bells 129; scratch dial 17, 121, 123; sundial 121, 123-4

Brede Church, Sussex, bells 38

Brede mark 38, 210

Brooke Church
bellringing customs 130; bells 37, 42, 49, 53-4, 96, 125-30

Browne's Hospital, Stamford, Lincs clock 68, 87, 90

Burghley House, clock 71, 90

Burley Church
bell 56, 130, 131-2; bellringing customs 132, 376; clock 75, 82, 130, 132-5; clock bell 131; cupola 131; scratch dial 18, 130, 132; sundial 132

Burley on the Hill 131-2

Buttercross, Oakham, sundial 20, 265-6

Caldecott Church
bellcote 13; bellringing customs 138, 378, 380; bells 42, 55-6, 135-8; clock 60, 67, 69, 74, 136, 140-2; Sanctus Bell 138, 379; Sanctus bellcote 13; scratch dial 17-18, 136, 138-40; sundial 20, 136, 140

Caldecott, Old Sun Inn, sundial 142

Cambridge, All Saints' Church, bells 55

Cambridge, St Mary's Church 262

Cambridge, St Peter's Chapel, bells 57

Castle Bytham Church, Lincs, clock 87

Chacombe Church, Oxon 28

Clipsham Church
bellringing customs 144, 376; bells 54-5, 142-4; clock 64, 73-4, 92, 143, 144-7; Ellacombe chiming frame 142-3; scratch dials 18, 143-5; sundial 143-4

Clipsham Hall
   sundials 147; stables clock 65
Cottesmore Church
   bellringing customs 151, 376; bells 14, 41-2,
   46, 54-6, 148-51; chime barrel 64, 66, 78,
   80-1, 152, 261; clock 64, 66, 78, 80, 81-3,
   148, 151-3
Cottesmore Hall 83
Cottesmore Hunt Kennels, clock 68, 76, 101
Daglingworth, Glos, Saxon sundial 15
Deene Park, Northants, clock 67
Donnington Church, Lincs 245
East Farndon Church, Northants, bells 30
Edith Weston
   Edith Weston Hall sundial 160; Sundial
   House sundial 160
Edith Weston Church
   bellringing customs 156, 376, 379; bells 21,
   27, 42, 48, 52-3, 56, 153-6, 245; clocks 61,
   65, 70, 73-4, 76, 82, 85, 87, 92, 145, 154,
   156-60, 314
Egleton Church
   bellringing customs 162, 164, 376, 379;
   bells 56, 161-4; Priest's Bell 161, 162, 379;
   scratch dials 17-18, 161, 162-5; sundial 161,
   165
Empingham Church
   bellringing customs 169, 376, 379; bells 14,
   38, 40, 42, 54-6, 166-70; clock weights 171;
   clocks 77, 82, 85, 87, 145, 166, 170-2; Ellacombe chiming frame 166; handbells 169;
   Ringers' Rules 167; scratch dials 166, 170
Essendine Church
   bellcote 12, 172-3; bellringing customs 173;
   bells 39, 172-3
Exton Church
   bellringing customs 176-7, 376; bells 42, 49,
   55-6, 174-7; clock 83, 174, 177-8;
   Ellacombe chiming frame 174, 176;
   handbells 177; Ringers' Rules 177
Exton Hall
   chapel bellcote 49; chapel bell 49, 178;
   sundial 178
Exton House, sundials 11
Exton School, bell 178
Gaulby Church, Leics, clock 61, 184
Geeston House, Ketton, sundial 207
Glaston Church
   bellringing customs 182, 380; bells 34-5, 42,
   53, 179-83; clock bell 67, 131, 182, 253;
   clocks 60, 67, 69, 73, 82-3, 179, 183-6,
   240; Ellacombe chiming frame 179-80;
   handbells 182-3; scratch dial 179, 183;
   sundial 179, 183
Glaston, Main St, sundial 186
Glaston Railway Mission Chapel, bell 186
Gloucester Cathedral 34
Great Casterton Church
   bellcote 12, 187, 189; bellringing customs
   188-9, 376; bells 12, 14, 41, 48, 186, 187-9;
   old bell clappers 187; sundial 186, 189
Great Casterton Primary School
   bell 189; bellcote 189
Greetham
   Chapman's Cottage sundial 196; Greetham

House sundial 195, 196; Ram Jam Inn
   sundial 16, 196; School clock 79, 196;
   Vicarage sundial 195
Greetham Church
   bellringing customs 192, 376; bells 14, 29,
   54-5, 56, 190-3; clocks 82-3, 85, 145, 190,
   194-5; scratch dials 18, 190, 193-4; sundial
   190, 194
Gretton Church, Northants, clock 67, 69-70,
   80, 323
Haddon Church, Hunts, bells 33
Hambleton
   Hilltop sundial 201; Home Farmhouse
   sundial 201; Manor House sundial 201; Old
   Hall sundial 201; Post Office clock 26, 200
Hambleton Church
   bellringing customs 198-9, 376; bells 42, 53-
   4, 196-9, 379; clock 61, 63, 71, 76, 80, 82,
   84, 91, 197, 199-200; Priest's Bell 196, 198,
   380; Sanctus Bell 198; sundial 197, 199
Hambleton Hall
   clock 200-1; sundial 20, 200-201
Hampton Lucy Church, Warks, clock 81
Harringworth Church, Northants
   clock 60-1, 71; Sanctus Bell 58
Henley in Arden Church, Warks, clock 81
Horn Church, bells 201
Horse Guards Parade, London, clock 83
Huntingdon, St Mary & St Benedict's Church
   49
Kenilworth, St Nicholas' Church, Warks, clock 81
Ketton
   Ketton Hall sundial 207; railway station bell
   206; school bell 206; St Mary's Home bell
   206; The Firs sundial 207; The Priory sundial
   207
Ketton Church
   bellringing customs 205, 376; bells 30, 34-5,
   42, 46, 55, 202-5; clock 205-6; Ellacombe
   chiming frame 202, 204; handbells 205;
   scratch dial 17, 202, 205-6
Ketton Grange
   stables clock 81, 206-7; Stables Cottage
   sundial 206
Kings Cliffe Church, Northants, clock 85
Kings Lynn, St Margaret's Church, Norfolk,
   bells 52
Kings Norton Church, Leics, chime barrel 61-
   62, 67, 261
Kirkdale, Yorks, Saxon sundial 15
Langham
   Langham Hall clock 82, 214; Manor House
   sundial 20, 213-14; Ranksborough Hall
   sundial 214
Langham Church
   bellringing customs 210-12, 375-6, 378-9;
   bells 14, 31, 38, 42, 47, 54-6, 208-12; clock
   65, 78, 81-2, 84, 208, 212-13; Peal Boards
   211; Sanctus Bell 208, 379
Leicester Cathedral 47, 62, 202
Leicester, All Saints' Church, bells 32, 34, 50
Leicester, St Martin's Church
   bells 47, clock 62
Leicester, St Mary de Castro's Church   37
Leicester, St Nicholas Church, bells 37

Leicester, St Peter's Church, bells 50
Little Bowden Church, Leics, bells 33, 251
Little Casterton Church
   bellcote 12, 215; bellringing customs 216;
   bells 53, 56, 215-16; Priest's Bell 380;
   scratch dial 216
Loddington Hall stables, Leics, clock 67, 184
Loughborough, All Saints' Church, bells 49
Luffenham Hall, sundial 251
Luffenham House, sundial 11, 251
Lyddington Church
   bellringing customs 218-19, 375-6, 380;
   bells 42, 55-6, 216-19; clock 60-2, 67, 69,
   70, 74, 79-80, 85, 216, 220-2; Priest's Bell
   217; Ringers' Rules 219; Sanctus Bell 217;
   scratch dial 216, 219-20
Lyddington, Manor House, sundial 222
Lyndon Church
   bellringing customs 225, 380; bells 27, 42,
   48, 52, 54-5, 222-5; cupola 227; Sanctus
   Bell 224; scratch dials 17, 222, 225-6
Lyndon Hall
   sundial 226; stables clock 82, 92, 226-7
Mancetter Church, Warks, bells 32
Manton Church
   bellcote 12, 13, 227; bellringing customs
   229-30, 376; bells 33, 42-4, 54, 227-30;
   Sanctus Bell 227-8, 379; Sanctus bellcote 13;
   scratch dials 17, 18, 227, 229-30
Market Overton Church
   bell repair 233; bellcote 13; bellringing
   customs 234, 376, 378, 380; bells 30-1, 42,
   55, 231-4; clocks 71, 73, 77, 231, 234-5;
   Ellacombe chiming frame 231; Peal Board
   231; Sanctus Bell 229, 231, 233, 379;
   Sanctus bellcote 13, 229, 231, 233; scratch
   dial 231, 234; sundials 231, 234
Market Overton, Manor House, sundial 234-35
Martinsthorpe Church, bells 18, 236
Martinsthorpe House, sundial 11, 236
Milwich Church, Staffs, bells 57
Morcott Church
   bellringing customs 239, 376, 379; bells 30-
   1, 34, 36, 54, 56, 236-9; clocks 70, 82, 85,
   237-8, 239-40; clock bell 11, 56, 57, 131,
   237, 238-9; Priest's Bell 237; Sanctus Bell
   236-7; sundial 237, 239
Morcott, Sundial House, sundial 240
Nassington Church, Northants, clock 85, 145
Newarke Houses Museum, Leicester 23, 62-3,
   81, 90
Newton Church, Lincs, bells 51
Normanton Church
   bells 49, 241-2; clock 241-2
Normanton Hall
   stables clock 67, 242-3; sundial 11, 242-3
North Luffenham Church
   bellringing customs 248-9, 378, 380; bells
   30-1, 34, 42-4, 46, 53, 244-9, 267, 268;
   clock 60-1, 67, 69, 74, 85, 244, 248, 249-
   50; clock bell 245, 248; medieval bellframe
   244-5; scratch dial 17, 244, 248-9; sundial
   244, 249
North Luffenham
   Manor Farm sundial 251; Sundial Cottage

sundial 251
North Luffenham, The Pastures
   clock 79, 250; clock bell 33, 251; sundial
   251
Norwich, St Lawrence's Church, Norfolk 83
Oakham
   Cemetery bell 264; High St sundial 20, 265;
   St Joseph's RC Church bell 264; The
   Cottage sundial 264
Oakham, Catmose House
   house bell 264; sundial 264
Oakham Church
   Angelus Bell 264; Bellman 260; bellringing
   customs 258-60, 264, 375-6, 378, 379, 380;
   bells 14, 29, 36, 38-9, 56, 252-60; chime
   barrel 61, 66, 80; clock 66, 78, 80, 81, 85,
   252, 254, 260-3, ; clock bell 254, 257-8;
   handbells 258-9; Priest's Bell 253-4, 256,
   258-9, 380; Ringers' Rules 260; Sanctus Bell
   258; Sermon Bell 259
Oakham School
   Old Stables clock 71, 265; School House bell
   turret 265
Oakham, St John & St Anne's Chapel
   bell 30-1, 263; bellringing customs 264;
   sundial 263
Oakham Union Workhouse
   bell 329; longcase clock 265
Orton Waterville Church, Lincs, bells 53
Oxford, St John's College, clock 75
Passenham Church, Northants, bells 28
Peterborough Cathedral
   bells 48; clock 85
Peterborough, St Joseph's Church 49
Peterborough, St Jude's Church, bells 49
Pickworth Church
   bell 39, 266; bellringing customs 266, 380;
   Spur Bell 380
Pickworth, Sundial Cottage, sundial 266
Pilton Church
   bellcote 12; bellringing customs 268; bells
   48, 56, 245, 267-8
Preston
   Main St sundial 273; Manor House sundial
   20, 273; Wings House sundial 273
Preston Church
   bellringing customs 271, 376, 378; bells 32,
   34-5, 42, 44, 48, 50, 269-71; clock 60-1,
   69, 70-1, 77, 268, 271-3; Gabriel Bell 269,
   270, 376; Sanctus Bell 50, 269-71, 379;
   scratch dial 18, 268, 271-2; sundial 20, 268,
   271, 273
Ridlington Church
   bellcote 12; bellringing customs 275, 376,
   380; bells 36, 42-4, 55, 273-5; clock 276;
   handbells 275; Sanctus Bell 275, 379; sundial
   274, 275-6
Ridlington sundial 276
Royal Exchange, London, clock 66
Royal Observatory, Greenwich 25
Rutland County Museum 9, 21, 61-2, 67, 73,
   76, 81, 85, 93, 156-7, 160, 265
Ryhall Church
   bellringing customs 279; bells 14, 37, 54-5,
   277-9; clock 62-4, 78, 82, 84, 277, 280-2;

clock bell 277, 279; Ringers' Rules 279;
   sundial 277, 280
Ryhall Vicarage, sundial 282
Saxby Church, Leics, bells 303
Seaton Church
   bellringing customs 285, 376; Bellrope Field
   283-4; bells 27, 28, 51, 52, 55, 283-5; clock
   79, 82, 283, 286-7; Parish Clerk's Allotment
   284; scratch dials 17, 283, 285-6; sundial
   283, 286
Seaton Railway Mission Chapel, bell 287
Seaton School, bellcote 287-8
Sherbourne Church, Warks, clock 81
South Luffenham
   South Luffenham Hall sundial 294: School,
   bellcote 294
South Luffenham Church
   Bellringers' Field Charity 288-9; bellringing
   customs 290, 375-6, 378-9; bells 36, 42, 45,
   54, 288-90; clock 60, 69, 71, 73-4, 79, 85,
   288, 291-4; Ellacombe chiming frame 288,
   290; Sanctus Bell 379; scratch dials 17, 18,
   288, 290-2
St Albans' Abbey, Herts 66
St Ives Church, Cambs, bells 49
St Paul's Cathedral, London clock 66
Stamford, Lincs
   All Saints' Church clock 87; Museum 64, 77,
   85, 170; St George's Church, bells 53-4; St
   John the Baptist's Church, bells 53; St
   Martin's Church 56
Stamford, St Mary's Church, Lincs
   bells 37, 56; chime barrel 62-4, 66; clock
   63-4
Stapleford Church, Leics
   bells 14, 303; chime barrel 61, 96, 261
Stocken Hall Coach House
   clock 24-5, 84, 300-2; clock bell 39, 300-2
Stoke Doyle Church, Northants, clock 67-8
Stoke Dry Church
   bell 15, 31, 294-5; bellcote 12, 294; bellman
   294-5; bellringing customs 295, 378; clock
   294-5, 297; Sanctus Bell 294; scratch dials
   17, 18, 294, 295-7
Stretton Church
   bellcote 12, 13, 48, 297; bellringing customs
   299; bells 15, 54, 297-9; Sanctus Bell 297,
   379; Sanctus bellcote 13, 297-8; scratch dials
   17, 298-300; sundial 298-300
Stretton School
   bell 39, 302; bellcote 302; clock 302
Surfleet Church, Lincs, bells 53
Swaton Church, Lincs, bells 51
Swynscombe Church, Oxon, bells 271
Teigh Church
   bellringing customs 305, 375-6, 378-9; bells
   30-1, 45, 56, 303-5; clavier 303; clock 62,
   78, 303, 306; scratch dial 18, 303, 305;
   Sermon Bell 305
Terrington St Clements Church, Norfolk, bells
   51, 52
Thistleton Church
   bell 47, 306-7; bellringing customs 307,
   375-6, 380; clock 71, 75, 82, 306, 307-8
Thorpe by Water, sundial 308

Tickencote Church
   bellcote 12, 309; bellringing customs 310,
   376; bells 38, 54-5, 309-10; Sanctus Bell
   309; Sanctus Bellcote 309
Tinwell Church
   bellringing customs 313, 376, 378-9; bells
   42, 54, 310-13; clocks 65, 73-4, 85, 92,
   158, 311, 313-14; Ellacombe chiming frame
   311; Ringers' Rules 279, 313
Tixover Church
   bell 38, 315; bellringing customs 315;
   scratch dial 17, 315-6; sundial 315-16
Tixover Grange, sundial 316
Uppingham
   Crown Hotel sundial 20, 330-1; Oddfellows
   Hall bellcote 330; Pleasant Terrace sundial
   331, Sundial Cottage sundial 330; Sundial
   House sundial 20, 330-1; Unicorn Inn
   sundial 330-1; Uppingham Hall sundial 329
Uppingham Church
   Angelus Bell 319-21, 380; bellringing
   customs 321-3, 375, 378-9; Bellrope Land
   316-17; bells 38-9, 42, 49, 55, 316-23; clock
   60-1, 69-71, 75, 79, 82, 91, 316-17, 321,
   323-5; Priest's Bell 380; Sanctus Bell 317-
   18, 380
Uppingham, National School
   bell 330; bellcote 330
Uppingham School
   Brooklands sundial; 329 Manor House
   sundial; 329 Masons' Lawn sundial 330
Uppingham School Chapel
   bellringing customs 327; bells 29, 325-7;
   clavier 326; Sermon Bell 327
Uppingham School, clock 327; clock bell 44,
   327-8; cupola 327; school bell 327-8;
   sundials 20;
Uppingham School, Johnson's Schoolroom
   bellcote 328; sundial 328
Uppingham School Library
   bell 330; cupola 330
Uppingham School Upper pavilion
   bell 328; clock 328; cupola 329
Uppingham Union Workhouse
   bell 230, 329; bellcote 230, 329
Wadham College, Oxford, clock 23, 75
Wakerley Church, Northants 242
Walgrave Church, Northants, Sanctus Bell 58
Wardley Church
   bellringing customs 332-3; bells 33, 55, 332-
   3; Sanctus Bell 332, 379; sundial 20, 332-3
Wellesbourne Church, Warks, clock 81
Wellingborough Church, Northants, clock 67
Westminster Palace, London, clock 26, 39, 66,
   91, 233, 262
Westminster, St John's Church, London, 241
Whissendine Church
   bellringing customs 336-7, 375-6, 378-9;
   bells 39, 42, 46, 96, 333-7; clock 63, 67, 82,
   334, 337-8; Ellacombe chiming frame 334-5;
   hourglass 337
   scratch dials 17, 334, 337-8
Whissendine Rectory, sundial 338
Whitwell Church
   bellcote 12, 49, 339; bellringing customs

339, 341, 376, 380; bells 49, 57, 339-41; scratch dials 17, 18, 339, 340-1; sundial 339, 341

Whitwell Rectory, sundial 341

Windsor Castle, Berks, clock, 75

Wing
  City Yard House sundial 20, 346; Sundial House sundial 20, 346
Wing Church
  bellcote 12, 341; bellringing customs 344-5; bells 33-4, 42-3, 49, 341-5; clock 82, 342,

345-46; handbells 344; Sanctus Bell 341, 344, 379; scratch dial 18, 342, 344-5
Wing Railway Mission Chapel, bell 346
Wing Wesleyan Chapel 342, 345
Wollaston Museum, Northants 67
Woolsthorpe Manor, Lincs 76-7

Surnames

Ablett 157
Adcock 261
Adkins 325
Aldred 33
Allen / Allin 115, 312, 318
Allsop 142
Almond 209, 253
Andrew 157
Angell 37, 236
Arnold 311, 312, 313
Ashby 345
Atkin 115
Atlay 108
Ayscough 76, 235
Baggallay / Bagley 198, 199, 211, 344, 376
Bagnall 318
Bailey 197
Baines / Bains 117, 197, 269, 273, 324
Baker 98, 176, 211, 345, 391
Bardwell 177
Barker 224
Barnes 149, 150
Barnett 142
Barrand 168, 169
Barratt / Barrett 346, 322
Barry 302
Barsby 223, 334
Bartram 211
Basset 247
Bateman 218
Baxter 67
Beadman 126, 128, 129
Beaumont 155
Beeson 133
Bell 95
Bellhouse 314
Bennitt 323
Benson 225
Beridge / Birridge 95, 199
Berry 343
Bertie 64, 176
Bevan 97
Billington 323, 328
Billins 303
Billows 262
Bindley 115
Bingham 184, 280
Birch 184
Blackband 303, 306
Blackburn 95, 253
Bland 169, 170
Bloodworth 153
Bolland 249
Bonner 311
Boon 70
Booth 220

Bosworth 192
Bowdell 58, 181, 182
Boyer 323
Bradley 318, 391
Bradshaw 60, 71, 311, 312
Brandon 231
Branston 219
Brasbridge 334
Brewster 219
Bridge 231
Brittayne / Britten 321, 316
Broom 140, 153, 269
Broughton 188, 284
Brown / Browne 27, 137, 176, 216, 217, 218, 224, 244, 245, 317, 318
Brudenell 181, 182
Brushfield 318
Bryan / Bryon 58, 122, 218
Buchan 177
Buckby 124
Bull 97, 156
Bullingham 204
Bullock 324
Bulstrode 276
Bunce 98
Burdett 212, 213
Burn 259
Burnaston 219
Burton 97, 260, 318
Bushell 201
Butler 42, 188, 253
Cadman 188
Canner 168
Canyng 87
Capendale 98, 99
Carrier 195
Carter 83, 197, 277, 309
Casey 231
Casterton 98
Catlin 319, 320
Cave 292
Cecil 64
Chamberlain, 97
Chapman 138, 180, 196, 295
Char...y 190
Charles 190, 191, 258
Christian 136, 151, 245, 269, 319, 324
Clarke / Clark 192, 260, 319, 337, 391, 393
Clarke Jervoise 209, 211
Clayton 224
Cleeve 257
Clephan 393
Cliffe 223
Cloxton 211
Coale / Cole 97, 208
Cobb 329

Cocks / Cox 137, 194, 233
Codrington 269, 270
Coke 208
Coldwell 309
Collins 113
Collwell 138, 216
Colly 181, 182
Conant 223, 225, 391
Cook / Cooke 323, 334, 337, 345
Cooper 391
Cort 209
Costall 232
Coulson 98
Coverley 83, 148, 151, 194, 195
Cradin 218
Crowden 283
Crowsforth 253
Crowson 187
Crowther-Beynon 155
Cuminge 336
Cundy 241
Cunningham 335
Dalby 211, 212, 263, 272, 393
Dale 272
Daniel-Tyssen 391, 393
Darnell 154, 197, 245
Davies 99, 154, 155, 231, 389, 391, 393
Day 319
de Aldesworth 236
Deane 147
Denham 231
Dennis 174, 244, 391
Dennison 174, 269
Dexter 108, 269
Dickinson 197
Digby 111, 247
Dolby 314
Drake 321, 378
Draper 232
Dunstone 260
Eastman 326, 327, 328
Eayre 142
Edgson 95, 99, 303, 338
Edwards 307
Elgood 393
Ellacombe 393
Ellingworth 260, 264
Elliott 71
Ellis 130
Emerson 107
English 198
Erle 77
Escritt 283
Evans-Freke 181
Exton 247
Fancourt 168, 328

Faulkner / Ffalkner 128, 322
Faulks 209
Field 179
Fenn 280
Fernyhough 280, 281
Ffleming 317
Ffowks 322
Field 120
Fielding / Ffielding 245, 317
Finch 132, 177, 256, 313, 314
Fleetwood-Hesketh 300, 302
Fletcher 393
Flint 342
Floar 335
Fludyer 103, 307, 308, 391
Forman 323
Foster 330
Fountain 152
Fowler 319, 334, 335
Fox 184
Franklin 48
Fraser 252
Frear 306
Freeman 234, 391
Freestone 204
French 324
Frisby 219
Fryer 200
Gale 134
Gatty 330
Gay 391
Geeson 136, 319
Gibson 212, 247
Gilham / Gillhan 137, 219
Gill 393
Glenn 212, 311
Goodacre 307
Goodlad 312
Goodwin 74
Gord 276
Gore Browne 180
Gore Marshall 200
Gough 108, 112, 114
Gouldin 311, 314
Gray 317
Green 174, 177, 318, 319, 323
Greenfield 336
Gregory 198
Gretton 335
Grinter 258
Grocock 108, 114, 115, 118
Guilford 150
Hack 338
Hackett 71, 73
Hall 188
Halliday 251, 391
Hamilton 314
Hanbury 123, 125, 134
Hand 184, 223
Handscombe 391
Hansworth 276
Harbottle 161
Hardy 151, 157, 158, 307, 308
Harris 179, 346, 393
Harrison 223

Hasson 213
Hazelrigg 275
Heathcote 168, 241
Heathcote-Drummond-Willoughby 169
Henton 93, 94, 141, 238, 239, 240
Herbert 219
Hewitt 7
Hickling 345
Hicks 249
Higgs 255, 258, 259
Hill / Hills 102, 134, 322, 324
Hilland 216
Hinman 245, 303
Hippesley 184, 323
Hirrup 318
Hoare 133
Hodges 74
Holden 290
Holland 98, 99
Hollis 41, 107, 149, 152
Holman 140
Holmes 279, 324
Hooke 23
Hooper 309
Hope 323
Horne 16, 334
Horsley 256
Hoskins 391
Hotchkin 223, 318
Houghton 74
How 318
Howard 309
Hoy 255, 337
Hubbard 208, 211, 212
Hubbert 208
Hudson 197, 323
Humphrey 168, 169
Hunt 224, 247, 248
Hurry 108
Hurst 176
Hutchings 95, 108, 109, 110, 151, 153
Hutton 181, 182
Ingram 221, 393
Ireland 175, 218
Islip 289
Jackson 202, 339
Jacques 180
James 211, 323
Jarman 115
Jeffs 136
Jewitt 393
Johnson 108, 247, 312, 319, 323
Joles 311
Jones 115, 130, 391, 393
Joyce 206
Judd 130
Keal 258
Kennedy 130
Kernick 319
Ketell 303
Killingback 129, 130
King 317, 180
Kirk 191, 192
Knox 174, 178
Lam 216

Lambard 239, 241
Lane 328
Lang 180
Langley 319, 393
Langton 319, 323
Latham 388, 391, 393
Law / Lawe 237, 245, 247
Laxton 70, 80, 149
Lee 88, 148, 150, 159, 170, 188, 258, 292
Lewin 120, 253
Lewis 231
Lightfoot 231
Line 199
Littledyke 174, 245, 249, 267, 297, 311
Long 337
Looe 313
Loseby 272
Lovick Cooper 168
Lowe 311, 339
Luck 388, 389, 393
Lupton 61
Lutyens 388
MacDonald 231
Mackerith 317, 319, 323
Mackworth 168
Maclachlan 108
Maidwell 257
Makey 157
Mandall 211
Mane 55
Mangan 221
Manning 318
Mantle 209
Manton 85, 220, 269
Marlow 115
Marriot 220, 221, 309, 324
Marshall 134, 256
Mason 98, 108
Matkin 265
Matthews 328
Maxwell 188
Mayhew 129
Mear / Mears 267, 330
Medmore 324
Meredith 95
Mills 169, 180, 324
Mirehouse 77
Mitchell 167
Moore 115
Morris 211, 337
Morton 81
Mosendew 324
Mould 84, 212
Muggleton 219
Munday 211
Munton 124, 247, 245
Murrey 211
Naylor 157
Needham 259
Nelson 259
Nettle 133
Nevinson / Nevison 222, 223, 225, 324
Newmarch 115
Newton 76
Noel 176

Norman 132
Norris 50, 51
North 388-93 and *passim*
Norton 129
Nutt 204, 318
Ogden 170, 339
Ohlsson 233
Ordish 393
Orland 121
Orme 317
Orton 129, 338
Ovens 159
Owen 58, 168, 169, 328
Paddy 341, 342
Padmore 389
Paget 392, 393
Pakeman 318, 331
Palmer 311, 318
Parker 192
Patteson / Pattison 108, 114, 328
Pawlett 197, 199
Paybury 62
Payne 157, 289
Pearson 133, 391
Peas 255
Peet 154
Penn 48
Penney 249
Pepper 99, 311, 331
Pepperday 180
Phillips 199, 252, 258, 260
Phillpott 306
Pilkington 108, 113, 114
Piper 339
Pitts 245
Plant 140
Pochin 62
Porter 232
Powell 274
Pretty / Pritty 142, 218
Price 148, 249
Pridmore 184, 292
Pulford 269, 323
Quenborough 272
Raven 388, 389
Rawlings 122, 269
Rawnsley 69
Read 261
Redlich 303, 306
Redmile 311
Reeve 115
Richardson 80, 217
Riley 209
Ringrose 67, 118
Robarts 321, 324
Robinson 269
Rogers 391
Rowlatt / Rowlet 322, 204
Royce 102, 339
Ruddell / Ruddle 123, 261
Rudkin 103, 258
Russell 157
Scotney 249
Scott 111, 180, 231, 253, 254, 258, 261, 376

Senescall 195
Sewell 211, 212, 209, 219, 317, 318
Sharman 80, 190, 219, 319
Sharpe 216, 269
Sheffield 284, 285
Sheldon 109, 192
Sherrard 303
Sherwood 317
Shillaker 211
Shortt 137
Simon 345
Simpson 180
Sims 303, 304
Singlehurst 142
Sissen / Sisson 204, 312
Sitton 317
Slater 183, 269
Sleath 115, 116, 217
Small 209
Smalley 318
Smallfield 377
Smith / Smythe 76, 113, 142, 158, 174, 197,
   211, 214, 228, 263, 305, 311
Sneath 69, 169
Snodin 338
Sparkes 83, 144, 146
Spreckley 391
Spring 292
Springthorpe 176
Squire 305, 306
Stafford 245, 247, 248, 318, 335, 336
Stainbank 392
Stanger 138, 205, 306
Stapleton 290
Steans 173
Steel 211
Stevens 180, 254, 256, 257, 262
Stevenson 136, 154, 219
Stimson / Stinson 81, 258
Stone 156, 157, 199
Storey / Story 231, 314
Stuart 391
Stubbs 195
Stuckley 76, 309
Sturges 334
Sutton 83, 194, 392
Swaby 166
Swann 108
Swayen 318
Swingler 213
Swire 204
Symmes / Symmys 276
Tailby 290
Tayler / Taylor 108, 222, 323, 389, 393
Tempest 194
Thompson / Thomson 26, 261, 307
Thring 325
Thurlby 99, 185
Tibbert 167
Tidd 100
Tilley 84
Tipping 315
Tiptaft 311, 312

Tires / Tyres 200
Tookey 289, 318
Toon 197, 253, 334
Towell 180, 258, 277, 307, 308
Traylen 107, 202
Turner 77, 269, 319
Tyler 60, 223, 324, 337
Tyssen-Amherst 393
Utting 392
Veasey 199
Venables 388, 393
Verney 114
Vickers 231
Vines 199
Voysey 79, 250, 251
Wade 120, 319, 320
Wadking 199
Wadland 219
Wales 391
Walker 199, 234
Wallace 175, 180
Wallbanks 289
Waltham 181
Wand 231
Ward 108, 112, 199, 258
Waring 218
Warren 220
Wartnaby 334
Waterfield 216
Watson 337
Watts / Wottse 85, 90, 179
Webster / Webstee 96, 177
Wells 18, 317
Westwood 188
Wheeler 258
Whiston 224
White 74, 330, 336, 337
Whitelaw 130
Wiggington 98, 264
Wilbon / Wilbron / Wilburn 95, 98, 99, 323
Willford 289
Willmott 154
Willoughby 169
Wilson 169
Wilton Hall 173
Wing 142, 234, 239, 324,
Wingfield 232, 391
Witt 179
Wood / Woods 231, 211, 254, 275, 290, 319,
   391
Woodcock 140, 197
Woolman 148
Wordsworth 182, 391
Worrall 343
Worsdale 83
Worth 114
Wotton 204
Wren 75
Wright 130, 136, 217, 218, 319
Wyare 303
Wymarke 316
Wymond 245
Yard 96, 97